Advances in Sol-Gel Derived Materials and Technologies

Series Editors

Michel Andre Aegerter
Michel Prassas

For further volumes:
http://www.springer.com/series/8776

The International Sol-Gel Society (ISGS)

Dear Readers,

The International Sol-Gel Society is proud to continue its long-lasting collaboration with Springer to offer this new series of books on sol-gel science. They are intended to inform novices and experts on the latest advances on the following topics: aerogels, sol-gel characterization, hybrids, nano-composites, sol-gel coatings for energy applications, and membranes.

These books have been written by world-class scientists who believe in the pivotal role played by the ISGS in the development of the sol-gel science community. Since its creation in 2003, the ISGS has tremendously evolved in its infrastructure but has kept the same missions: educate, communicate and coordinate.

The first mission of the ISGS is to educate scientists and technologists, from both academia and industry, through topical workshops and summer schools. The biannual ISGS summer school has become the reference for new Ph.D. students by providing comprehensive and practical teaching. It is also a unique opportunity for them to start networking in a relaxed atmosphere with more advanced researchers.

The second mission of the ISGS is to foster the communication between researchers from different fields and geographical regions through the organization of conferences and the publication and circulation of technical papers. The biannual International Sol-Gel Conference attracts more than 400 researchers coming from more than 40 different countries. The new ISGS website (www.isgs.org) is designed to provide a more interactive structure where members can participate in the life of the ISGS. In that regard, our quarterly newsletter encourages people to contribute by sharing their personal stories and opinions.

The third mission of the ISGS is to coordinate the promotion of sol-gel science and technology in the scientific and industrial communities. The official Journal of the ISGS, (The *Journal of Sol-Gel Science and Technology*) ranks 3rd in the category of Materials Science and Ceramics, with a 2010 Impact Factor of 1.525. The involvement of more industrial researchers in the functioning of the ISGS has become a top priority for the board.

Finally, the ISGS honors its commitment across the different missions by providing travel scholarship to students, and by sponsoring sol-gel related workshops, schools and local conferences.

The success of this society has been built on the forward thinking of the past presidents and boards. They have created the necessary infrastructure for an international society and they have shown a sustainable path forward. Today, our novel mission with you is to establish a stronger sense of global community in sol-gel science.

I wish you a very educative and pleasant reading.

Geraud Dubois
President of the International Sol-Gel Society

Mario Aparicio · Andrei Jitianu
Lisa C. Klein
Editors

Sol-Gel Processing for Conventional and Alternative Energy

Springer

Mario Aparicio
Institute of Ceramics and Glass
Campus of Cantoblanco
28049 Madrid
Spain

Lisa C. Klein
Rutgers University
Piscataway, NJ
USA

Andrei Jitianu
Department of Chemistry
City University of New York-Lehman College
250 Bedford Park Boulevard West
Bronx, NY 10468
USA

ISBN 978-1-4614-1956-3 e-ISBN 978-1-4614-1957-0
DOI 10.1007/978-1-4614-1957-0
Springer New York Heidelberg Dordrecht London

Library of Congress Control Number: 2012930130

Printed on acid-free paper

Springer is part of Springer Science+Business Media (www.springer.com)

Dedicated to Charo, Mihaela, and Tim
for their patience and understanding

Preface

Energy is on everyone's mind these days because of the expanding need for electricity, lighting, heating, cooling, and transportation fuels. Finding more efficient ways to produce energy is a global priority. In addition, the production of energy needs to be sustainable and clean. At the core of the energy infrastructure is the need for materials for energy conversion, energy storage, and energy distribution.

The focus of this book is on ways to synthesize, assemble, and modify materials that find use in systems designed for energy conversion and energy storage. The sol-gel process has been used for a variety of systems that serve the energy market. The singular advantage of the sol-gel process is its being a low-temperature route to simple and complex oxide materials. Low-temperature processing allows the combination of organic and inorganic materials, such as hybrid membranes and ion conductors. Another advantage of the sol-gel process is that it begins in liquid form, so that it is easily adapted to coatings and surface treatments, such as electrochromic windows and multilayer thin film stacks. Furthermore, the sol-gel process leads to high surface area materials and porous materials, meaning it has applications in catalysis, environmental remediation, and carbon dioxide sequestration.

Through the lens of sol-gel processing, this book is about the selection and preparation of materials for batteries, fuel cells, photovoltaics, thermoelectrics, energy harvesting ferroelectrics, and supercapacitors. The properties of these materials are explored in detail. Their applications and performance are described and evaluated. The interested reader can gain a sense of the state of the art and, at the same time, the need for further improvement.

Since new solutions are needed to solve the problem of increasing energy consumption, it would bring the co-editors great satisfaction if this book motivates some or all of its readers to double their efforts toward finding new ways to use the sol-gel process to address materials, needs. The co-editors wish to thank Mike Luby of Springer for his encouragement to take on this project. Merry Stuber is thanked for her editorial assistance. This project would not have come together if it were not for the outstanding response from the chapter authors and their wonderful contributions. Also, the co-editors appreciate the chance to work together on this project, which has led to further opportunities for research collaborations.

On behalf of my co-editors, Mario Aparicio and Andrei Jitianu,

Lisa C. Klein

Contents

Chapter 1
Conventional Energy Sources and Alternative Energy Sources and the Role of Sol-Gel Processing

Lisa C. Klein

Abstract Energy is an ever-present concern. Materials are in some cases a limiting factor in energy production. It is necessary to improve the available materials for energy storage, energy conversion and energy distribution. One way to address some of these materials issues is sol-gel processing. This chapter briefly introduces the role of sol-gel processing for energy applications.

Keywords Alternative energy · Biomass · Clean energy · Conventional energy · Energy conversion · Energy storage · Fossil fuels · Hydrogen · Non-renewable energy · Renewable energy

1.1 Introduction

This book is intended to be a comprehensive source of information about the use of sol-gel processing in materials in energy systems, energy conversion, energy storage and energy generation. Both conventional energy (e.g., nuclear) and alternative energy (e.g., solar) are included. The goal is to cover the present applications, such as nuclear fuel processing, electrolytes for fuel cells and electrochromics, and dye-sensitized solar cells (DSSC), among others. For each of the existing technologies, the plan is to cover the processing, properties and industrial applications that are known today. Related topics such as carbon sequestration, wastewater treatment and photocatalysis are covered as well.

The authors in reviewing and assessing the methods and applications are able to identify materials challenges presented by conventional and alternative energy

L. C. Klein (✉)
Department of Materials Science and Engineering, Rutgers University,
607 Taylor Road, Piscataway, NJ 08854, USA
e-mail: licklein@rci.rutgers.edu

M. Aparicio et al. (eds.), *Sol-Gel Processing for Conventional and Alternative Energy*, Advances in Sol-Gel Derived Materials and Technologies,
DOI: 10.1007/978-1-4614-1957-0_1,

Table 1.1 2006 National Academy of Engineers "Engineering Challenges" [2]

Make solar energy affordable
Provide energy from fusion
Develop carbon sequestration methods
Manage the nitrogen cycle
Provide access to clean water
Restore and improve urban infrastructure
Advance health informatics
Engineer better medicines
Reverse-engineer the brain
Prevent nuclear terror
Secure cyberspace
Enhance virtual reality
Advance personalized learning
Engineer the tools for scientific discovery

generation that require new materials and innovative processing. Overall, these specialists provide state-of-the-art information about different aspects of materials for energy production, with a focus on their processing. The need for materials innovation, in general, to promote renewable energy has been reviewed recently [1].

1.2 Motivation

What is the motivation for surveying sol-gel processing in the context of energy? It is safe to say that energy, and especially alternative energy, is on the minds of everyone, including no doubt materials scientists. In the compilation of "Engineering Challenges" by the National Academy of Engineers (NAE) the first on the list is "make solar energy affordable" (see Table 1.1). Sol-gel processing is one of the ways that this challenge is being met. This book will go beyond this challenge to address other ways to make conventional energy and alternative energy affordable and clean.

1.3 Conventional Versus Alternative Energy Sources

To be clear about what is meant by the many terms used to classify and describe energy, a few definitions are in order. In discussing alternative and renewable energy, there are several terms that are frequently used to describe the different technologies. In general, alternative energy, renewable energy, clean energy and sustainable energy broadly are used interchangeably to include all forms of energy that are not classified as conventional fossil fuels, meaning coal, oil and

Table 1.2 Non-renewable versus renewable

Non-renewable	Renewable
Fossil fuels	Solar energy
Petroleum	Wind power
Natural gas	Hydropower
Coal	Geothermal
Nuclear	Tidal power
	Biomass

natural gas. Alternative energy includes nuclear energy, clean coal, hydrogen, recycled energy and other non-traditional fossil fuel energy sources such as landfill gas. Energy sources are classified as non-renewable or renewable in Table 1.2.

Nuclear energy is sometimes classified in a category by itself, because it has a large role in the global energy picture, representing nearly 20% of electrical power generation in the United States alone. Notwithstanding recent events, nuclear energy will likely play a significant role in reducing greenhouse gases and achieving energy independence goals. Currently, for example, the use of nuclear power avoids ~2.5 billion metric tons of carbon dioxide emissions annually on a global basis.

Clean coal refers to a set of technologies that are aimed at reducing the level of pollution associated with the burning of coal and minimizing the amount of carbon dioxide that is emitted into the atmosphere. It includes technologies such as coal gasification, coal liquefaction, and carbon capture and storage (CCS). Given the worldwide abundance of coal, these technologies are expected to be central to any energy future involving coal.

Hydrogen is a carrier of energy and not an energy source. It is considered an alternative energy because it does not generate CO_2 during combustion. Today, however, hydrogen is mainly produced from fossil fuel and to a lesser extent by electrolysis of water using energy produced from fossil fuels. In the future, hydrogen could be a stored form of energy produced from other forms of renewable energy such as wind.

Recycled energy refers to high quality waste heat from industrial processes that can be captured and then used to generate electricity through the use of a steam turbine generator or to produce steam or hot water for heating applications. Thermoelectrics are a material class aimed at using waste heat to generate electricity directly.

Landfill gas is another source of alternative energy. It is a natural by-product of the decomposition of organic waste in landfills, and typically it comprises primarily methane. Since methane is a greenhouse gas that is about 20 times more potent than carbon dioxide at trapping the sun's heat within the earth's atmosphere, it is critical to capture landfill gases (LFG) and to prevent their escape into the atmosphere.

1.4 Renewable Energy Resources

In general, renewable energy includes a broad range of technologies that generate energy from non-fossil fuel resources, including wind, sun, rivers, oceans and deep drilling underground in the case of geothermal resources. Wind energy is converted into electrical energy when the wind turns propeller-like blades, which spin a shaft that is connected to a generator.

Solar energy is the sun's energy that reaches the surface of the earth. In energy terms it represents approximately 1 $KW/m^2/h$ depending on the time of year, weather, longitude and altitude. Solar energy systems can be divided into two major types. Concentrating solar power technologies use mirrors to reflect and concentrate sunlight onto receivers that heat a fluid, such as water, which in turn drives a steam turbine.

Photovoltaic solar energy systems, which are referred to as PV solar, use semiconductor materials to convert light energy into electrical energy. Some PV solar energy systems are already considered cost competitive in the 10 most expensive energy markets in the US with some experts predicting that PV may become cost competitive with coal within the next 10 years.

Hydroelectric power is produced from flowing water, typically rivers and streams. It is estimated that approximately 20% of the world's electricity is generated through hydro power, which currently accounts for over 80% of electricity derived from renewable sources.

Biomass refers to all forms of organic matter that can be converted into energy through combustion, directly or as a liquid fuel. There are enormous biomass resources, such as wood wastes, primary mill residues, forest residues, agricultural waste and dedicated energy crops. Other biomass resources exist, but are more distributed and not easy to quantify.

Geothermal energy is a form of renewable energy derived from heat deep within the earth's crust. Geothermal energy has wide applicability for heating and cooling residential homes and commercial facilities, and it currently represents about 0.5% of US generation capacity.

Ocean energy involves harnessing the power of the tides, currents or waves. These forms of energy hold significant potential but they are still in the early stages of development with typically only prototype units in operation.

1.5 Energy Storage and Energy Conversion

For an excellent review of electrochemical systems, including batteries, fuel cells and supercapacitors, the review by Winter and Brodd [3] is recommended. For a progress report on sol-gel processing in batteries and fuel cells, as of 2004, there is a chapter in Sakka's comprehensive three-volume set on sol-gel technology [4]. Since that time, there has been an explosion in research on sol-gel processing for fuel cells, especially proton-exchange membrane fuel cells (PEMFC) [5–7] and direct-methanol fuel cells (DMFC) [8]. The development of DSSC has been a major achievement for sol-gel processing, due in large part to the success of the titania-based work of Grätzel [9].

In summary, energy, its sources, its storage and its conversion is a worldwide concern. Many ways to address this concern are dependent on materials and their processing. This book collects the present-day approaches using sol-gel gel processing. A positive result of this book is that it catalyzes new ideas to address energy needs.

Acknowledgments Tim Kinsella is thanked for sharing his vast knowledge of the energy industry and for working to promote alternative energy.

References

1. Baxter J, Bian Z, Chen G, Danielson D, Dresselhaus MS, Federov AG, Fisher TS, Jones CW, Maginn E, Kortshagen U, Manthiram A, Nozik A, Rolison DR, Sands T, Shi L, Sholl D, Wu Y (2009) Nanoscale design to enable the revolution in renewable energy. Energy Environ Sci 2:559–588
2. http://www.engineeringchallenges.org
3. Winter M, Brodd RJ (2004) What are batteries, fuel cells, and supercapacitors? Chem Rev 104:4245–4269
4. Klein LC, Aparicio M, Damay F (2004) Sol-gel processing for battery and fuel cell applications. In: Sakka S (ed) Sol-gel technology volume III. Kluwer Publishers, Boston, pp 311–328 (Chapter 15)
5. Aparicio M, Klein LC, Adjemian K, Bocarsly A (2002) SiO_2-P_2O_5-ZrO_2 Sol-gel/NAFION^{TM} composite membranes for PEMFC. In: Manthiram A, Kumta P, Sundaram SK, Ceder G (eds) Ceramic transactions 127: materials for electrochemical energy conversion and storage. American Ceramic Soc., Westerville, pp 167–176
6. Damay F, Klein LC (2003) Transport properties of NAFION/SiO_2-P_2O_5-ZrO_2 composite membranes for proton exchange membrane fuel cells. Solid State Ion 162–163:261–267
7. Aparicio M, Klein LC (2005) Synthesis and characterization of NAFION^{TM}/60SiO_2-30P_2O_5-10ZrO_2 sol-gel composite membranes for proton exchange membrane fuel cells. J Electrochem Soc 152:A493–A496
8. Daiko Y, Klein LC, Kasuga T, Nogami M (2006) Hygroscopic-oxides/Nafion® hybrid electrolyte for direct methanol fuel cells. J Membr Sci 281:619–625
9. Grätzel M (2004) Solar cells based on sol-gel films. In: Sakka S (ed) Sol-gel technology volume III. Kluwer Publishers, Boston, pp 295–310 (Chapter 14)

Chapter 2
Solid Oxide Fuel Cells

Chendong Zuo, Mingfei Liu and Meilin Liu

Abstract Solid oxide fuel cells (SOFCs) have potential to be the most efficient and cost-effective system for direct conversion of a wide variety of fuels to electricity. The performance and durability of SOFCs depend strongly on the microstructure and morphology of cell components. As a unique synthesis and processing technique with easy control of composition, structure, morphology, and microstructure, sol-gel processes have been widely used for fabrication of key SOFC materials or critical components with desired properties or functionalities unattainable otherwise. In this chapter, we provide an overview on sol-gel processes applied for preparation of homogeneous and fine powders of electrolyte, electrode, and ceramic interconnect materials, for deposition of dense electrolyte membranes or porous electrode films, and for modification of electrode or metallic interconnect surface or interface to enhance catalytic activity, to improve tolerance to coking or contaminant poisoning, and to increase stability against oxidation, reduction, or other degradation mechanisms. While significant progress has been made in controlling and tailoring the composition, microstructure, morphology, and hence functionality of materials and components, many challenges still remain to make sol-gel processes cost-effective and versatile for many applications. The development of novel sol-gel processes as well as the exploration of the existing ones to new applications continues to be an intriguing research pursuit.

C. Zuo · M. F. Liu · M. L. Liu (✉)
School of Materials Science and Engineering,
Center for Innovative Fuel Cell and Battery Technologies,
Georgia Institute of Technology, 771 Ferst Drive,
Atlanta, GA 30332-0245, USA
e-mail: meilin.liu@mse.gatech.edu

C. Zuo
e-mail: zuochd@hotmail.com

M. F. Liu
e-mail: mingfei.liu@mse.gatech.edu

M. Aparicio et al. (eds.), *Sol-Gel Processing for Conventional and Alternative Energy*,
Advances in Sol-Gel Derived Materials and Technologies,
DOI: 10.1007/978-1-4614-1957-0_2,

Keywords Anode · Cathode · Coatings · Electrode/electrolyte interface · Electrolyte · Gadolinia-doped ceria (GDC) · ILTSOFC · Lanthanum strontium manganate (LSM) · Magnesium doped lanthanum chromate · NiO/YSZ · Sol-gel · Solid oxide fuel cell (SOFC) · Surface modification · Yttria-stabilized Zirconia (YSZ)

2.1 Introduction

A fuel cell is a system that converts the chemical energy of a fuel directly to electricity. Compared to other types of fuel cells, a solid oxide fuel cell (SOFC) is an all-solid-state fuel cell based on a solid oxide electrolyte [1], which potentially offers the highest energy efficiency with minimum emissions and hold promise for direct utilization of a wide variety of fuels, from hydrogen to natural gas, coal gas, reformed gasoline or diesel, and gasified carbonaceous solids (e.g. municipal solid waste and biomass) [2–4]. SOFCs are simple, reliable, environmentally benign, and highly efficient (up to ~85% energy efficiency when combined with gas turbine) compared to engines and modern thermal power plants (~30%) [5].

2.1.1 Configuration of a SOFC

A single SOFC consists of an anode and a cathode separated by a solid oxide electrolyte (an ionic conductor), as schematically shown in Fig. 2.1a [6]. The solid electrolyte can be an *oxygen ion*, a *proton*, or a *mixed oxygen ion–proton* conductor, but it must be an electronic insulator (prohibiting the conduction of electrons or electron holes) and gas impermeable (in a dense membrane form). While SOFCs based on proton conductors (e.g., $BaZr_{0.1}Ce_{0.7}Y_{0.2}O_{3-\delta}$—based electrolytes) have attracted much attention in recent years, the most studied SOFC systems to date are based on oxygen ion conductors such as yttria-stabilized zirconia electrolyte (YSZ, with a composition of 8 mol.% Y_2O_3-92 mol.% ZrO_2, sometimes referred as 8YSZ); the anode is a porous nickel-YSZ cermet; and the cathode is a porous composite that usually contains YSZ and $La_{1-x}Sr_xMnO_{3-\delta}$ (LSM, usually x varies from ~0.15 to ~0.20). YSZ-based SOFCs usually operate at high temperatures (750–1,000°C) to be efficient because of the limited transport and catalytic properties of the SOFC materials at low temperatures. To reduce the operating temperature, doped ceria (such as $Ce_{0.9}Gd_{0.1}O_{2-\delta}$ or GDC and $Ce_{0.9}Sm_{0.1}O_{2-\delta}$ or SDC) have been used as the electrolyte and $La_{0.6}Sr_{0.4}Co_{0.2}$-$Fe_{0.8}O_{3-\delta}$ (LSCF) as the cathode for SOFCs to be operated at low temperatures (<700°C). Figure 2.1b is a schematic for a typical single SOFC based on an electrolyte of an oxygen ion conductor using hydrogen as the fuel and oxygen as the oxidant.

During operation, oxygen molecules are adsorbed, dissociated, and reduced on the cathode surface to ionic oxygen species before incorporated into the lattice as

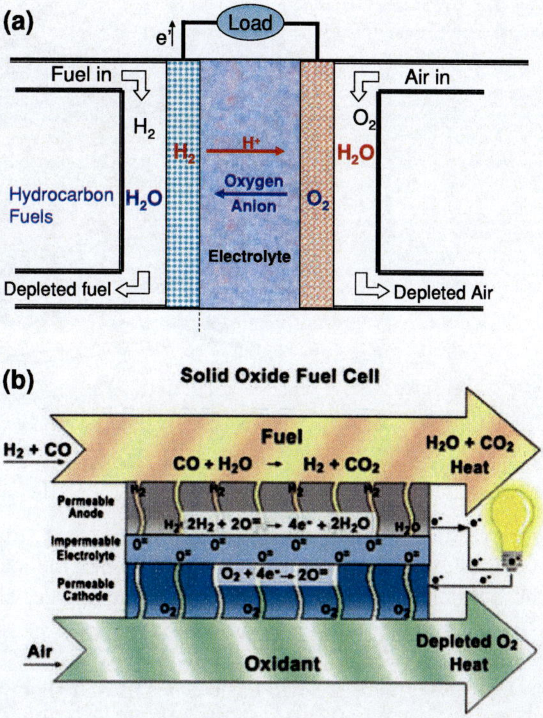

Fig. 2.1 Schematics of (**a**) SOFC based on different types of solid electrolyte [6] and (**b**) operating concept of an SOFC based on an oxygen ion conducting electrolyte [7]

oxygen ions, which then move through the electrolyte to the anode and combine with fuel molecules to form water and carbon monoxide/dioxide (if a hydrocarbon fuel is used). Outside the cell, electrons move from the anode to the cathode through an external circuit, converting chemical energy of the fuel to electrical energy. In Kroger's notation, oxygen reduction on the cathode can be described as follows:

$$\frac{1}{2}\mathrm{O}_2 + 2\mathrm{e} + V_O^{\bullet\bullet} = \mathrm{O}_O^X \tag{2.1}$$

At the same time, fuel molecule (e.g., hydrogen) is oxidized on the anode by combining with oxygen ions and release electrons:

$$\mathrm{H}_2 + \mathrm{O}_O^X = \mathrm{H}_2\mathrm{O} + 2\mathrm{e} + V_O^{\bullet\bullet} \tag{2.2}$$

The combination of the reactions (2.1) and (2.2) yields the overall reaction of the fuel cell,

$$\mathrm{H}_2 + \frac{1}{2}\mathrm{O}_2 = \mathrm{H}_2\mathrm{O} \tag{2.3}$$

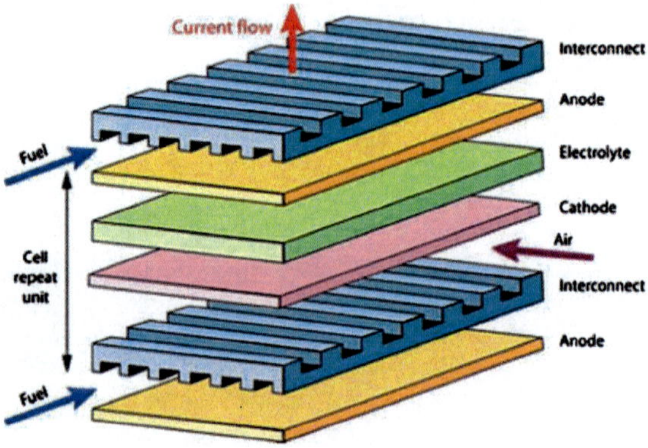

Fig. 2.2 Schematic for a planar SOFC design [8]

2.1.2 Solid Oxide Fuel Cell Structure

For practical applications, SOFCs have different structures, and each of them has its own advantages and disadvantages, though the materials for cell components in these different designs are either the same or very similar in nature.

- Planar versus tubular solid oxide fuel cells

Currently, there are two basic types of SOFCs in terms of cell structure: one is planar and the other is tubular. For planar SOFCs, each cell is made into a flat disk, square, or rectangular plate. The cells are put in series and connected by the interconnect plates, as schematically shown in Fig. 2.2. For tubular SOFCs, usually the electrode (either cathode or anode) is made into a long-tube with a porous wall. Outside the electrode tube are the electrolyte and then another electrode. Cells are also connected in series through interconnects, as schematically shown in Fig. 2.3. Earlier studies in SOFCs were focused on high temperature tubular SOFC systems; since the late 1990s, accompanied with the reduction of electrolyte thickness in the planar SOFC technology, the development of planar SOFC systems has drawn great interest due to its apparent advantages in power density and the ease of fabrication. However, tubular SOFC is still favorable for portable applications where rapid start-up and cool-down are required. The comparisons of planar and tubular structures are summarized in Table 2.1 [1].

- Electrolyte-supported versus electrode-supported solid oxide fuel cells

As illustrated in Fig. 2.1, a single SOFC has three layers: a porous anode and a porous cathode separated by a dense electrolyte membrane. The relative thickness for each cell component depends on the cell structure, electrolyte-supported or electrode-supported SOFCs, as illustrated in Fig. 2.4. In terms of processing technique, electrode-supported cells are more demanding than electrolyte-supported cells. However, electrode-supported structures are now more widely used. For example, for the tubular structure, Siemens adopts the cathode-supported

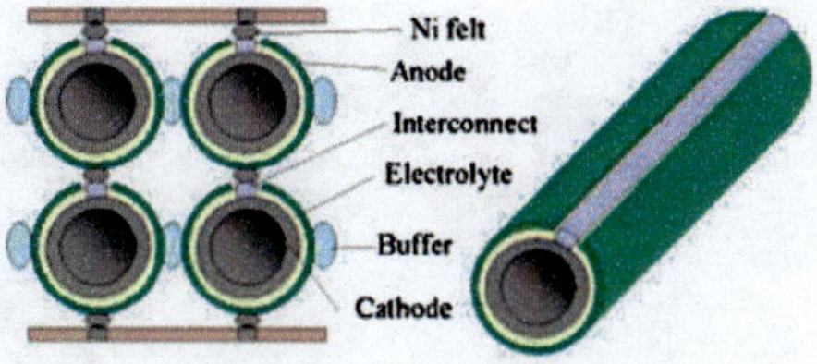

Fig. 2.3 Schematic for a tubular SOFC design [9]

Table 2.1 Comparison of planar and tubular structure for solid oxide fuel cells [1]

	Planar	Tubular
Power per unit area	Higher	Lower
Power per unit volume	Higher	Lower
Ease of fabrication	Easier	Difficult
Cost of fabrication	Higher	Lower
Ease of sealing	Difficult	Easy
Long-term stability	Fair	Excellent
Thermo-cycling stability	Fair	Good

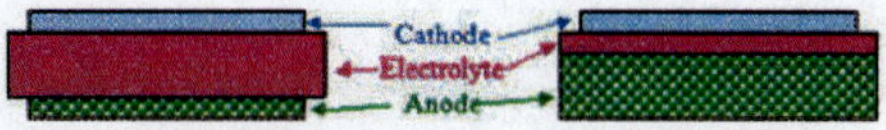

Fig. 2.4 Schematic for the cross-section of **(a)** an electrolyte-supported SOFC and **(b)** an anode supported SOFC

structure, and for the planar structure, most industrial teams adopt the anode-supported structure. The main advantage of the electrode-supported structure is that it provides thinner electrolyte and thus lower electrolyte ohmic resistance, which enables the operation of SOFC at lower temperatures, especially for the anode-supported planar structure [1].

2.1.3 Advanced Intermediate- and Low-Temperature SOFCs (ILT-SOFCs)

SOFC is currently attracting tremendous interest because of its huge potential for power generation in stationary, portable, and transport applications and of the increasing need for sustainable energy resources. The major current impediment of commercializing SOFCs is the high cost which results from high operating temperature (800–1,000°C). By lowering the operating temperature (500–750°C), the advanced intermediate- and low-temperature SOFCs (ILT-SOFCs) have the potential to greatly reduce the cost of interconnect, manifolding, and sealing materials, in addition to improved reliability, portability, and operational life [10]. However, the interfacial polarization resistances between electrolyte and

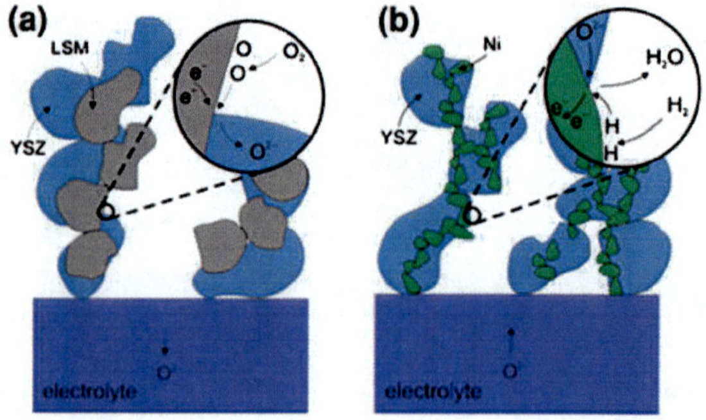

Fig. 2.5 Schematic of (**a**) Reduction reaction on the TPB of a cathode made of LSM-YSZ and (**b**) Oxidation reaction on the TPB of an anode made of Ni–YSZ [13]

electrodes increase dramatically as the operating temperature is reduced [11]. Thus, the development of novel electrode materials and/or unique microstructure is one of the critical issues in development of new generation SOFCs.

As shown in Fig. 2.5, a critical part of most fuel cells is often referred to as the triple-phase boundary (TPB), where the actual electrochemical reactions take place, are found where reactant gas, electrolyte, and electrode meet each other. For a site or area to be active, it must be exposed to the reactant, be in electrical contact with the electrode, be in ionic contact with the electrolyte, and contain sufficient electro-catalyst for the reaction to proceed at a desired rate. The density of these regions and the microstructure of these interfaces play a critical role in the electrochemical performance of SOFCs [12]. Thus porous electrodes with fine particle size are preferred in SOFCs to achieve high surface area which significantly increases the length of TPB. In order to reduce the operating temperature, the resistive loss occurring in the electrolyte should also be minimized. One of the solutions is to decrease the thickness of the solid electrolyte from several hundred micrometers, the usual thickness in conventional electrolyte-supported cell, to a range close to ten micrometers (electrode-supported cell). Therefore, advanced fabrication processes are desirable in order to create dense, thin electrolyte and porous electrode films with fine microstructures so that electrochemical performance can be enhanced at lower operating temperatures.

2.1.4 Applications of Sol-Gel Process in SOFCs

Sol-gel processes have been widely used for synthesis of an inorganic network through a chemical reaction in a solution at low temperatures. The most obvious features of this reaction, the transition from a liquid into a solid (di- or multiphasic gel), led to the expression "sol-gel process" [14]. An overview of the sol-gel process is presented in a simplified chart (Fig. 2.6).

Sol-gel chemistry is a remarkably versatile approach for fabricating materials and components. The advantages of the sol-gel processes include excellent control

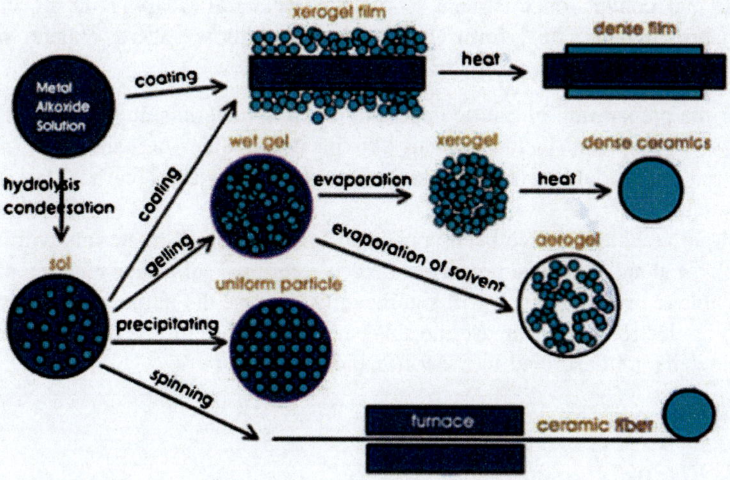

Fig. 2.6 Simplified Chart of sol-gel processes [14]

of microstructure and ease of compositional modification at relatively low temperatures by using simple and inexpensive equipment [15]. Sol-gel techniques have been used not only for powder synthesis or thin-film coating but also for modification of electrode surfaces or electrode/electrolyte interfaces.

2.2 SOFC Material Powders Derived from a Sol-Gel Process

The sol-gel method of obtaining ceramic oxide materials has been extensively developed and consists essentially of three steps [15, 16]:

1. Preparation of a starting solution. It involves mixing low viscosity (to ensure homogenization at molecular level) solutions of suitable 'precursors', i.e., metal derivatives; these precursors can in some instances be the 'metal oxide' sols themselves
2. Gelling stage: consists of forming a uniform sol and causing it to gel; this is the key step in the process to endow chemical homogeneity on the ceramic product during desiccation. The transition from the sol to the gel state can be achieved in three different ways:

   - growth of polymeric molecules (which crosslink randomly to a three dimensional network)
   - growth of individual particles (which grow together as they become larger)
   - stabilization of colloids by surface charges (change of the zeta potential and a following interparticle condensation process leads to gelation).

3. Thermal conversion treatment to the final material: shaping during or after gelation into the final form (bulk materials, hollow spheres, fibers, surface coatings, etc.) before firing.

For the preparation of multicomponent ceramic materials, the most commonly employed precursors include metal alkoxides, and other derivatives of metals, such as, metal oxides (sols or solids), nitrates, acetates (carboxylates), and β-diketonates.

Sol-gel techniques have been applied to fabricate SOFC materials with favorable physical and chemical properties that conventional solid state reaction method are unable to provide, which will significantly increase the number of reaction site (TPB) in electrodes or decrease the sintering temperature of the dense electrolyte. More details are described in the following sections.

2.2.1 Electrolyte Materials

- Synthesis of YSZ electrolyte powders

The most widely used electrolyte material for SOFCs is yttria-stabilized zirconia (YSZ) with a typical composition of $(ZrO_2)_{0.92}(Y_2O_3)_{0.08}$ or 8YSZ. The ionic conductivity of YSZ varies with dopant concentration and increases exponentially with temperature [1].

Figure 2.7 illustrates a flow chart for the synthesis of ultrafine YSZ powders through a sol-gel process [17]. The precursors were the zirconium propoxide ($Zr(OPr)_4$) and yttrium nitrate hexahydrate in 1-propanol. After the gelation step at 50°C, samples were dried at 80°C for a minimum of 24 h to obtain a xerogel. Then, they were calcined at 950°C for 2 h. XRD analysis indicates that the obtained powders have a single phase [18]. Figure 2.8 illustrates these nano-structured YSZ powders form an agglomerate with an average primary particle size of 50–100 nm and have nearly spherical morphology. These nano-engineered YSZ particles with controlled morphology and particle size can aid in the packing of particles in the green films and hence enhance sintering kinetics. This, in turn, assists densification of electrolyte films at lower firing temperatures.

- Synthesis of doped ceria electrolyte powders

With high ionic conductivity between 500 and 700°C, doped cerias have been extensively studied as electrolytes in reduced-temperature SOFCs. Gadolinia-doped ceria (GDC, $Ce_{0.9}Gd_{0.1}O_{1.95}$) is considered to be one of the most promising electrolytes for SOFCs to be operated below 650°C [19]. Further, doped cerias have also been successfully used as part of anodes for SOFCs, especially those using hydrocarbon fuels [20, 21]. Nano-crystalline GDC powder has been prepared by a sol-gel thermolysis method [20]. After the GDC gel precursors calcined at 400°C, the powders showed cubic fluorite structure with an average crystallite size

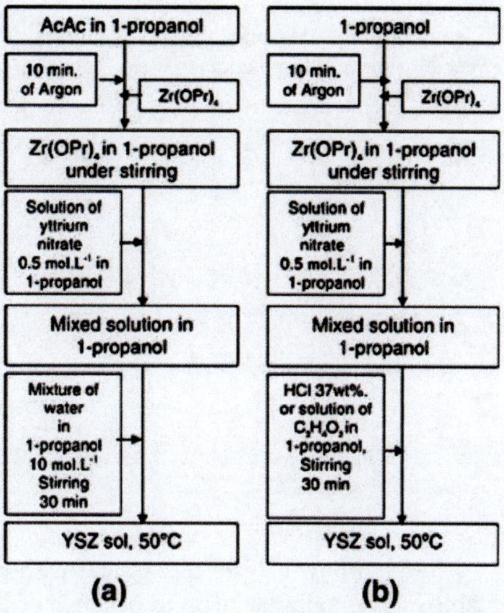

Fig. 2.7 **a** Procedure of synthesis with AcAc; **b** Procedure of synthesis in acid environment [17]

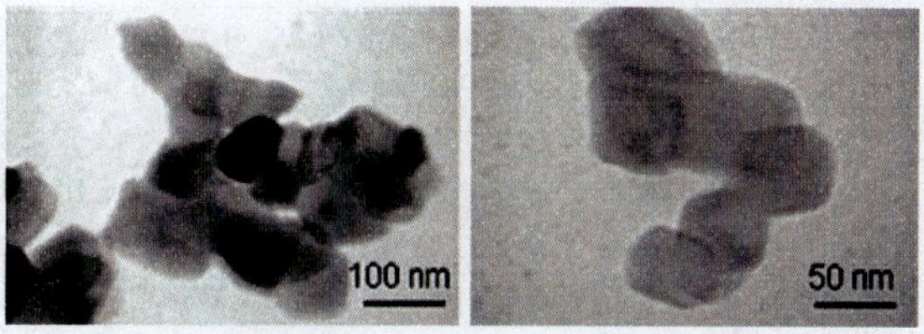

Fig. 2.8 TEM features of a micrograph of gel calcinated at 400°C with a CA/EG RATIO = 2.4 [18]

of 10 nm. The powders calcined at lower temperatures showed better sinterability and higher ionic conductivity ($\sim 2.21 \times 10^{-2}$ S cm^{-1} at 700°C in air).

2.2.2 *Cathode Materials*

Sol-gel process has been widely used for synthesis of cathode materials, such as perovskite LSM [22], $(Pr_{0.7}Ca_{0.3})_{0.9}MnO_{3-\delta}$ [23], Fe-doped LSM [24], LSCF [25, 26], $Pr_{1-x}Sr_xCo_{0.8}Fe_{0.2}O_{3-\delta}$ [27], La, Pr, Sm and/or Ba co-doped $Ln_{0.58}Sr_{0.4}Fe_{0.8}Co_{0.2}O_3$ [28, 29], Sc doped $SrCoO_3$ [30], BSCF [31], $La_{1-x}Sr_xCuO_{3-\delta}$ [32]; layered perovskite oxide of $PrBaCo_2O_{5+\delta}$ [33], PrBaCu-$CoO_{5+\delta}$ [34], $GdBaCo_2O_{5+\delta}$ [35], and K_2NiF_4 type cathode materials La_2Ni_{1-x}-$Cu_xO_{4+\delta}$ [36], $La_{2-x}NiO_{4+\delta}$ [37] etc. In order to reduce interfacial polarization resistance, an important factor to consider is the grain size of the porous cathode. The TPBs can be increased by decreasing grain size or increasing the surface to volume ratio. Sol-gel process has been studied for the preparation of highly homogeneous and fine cathode powders [38, 39].

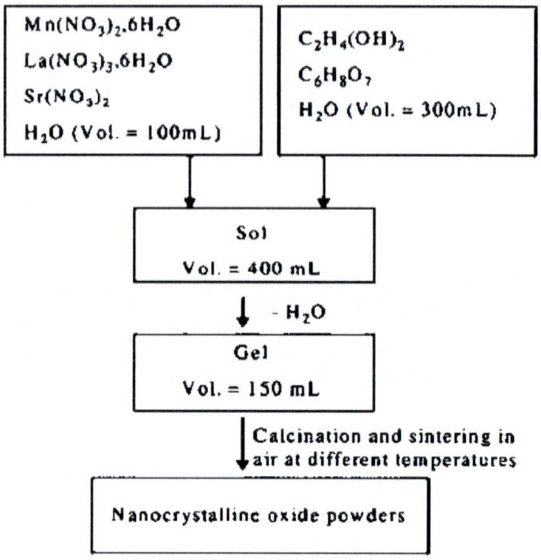

Fig. 2.9 A flow chart illustrating the processing procedure for LSM*x* powders preparation [22]

LSM is known to be a classical cathode material for SOFCs based on yttria-stabilized zirconia electrolyte because of its high electrical conductivity, excellent thermal, chemical stability and compatibility with the zirconia-based electrolyte at the working temperatures. Shown in flow chart of Fig. 2.9 is a typical sol-gel process for LSM powders [22]. In this process, the $La(NO_3)_3 6H_2O$, $Mn(NO_3)_2 6H_2O$ and $Sr(NO_3)_2$ precursors were dissolved into water. Ethylene glycol and citric acid were used as polymerization/complexation agents, respectively. The stable solution was then heated on a thermal plate, where polymerization occurs in the liquid solution and leads to a homogeneous sol. When the sol is further heated to remove the excess solvent, an intermediate resin is formed. Calcination of the resin at 400°C in air was performed before final sintering at various temperatures (600–1,000°C). LSM powder has a cubic crystalline structure with particle sizes as small as 40 nm.

Recently, Zhou et al. [40] developed a simple in situ sol-gel derived carbon templating process to synthesize nano-sized $La_{0.6}Sr_{0.4}Co_{0.2}Fe_{0.8}O_{3-\delta}$ (LSCF) and $La_{0.8}Sr_{0.2}MnO_{3-\delta}$ (LSM). These perovskite oxides were prepared using an EDTA-citrate complexing process to facilitate homogeneous mixing of the metal ions in the molecule level. After calcination of the dried gel under a reducing condition at high temperatures, perovskite oxide and carbon particles were formed simultaneously, producing nano-sized LSCF–carbon and LSM–carbon composites with a grain size of 20–30 nm. Further calcination of the obtained composites in air removes the carbon, resulting in nano-sized LSCF and LSM with a crystalline size of 14 nm, which is smaller than that prepared by the calcination of the solid precursor in air directly (18–22 nm). Such a decrease in crystalline size of perovskite via the indirect calcination process was ascribed to the suppressing effect of carbon in the grain growth of perovskite. Furthermore, when the in situ created carbon was used as a template for pore forming, a highly porous microstructure was obtained. Figure 2.10 shows the typical surface morphology (SEM image) of a sintered pellet [40] with porosity of ~75%, as determined by the

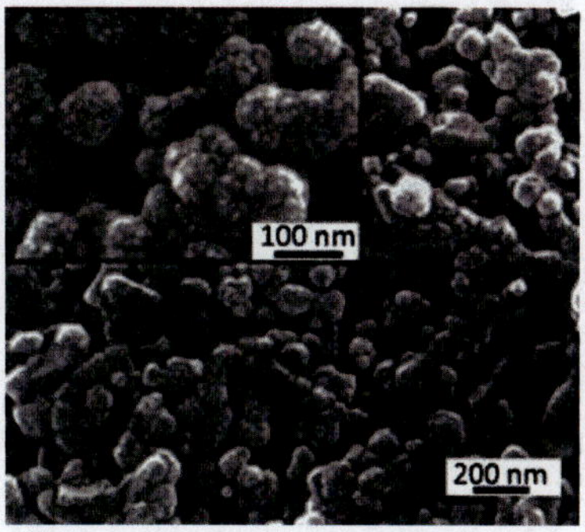

Fig. 2.10 Typical surface morphologies of a porous LSCF pellet prepared using the in situ carbon templating process after firing at 1,000°C in air for 5 h [40]

Archimedes method, which is much higher than those obtained from other processes (30–40%). The pores size with irregular pore shape was varied from several μm to nm while the particle size was in the range of 10–20 nm. The high porosity and high surface area make this in situ carbon templating process very promising for the preparation of porous electrode and catalysts for chemical and energy transformation applications.

2.2.3 Anode and Interconnect Materials

Traditionally anode materials are binary mixtures of nickel (Ni) and yttria-stabilized zirconia (YSZ) particles with a porosity of above ~30%. Such cermets meet all of the requirements for the anode: the interconnecting Ni network acts as a good catalyst for the electrochemical oxidation of hydrogen and provides an excellent conduction path for electrons released from the electrochemical oxidation of hydrogen; the interconnecting electrolyte material network provides a path for the oxygen ions from the electrolyte and also acts as a constraint for the growth and coalescence of nickel particles so that the fine nickel network could be maintained after long-term operation at elevated temperatures. The performance of the anode, in terms of minimal electrode polarization loss and minimal degradation during operation, depends strongly on its microstructure, and therefore on the precursor powders [41, 42]. It is important that both an ionic conducting network is formed by the YSZ particles and an electronically conducting one is formed by the Ni particles, and that the TPB between the two networks and the gas-phase is long. Furthermore, the relative size of the particles is important, the YSZ particles need to be much smaller than the Ni particles to minimize degradation due to Ni-particle sintering and coarsening during operation [41, 43].

- Synthesis of NiO–YSZ anode powders

The standard way of producing the anode precursor powder mixture is to produce YSZ and NiO particles separately and mix them in a mill [44]. However,

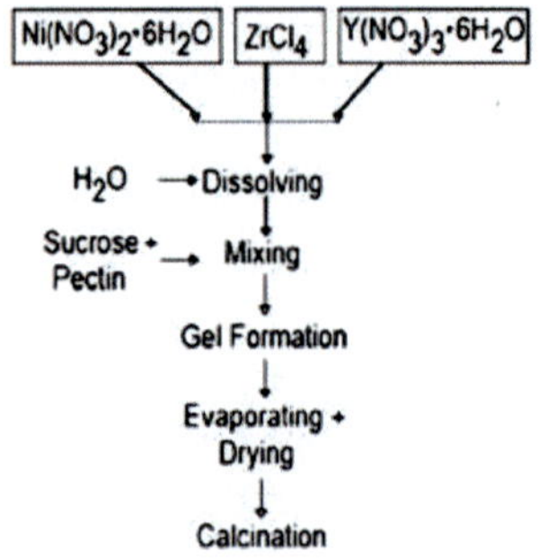

Fig. 2.11 A flow chart illustrating the preparation of samples [41]

it would be advantageous to produce the powder mixture in one single processing step via a sol-gel process (Fig. 2.11) because the mixing of species occurs on the atomic scale [41], Zirconium tetrachloride ($ZrCl_4$), yttrium nitrate hexahydrate ($Y(NO_3)_36H_2O$), and nickel nitrate hexahydrate ($Ni(NO_3)_26H_2O$) were used as precursors. The obtained xerogel is then calcined to obtain NiO/YSZ powder. During calcination this polymeric metal ion complex is decomposed into CO_2 and H_2O, and their escape from the reaction mixture prevents agglomeration by ensuring that the mixture remains porous. Thus fine particles, which only moderately agglomerated, are formed as the final powder. The Fig. 2.12 illustrated that the agglomerates of NiO/YSZ samples are quite similar in all final powders with different NiO/YSZ ratios. The mean particle sizes of the powders ranging from 28 to 31 nm and a well-crystallized NiO and YSZ powder mixture is formed even for the samples calcined at 800°C, as indicated by XRD analysis.

- Synthesis of LSCM anode powders

Significant effort has been devoted to the development of anode catalysts for SOFCs run on carbon-containing fuels [45, 46]. For example, $La_{0.75}Sr_{0.25}Cr_{0.5}Mn_{0.5}O_3$ (LSCM) perovskite oxide exhibited excellent redox stability and catalytic activity in both methane and hydrogen SOFCs [47–49]. LSCM powders were prepared using a sol-gel combustion method [50]. $La(NO_3)_3$, $Sr(NO_3)_2$, $Cr(NO_3)_36H_2O$ and $Mn(NO_3)_26H_2O$ salts were first dissolved in deionized water. Then citric acid was added as a chelating agent. After being heated on a hot plate to evaporate water to become dry gel, the resulting powder was calcined at 1,000°C to prepare LSCM powders. The XRD pattern showed that pure LSCM perovskite oxide was obtained and the TEM image (Fig. 2.13) showed that the average particle size was about 200 nm. Thus, when compared to the powders derived from a solid-state reaction method, the LSCM powders derived from the sol-gel combustion method at a relatively low calcination temperature had uniform morphology [51].

- Synthesis of LMC anode powders

Magnesium doped lanthanum chromate ($La_xMg_{1-x}CrO_3$) is considered as interconnect material for SOFC applications. The nano powder of this interconnect

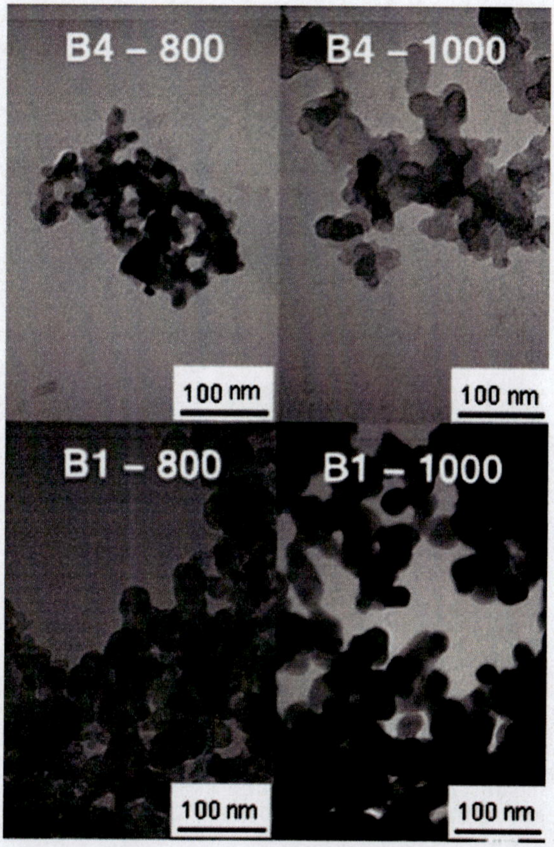

Fig. 2.12 TEM images of the B samples calcined at 800 and 1,000°C [41]

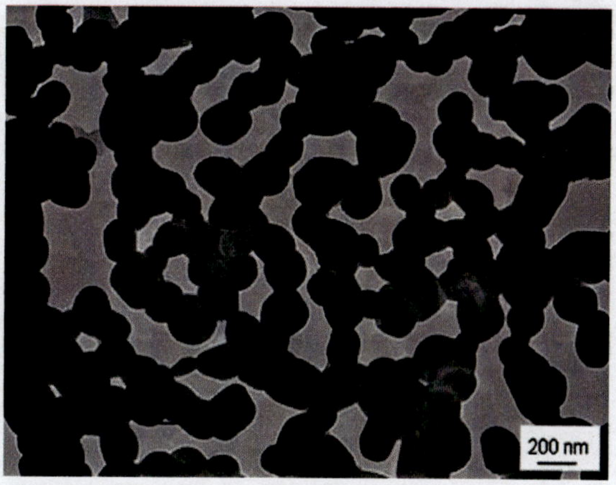

Fig. 2.13 TEM image of as-prepared LSCM powders [50]

material was prepared using sol-gel process [52]. The precursors were $La(NO_3)_36H_2O$, $4MgCO_3Mg(OH)_25H_2O$, $(NH_4)_2Cr_2O_7$ and $Cr(NO_3)_39H_2O$ and the chelating agent was citric acid and the dispersant agent was ethylene glycol. The SEM images reveal that the particle size of $La_{0.7}Mg_{0.3}CrO_3$ powders achieved by sol-gel process is in the range of ~50–200 nm. The powders prepared using $(NH_4)_2Cr_2O_7$ is more cost-effective than using $Cr(NO_3)_39H_2O$ since it also acts as fuel. The TGA plots depict that there is no further weight loss after reaching 350 and 575°C for the LMC gels prepared using AD and Cr–N, respectively, an indication of complete combustion of the precursors and formation of oxide phases.

2.3 Fabrication of SOFC Components Using a Sol-Gel Process

The performance of a SOFC depends strongly on the microstructures of its components. For a single cell, each component should have specific microstructures (in addition to the requirements on intrinsic properties of the materials) in order to achieve high efficiency: a porous, gas-permeable electrode (cathode and anode) with high specific surface area to increase the TPB length and thin, gas-tight electrolyte membrane to reduce the ohmic resistance. Sol-gel process has unique advantages for the fabrication of this kind of ceramic films because it does not require costly equipment, allows a lower processing temperature, and can control microstructure and chemical composition easily. The sol-gel process has been used not only for fabrication of thin electrolyte membranes but also for modification of electrode surfaces and electrode/electrolyte interfaces to improve the microstructure and electrochemical performance of cell components. More details of the sol-gel process for fabrication of cell components are described in the following sections.

2.3.1 Electrolyte

For efficient operation of a SOFC at low or intermediate temperatures, the electrolyte must be used in a thin-film form to reduce the ohmic losses. Further, the electrolyte membranes must be continuous and crack-free in order to prevent gas leakage, and must rest between two porous electrodes through which gases can pass freely to or away from the active sites for electrochemical reactions near the electrode–electrolyte interface. Thus, the fabrication of defect-free, dense electrolyte films on porous electrodes assumes significant importance.

There are a number of techniques for depositing thin films of ceramics on dense and porous substrates, including chemical vapor deposition, electrochemical vapor deposition, and various sputtering processes using ion beam, magnetron, electron beam, and so forth [53]. The drawbacks of these physical deposition techniques include difficulties in obtaining good compositional homogeneity [54, 55] and high costs due to the requirements of vacuum conditions [55]. Sol-gel techniques, on the other hand, overcome these problems and offer many additional advantages [56, 57]. Certainly one of the most technologically important aspects of sol-gel process is that, prior to gelation, the fluid sol or solution is ideal for preparing thin films by such common processes as dipping, spinning, or spraying. Both dense and porous structures can be easily tailored through the precursors and there is no limitation on the shape or size of the substrate surface. However, the most important advantage of sol-gel process over conventional coating methods is the ability to control precisely the microstructure (pore volume, pore size, and surface area) and composition (stoichiometry) of the film [58–60]. The disadvantages of sol-gel process include the cost of the raw materials (especially alkoxide precursors), shrinkage that accompanies drying and sintering, and processing times. Thin-film coating is one of the well-known applications of sol-gel process which benefits from most of the advantages of sol-gel process just cited while avoiding these disadvantages [58, 61].

Starting from a molecular precursor, the polymeric sol can be prepared by partial hydrolysis of corresponding metal alkoxide. If the rate of hydrolysis or condensation is very fast, then some kinds of organic acids, beta-dicarbonyls, and alkanolamines have been used as chelating agent in sol-gel processes to control the extent and direction of the hydrolysis-condensation reaction by forming a strong complex with alkoxide. For this reason, extremely thin, dense, and well-defined electrolyte films can be derived from sol-gel coatings on a porous electrode to form an electrode supported SOFC and the thickness can be controlled by the number of applied coatings. Purity and uniformity can also be controlled by the solution chemistry of the sol-gel process.

YSZ films with different thicknesses, such as ultra-thin film (0.1–10 μm) [53–56] and thick YSZ (8–100 μm) films [58–60], have been successfully deposited onto dense or porous electrode using a sol-gel process. Since the internal resistance of the cell decreased with decreasing the electrolyte thickness, high performance of 477 mW/cm^2 at 600°C and 684 mW/cm^2 at 800°C were achieved in an anode supported SOFC, which consists of a Ni–YSZ anode, a 4YSZ electrolyte (0.5 μm), and a Pt/Pd cathode Cathode [62]. Other electrolyte materials, such as SDC [61], BCG [63], zirconia films doped with Sc_2O_3 and Al_2O_3 (85 mol.% ZrO_2-11 mol.% Sc_2O_3-4 mol.% Al_2O_3) [64], have also been successfully fabricated via a sol-gel process.

Sol-gel process was also used for preparing an electronic blocking layer, such as thin YSZ layer, on doped ceria electrolyte to enhance the OCV [65–67]. Kim et al. [67] successfully fabricated 2 μm thick YSZ layer on YDC electrolyte surface using a sol-gel spin coating method. The open circuit voltage (OCV) of a single cell based on this composite electrolyte was ~0.5 V higher than that for an uncoated YDC electrolyte at temperatures from 700 to 1,050°C. The maximum power density of the single cell with the bi-layer electrolyte was 122 mW/cm^2 at 800°C, comparable to that of an YSZ single cell with the same electrolyte thickness at 1,000°C (~144 mW/cm^2).

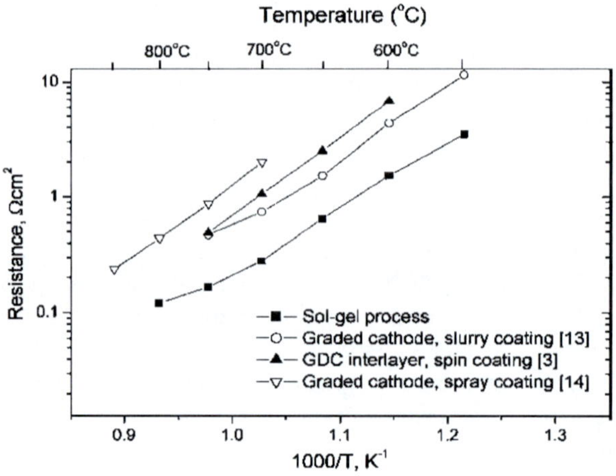

Fig. 2.14 Interfacial polarization resistances of LSM-GDC electrodes on a YSZ electrolyte prepared by different processes [71]

Another application of sol-gel process is to enhance the density of the YSZ layer through sol-gel coating on the prepared YSZ layer [68, 69] using other ceramic process. Burgess et al. [68] found that the sol-gel coated YSZ layer reduced the gas permeation through the electrolyte layers by up to 37% after firing to 500 or 650°C compared to that of YSZ layer (prepared by plasma spray YSZ coating). The sol-gel coatings led to a substantial increase in OCV and a small decrease in series resistance due to the decreased porosity of the electrolytes and a large decrease in polarization resistance due to increased interfacial surface area. Sol-gel coating has been also tried to remove defects produced after the slurry coating of YSZ on a porous anodeAnode [69].

2.3.2 Cathode

One of the most important goals in SOFC research is the reduction of the working temperature from high temperature (such as 900–1,000°C) to intermediate temperature (500–800°C) for future applications. However, at intermediate temperature, one of the main limiting factors is the high polarization resistance of the cathode, which limits SOFC power densities. In order to improve the cathode performance, the cathode has to be porous and have a small grain size to optimize the active surface area and the TPB length [70]. Sol-gel process is ideally suited for assembling such microstructures. The advantages of a sol-gel process over other technologies include: (i) the microstructure and composition of electrode materials can be controlled with relative ease, (ii) the electrode and electrolyte adherence is strong, and (iii) low-temperature processing is possible.

Xia et al. [71] developed a modified sol-gel process to fabricate LSM/GDC composite cathode for honeycomb SOFCs based on stabilized zirconia electrolytes. The sol-gel process derived cathodes show much lower interfacial polarization resistances than those prepared by slurry coating or spray deposition (shown in Fig. 2.14). The interfacial polarization resistances were 0.65 and 0.16 Ω cm^2 at 650 and 750°C, respectively, as determined using impedance spectroscopy. The high performance suggested that the sol-gel process has the advantage of developing strong bonding between the electrode and electrolyte at low temperatures and achieving desirable microstructures for fuel cell electrodes: small grains, high porosity, large surface area, and long TPB.

It is believed that cathode performance can be improved by using a thin layer with flat surface and small grains diameter at the cathode/electrolyte interface, covered by a thicker and porous layer with larger grains. This duplex microstructure design will increase the reaction zone and ensures the current collection by the additional porous layer [37]. Under this guidance, Fontaine et al. [37] developed thin graded porosity and composition of $La_{2-x}NiO_4$ ($x \geq 0$) cathode interlayer onto YSZ substrates using a sol-gel process. Through careful control of organic concentration, the ratio of transition metals, the number of coats, and the withdrawal speed, homogeneous, crack-free, and porous $La_{2-x}NiO_4$ ($x \geq 0$) films were successfully fabricated by dip-coating the sol onto YSZ substrates. The results indicated that low organic concentration allows the synthesis of cathode interlayer with fine microstructure and spherical grain of about 50 nm, while coarse microstructure with platelets is obtained for higher concentration. The film thickness was influenced by the number of coatings and the withdrawal speed. Controlled both film microstructure and composition can be easily achieved by changing processing parameters. Unfortunately, the authors did not report the electrochemical properties of the interlayer derived from the sol-gel process.

Sol-gel process is an ideal tool for functionally graded multilayer structure formation. Functionally graded 4-layer cathode was prepared by sol-gel/slurry coating [72]. The interlayer in contact with the electrolyte consisted of 50% LSM and 50% GDC and the composition was changed from a catalytically active layer to a current collection layer. The particulate sol of GDC was prepared and then the sol was mixed with LSM or other current collecting materials to produce the slurry for multilayer coating. The cathode interfacial polarization resistances for this graded cathode fired at 900°C are 0.21 and 0.10 Ω cm^2 at 700 and 800°C, respectively.

Another advantage of the sol-gel process is lower processing temperature, avoiding undesirable reaction between electrolyte and cathode materials. The cathode materials, such as lanthanum strontium cobalt ferrite (LSCF), strontium-doped samarium cobaltite (SSC), have attracted increasing attention for intermediate temperature SOFC due to their mixed-conduction characteristics and relatively high ionic conductivity. However, those materials have a strong tendency to react with YSZ to form insulating phases, like $La_2Zr_2O_7$ or $SrZrO_3$, at the cathode-electrolyte interface at high temperatures. Sol-gel can significantly reduce the sintering temperature of the cathode. Tang et al. [73] successfully prepared

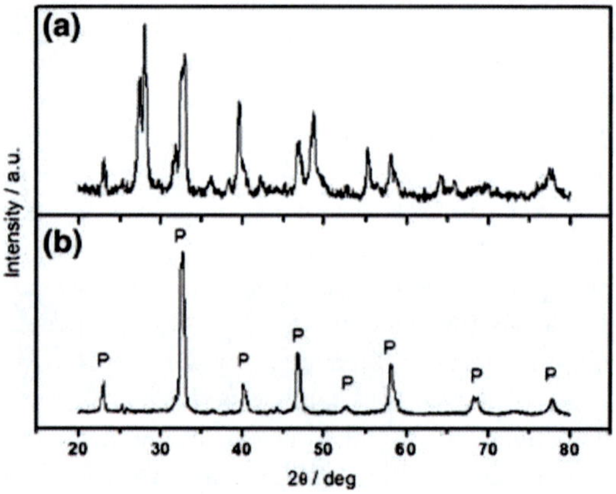

Fig. 2.15 XRD patterns of the decomposition products from LSM precursors (**a**) without and (**b**) with Triton X-100, sintered at 800°C for 1 h. [75]

200 nm thick SSC film on YSZ using dip-coating from the alkoxide precursor solution modified with an organic polymer (polyethylene glycol (PEG)). They found single-perovskite phase SSC films were formed when the films were heated at 700°C, and the crystalline size of the SSC film was around 50–100 nm when PEG was added. Other materials, like strontium-substituted lanthanum ferromanganites, $La_{0.8}Sr_{0.2}Mn_{1-y}Fe_yO_Z$ ($y = 0, 0.2, 0.5, 0.8, 1$), LSMF2Y ($Y = 10y$) films have been synthesized by a polymeric sol-gel route and deposited on YSZ substrates by a dip-coating process [24].

Sol-gel process has also been widely used for cathode infiltration, such as YSZ or SDC infiltrated LSM [74], LSM infiltrated YSZ [75], copper manganese spinel infiltrated YSZ [76]. Sholklapper et al. [75] carefully studied the phase formation for the LSM infiltrated YSZ. They found that the addition of the commercial polymeric dispersant to the nitrate precursors facilitated the formation of perovskite LSM particles at low temperatures. Figure 2.15 displays XRD patterns for the nanoscale synthesis of $La_{0.85}Sr_{0.15}MnO_3$, with and without the surfactant Triton X-100. As indicated in Fig. 2.15a, directly decomposing the nitrate precursors at 800°C did not yield a phase-pure LSM perovskite. In contrast, in the presence of the surfactant, the majority of characteristic peaks in Fig. 2.15b correspond to the LSM perovskite phase. Presumably, the surfactant has the effect of complexing the metal ions, so that the individual oxides do not form prematurely. Such complexing effects are well known in a sol-gel process.

2.3.3 Anodes

More recently, metal supported cells, which uses metal mechanical support deposited with ceramic active materials (anode, electrolyte and cathode), have been developed for a new generation of fuel cells [77, 78]. However, the use of metallic material as cell support induces thermal limitations such as the decrease in

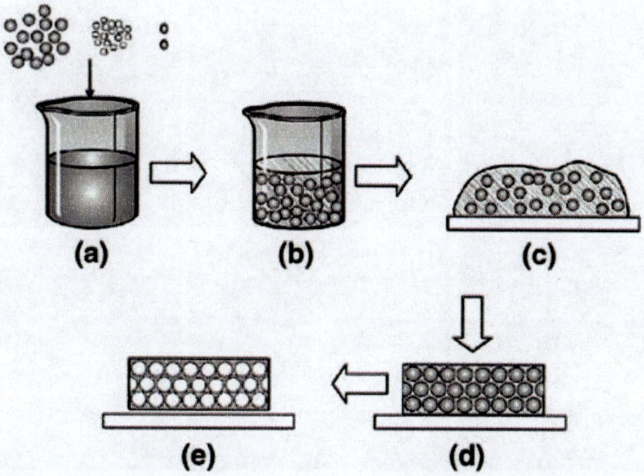

Fig. 2.16 Schematic diagram of the preparation of thin layer materials with macroporous microstructure. Starting metal salts are dissolved in distilled water, citric acid was used as complexing agent and PMMA microspheres as pore former (**a**). The homogeneous suspension (**b**) is deposited on a substrate (**c**) rendering an ordered template of PMMA microspheres with an infiltrated citrate gel of stoichiometric cations (**d**). An inorganic porous microstructure is obtained after calcination (**e**) [81]

anodic material sintering temperature. Indeed, a heat thermal treatment above 800°C would damage metal support, either mechanically or by corrosion. Rieu et al. [79, 80] successfully prepared both thin (around 100 nm) and thick (10–20 μm) NiO–YSZ anodic film on dense and porous metallic supports. A low thermal treatment at only 800°C leads to a coherent anodic duplex stacking which is continuous, homogeneous and adherent.

2.3.4 Other Components Derived from a Sol-gel Process

Sol-gel process is also employed to prepare thin layer oxides with controlled macroporous microstructure for SOFC applications [81]. Shown in Fig. 2.16 is the schematic diagram of the preparation of the thin layer materials. The method uses aqueous stoichiometric cation solutions of metal nitrates, citric acid as complexing agent and polymethyl methacrylate (PMMA) microspheres as pore former. They have successfully prepared several mixed oxides with fluorite and perovskite-type structures, i.e. doped zirconia, ceria, ferrites, manganites, and NiO–YSZ composites. The synthesized materials are nanocrystalline and have a homogeneous pore distribution and relatively high specific surface area. Some of the microstructures are presented in Figs. 2.17 and 2.18. Homogeneous and ordered pore arrangement with uniform wall thickness were achieved for all of the samples. The specific structure of the macroporous microstructure makes them interesting for SOFC and catalysis applications in the intermediate temperature range.

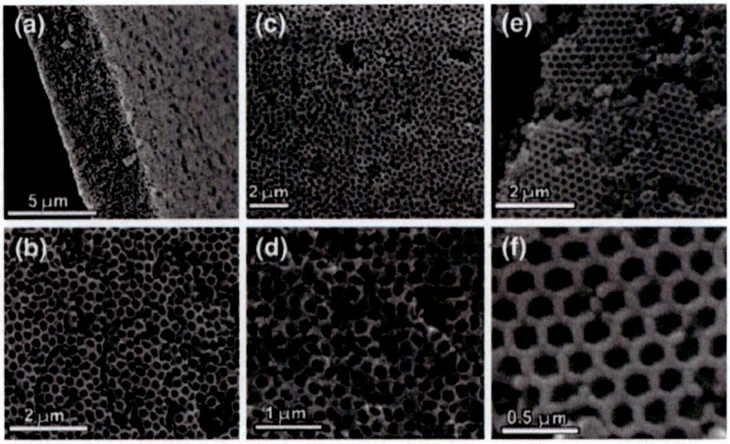

Fig. 2.17 SEM images of the different electrolytes: **a** and **b** cross-section and surface view, respectively of the YSZ layers; **c** and **d** surface of LMO at different magnifications; **e** and **f** surface of CGO at different magnifications [81]

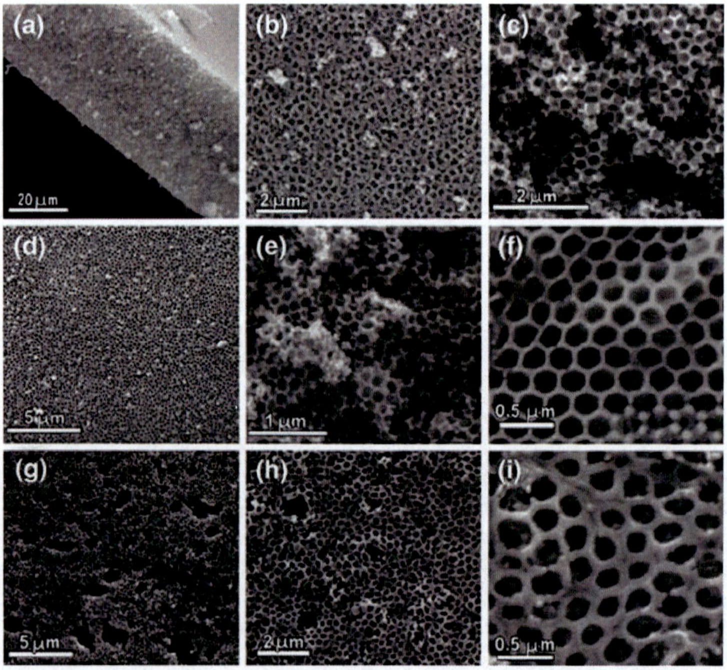

Fig. 2.18 SEM images revealing the microstructural details of the surface and cross-section view of different electrodes: **a–c** LSM, **d–f** LSF and **g–i** LSTMG [81]

In summary, the sol-gel technique has been widely used in recent years to form fuel cell components with enhanced microstructures for SOFCs. Both dense and porous structure can be easily tailored through the precursors and there is no limitation in the shape or size of the substrate surface. However, the sol-gel method also has some disadvantages such as a complex synthesis route; the quality of the coating was influenced much by many process parameters. And the typical gels show inherently large shrinkage, usually 20–25 vol%, which will introduce huge shrinkage stresses during drying/heating process. For a thick film application, in order to avoid crack formation, coating and drying/heating processes have to be repeated several times to get the designed thickness, which is time-, labor- and energy-intensive process. The sol-gel process also needs further development to deposit micro-structured materials in a selective area such as the triple-phase boundary of a fuel cell.

2.4 Surface Modification by a Sol-Gel Process

As mentioned in Sect. 2.3, ultra-thin coating can be easily achieved using a sol-gel process and the coating process is not limited by the shape and size of the substrates. Up to now, sol-gel processes have been successfully employed in SOFCs to enhance the performance and stability of the components.

2.4.1 Enhance Cathode Performance and Stability Through Surface Medication

In recent years, $La_xSr_{1-x}Co_yFe_{1-y}O_{3-\delta}$ (LSCF)-based cathodes have attracted much attention due to its much higher ionic and electronic conductivity than the conventional LSM cathode [82–84] and higher performance have been achieved in the temperature range of 600–800°C. However, the electrochemical activity of the stand-alone LSCF cathode is likely to be limited by the surface catalytic properties [85, 86]. The dominant oxygen reduction reaction is associated with the surface exchange process [87]. Inadequate long-term stability of LSCF is also a primary factor impeding the widely practical commercial application [88–90]. In order to enhance the performance and/or stability of LSCF cathode, Liu et al. proposed to improve the performance and stability of a porous LSCF cathode by the application of a thin, catalytically active coating though infiltration [91], as schematically shown in Fig. 2.19. The new electrode structure can make effective use of desirable properties of two different materials: fast ionic and electronic transport in the backbone of LSCF cathode and facile surface kinetics on the thin surface coating. The coating could be a discontinuous porous layer or a continuous dense film.

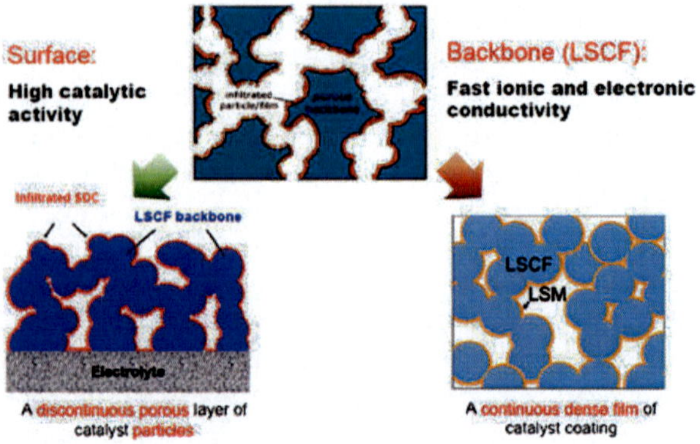

Fig. 2.19 Schematics of catalyst-infiltrated LSCF cathodes

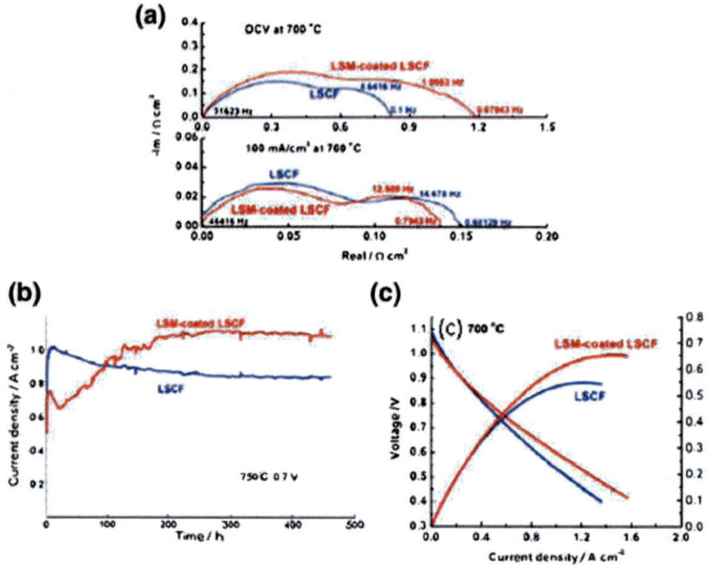

Fig. 2.20 **a** Interfacial impedance $\{Z_{cell}-Z_b\}$ of fuel cells with and without infiltration of LSM measured at OCV and at 100 mA cm^{-2}. **b** Current density of two test cells with and without infiltration as a function of time at a constant voltage of 0.7 V and approximate cathodic overpotential of −0.12 V. **c** Cell voltages and power densities as a function of current density for full cells with and without infiltration of LSM after long-term testing [92]

Improved performance and stability has been reported by using thin catalytic coating of LSM [92], SSC [93] and SDC [94] on LSCF surface. For example, Matt et al. [92] systematically studied the electrochemical behavior of LSM infiltrated LSCF cathode. The thin coating of LSM on the LSCF surface significantly enhanced the long-term stability and performance, as shown in Fig. 2.20. The thin

LSM film formation on LSCF surface and stability of LSM/LSCF were carefully discussed by Choi et al. [95]. In the SDC infiltrated LSCF cathode, Nie et al. [94] found that the microstructures and performances of SDC infiltrated LSCF cathodes depend sensitively on the amount of SDC introduced and the microstructure of the SDC coatings. With 10 μL of 0.25 mol/L SDC infiltrated into LSCF cathode, the interfacial resistance reduced to about half of those for the blank cathode, implying 50% improvement of the electrochemical performance. Short term stability (about 100 h tests) in the anode supported cell indicated that the SDC infiltrated LSCF can significantly improve the performance and stability; however, the detailed mechanism is yet to be determined. Yoon et al. [74] deposited a YSZ or samaria-doped ceria ($Sm_{02}Ce_{08}O_2$, SDC) film at the triple-phase boundary (TPB) of LSM/YSZ/gas to increase the number of electrochemical reaction sites, which resulted in a decrease of cathode polarization and increase of cell performance. A maximum power densities of the cell modified by the SDC sol-gel coating were about 0.53 W/cm^2 at 750°C and about 0.19 W/cm^2 at 650°C [74].

2.4.2 Modification of Porous Anodes for Sulfur and Carbon Tolerance

One of the major advantages of SOFCs is the fuel flexibility. A wide variety of practical hydrocarbons such as methane, propane, gasoline, diesel and kerosene can be direct utilized as the fuels in SOFCs [2, 3, 96–98]. The direct utilization of hydrocarbon fuels will increase the operating efficiency and reduce system costs, which will accelerate substantially the use of SOFCs in transportation, residential, and distributed-power applications. The conventional anode for a SOFC, a composite consisting of nickel and yttria-stabilized zirconia, has excellent catalytic activity for fuel oxidation, good conductivity for current collection, and matched compatibility with YSZ electrolyte for easy cell fabrication, but it is highly susceptible to carbon buildup (coking) and deactivation (sulfur poisoning) by contaminants commonly encountered in readily available fuels [2].

To overcome these problems, substantial effort has been devoted to the development of new anode materials and novel electrode structures. Though some new anode materials have been developed with improved tolerance, such as Cu-based cermet [99–101], $La_{0.75}Sr_{0.25}Cr_{0.5}Mn_{0.5}O_3$ [47], $Sr_2Mg_{1-x}Mn_xMoO_6$ ($0 \leq x \leq 1$) [102], doped (La,Sr)(Ti)O_3 [103, 104], and $La_{0.4}Sr_{0.6}Ti_{1-x}MnxO_3$ [105], their practical application is stalled by other problems such as low electronic conductivity, poor electrochemical activity, and limited physical, chemical, and thermal compatibility with YSZ electrolyte. An effective approach is using a catalytic coating to modify the Ni–YSZ surface to enhance the coking and sulfur tolerance. Yoon et al. [106] used a coating of SDC sol to modify the Ni–YSZ anode for hydrocarbon fuel. The surface modification of Ni/YSZ anode resulted in an increase of structural stability and enlargement of the TPB, which served as a catalytic reaction site for oxidation of carbon or carbon monoxide. Maximum power density of 0.3 W/cm^2 was achieved at 700°C using the mixture of methane (25%) and air (75%) as the fuel and air as the oxidant and the cell was operated for 500 h without significant degradation of cell performance [106]. Yang et al. [2] developed a catalyst material of BZCYYb, which shows high water adsorption capability at high temperatures. With BZCYYb infiltrated into NiO–YSZ anode, the cell showed a stable power output for 1,000 h in wet hydrogen contaminated with 10 ppm H_2S, implying that BZCYYb exhibits considerable stability for long-term sulfur tolerance, and the sulfur poisoning can be fully suppressed in the presence of a small amount of steam.

2.4.3 Protective Coatings for Metallic Interconnect

Metallic interconnects are preferred over the conventional ceramic interconnects, for planar-type SOFC, due to their superior electronic and heat conductivity and low-cost. Among these alloys, Cr_2O_3-forming ferritic stainless steel were one of the most promising candidates due to close match of their coefficients of thermal expansion (CTE) to those of the cell components. However, the long-term stability of the metallic interconnects and the compatibility issues are still a concern. For example, over long-term exposure during SOFC operation, Cr_2O_3 oxide will thermally grow on the surface and increase the electrical resistance. Additionally, the volatile Cr species, primarily in the form of CrO_3 or $CrO_2(OH)_2$, generated from Cr_2O_3 tends to deposit on the surface of cathode and/or the interface between cathode and electrolyte, forming a double-layer oxide scale structure of $(Mn, Cr)_3O_4$ spinel on top of Cr_2O_3, which leads to a significant decrease in cathode activity and subsequent the stack performance [107, 108]. To solve these potential problems, using surface modification via application of a protective coating of conductive oxides is an effective approach to enhance oxidation resistance, surface stability, scale adhesion and conductivity as well as alleviate Cr vaporization [107–109]. Zhu et al. [107] successfully applied a $LaCrO_3$ thin film on a ferritic stainless steel substrate via sol-gel coating process. The $LaCrO_3$ coating provided effective protection for the interconnect steel during oxidation of twelve 100 h cycles at 800°C in air, whereas significant spallation and weight loss were observed for the uncoated steel. Pu et al. [108] investigated the oxidation behavior and the microstructure of the SUS 430 ferritic stainless steel in air at 800°C for 200 h. They found that a $NiCo_2O_4$ protective coating prepared by sol-gel process can significantly increase the oxidation resistance of the SUS 430 alloy by limiting the access of O_2 in air to the outwardly diffused cations, while the electrical conductivity is considerably improved due to inhibited growth of resistive Cr_2O_3 and the formation of conductive spinel phases in the oxide scale. The parabolic rate constant of the oxidation kinetics is 8.1×10^{-15} g^2 cm^{-4} s^{-1} for the coated specimen, in comparison with 8.3×10^{-14} g^2 cm^{-4} s^{-1} of the uncoated. Hua et al. [110] successfully prepared a $MnCo_2O_4$ protective coating on SUS 430 alloy by a sol-gel process. They found that the best technique for forming the $MnCo_2O_4$ protective coating was calcined in reducing atmosphere followed by pre-oxidation in the air, which enhances the oxidation resistance, and improves the electrical conductivity and adherence of coated SUS 430 alloy significantly. As a result, the $MnCo_2O_4$ spinel is the most potential candidate for SOFC metallic interconnect protective coating application [109].

2.5 Summary

SOFCs have attracted worldwide attention because of their high energy efficiency and excellent fuel flexibility. The performance of SOFCs depends sensitively on the microstructures of cell components. For example, the electrodes (both cathode and anode) must be porous and gas-permeable to minimize resistance to gas transport and have high specific surface area to increase the number of active sites (such as TPBs) for electrochemical reactions. In contrast, the electrolyte membrane must be thin to reduce the ohmic loss and gas-tight (or dense) to prevent gas leakage. A sol-gel process has unique advantages for the fabrication of SOFC materials and components with desired microstructures. Further, it is cost-effective because it does not require costly equipment, allows lower processing temperature, and can easily control the chemical composition. Both dense and porous structures can be easily tailored and there is no limitation on the shape or size of the substrate surface. Sol-gel processes have also been successfully applied to preparation of highly homogeneous and fine powders and to modification of electrode surface or electrode/electrolyte interfaces.

However, sol-gel techniques also have some disadvantages: some synthesis routes may be complex, some processes can be time-consuming, precursors are often expensive and sensitive to moisture; and the quality of the coating may be influenced sensitively by many processing parameters. Sol-gel processes also need further development for deposition of nanostructured materials in a selective area such as the triple-phase boundary of a fuel cell. Finally, the problems of dimensional change on densification, and of shrinkage and stress cracking on drying, although not insurmountable, do require careful attention. These significant limitations underline the need for sol-gel process optimization to exploit their advantages to the maximum in applications where they can provide properties not attainable by other methods, including fabrication of thin-film electrolyte membranes and functionally graded multilayer electrode structures, infiltration of catalyst/electrode coatings onto scaffolds, and surface modification of electrodes for enhanced performance and stability.

References

1. Minh NQ, Takahashi T (1995) Science and technology of ceramic fuel cells. Elsevier, Amsterdam
2. Yang L, Wang SZ, Blinn K, Liu MF, Liu Z, Cheng Z, Liu ML (2009) Enhanced sulfur and coking tolerance of a mixed ion conductor for SOFCs: BaZr0.1Ce0.7Y0.2−*x*Yb*x*O3−*δ*. Science 326(5949):126–129

3. Yang L, Choi Y, Qin W, Chen H, Blinn K, Liu M, Liu P, Bai J, Tyson TA, Liu M (2011) Promotion of water-mediated carbon removal by nanostructured barium oxide/nickel interfaces in solid oxide fuel cells. Nat Commun 2:357
4. Zhe C, Wang JH, Choi YM, Yang L, Lin MC, Liu M (2011) From Ni-YSZ to sulfur-tolerant anodes: electrochemical behavior, modeling, in situ characterization, and perspectives. Energy Environ Sci Perspect Rev 4:4380–4409
5. Singhal SC (2000) Science and technology of solid-oxide fuel cells. MRS Bull 25(3):16–21
6. Liu M, Lynch ME, Blinn K, Alamgir F, Choi Y (2011) Rational SOFC material design: new advances and tools. Mat Today Invited Rev 14:534–546
7. http://science.nasa.gov/headlines/y2003/18mar_fuelcell.htm
8. http://www.doitpoms.ac.uk/tlplib/fuel-cells/high_temp_sofc.php
9. http://www.aki.che.tohoku.ac.jp/~koyama/html/research/SOFC.html
10. Ormerod RM (2003) Solid oxide fuel cells. Chem Soc Rev 32(1):17–28
11. Xia CR, Liu ML (2002) Novel cathodes for low-temperature solid oxide fuel cells. Adv Mat 14(7):521
12. (2005) 2004 fuel cell handbook: advanced technology for generating electricity series on renewable energy, biofuels, bioenergy, and biobased products, US Department of Energy, 7th edn. Progressive Management
13. http://www.doitpoms.ac.uk/tlplib/fuel-cells/sofc_electrode_materials.php
14. Pierre AC (1998) Introduction to sol-gel processing. Springer, London
15. Brinker CJ, Scherer GW (1990) Sol-gel science: the physics and chemistry of sol-gel processing, 1st edn. Academic Press
16. Mehrotra RC (1989) In: Aegerter MA, Souza Jr., DF, Zanotto ED (eds) Sol-gel science and technology. World Scientific Publishing Company, Singapore
17. Viazzi C, Deboni A, Ferreira JZ, Bonino JP, Ansart F (2006) Synthesis of Yttria Stabilized Zirconia by sol-gel route: Influence of experimental parameters and large scale production. Solid State Sci 8(9):1023–1028
18. Laberty-Robert C, Ansart F, Deloget C, Gaudon M, Rousset A (2001) Powder synthesis of nanocrystalline ZrO2–8%Y2O3 via a polymerization route. Mat Res Bull 36(12): 2083–2101
19. Steele BCH (2000) Appraisal of $Ce_{1-y}Gd_yO_{2-y/2}$ electrolytes for IT-SOFC operation at 500 degrees C. Solid State Ion 129(1–4):95–110
20. Prasad DH, Son JW, Kim BK, Lee HW, Lee JH (2008) Synthesis of nano-crystalline $Ce_{0.9}Gd_{0.1}O_{1.95}$ electrolyte by novel sol-gel thermolysis process for IT-SOFCs. J Eur Ceram Soc 28(16):3107–3112
21. Mogensen M, Sammes NM, Tompsett GA (2000) Physical chemical and electrochemical properties of pure and doped ceria. Solid State Ion 129(1–4):63–94
22. Gaudon M, Laberty-Robert C, Ansart F, Stevens P, Rousset A (2002) Preparation and characterization of $La_{1-x}Sr_xMnO_{3+\delta}$ ($0 <= x <= 0.6$) powder by sol-gel processing. Solid State Sci 4(1):125–133
23. Xiong L, Wang SR, Wang YS, Wen TL (2008) $(Pr_{0.7}Ca_{0.3})_{(0.9)}MnO_{3-\delta}$-SDC cathode for IT-SOFC. J Alloy Compd 453(1–2):356–360
24. Lenormand P, Castillo S, Gonzalez JR, Laberty-Robert C, Ansart F (2005) Lanthanum ferromanganites thin films by sol-gel process. Influence of the organic/inorganic R ratio on the microstructural properties. Solid State Sci 7(2):159–163
25. Ghouse M, Al-Yousef Y, Al-Musa A, Al-Otaibi MF (2010) Preparation of $La_{0.6}Sr_{0.4}$ $Co_{0.2}Fe_{0.8}O_3$ nanoceramic cathode powders for solid oxide fuel cell (SOFC) application. Int J Hydrogen Energy 35(17):9411–9419
26. Ding C, Lin H, Sato K, Hashida T (2008) Synthesis and characterization of $La_{0.8}Sr_{0.2}$ $Co_{0.8}Fe_{0.2}O_3$ nanoparticles for intermediate-low temperature solid oxide fuel cell cathodes. Water Dyn 987:35–38
27. Meng XW, Lu SQ, Ji Y, Wei T, Zhang YL (2008) Characterization of $Pr_{1-x}Sr_xCo_{0.8}Fe_{0.2}$ $O_{3-\delta}$ ($0.2 <= x <= 0.6$) cathode materials for intermediate-temperature solid oxide fuel cells. J Power Sources 183(2):581–585

28. Vert VB, Serra JM (2009) Influence of Barium incorporation on the electrochemical performance of $Ln_{0.58}Sr_{0.4}Fe_{0.8}Co_{0.2}O_{3-\delta}$) (Ln = La, Pr, Sm) Perovskites for oxygen activation at intermediate temperatures. Fuel Cells 9(5):663–678
29. Vert VB, Serra JM (2010) Improvement of the Electrochemical Performance of $Ln_{0.58}Sr_{0.4}$ $Fe_{0.8}Co_{0.2}O_{3-\delta}$ IT-SOFC Cathodes by Ternary Lanthanide Combinations (La-Pr-Sm). Fuel Cells 10(4):693–702
30. Zeng PY, Ran R, Zhihao CAH, Zhou W, Gu HX, Shao ZP, Liu SM (2008) Efficient stabilization of cubic perovskite $SrCoO_{3-\delta}$-delta by B-site low concentration scandium doping combined with sol-gel synthesis. J Alloy Compd 455(1–2):465–470
31. Shao ZP, Haile SM (2004) A high-performance cathode for the next generation of solid-oxide fuel cells. Nature 431(7005):170–173
32. Zheng MZ, Liu XM, Su WH (2005) Preparation and performance of $La_{1-x}Sr_xCuO_{3-\delta}$ as cathode material in IT-SOFCs. J Alloy Compd 395(1–2):300–303
33. Zhao L, He BB, Lin B, Ding HP, Wang SL, Ling YH, Peng RR, Meng GY, Liu XQ (2009) High performance of proton-conducting solid oxide fuel cell with a layered $PrBaCo_2O_{5+\delta}$ cathode. J Power Sources 194(2):835–837
34. Zhao L, Nian Q, He BB, Lin B, Ding HP, Wang SL, Peng RR, Meng GY, Liu XQ (2010) Novel layered perovskite oxide $PrBaCuCoO_{5+\delta}$ as a potential cathode for intermediate-temperature solid oxide fuel cells. J Power Sources 195(2):453–456
35. Pena-Martinez J, Tarancon A, Marrero-Lopez D, Ruiz-Morales JC, Nunez P (2008) Evaluation of $GdBaCo_2O_{5+\delta}$ as Cathode Material for Doped Lanthanum Gallate Electrolyte IT-SOFCs. Fuel Cells 8(5):351–359
36. Ferkhi M, Khelili S, Zerroual L, Ringuede A, Cassir M (2009) Synthesis, structural analysis and electrochemical performance of low-copper content $La_2Ni_{1-x}Cu_xO_{4+\delta}$ delta materials as new cathodes for solid oxide fuel cells. Electrochim Acta 54(26):6341–6346
37. Fontaine ML, Laberty-Robert C, Ansart F, Tailhades P (2006) Composition and porosity graded $La_{2-x}NiO_{4+\delta}$ (x > = 0) interlayers for SOFC: Control of the microstructure via a sol-gel process. J Power Sources 156(1):33–38
38. Livage J, Henry M, Sanchez C (1988) Sol-Gel Chemistry of Transition-Metal Oxides. Prog Solid State Chem 18(4):259–341
39. Shimizu Y, Murata T (1997) Sol-gel synthesis of perovskite-type lanthanum manganite thin films and fine powders using metal acetylacetonate and poly(vinyl alcohol). J Am Ceram Soc 80(10):2702–2704
40. Zhou W, Ran R, Shao ZP, Jin WQ, Xu NP (2010) Synthesis of nano-particle and highly porous conducting perovskites from simple in situ sol-gel derived carbon templating process. Bull Mat Sci 33(4):371–376
41. Suciu C, Hoffmann AC, Dorolti E, Tetean R (2008) NiO/YSZ nanoparticles obtained by new sol-gel route. Chem Eng J 140(1–3):586–592
42. Jiang SP, Chan SH (2004) A review of anode materials development in solid oxide fuel cells. J Mater Sci 39(14):4405–4439
43. Wilson JR, Kobsiriphat W, Mendoza R, Chen HY, Hiller JM, Miller DJ, Thornton K, Voorhees PW, Adler SB, Barnett SA (2006) Three-dimensional reconstruction of a solid-oxide fuel-cell anode. Nat Mater 5(7):541–544
44. Wilkenhoener R, Vassen R, Buchkremer HP, Stover D (1999) Mechanically alloyed Ni/8YSZ powder mixtures: preparation, powder characterization and sintering behavior. J Mat Sci 34(2):257–265
45. Sun CW, Stimming U (2007) Recent anode advances in solid oxide fuel cells. J Power Sources 171(2):247–260
46. Jacobson AJ (2010) Materials for solid oxide fuel cells. Chem Mat 22(3):660–674
47. Tao SW, Irvine JTS (2003) A redox-stable efficient anode for solid-oxide fuel cells. Nat Mater 2(5):320–323
48. Wan J, Zhu JH, Goodenough JB (2006) La0.75Sr0.25Cr0.5Mn0.5$O_{3-\delta}$ + Cu composite anode running on H2 and CH4 fuels. Solid State Ion 177(13–14):1211–1217

49. Zhu XB, Zhe L, Bo W, Chen KF, Liu ML, Huang XQ, Su WH (2010) Fabrication and performance of membrane solid oxide fuel cells with $La_{0.75}Sr_{0.25}Cr_{0.5}Mn_{0.5}O_{3-\delta}$ impregnated anodes. J Power Sources 195(7):1793–1798
50. Fu XZ, Melnik J, Low QX, Luo JL, Chuang KT, Sanger AR, Yang QM (2010) Surface modified Ni foam as current collector for syngas solid oxide fuel cells with perovskite anode catalyst. Int J Hydrogen Energy 35(20):11180–11187
51. Chen XJ, Liu QL, Chan SH, Brandon NP, Khor KA (2007) Sulfur tolerance and hydrocarbon stability of $La_{0.75}Sr_{0.25}Cr_{0.5}Mn_{0.5}O_3/Gd_{0.2}Ce_{0.8}O_{1.9}$ composite anode under anodic polarization. J Electrochem Soc 154(11):B1206–B1210
52. Ghouse M, Al-Musa A, Al-Yousef Y, Al-Otaibi MF (2010) Synthesis of Mg doped $LaCrO_3$ nano powders by sol-gel process for solid oxide fuel cell (SOFC) application. J New Mater Electrochem Syst 13(2):99–106
53. Stover D, Buchkremer HP, Uhlenbruck S (2004) Processing and properties of the ceramic conductive multilayer device solid oxide fuel cell (SOFC). Ceram Int 30(7): 1107–1113
54. Kueper TW, Visco SJ, De Jonghe LC (1992) Thin-film ceramic electrolytes deposited on porous and non-porous substrates by sol-gel techniques. Solid State Ion 52(1–3):251–259
55. Van Gestel T, Sebold D, Meulenberg WA, Buchkremer H-P (2008) Development of thin-film nano-structured electrolyte layers for application in anode-supported solid oxide fuel cells. Solid State Ion 179(11–12):428–437
56. Pan Y, Zhu JH, Hu MZ, Payzant EA (2005) Processing of YSZ thin films on dense and porous substrates. Surf Coat Technol 200(5–6):1242–1247
57. Peshev P, Slavova V (1992) Preparation of Yttria-stabilized Zirconia thin-films by a sol-gel procedure using alkoxide precursors. Mat Res Bull 27(11):1269–1275
58. Gaudon M, Laberty-Robert C, Ansart F, Stevens P (2006) Thick YSZ films prepared via a modified sol-gel route: thickness control (8–80 μm). J Eur Ceram Soc 26(15):3153–3160
59. Egger P, Soraru GD, Dire S (2004) Sol-gel synthesis of polymer-YSZ hybrid materials for SOFC technology. J Eur Ceram Soc 24(6):1371–1374
60. Lenormand P, Rieu M, Cienfuegos RF, Julbe A, Castillo S, Ansart F (2008) Potentialities of the sol-gel route to develop cathode and electrolyte thick layers Application to SOFC systems. Surf Coat Technol 203(5–7):901–904
61. Vo NXP, Yoon SP, Nam SW, Han J, Lim TH, Hong SA (2005) Fabrication of an anode-supported SOFC with a sol-gel coating method for a mixed-gas fuel cell. On the Convergence of Bio-Information,- Environmental-, Energy- Space- and Nano-Technologies, Pts 1 and 2, 277(–279):455–461
62. Chen YY, Wei WCJ (2006) Processing and characterization of ultra-thin yttria-stabilized zirconia (YSZ) electrolytic films for SOFC. Solid State Ion 177(3–4):351–357
63. Agarwal V, Liu ML (1997) Preparation of barium cerate-based thin films using a modified Pechini process. J Mater Sci 32(3):619–625
64. Chiba R, Yoshimura F, Yamaki J, Ishii T, Yonezawa T, Endou K (1997) Ionic conductivity and morphology in Sc_2O_3 and Al_2O_3 doped ZrO_2 films prepared by the sol-gel method. Solid State Ion 104(3–4):259–266
65. Mehta K, Xu R, Virkar AV (1998) Two-layer fuel cell electrolyte structure by sol-gel processing. J Sol-Gel Sci Technol 11(2):203–207
66. Jang WS, Hyun SH, Kim SG (2002) Preparation of YSZ/YDC and YSZ/GDC composite electrolytes by the tape casting and sol-gel dip-drawing coating method for low-temperature SOFC. J Mater Sci 37(12):2535–2541
67. Kim SG, Yoon SP, Nam SW, Hyun SH, Hong SA (2002) Fabrication and characterization of a YSZ/YDC composite electrolyte by a sol-gel coating method. J Power Sources 110(1): 222–228
68. Rose L, Kesler O, Tang ZL, Burgess A (2007) Application of sol gel spin coated yttria-stabilized zirconia layers for the improvement of solid oxide fuel cell electrolytes produced by atmospheric plasma spraying. J Power Sources 167(2):340–348

69. Kim SD, Hyun SH, Moon J, Kim JH, Song RH (2005) Fabrication and characterization of anode-supported electrolyte thin films for intermediate temperature solid oxide fuel cells. J Power Sources 139(1–2):67–72
70. Moon J, Song HS, Kim WH, Hyun SH, Kim J, Lee HW (2007) Effect of starting particulate materials on microstructure and cathodic performance of nanoporous LSM-YSZ composite cathodes. J Power Sources 167(2):258–264
71. Xia CR, Zhang YL, Liu ML (2003) LSM-GDC composite cathodes derived from a sol-gel process - Effect of microstructure on interfacial polarization resistance. Electrochem Solid State Lett 6(12):A290–A292
72. Zha SW, Zhang YL, Liu ML (2005) Functionally graded cathodes fabricated by sol-gel/slurry coating for honeycomb SOFCs. Solid State Ion 176(1–2):25–31
73. Tang ZL, Xie YS, Hawthorne H, Ghosh D (2006) Sol-gel processing of $Sr_{0.5}Sm_{0.5}CoO_3$ film. J Power Sources 157(1):385–388
74. Yoon SP, Han J, Nam SW, Lim TH, Oh IH, Hong SA, Yoo YS, Lim HC (2002) Performance of anode-supported solid oxide fuel cell with La0.85Sr0.15MnO_3 cathode modified by sol-gel coating technique. J Power Sources 106(1–2):160–166
75. Sholklapper TZ, Lu C, Jacobson CP, Visco SJ, De Jonghe LC (2006) LSM-infiltrated solid oxide fuel cell cathodes. Electrochem Solid State Lett 9(8):A376–A378
76. Zhang Q, Martin BE, Petric A (2008) Solid oxide fuel cell composite cathodes prepared by infiltration of copper manganese spinel into porous yttria stabilized zirconia. J Mat Chem 18(36):4341–4346
77. Matus YB, De Jonghe LC, Jacobson CP, Visco SJ (2005) Metal-supported solid oxide fuel cell membranes for rapid thermal cycling. Solid State Ion 176(5–6):443–449
78. Tucker MC (2010) Progress in metal-supported solid oxide fuel cells: A review. J Power Sources 195(15):4570–4582
79. Rieu M, Lenormand P, Panteix PJ, Ansart F (2008) New route to prepare anodic coatings on dense and porous metallic supports for SOFC application. Fuel Cells Bull 2008(12): 12–15
80. Rieu M, Lenormand P, Ansart F, Mauvy F, Fullenwarth J, Zahid M (2008) Preparation of Ni–YSZ thin and thick films on metallic interconnects as cell supports. Applications as anode for SOFC. J Sol-Gel Sci Technol 45(3):307–313
81. Marrero-Lopez D, Ruiz-Morales JC, Pena-Martinez J, Canales-Vazquez J, Nunez P (2008) Preparation of thin layer materials with macroporous microstructure for SOFC applications. J Solid State Chem 181(4):685–692
82. Jiang SP (2002) A comparison of O-2 reduction reactions on porous (La, Sr)MnO3 and (La, Sr)(Co, Fe)O-3 electrodes. Solid State Ion 146(1–2):1–22
83. Murray EP, Sever MJ, Barnett SA (2002) Electrochemical performance of (La, Sr) (Co, Fe)O-3-(Ce, Gd)O-3 composite cathodes. Solid State Ion 148(1–2):27–34
84. Yang L, Liu Z, Wang SZ, Choi YM, Zuo CD, Liu ML (2010) A mixed proton, oxygen ion, and electron conducting cathode for SOFCs based on oxide proton conductors. Journal of Power Sources 195(2):471–474
85. Lane JA, Benson SJ, Waller D, Kilner JA (1999) Oxygen transport in $La_{0.6}Sr_{0.4}Co_{0.2}Fe_{0.8}O_{3-\delta}$ -delta. Solid State Ion 121(1–4):201–208
86. Prestat M, Koenig JF, Gauckler LJ (2007) Oxygen reduction at thin dense $La_{0.52}Sr_{0.48}Co_{0.18}Fe_{0.82}O_{3-\delta}$ $_{\text{-delta}}$ electrodes. Part I: Reaction model and faradaic impedance. J Electroceram 18(1–2):87–101
87. Lee JW, Liu Z, Yang L, Abernathy H, Choi SH, Kim HE, Liu ML (2009) Preparation of dense and uniform $La_{0.6}Sr_{0.4}Co_{0.2}Fe_{0.8}O_{3-\delta}$ -delta (LSCF) films for fundamental studies of SOFC cathodes. J Power Sources 190(2):307–310
88. Simner SP, Anderson MD, Engelhard MH, Stevenson JW (2006) Degradation mechanisms of La-Sr-Co-Fe-O_3SOFC cathodes. Electrochem Solid State Lett 9(10):A478–A481
89. Kim JY, Sprenkle VL, Canfield NL, Meinhardt KD, Chick LA (2006) Effects of chrome contamination on the performance of $La_{0.6}Sr_{0.4}Co_{0.2}Fe_{0.8}O_{3-\delta}$ cathode used in solid oxide fuel cells. J Electrochem Soc 153(5):A880–A886

90. Benson SJ, Waller D, Kilner JA (1999) Degradation of $La_{0.6}Sr_{0.4}Fe_{0.8}Co_{0.2}O_{3-\delta}$ in carbon dioxide and water atmospheres. J Electrochem Soc 146(4):1305–1309
91. Liu M, Liu Z, Liu MF, Nie LF, Mebane DS, Wilson DS, Surdoval W (2010) Solid oxide fuel cells having porous cathodes infiltrated with oxygen-reducing catalysts, US Patent Application No. 12/837,757
92. Lynch ME, Yang L, Qin W, Choi J-J, Liu M, Blinn K, Liu M (2011) Enhancement of $La_{0.6}Sr_{0.4}Co_{0.2}Fe_{0.8}O_{3-\delta}$ durability and surface electrocatalytic activity by $La_{0.85}Sr_{0.15}MnO_{3\pm\delta}$ investigated using a new test electrode platform. Energy Environ Sci 4(6):2249
93. Lou XY, Wang SZ, Liu Z, Yang L, Liu ML (2009) Improving $La_{0.6}Sr_{0.4}Co_{0.2}Fe_{0.8}O_{3-\delta}$ cathode performance by infiltration of a $Sm_{0.5}Sr_{0.5}CoO_{3-\delta}$ coating. Solid State Ion 180(23–25):1285–1289
94. Nie LF, Liu MF, Zhang YJ, Liu ML (2010) $La_{0.6}Sr_{0.4}Co_{0.2}Fe_{0.8}O_{3-\delta}$ cathodes infiltrated with samarium-doped cerium oxide for solid oxide fuel cells. J Power Sources 195(15): 4704–4708
95. Choi J-J, Qin W, Liu M, Liu M (2011) Preparation and characterization of $(La_{0.8}Sr_{0.2})_{0.95}MnO_{3-\delta}$ (LSM) thin films and LSM/LSCF interface for solid oxide fuel cells. J Am Ceram Soc 94(10):3340–3345
96. Singhal SC (2000) Advances in solid oxide fuel cell technology. Solid State Ion 135(1–4): 305–313
97. Gong MY, Liu XB, Trembly J, Johnson C (2007) Sulfur-tolerant anode materials for solid oxide fuel cell application. J Power Sources 168(2):289–298
98. Atkinson A, Barnett S, Gorte RJ, Irvine JTS, Mcevoy AJ, Mogensen M, Singhal SC, Vohs J (2004) Advanced anodes for high-temperature fuel cells. Nat Mater 3(1):17–27
99. Park SD, Vohs JM, Gorte RJ (2000) Direct oxidation of hydrocarbons in a solid-oxide fuel cell. Nature 404(6775):265–267
100. Gorte RJ, Vohs JM (2003) Novel SOFC anodes for the direct electrochemical oxidation of hydrocarbons. J Catal 216(1–2):477–486
101. Lu C, Worrell WL, Gorte RJ, Vohs JM (2003) SOFCs for direct oxidation of hydrocarbon fuels with samaria-doped ceria electrolyte. J Electrochem Soc 150(3):A354–A358
102. Huang YH, Dass RI, Xing ZL, Goodenough JB (2006) Double perovskites as anode materials for solid-oxide fuel cells. Science 312(5771):254–257
103. Pillai MR, Kim I, Bierschenk DM, Barnett SA (2008) Fuel-flexible operation of a solid oxide fuel cell with $Sr_{0.8}La_{0.2}TiO_3$ support. J Power Sources 185(2):1086–1093
104. Pillai MR, Jiang Y, Mansourian N, Kim I, Bierschenk DM, Zhu HY, Kee RJ, Barnett SA (2008) Solid oxide fuel cell with oxide anode-side support. Electrochem Solid State Lett 11(10):B174–B177
105. Fu QX, Tietz F, Stover D (2006) $La_{0.4}Sr_{0.6}Ti_{1-x}Mn_xO_{3\text{-delta-}\delta}$ perovskites as anode materials for solid oxide fuel cells. J Electrochem Soc 153(4):D74–D83
106. Yoon SP, Han J, Nam SW, Lim TH, Hong SA (2004) Improvement of anode performance by surface modification for solid oxide fuel cell running on hydrocarbon fuel. J Power Sources 136(1):30–36
107. Zhu JH, Zhang Y, Basu A, Lu ZG, Paranthaman M, Lee DF, Payzant EA (2004) $LaCrO_3$-based coatings on ferritic stainless steel for solid oxide fuel cell interconnect applications. Surf Coat Technol 177:65–72
108. Pu JA, Hua B, Zhang WY, Wu JA, Chi B, Jian L (2010) A promising $NiCo_2O_4$ protective coating for metallic interconnects of solid oxide fuel cells. J Power Sources 195(21): 7375–7379
109. Fang Y, Wu C, Duan X, Wang S, Chen Y (2011) High-temperature oxidation process analysis of $MnCo_2O_4$ coating on Fe–21Cr alloy. Int J Hydrogen Energy 36(9):5611–5616
110. Hua B, Kong YH, Lu FS, Zhang JF, Pu JA, Li JA (2010) The electrical property of MnCo2O4 and its application for SUS 430 metallic interconnect. Chin Sci Bull 55(33): 3831–3837

Chapter 3
Inorganic-Based Proton Exchange Membranes for H_2/O_2 Fuel Cells

Masayuki Nogami and Lakshminarayana Gandham

Abstract The sol-gel method was applied to prepare proton exchange membranes. The proton exchange membrane fuel cells have attracted significant attention because of their ability to produce high energy without emission of harmful pollutants. The main technology of the fuel cells is to prepare the proton exchange membranes and Nafion-based perfluosulfonated ionomers have received consideration as possible electrolytes due to their high proton conductivity at room temperature. Despite this, some problems arise because of their thermal and chemical degradation, and stable materials are desired to be developed. We have proposed the sol-gel technique for the preparation of high proton-conducting membranes and successfully prepared inorganic and inorganic–organic hybrid membranes exhibiting high proton conductivities compared to that of perfluosulfonated ionomers. Metal alkoxides such as $Si(OC_2H_5)_4$ and $PO(OCH_3)_3$ are hydrolyzed, followed by heating at 400–600°C to form porous glasses, in which the water molecules are absorbed. The proton conductivities were investigated relating with the pore structure, adsorbed water molecules and glass compositions. The conductivities follow the Arrhenius equation and the activation energy decreases with increasing the logarithm of the product of proton and water concentration. The proton conductivities increase with increasing the content of water and hydroxyl groups and reach $\sim 10^{-2}$ S/cm at room temperature. Further inorganic–organic hybrids doped with the heteropoly acids and ionic liquids are prepared to increase the proton conductivities at high temperature under anhydrous condition.

M. Nogami (✉)
Department of Materials Science and Engineering, Nagoya Institute of Technology, Showa, Nagoya 466-8555, Japan
e-mail: mnogami@mtj.biglobe.ne.jp

L. Gandham
Materials Science and Technology Division (MST-7), Los Alamos, NM 87545, USA
e-mail: gandham@lanl.gov

M. Aparicio et al. (eds.), *Sol-Gel Processing for Conventional and Alternative Energy*, Advances in Sol-Gel Derived Materials and Technologies,
DOI: 10.1007/978-1-4614-1957-0_3,

Keywords Sol-gel method · Hybrid membranes · Electrolytes · Proton conductivity · Fuel cell

3.1 Introduction

Now-a-days, fuel cells have received considerable attention as an alternative power source with respect to the conventional fossil fuels, as conventional fossil fuels contributing to environmental pollution and decreasing the overall oil supply. In particular, proton exchange membrane fuel cells (PEMFCs), which electrochemically convert the chemical energy of a fuel, e.g., hydrogen, directly into electrical energy are one of the most promising clean energy technologies and suitable primary power source for transportation and stationary applications due to their high conversion efficiency, high power density, and benign environmental impact [1–4]. The heart of the fuel cell is the proton exchange membrane, which ensures the space separation between anode and cathode and mediates the electrochemical reactions by transporting protons. Among various technological requirements for PEMs, the most essential property is proton conductivity. Nafion-based perflourosulfonated ionomer membranes are among the commonly employed candidates for PEMFC due to their high and selective permeability for small cations, especially for protons [5, 6]. But they have severe drawbacks, namely cost and environmental issues and limitation of the operation temperature to below 100°C. In addition, Nafion and other perfluorinated sulfonic acid membranes suffer from low conductivity at low water contents or high temperatures, relatively low mechanical strength, low stability and high methanol crossover at higher temperatures [7–9]. Stable inorganic membranes with high proton conductivities, if developed, would extend beyond the limitation of the polymer membranes. Therefore, in recent years there has been a great deal of research in the development of anhydrous electrolyte membranes with high proton conductivities at temperatures above 100°C [10–13]. Operation of PEMFCs at high temperatures (>100°C) provides a number of technological benefits, including higher tolerance to significant quantities of CO, fast electrode kinetics, and no cathode flooding [14–16]. Also, PEMFC system complexity could be reduced by the development of 'water-free' electrolytes that do not require hydration. It also enables the PEMFC to be operated under 'warm' conditions (i.e. above 100°C) thus further improving its efficiency.

Nogami's research group applied the sol-gel process for these PEMFCs preparation and successfully developed new inorganic glass and inorganic–organic hybrid membranesHybrid membranes [17–28]. The strategy of the sol-gel process involves the formation of network structure similar to that exhibited in glasses through a hydrolysis of metal alkoxides at room temperature. After gelling the sol, the gels are heated to transform into the glasses at near the glass-transition temperature. One characteristic of this method is to contain the pores and water in the obtained glass, which enables to bring the proton conductivities. Further interesting of the sol-gel process is to make it possible to incorporate organic molecules into an inorganic network, so that they can be combined virtually at any ratio on the molecular level with the formation of hybrid organic–inorganic materials [29]. The mechanical properties of these materials can be varied from brittle and hard to rubbery depending on the content of organic modifiers in the inorganic network. Moreover, many factors such as temperature, condensation medium (acid/base), nature of the solvent and the type of alkoxide precursor, aging, drying, stabilization and densification may influence the sol-gel structure since they are related to the rates of hydrolysis and condensation that determine the structure of the gel. In this chapter, the sol-gel process is reviewed to prepare inorganic glasses and inorganic–organic hybrids for the PEMFCs electrolyte membranes.

3.2 Inorganic Proton Exchange Membranes

3.2.1 Sol-Gel Preparation of Porous Membranes

Porous silicate glasses containing P_2O_5, TiO_2, or ZrO_2 were prepared through the hydrolysis of metal alkoxides such as $Si(OC_2H_5)_4$ (TEOS), $Si(OCH_3)_4$ (TMOS), $PO(OCH_3)_3$, $Ti(OC_3H_7)_4$, and $Zr(OC_4H_9)_4$. Si-alkoxide was first hydrolyzed at room temperature with a solution of H_2O, C_2H_5OH, and HCl as a catalyst. After reacting $PO(OCH_3)_3$, $Ti(OC_3H_7)_4$, and $Zr(OC_4H_9)_4$, the solutions were further hydrolyzed using water and catalysts. In particular case of containing $Ti(OC_3H_7)_4$ and $Zr(OC_4H_9)_4$ the solutions were kept in air atmosphere instead of adding water. The hydrolyzed solutions were dried at room temperature to form a stiff gel about 0.1 mm thick, followed by heating in air at 400–600°C. Synthesized gels are porous, the surfaces of which are composed of the hydroxyl bonds. On heating the gel, The sequence of dehydration is removal of the physically adsorbed water at around 100°C, followed by the removal of water molecules hydrogen-bonded with the surface hydroxyl bonds, and finally by loss of water produced by a dehydration-condensation reaction taking above 400°C between the surface hydroxyl bonds; Me(Si, P, Ti, or Zr)OH + HOMe = Me-O-Me + H_2O. These dehydration-condensation reactions lead the pores to decrease. Nitrogen gas adsorption experiments were done to determine pore properties. Typical example of the nitrogen gas adsorption curves are shown in Fig. 3.1. The prepared glasses exhibit different dependence of the adsorption isotherms on the partial gas pressure. The specific surface area and pore volume of glasses (1), (2) and (3) shown in Fig. 3.1 are 1160, 726, 197 m^2/g and 0.59, 0.70, 0.70 cm^3/g, respectively. The pore size and the pore structure are determined by using acid or base catalysts and surfactant such as $CH_3(CH_2)_{15}N^+(CH_3)_3Br^-$ (CTAB). The pore size calculated from desorption branch of the isotherm is plotted in Fig. 3.2. It is evident that the prepared samples exhibit narrow pore size distribution and the average size of the pores in the glasses (1), (2) and (3) is 2, 4, and 15 nm, respectively. Thus, the pore

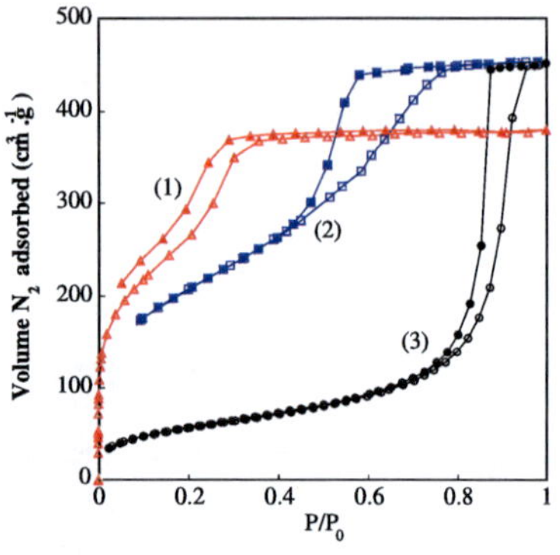

Fig. 3.1 Nitrogen gas adsorption–desorption isotherms of porous glasses

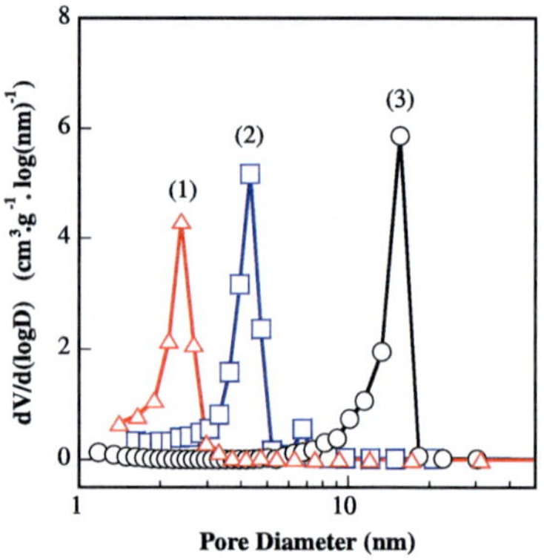

Fig. 3.2 Distribution curves of pore size determined from Nitrogen desorption isotherms

properties are easily controlled by choosing the raw materials and the synthesis conditions. In particular, the CTAB used as the template of the pore structure effectively produces small pores. The formation process of the porous glass structure in this system has been reported in elsewhere [22, 30]. In the sol-gel synthesis process the phosphorus ions react with the polymerized-silica network structure, forming the Si–O–P–OH bonds. The surface of the polymerized phosphorous-silicate structure further reacts with the micelles of the CTAB surfactant [23, 31]. This highly cross-linking of the structure results into formation of small pores after removing the CTAB.

When heating these porous gels, poly-condensation reaction takes place, forming the densified materials. The porous structure remains unchanged in glass heated below 800°C, pore surfaces of which are terminated with hydroxyl bonds and sensitive to air humidity. The nature of water and hydroxyl bonds in silicate glasses is investigated from the infrared (IR) spectra ranging from 2,500 to 4,000 cm^{-1}, which is strongly dependent on the glass compositions. As a example, the IR spectra of SiO_2 and P_2O_5-SiO_2 porous glasses are shown in Fig. 3.3. Two absorption bands peaking at $\sim$3,700 and 3,300 cm^{-1} with shoulder around 2,900 cm^{-1} are observed, which are both assigned to the OH stretching bonds. The position of the OH-absorption bands depends on the degree of strength of hydrogen bonding, and shifts to lower

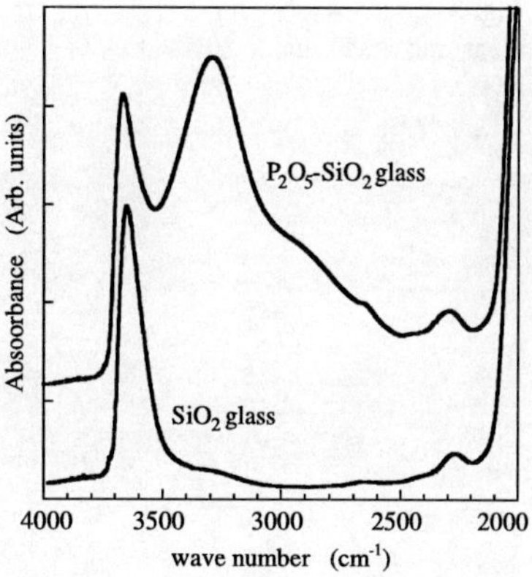

Fig. 3.3 Infrared spectra of SiO_2 and P_2O_5-SiO_2 glasses

wave number with increasing strength of the hydrogen bonding. Among the absorption bands, the band at 3,700 cm^{-1} is assigned to the hydrogen-bonding free OH groups bound with Si^{4+} ions. On the other hand, the bands at 3,300 and 2,900 cm^{-1} observed in the glasses containing P_2O_5 are assigned to the OH groups bound with the P^{5+} ions. When these porous glasses are exposed into an ambient air atmosphere, a pronounced broad band is induced at around 3,500–2,700 cm^{-1} region. These bands are assigned with molecular water hydrogen-bonded with hydroxyl bonds of the pore surfaces, the intensity of which increases with increasing the exposing time, indicating large quantities of absorbed-waters. These adsorbed-waters are reversibly removed by heating at 100–350°C. These experimental results indicate that the amount of water bound with hydroxyl groups of pore surfaces can be determined by exposure in a humid atmosphere, followed by a careful heat treatment at these temperatures. On the other hand, the intensity of the bands of SiOH and POH bonds at 3,700 and 3,300 cm^{-1}, respectively, remains unchanged during these low temperature-heating treatments. In our experiments, it is evident that heating at 600–800°C is needed to decrease the concentration of hydroxyl bonds.

When the glasses are exposed in a high humid atmosphere, water molecules are absorbed in pores. The water molecule is polar and chemically bonded with the hydroxyl groups such as POH or SiOH bonds at the pore surface. Further absorption of water molecules leads to the multilayer of water molecules, which are physically bound in pores. Therefore, the glasses having small size of pores are filled with the water at low humidity. On the other hand, for the glasses with the large size of pores, it is necessary for glasses to be exposed in high humidity for filling the pores with water.

3.2.2 Mechanism of Proton Conduction in Porous Glasses

Electrical conductivities of glasses are measured by changing the water contents in pores. The electrical conductivities are determined by direct and alternative-current

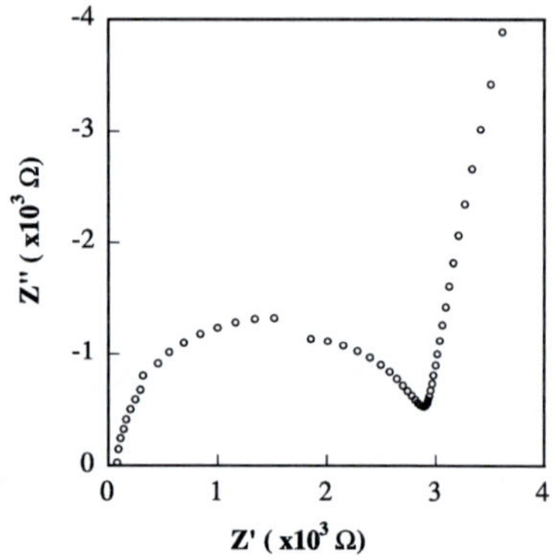

Fig. 3.4 Cole-Cole plot measured at 50 under 80% RH

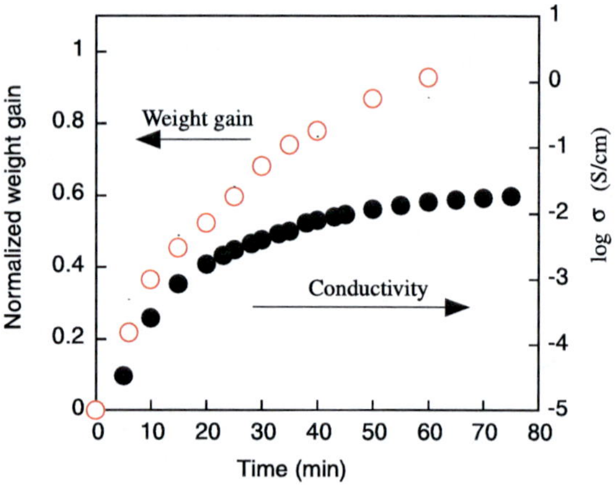

Fig. 3.5 Weight change and conductivity of porous glass during exposing to ambient atmosphere of 80% RH at room temperature

(dc and ac) methods, which are selected corresponding to the conductivities. The dc conductivity technique is useful in the small water-content range, which is measured with a vibrating reed electrometer, where the current was monitored under dc voltage, and the steady-state current was measured to determine the dc conductivity. On the other hand, the ac conductivity is determined from cole-cole plots by impedance measurements over the frequency range of 1–10 MHz, the typical plot of which is shown in Fig. 3.4. The conductivity (σ) of the sample is obtained from the intersection (R) of the semicircle on the real axis and the sample dimensions; $\sigma = d/(S\,R)$, where S is the area of cross section and d is the thickness of the sample. During these measurements, the relative humidity was controlled to be constant within 30–80% RH. Figure 3.5 shows the dependence of conductivity and weight gain on the exposing period in constant 70% RH. It is evident that the increase of conductivity is due to the water absorbed in pores. The conductivities are measured for glasses containing various amount of water, where the water contents are changed by exposing to constant relative humidity. An example of results is shown in Fig. 3.6, where the water content is related with the ratio of

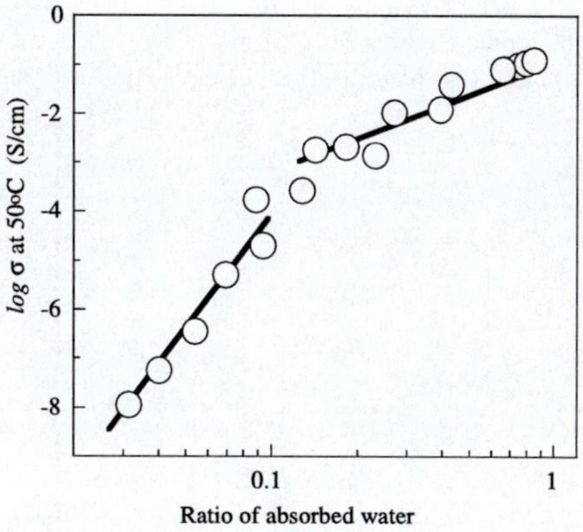

Fig. 3.6 Plotting of conductivity against ratio of pore volume absorbed by water molecules

pore volume absorbed by water. It is interesting to notice that the conductivities are well represented to be proportional to the water content $[H_2O]$ in log–log scale as follows.

$$\log \sigma = \ n \log[H_2O] \tag{3.1}$$

where n is a constant. In this high water-content region, that is larger than ~0.1, the absorbed water molecules are physically bonded. On the other hand, for the measurement of conductivities in the small absorbed-water region, that is less than ~0.1, the glasses exposed in high humid atmosphere are heated at 100–200°C to remove the water. In these treatments, the concentration of proton is kept to be constant to be compared with the number of protons of the surface MeOH groups, meaning that all the adsorbed water are hydrogen-bonded with MeOH groups. The result is also shown in Fig. 3.6. Evidently the conductivity is proportional to the water content, though having large n-value compared with that for large water-containing glasses.

The conductivities are measured as the function of temperature, which are shown in Fig. 3.7. It is evident that the conductivity, σ, varies exponentially with reciprocal temperature over the temperature range measured:

$$\sigma = \sigma_0 \exp(-E/RT), \tag{3.2}$$

where E is the activation energy, T the temperature, R the gas constant, and σ_0 the preexponential term. The charge carrier is an only proton dissociated from the hydroxyl bonds because of containing no carrier other than protons. The dissociated protons move from the initial site to a neighboring site under an electrical field, resulting in the electrical conductivity. The activation energies for conduction are related to the energy for the dissociation of proton from the hydroxyl bonds and/or the barrier for hopping between hydroxyl and water molecules. The conductivity measurements were done for the glasses containing various amounts of the proton and water, the activation energies (E) in which were determined and plotted in Fig. 3.8 as a function of the logarithm of concentration of the proton

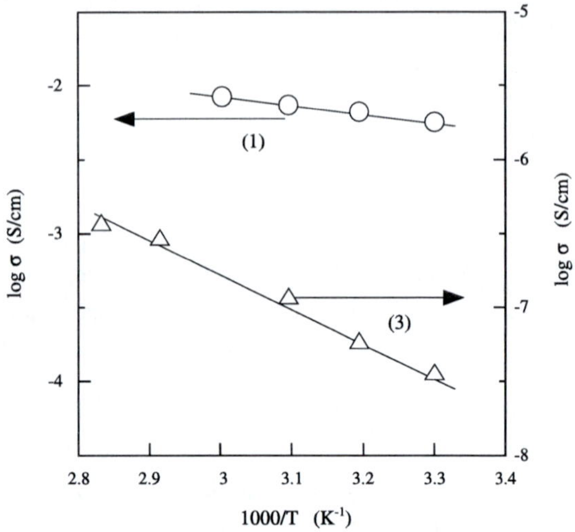

Fig. 3.7 Arrhenius plot of the conductivities for glasses (1) and (3) shown in Fig. 3.1

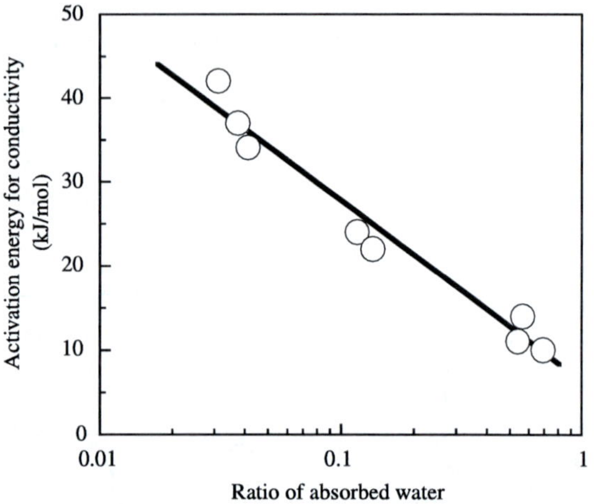

Fig. 3.8 Relation between the activation energy for proton conduction and ratio of pore volume absorbed by water molecules

[H^+] and water [H_2O] [18, 19]. In these experiments, the concentration of proton was estimated from the intensity of the 3,700 and 3,300 cm^{-1} bands of the glasses heated at 600–800°C. On the other hand, the water contents were determined by weighing the sample before and after exposing in the constant humidity. It is interesting to notice that the linear relation between E and $\log\{[H^+]\cdot[H_2O]\}$is satisfied;

$$E = E_0 - k \log_{10}\{[H^+]\cdot[H_2O]\}. \quad (3.3)$$

This relation strongly suggests that the activation energy for conduction in glasses containing water molecules bound with MeOH bonds is related to the dissociation of proton from the MeOH bonds and the dissociation equilibrium between the proton and water molecule. The first term in Eq. 3.3 corresponds to the energy required to dissociate a proton from the MeOH bonds. The dissociated proton moves to a water molecule bound with the MeOH bonds, forming the activated $H_2O{:}H^+$ state:

$$\mathrm{MeOH} \cdots \mathrm{H_2O} \rightarrow \mathrm{MeO^-} + \mathrm{H^+} : \mathrm{H_2O}, \tag{3.4}$$

where the dotted line represents the hydrogen bonding between the proton and water molecule. The second stage of the proton hopping is the dissociation of the proton from the activated $H_2O{:}H^+$ state. The second term in Eq. 3.3 is considered due to the dissociation of the proton from the activated $H^+{:}H_2O$ state.

In the region of small amount of the absorbed-water molecules; less than the number of MeOH bonds, all the water molecules are strongly hydrogen bonded with the MeOH groups and the dissociated proton moves by hopping between MeOH and H_2O. In high water content range, the conductivity also varies exponentially with reciprocal temperature, following the Arrhenius equation (Fig. 3.7). The obtained activation energies are followed the Eq. 3.3. This strongly suggests that the proton conduction is basically controlled by the same mechanism as for glasses containing a small amount of absorbed water. The dissociated proton moves to a neighboring H_2O molecule, forming the activated $H_2O{:}H^+$ state. The second stage of the proton hopping is the dissociation of the proton from the activated $H_2O{:}H^+$ state and hopping of proton between the water molecules:

$$\mathrm{H_2O_{(I)}} : \mathrm{H^+} \rightarrow \mathrm{H_2O_{(I)}} + \mathrm{H^+} \rightarrow \mathrm{H_2O_{(II)}} : \mathrm{H^+}, \tag{3.5}$$

where $H_2O_{(I)}$ and $H_2O_{(II)}$ represent the water molecules that between the proton is hopping. In this dissociation equilibrium, the dissociation energy is decreased with increasing $[H^+]$ and $[H_2O]$. Thus, it is concluded that the activation energy for proton conduction is reduced as the water content increases, resulting into high conductivity. These results are strongly useful how to design the high proton-conducting glasses by the sol-gel process.

3.2.3 Proton Conductivities of Sol-Gel P_2O_5-SiO_2 Membranes

The water absorption is strongly affected by the pore structure and glass compositions. First the role of the pore size on the proton conductivity is discussed. The gaseous water condenses to liquid corresponding to Kelvin's equation at a constant relative humidity. As the pore size decreases, the condensation of water vapors occurs at lower relative humidity. For example, in the glasses having small pores of 5 nm size, the pores are almost filled with water molecules at low humidity, resulting in the increased conductivity in the low humidity. On the other hand, in the glass with large pores of 15 nm size, only about 10% of the pore volume is occupied by water (see Fig. 3.1). Therefore, these small amounts of water

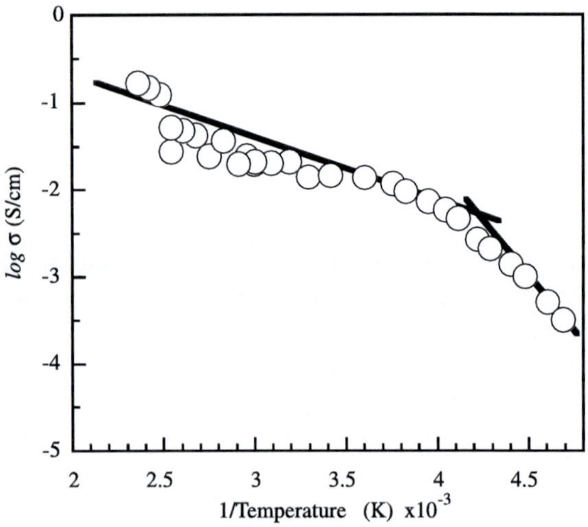

Fig. 3.9 Relation between proton conductivity and reciprocal temperature for P_2O_5-SiO_2 glass. Conductivities were measured under 100% RH

molecules are chemically bonded with the hydroxyl groups on the pore surfaces, resulting in not so much increased in the conductivity.

Glass composition is also very important to determine the proton conductivity. As previously mentioned, the activation energy for conduction in glasses is related to the dissociation of proton from the MeOH bonds and the dissociation energy is strongly determined by the metal ions of the hydroxyl bonds. For example, in the IR spectral experiments, the band at 3,700 cm^{-1} is assigned to the hydrogen-bonding free OH groups bound with Si^{4+} ions, indicating that the protons in SiOH groups are strongly bound with the oxygens to increase the dissociation energy of proton. This results in the low conductivity for only SiO_2 glass. On the other hand, the glasses containing P_2O_5 have the bands at 3,300 and 2,900 cm^{-1} which are assigned to the OH groups bound with the P^{5+} ions. Compared with the SiOH bonds, the POH bonds are strongly hydrogen-bonded with water molecules, resulting in the increased conductivity. According to our experimental experiences, the conductivities are increased by two order magnitudes higher than SiO_2 porous glass by adding P_2O_5 of 5 mol%. Figure 3.9 shows the Arrhenius plot of the conductivity, measured under 100% RH, for the $5P_2O_5{\cdot}95SiO_2$ (in mol%). The conductivity is ~30 m S/cm at around 30°C, which increases with increasing temperature, reaching ~170 m S/cm at 150°C. Considering the chemical stability of the inorganic glasses, the sol-gel derived-glasses would extend functionality beyond the limitation of perfluorosulfonate ionomers and are expected to have potential for the electrolyte of the fuel cell operating above 100°C, even though the highly humidified-condition is needed.

Of further interest in Fig. 3.9 is that the conductivities are well represented in one Arrhenius equation with the activation energy of ~10 kJ/mol at temperature down to −20°C. This result indicates the possibility of glasses used in the fuel cell at low temperature. The freezing temperature of the absorbed-water was determined to be −20°C from the differential scanning calorimetric experiment of the water-absorbed glass, which is much lower than that of bulk water. It is well known that the freezing temperature decreases with decreasing the size. The motion of water molecules which are confined in the small pores are restricted by the small-sized pores in the cross section, though free in the vertical direction. Thus, the restricted motion of the water confined in the small pores brings the freezing temperature to lower than 0°C.

3.2.4 Glass Membranes Impregnated with Heteropoly Acids

Heteropoly acids (HPA) have much effect to enhance the proton conductivities of silicate membranes, which are impregnated under humidified conditions. In our first work [32], a series of novel proton conductive inorganic–organic hybrid membranes doped with phosphotungstic acid (PWA)/phosphomolybdic acid (PMA) and trimethylphosphate $PO(OCH_3)_3$ have been prepared by sol-gel process with 3-glycidoxypropyltrimethoxysilane (GPTMS), and tetraethoxysilane (TEOS) as precursors. Thermal analysis including TG and DTA confirmed that the membranes are thermally stable up to 200°C. Thermal stability of membranes was significantly enhanced by the presence of SiO_2 framework. Proton conductivity of 1.59×10^{-2} S/cm with composition of 50TEOS-5PO$(OCH_3)_3$-35GPTMS-10PWA was obtained (1.15×10^{-2} S/cm for 10 mol% PMA) at 90°C under 90% relative humidity. The proton conductivity of the hybrid membranes was due to the proton-conducting path through the GPTMS-derived "polyethylene oxide (PEO)" networks in which the trapped solid acid (PWA/PMA) as a proton donor is contained. The molecular water absorbed in the polymer matrix was also presumed to provide high proton mobility, resulting in an increase of proton conductivity with increasing relative humidity.

Based on the above encouraging results, we have further investigated the above phosphosilicate matrix with incorporation of both PWA and PMA in addition with 3-aminopropyltriethoxysilane (APTES), using phosphorous precursor as H_3PO_4 [33]. Relatively high proton conductivity of 3×10^{-2} S/cm with composition of 50TEOS -25GPTMS-20H_3PO_4-5APTES-3PMA-6PWA was obtained (6.35×10^{-3} S/cm at 150°C, 50% RH) at 120°C under 90% relative humidity. Also, we have fabricated hybrid films doped with a mixture of PWA and PMA by sol-gel process with GPTMS, TMOS, and H_3PO_4 precursors [34]. Relatively high proton conductivity of 9.8 x 10^{-2} S/cm with composition of 20TMOS-50GPTMS-30H_3PO_4-3PMA-6PWA (mol%) was obtained at 130°C and 90% relative humidity as shown in Fig. 3.10. The hydrogen permeability was found to decrease from 1.12×10^{-10} to 8.5×10^{-11} cm^2/s in the temperature range 30–130°C for, 3PMA/6PWA (mol%) doped hybrid film.

The membrane electrode assemblies were prepared and a maximum power density of 68 mW/cm^2 at 140 mA/cm^2 as well as a current density of 262 mA/cm^2 were obtained at 130°C and 50% RH when utilized in a H_2/O_2 fuel cell (see Fig. 3.11). Further, the hybrid membranes prepared by sol-gel process using

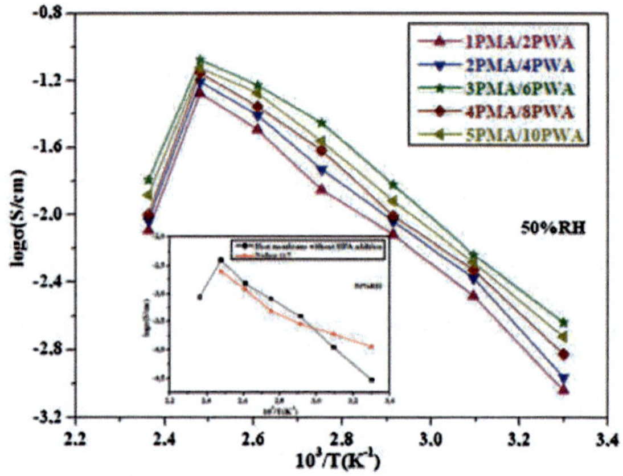

Fig. 3.10 Temperature dependence of conductivity of 20TMOS-50GPTMS-30H_3PO_4-xPMA/yPWA(x = 1,y = 2; x = 2,y = 4; x = 3,y = 6; x = 4,y = 8; x = 5,y = 10 mol%) hybrid membranes. Conductivity was obtained under 50% R.H. Inset shows the proton conductivity of host membrane without HPA addition and Nafion® 117 under 50% RH [34]

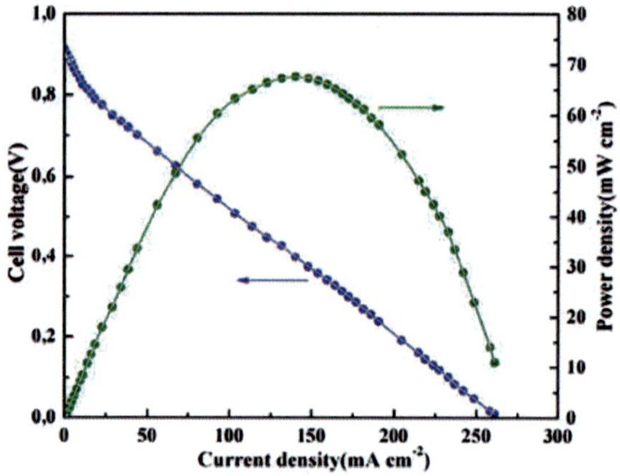

Fig. 3.11 Polarization curve displaying the current density and power density for the 3PMA/6PWA (mol%) doped hybrid membrane at 130°C and 50% RH [34]

TEOS, GPTMS, H_3PO_4, APTES, and STA have shown thermal stability up to 350°C. Conductivity of 2.85 × 10^{-2} S/cm has been obtained at 120°C under 90% RH for 10 mol% STA doped membrane [35].

3.3 Inorganic–Organic Hybrid Proton-Exchange Membranes

Operation of PEMFCs at high temperatures (>100°C) provides a number of technological benefits, including higher tolerance to significant quantities of CO, fast electrode kinetics, and no cathode flooding. Also, PEMFC system complexity could be reduced by the development of 'water-free' electrolytes that do not require hydration. To further operate the PEMFCs at high temperatures (>100°C), the development of 'thermally stable' and 'water-free' electrolyte membranes with high proton conductivities is required [10–13]. In these directions, we have investigated

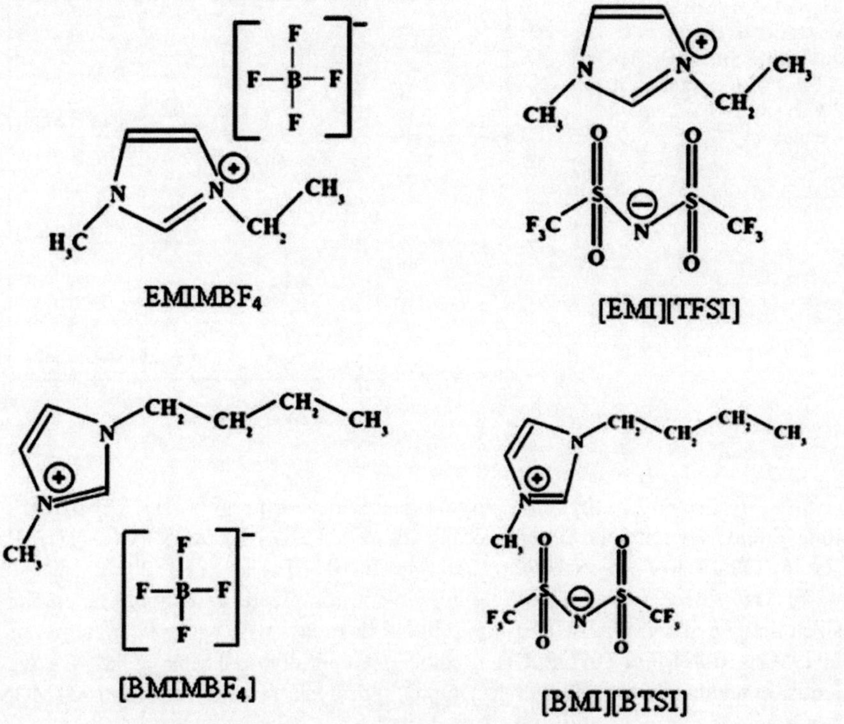

Fig. 3.12 Chemical structure of ionic liquids applied in this study

the preparation of inorganic–organic hybrids, where a substitute charge carrier such as room temperature ionic liquid (IL) is used in place of water. The chemical structures of some IL used in our study are shown in Fig. 3.12. Organosiloxane-based inorganic–organic hybrid is one of the promising materials for the proton-conducting membrane because of its high thermal stability and flexibility of organosiloxane networks. The inorganic–organic composite membranes based on organosiloxanes are a combination of siloxane linkages, organic chains, and mixed acid species. The thermal stability of membrane could be improved by the higher dissociation energy of Si–O bond than that of C–C bond [36].

3.3.1 Inorganic–Organic Hybrid Proton Exchange Membranes Based on Tetramethoxysilane/Vinyltrimethoxysilane/Trimethylphosphate and Ionic Liquid

The hybrid membranes were prepared by using tetramethoxysilane ($Si(OCH_3)_4$, TMOS), vinyltrimethoxysilane ($C_5H_{12}O_3Si$, VTMOS), trimethyl phosphate

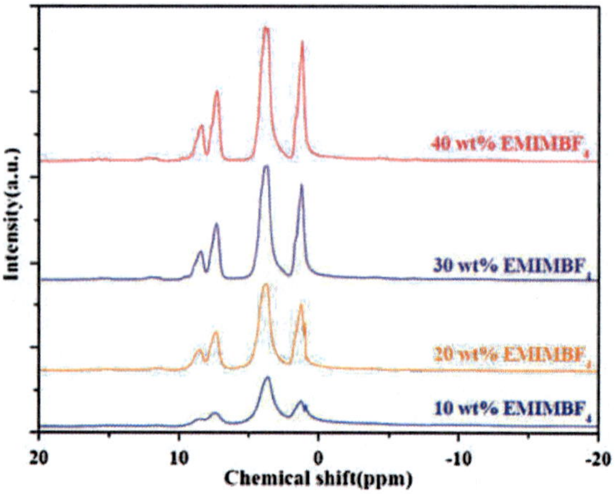

Fig. 3.13 ^{1}H MAS NMR spectra of 60TMOS-30VTMOS-10 $PO(OCH_3)_3$ -x [$EMIMBF_4$](x = 0, 10, 20, 30, and 40 wt%) composite membranes

($PO(OCH_3)_3$), and [1-ethyl- 3-methylimidazolium tetrafluoroborate] ($EMIMBF_4$) ionic liquid as precursors. Several compositions of TMOS-VTMOS-$PO(OCH_3)_3$-IL (i.e. 60TMOS-30VTMOS-10$PO(OCH_3)_3$-x [$EMIMBF_4$] (x = 10, 20, 30, and 40 wt%) were selected for the ionic liquid content optimization. It should be emphasized that we added [$EMIMBF_4$] ionic liquid in excess wt% to the host membrane 60TMOS-30VTMOS-10$PO(OCH_3)_3$ (mol%). Generally, in TMOS/VTMOS covalently bonded network, TMOS has rapid hydrolysis reaction rate and VTMOS possess good thermal stability.

Both of them can polymerize through hydrolysis and condensation reactions to produce a Si–O-Si network. The latter one has a vinyl group in its structure that permits an organic polymerization in order to produce a more flexible material. The structural study of the prepared hybrids is done using FT-IR, NMR, and TG-DTA experiments. Among these data, Fig. 3.13 shows the ^{1}H-NMR spectra, where the chemical shifts at 8.578, 7.425, 3.612, 1.237, and 0.886 ppm are found for 10 wt% [$EMIMBF_4$] ionic liquid doped hybrid membrane. Similarly, for 20, 30, and 40 wt% [$EMIMBF_4$] doped composites, signals at 8.495, 7.274, 3.762, 1.170, and 0.953 ppm; 9.515, 8.428, 7.273, 3.760, and 1.171 ppm; and 9.732, 8.361, 7.271, 3.679, 1.170, and 0.820 ppm are found, respectively. In these prepared hybrid membranes, several ^{1}H chemical shifts shifting compared with pure [$EMIMBF_4$] ionic liquid i.e.; [^{1}H NMR spectrum (δ/ppm relative to $CDCl_3$): 8.403 (s, 1H, H(2)), 7.358 (s, 1H, H(4)), 7.287 (s, 1H, H(5)), 4.034 (q, 2H, and $J = 21$ Hz, H(6)), 3.712 (s, 3H, H(8)), and 1.250 (t, 3H, and $J = 14$ Hz, H(7))] could be due to the strong interaction between phosphosilicate content and ionic liquid.

Thermal properties of all the prepared membranes are investigated by both thermogravimetry (TG) and differential thermal analysis (DTA), respectively. Figure 3.14(a) and (b) show the TG and DTA profiles of all the prepared hybrid membranes, including host phosphosilicate matrix. An initial loss, 5 wt% was found at 88°C for host matrix. This first weight loss region below 100°C is due

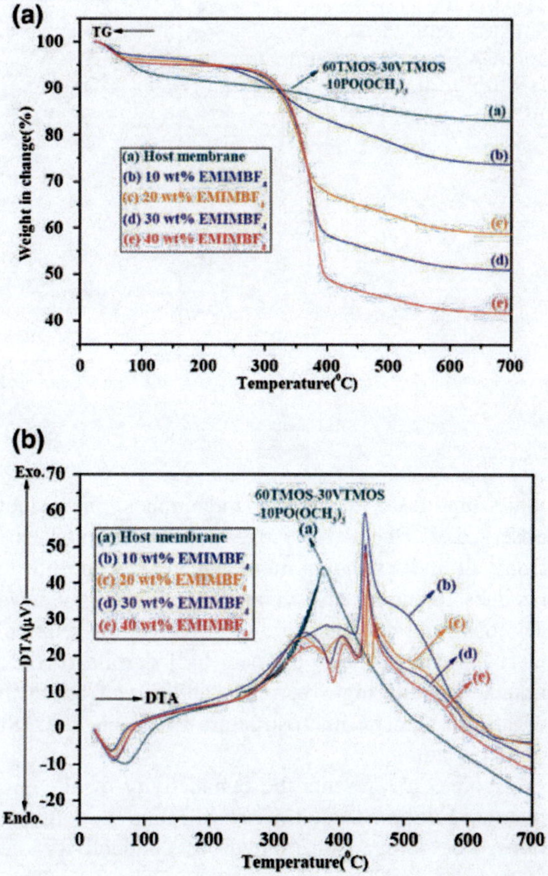

Fig. 3.14 **a** TG **b** DTA profiles of 60TMOS-30VTMOS-10PO$(OCH_3)_3$ -x [EMIM BF_4] (x = 0, 10, 20, 30, and 40 wt%) membranes

to the evaporation of physically weak and chemically strong bound water molecules. Also 10 and 15 wt% losses were observed around at 346 and 500°C due to the condensation of structural hydroxyl groups, respectively. On the other hand, the weight loss of the 10–40 wt% [$EMIMBF_4$] ionic liquid-based hybrid membranes below 100°C was 3.0 wt% only due to the high amount of added ionic liquid. However, all the hybrid membranes studied are found to be thermally stable up to temperatures of 280°C and can be used at temperatures within the 100–200°C range for PEMFCs. From the host matrix DTA profile, a broad exothermic peak at 375°C is identified, including a smaller one at 248°C along with one endothermic peak at 68°C. All the observed exothermic peaks suggest the decomposition or burning of the residual organic groups in the phosphosilicate matrix. Similarly, for 10–40 wt% [$EMIMBF_4$] ionic doped hybrid membranes, exothermic peaks at 388, 440, 498, and 623°C; 342, 402, 443, 456, and 526°C; 348, 405, 441, and 531°C; and 363, 409, 440, and 556°C are found due to decomposition of the doped ionic liquid. Also, from the all

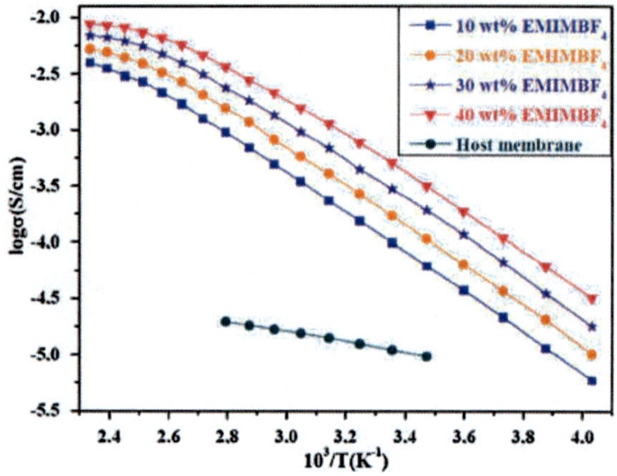

Fig. 3.15 Conductivity of 60TMOS- 30VTMOS-10PO$(OCH_3)_3$-x [$EMIMBF_4$] (x = 0, 10, 20, 30, and 40 wt%) hybrid membranes

ionic liquid doped hybrid membranes one endothermic peak at 54°C was observed. Further, above 625°C, no additional exothermic peaks are identified from all the prepared membranes DTA profiles suggesting that the organic residues including any volatile species removal were completed below 625°C. The observed decomposition temperature is relatively high with respect to other C–H bonding-related polymers [42] despite of the high amount of added ionic liquid, and this high thermal stability can be attributed to both the inorganic–organic composite microstructure based on Si–O–Si backbones and the stability of [BF_4] anion.

Figure 3.15 presents the conductivity of all the prepared hybrid membranes, measured within the temperature range −25 to 155°C under anhydrous conditions. For comparison, anhydrous conductivity of host phosphosilicate matrix from 15 to 85°C is also presented in Fig. 3.15. The thickness of all the prepared hybrid membranes for the conductivity measurements was about 0.2 mm. The maximum conductivity of host membrane at 85°C was 1.97×10^{-5} S/cm only due to the loss of water molecules that exist within the membrane matrix with an increase in the temperature under non-humidified conditions. In order to increase the conductivity of host matrix within the temperature range (−25) to 155°C, [$EMIMBF_4$] ionic liquid is added because ILs are able to act as protonic charge carriers at medium temperatures (100–200°C), even under anhydrous conditions. For all the ionic liquid doped hybrid membranes conductivity was increased with an increase in the ionic liquid weight percent ratio from 10–40 wt%. A maximum conductivity of 8.9×10^{-3} S/cm at 155°C was obtained for 40 wt% [$EMIMBF_4$] ionic liquid doped membrane under anhydrous conditions. At 85°C, the 40 wt% IL doped membrane has shown conductivity value of 3.7×10^{-3} S/cm which is two orders of magnitude higher than that of host matrix conductivity. In the [BF_4] anion, the highly electronegative fluorine atom contributes to the distribution of the anionic charge of borate. In addition to the anionic charge distribution, the effect of the surface covering of the anion

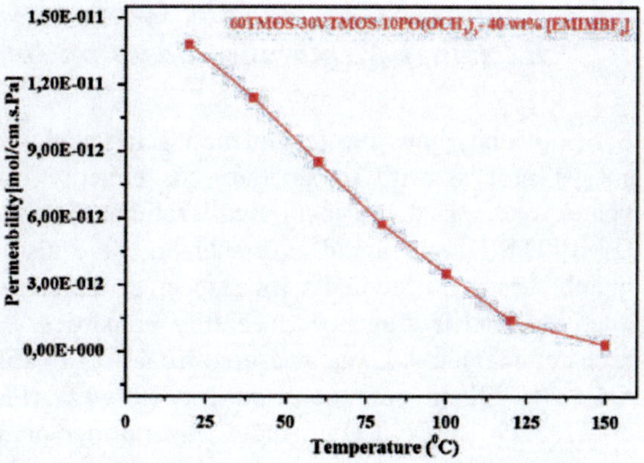

Fig. 3.16 Hydrogen permeation rate as a function of temperature for 40 wt% [EMIMBF$_4$] ionic liquid doped hybrid membrane

backbone by fluorine atoms may be a significant factor for weak interaction with the [EMIM] cation. It should be emphasized that even though the ionic liquid contribute to the anhydrous conductivity, the all available protonic carriers would be solely responsible for the observed higher conductivity of the ionic liquid doped membranes. Also, from Fig. 3.15, one can see that the plots of conductivity versus reciprocal temperature are not linear within the temperature range −25 to 155°C indicating that the conduction does not exactly follow an Arrhenius-type relation, but also exhibits the Vogel-Tamman-Fulcher (VTF) type behavior. Previously, this VTF type behavior in temperature dependence of electrical conductivity measurements was also observed for EMIM-based ionic liquids [37].

Figure 3.16 presents the measured hydrogen permeability values for 40 wt% [EMIMBF$_4$] ionic liquid doped hybrid membrane within the temperature range 20–150°C (thickness = 0.5 mm). The hydrogen permeability value was decreased from 1.38×10^{-11} to 0.3×10^{-12} mol/cm s Pa with increasing temperature from 20°C up to 150°C. The reason for the observed high H_2 permeability values of 40 wt% [EMIMBF$_4$] ionic liquid-based hybrid membrane might be due to the Knudsen diffusion process that exists in phosphosilicate matrices when the H_2 gas permeation decreases with temperature increment. Knudsen diffusion is important when the mean free path of the gas molecules is greater than the pore size. It is well known that the permeability coefficient of any pure gas through a membrane is directly related to the dimensions (size and thickness etc.) of the membrane including applied gas pressure [38–43]. It is possible by the selected sol-gel method to prepare thin membranes with desired thickness just by varying the volume of sol solutions of the studied membranes, in the Petri dishes. To increase the power density, decrease of the membrane resistance by means of reduction of the thickness of the electrolyte membrane is required without loss of mechanical strength. However the hydrogen permeability property of ionic liquids doped inorganic–organic hybrid membranes required deeper investigations and will be a subject of our future work.

3.3.2 Another Example of the Inorganic–Organic Hybrid Proton Exchange Membranes Based on Ionic Liquid

Series of anhydrous proton-conducting inorganic–organic hybrid membranes by sol-gel process with tetramethoxysilane/methyl-trimethoxysilane/trimethylphosphate and 1-ethyl-3-methylimidazolium-bis(trifluoromethanesulfonyl) imide [EMI][TFSI] ionic liquid as precursors were also prepared [36]. These hybrid membranes were studied with respect to their structural, thermal, proton conductivity, and hydrogen permeability properties. Proton conductivity of all the prepared membranes was measured from −25 to 155°C, and high conductivity of 5.4×10^{-3} S/cm was obtained for 40 wt% [EMI][TFSI] doped 40TMOS–50MTMOS–10PO(OCH$_3$)$_3$ (mol%) hybrid membrane, at 150°C under anhydrous conditions. The hydrogen permeability was found to decrease from 1.61×10^{-11} to 1.39×10^{-12} mol/cm s Pa for 40 wt% [EMI][TFSI] doped hybrid membrane as the temperature increases from 20 to 150°C. For 40 wt% ionic liquid doped hybrid membrane, membrane electrode assemblies were prepared and a maximum power density value of 0.22 mW/cm^2 at 0.47 mA/cm^2 as well as a current density of 0.76 mA/cm^2 was obtained at 150°C under non-humidified conditions when utilized in a H_2/O_2 fuel cell. Further, anhydrous hybrid membranes based on tetramethoxysilane/methyl-trimethoxysilane/diisopropyl phosphite and 1-butyl-3-methylimidazolium tetrafluoroborate (BMIMBF$_4$) ionic liquid were fabricated through sol-gel process and characterized [38]. In this case, the average pore size was increased proportionally with the ionic liquid weight percent ratio in the host phosphosilicate matrix from 2.59 to 11.71 nm, respectively. Thermogravimetric analysis and differential thermal analysis measurements confirmed that the hybrid membranes were thermally stable up to 260°C. Thermal stability of the hybrid membranes was significantly enhanced by the presence of inorganic SiO_2 framework and high stability of [BF$_4$] anion. For all the composite membranes, the conductivities were measured within the temperature range (−30–150°C), and a maximum conductivity of 7×10^{-3} S/cm at 150°C was achieved for 40 wt% ionic liquid-based composite membrane under nonhumidified conditions.

Also, proton-conducting hybrid membranes were prepared by sol-gel process with tetraethoxysilane/methyltriethoxysilane/trimethyl phosphate and 1-butyl-3 methylimidazolium tetrafluoroborate (BMIMBF$_4$) ionic liquid as precursors [39]. For all the prepared composite membranes the conductivities were measured within the temperature range (−30–150°C), and a maximum conductivity of 6.74×10^{-3} S/cm at 150°C was achieved for 40 wt% ionic liquid-ased hybrid membrane, under non-humidified conditions. The hydrogen permeability values were decreased with the temperature increment from 20 to 150°C and the measured H_2 permeability value for 40 wt% ionic liquid-based hybrid membrane was 0.5×10^{-12} mol/cm s Pa at 150°C. For 40 wt% [BMIMBF$_4$] doped hybrid membrane, membrane electrode assemblies were prepared and a maximum power density value of 2 mW/cm^2 at 4.73 mA/cm^2 as well as a current density of 7.4 mA/cm^2 was obtained at 150°C under non-humidified conditions when utilized

Table 3.1 Conductivities of Ionic liquid doped inorganic membranes

Ionic liquid	Matrix composition	Conductivity at 150°C under anhydrous condition (S/cm)
[EMI][TFSI]	TMOS-MTMOS-TMP	5.4×10^{-3}
$BMIMBF_4$	TMOS-MTMOS-DPP	7.0×10^{-3}
$BMIMBF_4$	TEOS-MTMOS-TMP	6.74×10^{-3}
$BMIMBF_4$	TMOS-TMPS-TEP	6.4×10^{-3}
[BMI][BTSI]	TEOS-ETEOS-TMP	5.0×10^{-3}
[EMI][TFSI]	TEOS-PDMS-TMP	4.87×10^{-3}
$EMIMBF_4$	TMOS-TEOS-MTEOS-TMP	1.0×10^{-2}
$EMIMBF_4$	TMOS-VTMOS-TMP	8.9×10^{-3}

in a H_2/O_2 fuel cell. Further, anhydrous proton-conducting hybrid membranes were prepared by the sol-gel process with trimethoxysilane (TMOS)/trimethoxy propylsilane (TMPS)/triethyl phosphate (TEP) and $BMIMBF_4$ ionic liquid as precursors [40]. Relatively maximum anhydrous conductivity of 6.4×10^{-3} S/cm was obtained for 40 wt% ionic liquid-based hybrid membrane, at 150°C, with a hydrogen permeability value of 0.8×10^{-12} mol/cm s Pa. Also we reported on membranes with anhydrous proton conduction prepared by the sol-gel process with tetraethoxysilane/ethyl-triethoxysilane/trimethylphosphate and 1-butyl-3-methyl imidazoliumbis (trifluoromethylsulfonyl) imide ionic liquid as precursors [41]. The conductivity of all the fabricated membranes was measured from -30 to 150°C, and a high conductivity of 5×10^{-3} S/cm was measured for the 40 wt% [BMI][BTSI] ionic liquid doped membrane at 150°C under anhydrous conditions. The hydrogen permeability was decreased from 1.84×10^{-11} to 2.7×10^{-12} mol/cm s Pa for the 40 wt% [BMI][BTSI] ionic liquid doped hybrid membrane during temperature increment from 20 to 150°C. For 40 wt% ionic liquid doped hybrid membrane, membrane electrode assemblies were fabricated, and a maximum power density value of 0.17 mW/cm^2 at 0.42 mA/cm^2 as well as a current density of 0.67 mA/cm^2 were achieved at 150°C under nonhumidified conditions. Recently, we have prepared proton-conducting membranes using tetraethoxysilane/poly (-dimethylsiloxane)/trimethyl phosphate and 1-ethyl-3 methylimidazolium-bis (trifluoromethanesulfonyl) imide ionic liquid as sol-gel precursors [42]. Conductivity of all the fabricated hybrid membranes was measured under anhydrous conditions within the temperature range −20 to 150°C, and a value of 4.87×10^{-3} S/cm at 150°C was achieved for 40 wt% [EMI][TFSI] ionic liquid doped 72TEOS-18PDMS-10PO $(OCH_3)_3$ (mol%) hybrid membrane. For 40 wt% ionic liquid doped composite membrane, the measured hydrogen permeability value at 150°C was 4×10^{-12} mol/cm s Pa. Very recently, we have prepared new composite membranes consisting of phosphosilicate (P_2O_5–SiO_2) contentand1-ethyl-3methylimi- dazolium tetrafluoroborate [$EMIMBF_4$] ionic liquid by sol–gel process [43]. Relatively a high conductivity of 1×10^{-2} S/cm was obtained for 40 wt% IL doped 30TMOS–30TEOS–30MTEOS–10PO$(OCH_3)_3$ hybrid membrane at 155°C under anhydrous conditions. The hydrogen permeability was found to decrease within the range 10^{-11} to 10^{-12} mol/cm s Pa for 40 wt% IL doped membrane as the temperature increases from 20 to 150°C. The conductivities of the ionic liquid-inorganic hybrids prepared in our research are summarized in Table 3.1.

References

1. Winter M, Brodd RJ (2004) What are batteries, fuel cells, and supercapacitors? Chem Rev 104:4245
2. Steele BCH, Heinzel A (2001) Materials for fuel-cell technologies. Nature 414:345
3. Carrette L, Friedrich KA, Stimming U (2000) Fuel cells: principles, types, fuels, and applications. Chem Phys Chem 1:162
4. Kim JH, Jo YY, Cho EA, Jang JH, Kim HJ, Lim TH, Oh IH, Ko JJ, Son IJ (2010) Effects of cathode inlet relative humidity on PEMFC durability during startup–shutdown cycling. J Electrochem Soc 157:B633
5. Mauritz KA, Moore RB (2004) State of understanding of Nafion. Chem Rev 104:4535
6. Kreuer KD, Paddison SJ, Spohr E, Schuster M (2004) Transport in proton conductors for fuel-cell applications: simulations, elementary reactions, and phenomenology. Chem Rev 104:4637
7. Kreuer KD (2001) On the development of proton conducting polymer membranes for hydrogen and methanol fuel cells. J Membr Sci 185:29
8. Harrison WL, Hickner MA, Kim YS, Mc Grath JE (2005) Poly (arylene ether sulfone) copolymers and related systems from disulfonated monomer building blocks: synthesis, characterization, and performance–a topical review. Fuel cells 5:201
9. Souzy R, Ameduri B (2005) Functional fluoropolymers for fuel cell membranes. Prog Polym Sci 30:644
10. Sen U, Celik SU, Ata A, Bozkurt A (2008) Anhydrous proton conducting membranes for PEM fuel cells based on Nafion/Azole composites. Int J Hydrogen Energy 33:2808
11. Itoh T, Hirai K, Tamura M, Uno T, Kubo M, Aihara Y (2008) Anhydrous proton-conducting electrolyte membranes based on hyperbranched polymer with phosphonic acid groups for high-temperature fuel cells. J Power Sources 178:627
12. Aslan A, Çelik SÜ, Bozkurt A (2009) Proton-conducting properties of the membranes based on poly (vinyl phosphonic acid) grafted poly (glycidyl methacrylate). Solid State Ionics 180:1240
13. Alabi CA, Chen Z, Yan YS, Davis ME (2009) Insights into the nature of synergistic effects in proton-conducting 4,4—1H,1H-Bitriazole-Poly (ethylene oxide) composites. Chem Mater 21:4645
14. Zhang J, Xie Z, Zhang J, Tang Y, Song C, Navessin T, Shi Z, Song D, Wang H, Wilkinson DP, Liu ZS, Holdcroft S (2006) High temperature PEM fuel cells. J Power Sources 160:872
15. Li Q, He R, Jensen JO, Bjerrum NJ (2003) Approaches and recent development of polymer electrolyte membranes for fuel cells operating above 100°C. Chem Mater 15:4896
16. Hu J, Luo J, Wagner P, Conrad O, Agert C (2009) Anhydrous proton conducting membranes based on electron-deficient nanoparticles/PBI-OO/PFSA composites for high-temperature PEMFC. Electrochem Commun 11:2324
17. Nogami M, Nagao R, Makita K, Abe Y (1997) Fast proton-conducting P2O5–ZrO2–SiO2 glasses. Appl Phys Lett 71:1323
18. Nogami M, Abe Y (1997) Evidence of water-cooperative proton conduction in silica glass. Phys Rev B 55:12108
19. Nogami M, Nagao R, Wong C (1998) Proton conduction in porous silica glasses with high water content. J Phys Chem B 102:5772–5775

20. Nogami M, Matsushita H, Goto Y, Kasuga T (2000) A sol–gel-derived glass as a fuel cell electrolyte. Adv Mater 18:1370
21. Nogami M, Daiko Y, Akai T, Kasuga T (2001) Dynamics of proton transfer in the sol−gel-derived P2O5−SiO2 glasses. J Phys Chem B 103:4653
22. Daiko Y, Kasuga T, Nogami M (2002) Proton conduction and pore structure in sol−gel glasses. Chem Mater 14:4624
23. Li H, Nogami M (2002) Pore-controlled proton conducting silica films. Adv Mater 14:912
24. Li H, Nogami M (2003) Ordered mesoporous phosphosilicate glass electrolyte film with low area specific resistivity. Chem Commun 236:236–237
25. Uma T, Nogami M (2006) High performance of H2/O2 fuel cells using Pt/C electrodes and P2O5-SiO2-PMA glasses as an electrolyte in low temperature. J Ceram Soc Jpn 114:748
26. Uma T, Nogami M (2007) Characterization and performance improvement of H2/O2 fuel cells based on glass membranes. J Electrochem Soc 154:B845
27. Uma T, Nogami M (2007) Structural and transport properties of mixed phosphotungstic acid/ phosphomolybdic acid/SiO2 glass membranes for H2/O2 fuel cells. Chem Mater 19:3604
28. Lakshminarayana G, Nogami M (2009) Synthesis, characterization and electrochemical properties of SiO2–P2O5–TiO2–ZrO2 glass membranes as proton conducting electrolyte for low-temperature H2/O2 fuel cells. J Phys D Appl Phys 42:215501/1
29. Mackenzie JD (2003) Sol-gel research—achievements since 1981 and prospects for the future. J Sol-Gel Sci Technol 26:23
30. Nogami M, Nagao R, Wong C, Kasuga T, Hayakawa T (1999) High proton conductivity in porous P2O5−SiO2 glasses. J Phys Chem B 103:9468
31. Ogawa M (1995) Incorporation of pyrene into an oriented transparent film of layered silica-hexadecyltrimethylammonium bromide nanocomposite. Langmuir 11:4639–4641
32. Lakshminarayana G, Nogami M (2009) Synthesis and characterization of proton conducting inorganic–organic hybrid nanocomposite membranes based on tetraethoxysilane/ trimethylphosphate/ 3-glycidoxypropyltrimethoxysilane/heteropoly acids. Electrochim Acta 54:4731
33. Lakshminarayana G, Nogami M (2009) Synthesis and characterization of proton conducting inorganic−organic hybrid nanocomposite membranes based on mixed PWA-PMA-TEOS-GPTMS-H3PO4-APTES for H2/O2 fuel cells. J Phys Chem C 113:14540
34. Lakshminarayana G, Nogami M (2009) Synthesis and characterization of proton conducting inorganic-organic hybrid nanocomposite films from mixed phosphotungstic acid/ phosphomolybdic acid/tetramethoxysilane/3-glycidoxypropyltrimethoxysilane/phosphoric acid for H2/O2 fuel cells. J Renew Sustain Energy 1:013106/1
35. Lakshminarayana G, Nogami M (2011) Proton-conducting hybrid membranes for medium temperature (>100°C) fuel cells. Ionics 17:287–291
36. Lakshminarayana G, Nogami M (2010) Inorganic–organic hybrid membranes with anhydrous proton conduction prepared from tetramethoxysilane/methyl-trimethoxysilane/ trimethylphosphate and 1-ethyl-3-methylimidazolium-bis (trifluoromethanesulfonyl) imide for H2/O2 fuel cells. Electrochim Acta 55:1160–1168
37. Vila J, Gin'es P, Pico JM, Franjo C, Jim'enez E, Varela LM, Cabeza O (2006) Temperature dependence of the electrical conductivity in EMIM-based ionic liquids: evidence of Vogel–Tamman–Fulcher behavior. Fluid Phase Equilibria 242:141
38. Lakshminarayana G, Tripathi VS, Tiwari I, Nogami M (2010) Anhydrous proton-conducting organic–inorganic hybrid membranes synthesized from tetramethoxysilane/methyltrimethoxysilane/ diisopropyl phosphite and ionic liquid. Ionics 16:385
39. Lakshminarayana G, Nogami M (2010) Proton conducting organic–inorganic composite membranes under anhydrous conditions synthesized from tetraethoxysilane/methyltriethoxysilane/trimethyl phosphate and 1-butyl-3 methylimidazolium tetrafluoroborate. Solid State Ionics 181:760
40. Lakshminarayana G, Vijayaraghavan R, Nogami M, Kityk IV (2011) Anhydrous proton conducting hybrid membrane electrolytes for high temperature (>100°C) proton exchange membrane fuel cells. J Electrochem Soc 158:B376

41. Lakshminarayana G, Nogami M, Kityk IV (2010) Anhydrous proton conducting inorganic–organic composite membranes based on tetraethoxysilane/ ethyl-triethoxysilane/ trimethylphosphate and 1-butyl- 3-methylimidazolium-bis (trifluoromethylsulfonyl) imide. J Electrochem Soc 157:B892
42. Lakshminarayana G, Nogami M, Kityk IV (2010) Synthesis and characterization of anhydrous proton conducting inorganic–organic composite membranes for medium temperature proton exchange membrane fuel cells (PEMFCs). Energy 35:5260
43. Lakshminarayana G, Nogami M, Kityk IV (2011) Novel hybrid proton exchange membrane electrolytes for medium temperature non-humidified fuel cells. J Alloys Compd 509:2238

Chapter 4
Sol-Gel Routes and Proton Conductors

Philippe Colomban

Abstract The sol-gel route offers the possibility of homogeneous preparation of materials, especially those with proton conductivity or able to host a proton conducting phase. Proton conductors are potential key materials for the development of CO_2 free energy production and conversion. The thermal stability of proton conductors in relation with the hydrogen bond strength is discussed. The most representative inorganic and organic proton conductors in their different forms are presented.

Keywords Grotthuss mechanism · Heteropolyacids (HPA) · Hybrid membranes · Hydrogen bonds · Inorganic membranes · Polymer membranes · Proton conduction mechanism · Proton conductivity · Proton exchange membrane fuel cell (PEMFC) · Sol-gel · Structure · Thermal stability

4.1 Introduction

Recently, in the twilight of the fossil fuel economy and in front of global warming problem, hydrogen appears as an alternative energy vector for a sustainable modern world. However two bottlenecks have to be successfully overcome [1, 2]:

i. its low cost and environment friendly (CO_2 free) production e.g. using the steam water electrolysers: electricity $+ H_2O \Leftrightarrow H_2 + O_2$

Ph. Colomban (✉)
Laboratoire de Dynamique, Interaction et Réactivité (LADIR) CNRS,
Université Pierre-et-Marie-Curie (UPMC), C49, 4 Place Jussieu,
75252, Paris Cedex 05, France
e-mail: philippe.colomban@upmc.fr

M. Aparicio et al. (eds.), *Sol-Gel Processing for Conventional and Alternative Energy*,
Advances in Sol-Gel Derived Materials and Technologies,
DOI: 10.1007/978-1-4614-1957-0_4,

ii. development of long life time high efficiency fuel cells (the reverse of the above equation) and CO_2/syngas converters:

$$4\ H_2 + CO_2 \Leftrightarrow CH_4 + 2H_2O\ (\text{or}\ CH_3OH)$$

These routes require a membrane, acting as an isolator for electrons but a conductor for the proton or the oxygen ion. Sensors controlling the H_2 level are also required.

The small size of proton makes that a high surface or volume concentration of proton at an electrode surface gives a huge capacitance. Consequently, proton conductors may have many other applications, for example as membrane in supercapacitors and in optical/solar electrochemical devices [3]. However, the expected better lifetime related to the small proton size may increase the reversibility of the intercalation/deintercalation process.

The proton, through hydrogen bonding, plays a very important role in the formation of gels. The sol-gel route offers the possibility of homogeneous preparation of the electrolyte, some electrode materials and/or the framework that will host the proton conducting phase. First, we will consider the gel route from its synthesis general point of view.

Sol-gel routes to prepare solid materials in their different forms (monolith, film and membrane, etc.) are used since decades [3–17]. Inorganic, organic and hybrid/mixed gels represent always one of the most promising lines of development [18–53] for low-current/intensity high capacity batteries, supercapacitors and electrochromic (EC) devices and even fuel cells. They consist of a conductive and electrochemically stable liquid electrolyte soaked within a polymer [18, 19], organic [20], hybrid [21–23] or porous [3, 24–44] matrix providing a "stable" and appropriate framework to retain the electrolyte. They are also potential materials as electrolyte for small size fuel cells [29–43], electrochromic [45, 46] and dye-sensitized photoelectrochemical cells (DSPEC) [23, 47]. Mixed electron/proton, organic and inorganic conductor gels may have specific use [48–50].

A gel is the association of a framework made of covalent-bonded species expanded by a continuous liquid phase. Consequently, framework elements are fixed giving rise to a heterogeneity at a very short scale. The interest of gel is to combine the expected advantages of a "solid" (a volume "stable" from the mechanical point of view in order to avoid dommageble leak requiring a very tight and costly container) and a liquid phase (high diffusion rates, good interface contact).

It is difficult to trace the beginning of Sol-gel history! Pionner works date from the 50 s for the preparation of ceramics by US Roy' group [4] and much earlier for glass compositions in Germany (see in ref [8]). However, the first comprehensive works on the relationships among Sol-gel synthesis, homogeneity, grain size control and low-temperature sintering occur with the need to prepare very homogeneous, transparent, large size ceramics [5]. Transparent PLZT perovskite (La_2O_3–$PbTiO_3$–$PbZrO_3$ solid solution) ceramics for advanced optical devices (optical memory, filter, Kerr' cell,...) were achieved by mixing Ti- and

Zr-alkoxides. The alkoxide mixture was hydrolyzed by acetic acid solution that simultaneously dissolves Pb oxide and La hydroxide powder to form a sol or a gel [5, 6], as a function of concentration and pH. This retains a large attention and gives impetus to focus the efforts of ceramists and glass makers with the help of metal–organic chemistry experts [9, 51] to develop new synthesis routes for film, monolithic ceramics, fibers and composites [52, 53]. Actually, most of the so-called chemical, coprecipitation, "chimie douce" routes ... are also partially or totally Sol-gel routes. Moreover the traditional clay route to prepare pottery is a Sol-gel route [13–15] that takes profit of the nanometric state of clay particles and their ability to form aqueous sol or gel as a function of pH/pK values. All these liquid ways offer the following advantages: (i) a good mixing at a very short scale (at least submicronic) of certain elements (those belonging to gel-former [13, 15, 53]), (ii) the small size of the particles promotes a good reactivity during the sintering and hence a grinding is not necessary, the later step requiring tools, time, energy and hence increasing significantly the cost and the duration of the synthesis - *this is the main reason of the success of Sol-gel routes in many academic studies !*, (iii) a good control of the viscosity that makes possible forming coating, fibers,... and allowing the dispersion of inert grains in the viscous medium [13], (iv) a low temperature sintering [11–13].

Another great advantage is the mixing of organic and inorganic state-of-the-art, e.g. silanes (Si–Xn, X being an organic branch) and alkoxides (Si–O–Xn) to give hybrid compounds [54].

In most of the cases the gel state is controlled by the formation of hydrogen bonds. The main inorganic gel formers are the silicon atoms and the atoms with rather similar properties (P, Al, Ti, Zr, etc. [13, 14]) that form strong covalent-bonded tetrahedron, octahedron, etc. and hence crystalline or amorphous inorganic polymers. They form 2 or 3 Dimension backbone of strong covalent-bonded entities with an easy orientation disorder and non-saturated oxygen atoms that may be acceptors for H-bond. Furthermore, the Si atom ability to establish both strong Si–C and Si–H bond makes a bridge with an organic chemistry [44–48]. Hydrolysis-polycondensation of convenient alkoxide mixtures leads to a gel, or a sol according to the synthesis parameters. Drying at room temperature leads to xerogel—also called a porous glass, a dried glass or just "glass"—that contains always water, hydroxyl ions and more complex protonic species [55–57]. Full densification is only achieved at much higher temperature, between ~500°C for poor Sol-gel former composition (i.e. ZrO_2 [55]) and up to 1,000°C for aluminosilicates [56–63], with the elimination of the residual protons. The gel can be used to form a membrane immobilizing the proton conductor or coated on another membrane in order to optimize the properties of the resulting material [22].

4.2 The Protonic Species and the Proton Conduction

The proton conduction is a kind of ionic conductivity but the very small size of the proton, in between that of an electron and of the lithium ion, the smallest 'standard' ion, makes that the proton conductivity is a very specific topic.

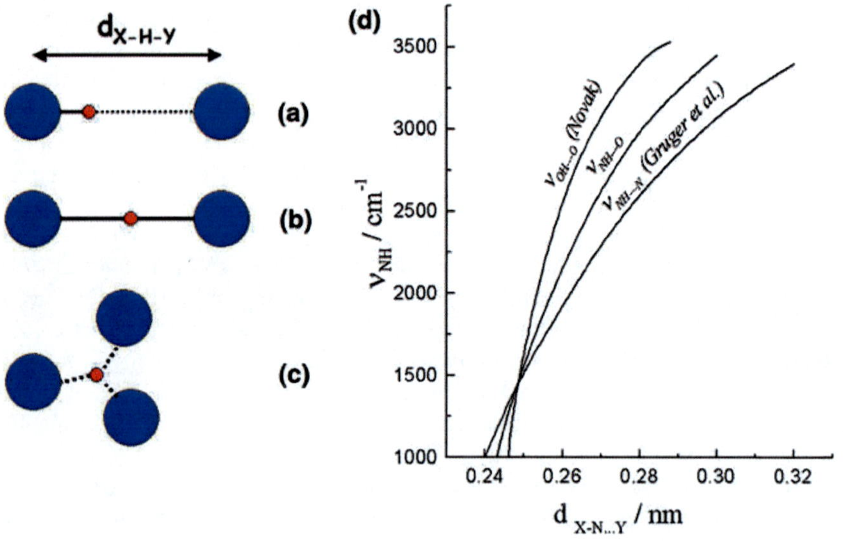

Fig. 4.1 Schematic of a weak asymmetric H-bond (**a**), of a strong symmetric H-bond (**b**) and of an ionic proton (**c**); in **d** the very large infrared wave numbers shift of the *ν*O–H or *ν*N–H band center- of -gravity versus the acceptor-donnor distance issued of diffraction data [2–4] gives illustration of the very large variation of the hydrogen bond strength; the N–H..O curve has been drawn by interpolation [79, 81]

The effect of very high electrical charge vs. size ratio makes that in most of the cases the proton penetrates the covalence shell of an acceptor species, such as oxygen ions, ammoniac, hydrazine, imidazole, … to form hydroxyl (OH^-), ammonium (NH_4^+), (H_3O^+), hydrazinium ($N_2H_5^+$), imidazollium ($C_3N_2H^+$) ion, etc. [3, 64–73]. A certain interaction, the so-called hydrogen bond is established between the proton bonded to the acceptor and the similar species in its vicinity [74–80]. Actually, the interaction versus two acceptors can be shifted quasi continuously and can be inversed (proton jump also called Grotthuss' mechanism). For instance considering the oxonium ion H_3O^+ interacting with a water molecule H_2O, the O–H$^+$….O distance can vary from d > 0.29 nm (no H-bond, Fig. 4.1a–c) to ~0.245 nm when the bond is fully symmetrical [3] and gives the Hydroxonium $H_5O_2^+$ ion (Fig. 4.1d). Thus the X–H bond (X = O or N) is the sole bond that may have a quasi continuous values from a very low value, less than 1 kcal/mole, i.e. a bond broken by the thermal agitation at room temperature, to a rather high value (>50 Kcal/mole).

Furthermore, the corresponding potential well is very anharmonic that has an important effect on the thermal expansion of the material and its vibrational signature.

The thermal stability of proton conducting material is thus directly related to the nature of the protonic species and to the strength of the H-bond.

Table 4.1 Typical conducting protonic species and their stability range

Protonic species	Typical stability range (°C)
H_2O, $H^+(H_2O)_n$ (gels)	<80
H_3O^+, NH_4^+	<400
OH^-	<600
H^+	<900

Thermal equilibrium between the species having different H-bonding strength is also common. On the contrary, other bonds have rather fixed distance/bond strength or show step-by-step variations (single C–C or C–N, double C = C or N = N and triple bond). The other main exception is the Π-bond in aromatic moieties. In rare cases the proton shares its interaction with more than 2 acceptors and became the ionic species [3, 64, 65]. The question if the stabilization of these new forms by electronic defect(s) is necessary remains open.

The variation of the bonding strength between a proton and its environment has important consequences, especially regarding the thermal stability of hydrated materials such as gels: weaker the H-bond, lower the thermal stability. For instance, phosphotungstic ($H_3PW_{12}O_{40}$, 29 H_2O, PWA) and phosphomolibdic ($H_3PMo_{12}O_{40}$, 30 H_2O) acid hydrates, two well-established crystalline proton conductors [3, 26, 73, 76] exhibit dehydration level steps at ~21, 12, 6 (> 60–80°C) and 0 H_2O molecules (>250°C), that have a strong effect on the nature of protonic species [26, 64, 76]. Similarly, for the poly(perfluorosulfonic) acid (hydrated) membrane, known by its commercial name Nafion®, a water loss (and conductivity, mechanical strength,...) starts over 80°C [82]. Polyalinine (PANI), a mixed proton–electron conductor, is also dehydrated above the same temperature and above ~200°C an irreversible degradation takes place [50, 82–84]. Ammonium and oxonium β-aluminas are stable up to ~400°C [24, 25] and above 500°C the evolving protons destroy the β-alumina framework leading to the high crystalline transition alumina. Note, in the lack of protons the thermal stability of β-alumina exceeds 1600°C. These transformations give a good illustration of the low stability of the different water and/or proton containing phases. In summary, the departure of weak H-bonded hydration water takes place below ~80–120°C, H_3O^+ ions lose their water at ~250–400°C as a function of the host framework and "ionic" protons and hydroxyl ions remain hosted in oxide frameworks above 500°C, sometimes up to 1,000°C [26–30, 85, 86, (Table 4.1)]. There is thus a direct relationship between the thermal stability and the strength of the proton bonding with its framework, in other words, with the Brönstedt acidity of the host framework [3]. Consequently, the thermal stability of proton conducting gels cannot exceed ~80°C at usual water pressure.

The solution to increase the thermal stability range of the material is to control the water pressure, in the vicinity of the protonic species. There are two alternatives applied to the required water partial pressure: (i) the entire device itself must work under pressure, *i.e.* autoclave, (ii) the proton conductor is encapsulated by a host framework that retains kinetically the water and hence increases the water partial pressure at the very local scale. Since the Jurin' law determines the local partial pressure, the size and the geometry (curvature radius) of the pores are also important parameters.

Some materials can be prepared by direct synthesis (e.g. HUP: $H_3OUO_2PO_4$, $3H_2O$, ZrHP: $Zr(HPO_4)_2$, nH_2O)) some others are only obtained after ionic exchange [3, 35, 82] in the liquid or gas state (hydrogen plasma [87]), with or without subsequent thermal treatments.

Crystalline proton conductors are good models to study the proton conductivity but they exhibit marked phase transitions and conductivity jumps. On the contrary, corresponding amorphous and/or gel phases do not generally exhibit conductivity jump [85]. However, the superionic conductivity requires static order that promotes the dynamic disorder [88]. Consequently, the conductivity of amorphous and gel materials is lower but can be spread out to larger temperature ranges. The gel is homogeneous without defined interfaces that promote a good contact with solid electrodes.

There are thus different routes to obtain a proton conducting gel: (i) forming a 3D acid framework that hosts some sites able to receive protons or to be grafted by protonic species, (ii) forming a micro/meso porous framework that retains acidic water, (iii) forming a framework in which a crystalline or amorphous proton conductor can be dispersed. First we will present a few compounds that can be considered as the model of different proton conductors, in their crystalline or gel form. Then we focus the chapter on the materials that retain attention to overcome some of the intrinsic weakness of proton conductors or that may be obtained in the gel form.

4.3 Model Proton Conductors

Different criteria can be used to classify the proton conductors.

The first one is their intrinsic stability range that depends on the nature of the protonic species as well as on the interaction between these species and the host framework: typically hydrates, frameworks hosting oxonium or ammonium ions and then protonated anhydrous compounds.

The second one is the geometry of the conducting pathway for the conducting species: surface or bulk conduction. The crystal and/or grain habits (fibers, ribbons, ...) play important role because the surface is always different from the bulk, especially regarding the possibility of interaction with water and protonic species.

Obviously, there is also another criteria related to the use, the conductivity at the working temperature and its activation energy (Table 4.2).

All protonic conductors where the conductivity results from the Grotthuss mechanism may have at least one couple of proton donor and acceptor: the proton jump from one ion to the neighboring one (inversion of the asymmetric H-bond), being easier if a reorientation of the donor and/or acceptor ions takes place. On the contrary, for ionic (interstitial) protonic species the diffusion does not need such coupled motions.

Table 4.2 Model proton solid conductors [3, 35]

Name	Synthesis	Stability range/°C	Species	Activation energy/eV
Ice	D	<0	H_3O^+/H_2O	0.6
PWA	D	0–30	$H_5O_2^+/H_3O^+/H_2O$	0.15/0.25
HUP	D	−50–80	H_3O^+/H_2O	0.3
HUP "gels"	D		$H_3O^+/H_2O/HPO_4^-/PO_4^{2-}/$	
ZrHP, n H_2O	D		$HPO_4^-/PO_4^{2-}/H_2O$	0.3–0.5
ZrHP gel	D		$HPO_4^-/PO_4^{2-}/H_2O$	
Oxide gels	D		$H_3O^+/H_2O/OH^-$	
Nafion®	Ex	−10–80	$HSO_4^-/SO_4^{2-}/H_2O$	0.2
Nafion ® gel	D		$HSO_4^-/SO_4^{2-}/H_2O$	
PANI	Ex	0–150	$H^+/H_3O^+/H_2O/e^-$	0.1
PANI gel	Ex	0–80	$H^+/H_3O^+/H_2O/e^-$	
PVA-H_3PO_4			$HPO_4^-/PO_4^{2-}/H_2O$	1
H_3O^+/NH_4^+ β alumina	Ex		H_3O^+/NH_4^+	0.2–0.6
$H^+(H_2O)n$ β alumina	Ex	−50–150	$H_5O_2^+/H_3O^+/H_2O$	0.2–0.5
$CsHSO_4$	D		HSO_4^-/SO_4^{2-}	0.35
H_xWO_3	Ex/D		H^+	
H_xPWA	D		H^+	0.4–0.6
Perovskites $H_zSrZr_{1-x}Ln_xO_{3-y}$	Ex	300–550/ 550–800	H^+	0.3/0.6
Aluminosilicate	Ex	500–900	H^+	n.m.

D direct synthesis; *Ex* protonated materials is obtained after ionic/plasma exchange; *n.m.* not measured. See text for the acronyms

4.4 Inorganic Polymers and Gels

Inorganic gels are observed for some inorganic polymers such as silicates, phosphates, arsenates, antimonates, titanates, vanadates…, i.e. structures built of strong molecular units: SiO_4, PO_4, AsO_4, VO_4, TiO_6,.. [3, 15, 16, 24, 48]. Crystalline phases and small crystals may be obtained for defined compositions but the presence of defects (formation of X_2O_7, X_3O_9, etc.... species) and protonation (e.g. $H_2PO_4^-$, HPO_4^{2-}, HSO_4^-) promote the disorder. Additionally, the bonding with water molecules leads to an amorphous phase and/or a gel. Gels may also result from the very small grain sizes, their specific shape (slabs, needles) and surface properties that permit strong bonding with water: clays, vanadium, tin or cerium oxides are good examples. Consequently, in many cases the differentiation among surface-, interface- and bulk-conductivity does not make sense.

The situation is very similar with anhydrous inorganic forms but in the lack of water molecules, glasses are formed and not gels. Rapid quenching is required to increase the orientational disorder and to avoid crystallization [85]. The gel composition (hydroxyl and water content) highly depend from the synthesis

Table 4.3 Examples of proton conductor gels and associated framework or dispersed phase

Backbone/host framework	Dispersed phase	Reference
$SiO_2-P_2O_5-ZrO_2$	PWA	Seo et al. [89]
SiO_2	PWA	Mioc et al. [26]
$SiO_2-Al_2O_3$	PWA	Mioc et al. [104]
ICS-SiO_2/PPG	PWA,SiWA	Lavrensic–Stangar et al. [44], id. [22], id. [105] Surca et al. [106], Surca Vuk et al. [46]
$SiO_2-Al_2O_3-P_2O_5-ZrO_2$ etc.	Acidic water	Blanchard et al. [25]
$SiO_2(-Al_2O_3-P_2O_5-ZrO_2$etc.)	Nafion (ionomer)	Mauritz [30], Deng et al. [31] Aparicio et al. [90], Jiang et al. [32]
DBSA:PEG/SiO_2	Nafion	Thangamuthu and Lin [91]
Silica gel	Nafion-like gel	Zoppi et al. [28, 103], Shao et al. [29]
Silica gel	Nafion & polianiline	Chen et al. [92]
PEG/SiO_2	PWA	Lin et al. [93]
Nafion®	ZrHP	Yang et al. [42]
ZrHP	Silica (xero)gel	Alberti et al. [27]
ZrHP	PWA	Helen et al. [94]
Chitosan	NH_4NO_3	Ng and Mohamad [95]
Chitosan- Al_2SiO_5	H_3PO_4-NH_4NO_3	Majid and Arof [96]
PAPS	H_3PO_4	Tezuka et al. [97]
PBI	H_3PO_4	Jalani et al. [18]
PEO	H_3PO_4	Qiaio et al. [98]
SiO_2	Bis(trifluoromethyl sulfonyl)imide + ethyl-3-imidizolio)butane-4-sulfonate	Mizumo et al. [36]
Silsesquioxane	Trifluoroacetic/acetic acid	Surca Vuk [46]
SiO_2	NH_4PO_3	Peng et al. [99]
SiO_2	H_3PO_4-fullerene	Li et al. [100]
SPEEK	$Zr(HPO_4)$ nH_2O	Krishnan et al. [101]
PVDF-hexafluoropropylene	Silica	Sel et al. [102]
PMMA		Agnihotry et al. [19]

parameters and a very little attention was given to the composition measurement [56–58].

4.5 Organic and Hybrids Organic–Inorganic Materials

Organic polymers are made of long chains more or less interconnected. Thus hydrophilic backbone are easily filled and expansed with water. The use of short oligomers in place of long macromolecules gives rise to gels that can be deposited at the surface or injected in the pores of an inorganic matrix. Hybrid materials were first developed in order to optimize process-related properties (viscosity, wettability, polymerization, porosity, …) [30, 54] and then for other properties including the proton conductivity [22]. Two main routes were followed (Table 4.3): (i) the hosting of a polymeric liquid or solid proton conductor in a micro/mesoporous inorganic framework, (ii) the mixing of liquid proton conductor with a polymer. The big difficuty is the optimization of both electrical and mechanical properties in the required thermal range in presence of a strong water pressure gradient through the electrolytic membrane.

References

1. Olah GA (2005) Beyond oil and gas: the methanol economy. Angew Chem Int Ed 44(18):2636–2639
2. Iwahara H, Esaka T, Uchida H, Maeda N (1981) Proton conduction in sintered oxides and its application to steam electrolysis for hydrogen-production. Solid State Ionics 3/4:359–363
3. Colomban Ph (1992) Proton conductors: solids, membranes and gel—materials and devices. Cambridge University Press, Cambridge
4. Roy R (1969) Gel route to homogeneous glass preparation. J Am Ceram Soc 52:344–347
5. Snow G (1973) Fabrication of transparent electronic PLZT ceramics by atmosphere sintering. J Am Ceram Soc 56:91–96
6. Colomban Ph (1976) Frittage de céramiques transparentes PLZT. L'Industrie Céramique 697:531–535
7. Yoldas BE (1977) Effect of variation in polymerized oxides on sintering and crystalline transformations. J Am Ceram Soc 65:387
8. Sakka S, Kamiya K (1980) Glasses from metal alcoholates. J Non-Cryst Solids 42:403–422
9. Bradley DC, Mehrotra RC, Gaur DP (1978) Metal alkoxides. Academic Press, London
10. Mazdiyasni KS (1982) Powder synthesis from metal-organic precursors. Ceram Int 8:42–56
11. Perthuis H, Colomban Ph (1984) Well densified NASICON-type ceramics elaborated using Sol-GelSol-Gel process and sintering at low temperatures. Mater Res Bull 19(5):621–631
12. Johnson DW Jr (1985) Sol-Gel processing of ceramic and glass. Am Ceram Soc Bull 64:1587–1602
13. Colomban Ph (1989) Gel technology in ceramics, glass-ceramics and ceramic–ceramic composites. Ceram Int 15:23–50
14. Colomban Ph (1996) Raman studies of inorganic gels and of their sol-to-gel, gel-to-glass and glass-to-ceramic transformation. J Raman Spectrosc 27:747–758
15. Colomban Ph, Boilot J-P (1985) Polymères inorganiques (xerogels et verres) dans les systèmes M2O-M'O2SiO2–P2O5X2O3. Rev Chimie Minérale 22:235–255
16. Boilot JP, Colomban Ph (1985) Sodium and lithium superionic gels and glasses. J Mater Sci Lett 4:22–24
17. Klein LC (1988) Sol-Gel technology. NOYES publications, Park Ridge
18. Jalani NH, Ramani M, Ohlsson K, Buelte S, Pacifico G, Pollard R, Staudt R, Datta R (2006) Performance analysis and impedance spectral signatures of high temperature PBI-phosphoric acid gel membrane fuel cells. J Power Sources 160:1096–1103
19. Agnihotry SA, Pradeep P, Sekhon SS (1999) PMMA based gel electrolyte for EC smart window. Electrochim Acta 44(18):3121–3126
20. Staiti P, Minutoli M, Lufrano F (2002) All solid electric double layer capacitors based on Nafion ionomer. Electrochem Acta 47:2795–2800

21. Honma I, Takeda Y, Bae JM (1999) Protonic conducting properties of sol-gel derived inorganic/organic nanocomposites membrane doped with acidic functional molecules. MRS Symp Proc Organic/inorganic Hybris Mater II 576:257–262
22. Lavrencic Stangar U, Orel B, Schmitz A, Colomban Ph (2001) Proton-conducting sol-gel hybrids containing heteropoly acids. Solid State Ionics 145:109–118
23. Lavrencic Stangar U, Orel B, Surca Vuk A, Sagon G, Colomban Ph, Stathatos E, Lianos P (2002) In-situ resonance Raman microscopy of a solid-state dye sensitied photoelectrochemical cell. J Electrochem Soc 149(11):E413–E423
24. Blanchard N, Boilot JP, Colomban Ph (1983) Formation of superionic gels and glasses by low temperature chemical polarization. Solid State Ionics 910:639–643
25. Blanchard N, Boilot JP, Colomban Ph, Pouxviel J-C (1986) New glasses from metal-organic precursors: preparation and properties. J Non-Cryst Solids 82:205–209
26. Mioc UB, Milonjic SK, Malovic D, Stamenkovic V, Colomban Ph, Mitrovic MM, Dimitrijevic R (1997) Structure and proton conductivity of 12-tungstophosphoric acid doped silica. Solid State Ionics 97:239–246
27. Alberti G, Casciola M, Costantino U, Peraio A, Rega T (1995) Proton-conducting solid dispersions of silica and zirconium phosphate pyrophosphate. J Mater Chem 5:1809–1812
28. Zoppi RA, Yoshida IVP, Nunes SP (1998) Hybrids of perfluorosulfonic acid ionomer and silicon oxide by sol-gel reaction from solution. Morphology and thermal analysis. Polymer 39:1309–1315
29. Shao PL, Mauritz KA, Moore RB (1995) [Perfluorosulfonate ionomer] [mixed inorganic oxide] nanocomposites via polymer in situ sol-gel chemistry. Chem Mater 7:192–200
30. Mauritz KA (1998) Organic-inorganic hybrid materials: perfluorinated ionomers as sol-gel polymerization templates for inorganic alkoxides. Mater Sci Eng C-Biomimetic Supramol Syst 6:121–133
31. Deng Q, Moore RB, Mauritz KA (1998) Nafion® (SiO2, ORMOSIL and dimethylsiloxane) hybrids via in situ sol-gel reactions; characterization of fundamental properties. J Appl Polymer Sci 68:747–763
32. Jiang RC, Kunz HR, Fenton JM (2006) Composite silica/Nafion (R) membranes prepared by tetraethylorthosilicate sol-gel reaction and solution casting for direct methanol fuel cells. J. Membr Sci 272:116–124
33. Wang HT, Holmberg BA, Huang LM, Wang ZB, Mitra A, Norbeck JM, Yan YS (2002) Nafion-bifunctional silica composite proton conductive membranes. J Mater Chem 12:834–837
34. Colomer MT, Anderson MA (2001) High porosity silica xerogels prepared by a particulate sol-gel route: pore structure and proton conductivity. J Non-Cryst Solids 290:93–104
35. Colomban Ph, Novak A (1988) Proton transfer and superionic conductivity in solids and gels. J Mol Struct 77:277–308
36. Mizumo T, Watanabe T, Ohno H (2008) Thermally stable abd proton conductive ionogel based on bronsted acidic ionic liquid with the support of silicate network. Polymer J 40:1099–1104
37. Surca-Vuk A, Gaberscek M, Orel B, Colomban Ph (2004) In situ resonance Raman spectroelectrochemical studies of a semi-solid redox (I3-/I-) electrolyte encapsulated in a hybrid electrochemical cell. J Electrochem Soc 151:E150–E161
38. Lavrencic-Stangar U, Orel B, Régis A, Colomban Ph (1997) Chromogenic WPA/TiO2 hybrid gels and films. J Sol-Gel Sci Technol 8(1–3):965–971
39. Herring AM (2006) Inorganic-polymer composite membranes for proton exchange membrane fuel cells. Polymer Rev 46:245–296
40. Miyake N, Wainright JS, Savinell RF (2001) Evaluation of a sol-gel derived Nafion/silica hybrid membrane for proton electrolyte membrane fuel cell application–I. Proton conductivity and water content. J Electrochem Soc 148:A898–A904
41. Damay F, Klein LC (2003) Transport properties of Nafion ™ composite membranes for proton-exchange membranes fuel cells. Solid State Ionics 162:261–267

42. Yang C, Srinivasan S, Bocarsly AB, Tulyani S, Benziger JB (2004) A comparison of physical properties and fuel cell performance of Nafion and zirconium phosphate/Nafion composite membranes. J Membr Sci 237:145–161
43. Holclajtner-Antunović I, Mioč UB, Todorović M, Jovanović Z, Davidović M, Bajuk-Bogdanović D, Laušević Z (2010) Characterization of potassium salts of 12-tungstophosphoric acid. Mater Res Bull 45:1679–1684
44. Lavrencic-Stangar U, Grosejl N, Orel B, Colomban Ph (2000) Structure of and interactions between P/SiWA Keggin nanocrystals dispersed in an organically modified electrolyte membrane. Chem Mater 12:3745–3753
45. Grandvist CG (1995) Handbook of inorganic electrochromic materials. Elsevier, Amtersdam
46. Surca Vuk A, Jovanovski V, Pollet-Villar A, Jerman I, Orel B (2008) Imidazolium-based ionic liquid derivatives for application in electrochromatic devices. Sol Energy Mater Sol Cells 92:126–135
47. Lavrencic Stangar U, Orel B, Groselj N, Colomban Ph, Stathatos E, Lianos P (2002) In situ resonance raman studies of a dye-sensitized photoelectrochemical cell with a Sol-Gel electrolyte. J New Mater Electrochem Syst 5:223–231
48. Livage J (1991) Vanadium pentoxide gels. Chem Mater 3(4):578–593
49. Vuk AS, Orel B, Drazic G, Colomban Ph (2002) Vibrational spectroscopy and analytical electron microscopy studies of Fe-V-O and In-V-O thin films. Monatshefte fur Chemie 133(6):889–908
50. El Khalki A, Gruger A, Colomban Ph (2003) Bulk-surface nanostructure and defects in polyaniline films and fibers. Synth Met 139:215–220
51. Okamura H, Bowen HK (1986) Preparation of alkoxide for the synthesis of ceramics. Ceram Int 12:161–171
52. Colomban Ph (2011) SiC from amorphous to nanosized materials, the example of SiC fibers issued of polymer precursors. In: Mukherjee M (ed) Silicon carbide, materials, processing and applications in electronic devices. INTECH, pp 161–186. ISBN: 978-953-307-968-4
53. Colomban Ph (2005) Nano/micro-structure and property control of single and multiphase materials, ch. 12. In: Komarneni S, Lee B (eds.) chemical processing of ceramics-second edition, CRC Press, Boca Raton pp. 303–339
54. Schubert U, Hüsing N, Lorenz A (1995) Hybrid inorganic-organic materials by sol-gel processing of organofunctional metal alkoxides. Chem Mater 7:2010–2027
55. Bruneton E, Colomban Ph (1992) Influence of hydrolysis conditions on crystallization, phase transitions and sintering of zirconias prepared by alkoxide hydrolysis. J Non-Crystall Solids 147, 148:201–205
56. Vendange V, Colomban Ph (1996) How to tailor the porous structure of alumina and aluminosilicate gels and glasses. J Mater Res 11:518–528
57. Vendange V, Colomban Ph (1996) Determination of the hydroxyl group content in gels and porous "glasses" issued of alkoxide hydrolysis by combined TGA and BET analysis. J Porous Mater 3:193–200
58. Vendange V, Colomban Ph (1994) Densification mechanisms of alumina, aluminosilicates and aluminoborosilicates gels, glasses and ceramics. J Sol-Gel Sci Technol 2:407–411
59. Touati F, Gharbi N, Colomban Ph (2000) Structural evolution in polyolysed hybrid organic-inorganic alumina gels. J Mater Sci 35:1565–1570
60. Colomban Ph (1989) Structure of oxide gels and glasses by IR and Raman scattering: I. aluminas. J Mater Sci 24:3002–3010
61. Colomban Ph (1989) Structure of oxide gels and glasses by IR and Raman scattering: II. mullites. J Mater Sci 24:3011–3020
62. Colomban Ph (1992) Hydrogen bonding in hydrogen substituted lithium aluminosilicates. J Mol Struct 270:407–416
63. Slodczyk A, Colomban Ph, Zaafrani, O, Lacroix B, Sala B (2011) What is the true nature of conducting proton in perovskite membranne: hydroxyl ion or intertitial proton? MRS 2010 Fall Meeting EE In Proceeding vol. 1309/mrsf10-1309-ee03-21

64. Colomban Ph, Tomkinson J (1997) Novels forms of hydrogen in solids : the ionic proton and the "free" proton. Solid State Ionics 97:123–134
65. Colomban Ph (1999) Latest developments in proton conductors. Ann Chimie Sci Matériaux-Paris 24:1–18
66. Colomban Ph, Lucazeau G, Novak A (1981) Vibrational study of hydrogen β alumina. J Phys C (Solid State Physics) 14:4325–4333
67. Pham-Thi M, Adet Ph, Velasco G, Colomban Ph (1986) Perfomance of a hydrogen uranyl phosphate-carbon double layer solid capacitor. Appl Phys Letts 48(20):1348–1350
68. Pham Thi M, Colomban Ph (1985) Morphological, X-ray and vibrational chemistry of various uranyl phosphate hydrates. J Less-Common Metal 108:747–751
69. Pham Thi M, Colomban Ph (1986) Acetone intercalation/desintercalation process in HUP (H3OUO2PO4.3H2O) framework. A model for electrochemical degradation. kinetic and conducting properlies. J Mater Sci 21:1591–1600
70. Colomban Ph, Boilot J-P, Kahn A, Lucazeau G (1978) Structural investigation of protonic conductors NH4 + β-Alumina. Nouv J Chimie 2:21–32
71. Colomban Ph, Boilot J-P, Chagnon P, Guilloteau G (1978) Stabilité des conducteurs protoniques de types βAl2O3. Bull Soc Fr Ceram 119:3–12
72. Colomban Ph, Novak A (1988) Proton transfer and superionic conductivity in solids and gels. J Mol Struct 177:277–308
73. Kremenovic A, Spasojevic-De Bire A, Bourrée F, Colomban Ph, Dimitrijevic R, Davidovic M, Mioc U.B (2002) Structural modifications of Dodecatungstophosphorichexahydrate induced by temperature in the (10–358 K) range. In situ high resolution neutron powder diffraction investigation. Solid State Ionics 150(3–4):431–442
74. Colomban Ph, Novak A (1989) Nature of protonic species and gel-crystal transition in hydrated zirconium phosphate. J Mol Struct 198:277–296
75. Colomban Ph, Doremieux-Morin C, Piffard Y, Limage M-H, Novak A (1989) Equilibrium between protonic species and conductivity mechanism in antimonic acid, H2Sb4O11.n H2O. J Mol Struct 213:83–96
76. Mioč UB, Todorović MR, Davidović M, Colomban Ph, Holclajtner-Antunović I (2005) Heteropolycompounds, from proton conductors to biomedical agents. Solid State Ionics 176(39–40):3005–3017
77. Novak A (1974) Hydrogen bonding in solids correlation of spectroscopic and crystallographic data. Struct Bonding 18:177–216
78. Blinc R, Hadzi D, Novak A (1960) The relation between the bridge length of short hydrogen bonds, the potential curve, and the hydroxyl stretching frequency. Zeitschrift Elektrochemie 64(5):567–571
79. Lautié A, Froment F, Novak A (1976) Relationship between NH stretching frequencies and N..O distances of crystal containing NH..O hydrogen bonds. Spectrosc Letts 9(5):289–299
80. Hadzi D, Orel B, Novak A (1973) Infrared and Raman-spectra of some acid salt containing symmetric hydrogen-bonds. Spectrochim Acta Part A 29:1745–1753
81. Colomban Ph, Gruger A, Novak A, Régis A (1994) Infrared and Raman study of polyaniline. Part I : hydrogen bonding and electronic mobility in emeraldine salts. J Mol Struct 317:261–271
82. Yao YFY, Kummer JT (1963) Ion exchange properties of and rates of ionic diffusion in beta-alumina. J Inorg Nucl Chem 29(9):2453–2503
83. Gruger A, Régis A, Schmatko T, Colomban Ph (2001) Nanostructure of nafion membranes at different states of hydration. An IR and Raman study. Vib Spectrosc 26:215–225
84. Gruger A, El Khali A, Colomban Ph (2003) Protonation, sol formation and precipitation of poly- and oligo.anilines. J Raman Spectrosc 34(6):438–450
85. Mihri T, Colomban Ph (1989) Spread of the conductivity jump in mixed alkali acid sulphates Csl-xMxHSO$_4$. Solid State Ionics 35:99–103
86. Mioc UB, Colomban Ph, Davidovic M, Tomkinson J (1994) Inelastic neutron scattering study of protonic species during the thermal dehydration of 12-tungstophosphoric hexahydrate. J Mol Struct 326:99–107

87. Schnell Ph, Velasco G, Colomban Ph (1981) Thin films of β-Alumina-like structure. Solid State Ionics 5:291–294
88. Colomban Ph (1986) Raman study of the polymer-superionic NASICON transformation. dynamics, static inorganic orientational disorder and superionic conductivity. J Mol Struct 143:191–194
89. Seo D, Park S, Shul YG (2009) Preparation of silica-based proton conductors for intermediate temperature fuel cells. Korean J Chem Eng 26:1016–1021
90. Aparicio M, Damay F, Klein LC (2003) Characterization of SiO2–P2O5-ZrO2 Sol-Gel/ Nafion ™ composite membranes. J Sol-Gel Sci Techn 26:1055–1059
91. Thangamuthu R, Lin CW (2005) DBSA-doped PEG/SiO2 proton-conducting hybrid membranes for low temperature fuel cell applications. Solid State Ionics 176:531–538
92. Chen CY, Garnica-Rodriguez JI, Duke MC, Dalla Costa RF, Dicks AL, da Costa JCD (2007) Nafion/polyaniline/silica composite membranes for direct methanol fuel cell application. J Power Sources 166:324–330
93. Lin CW, Thangamuthu R, Chang PH (2005) PWA-doped PEG/SiO2 proton-conducting hybrid membranes for fuel cell applications. J Membr Sci 254:197–205
94. Helen M, Viswanathan B, Murthy SS (2007) Synthesis and characterization of composite membranes based on alpha zirconium phosphate and silicotungstic acid. J Membr Sci 292:98–105
95. Ng LS, Mohamad AA (2008) Effect of temperature on the performance of proton batteries based on chitosan-NH4NO3-EC membrane. J Membr Sci 325:653–657
96. Majid SR, Arof AK (2008) FTIR studies of chitosan-orthophosphoric acid-ammonium nitrate-aluminosilicate polymer electrolyte. Mol Cryst Liquid Cryst 484:483–492
97. Tezuka T, Tadanaga K, Hayashi A, Tatsumisago M (2008) Proton conductive inorganic-organic hybrid membranes prepared from 3-aminopropyltriethoxysilane and phosphoric acid by the sol-gel method. Solid State Ionics 179:1151–1154
98. Qiaio J, Yoshimoto N, Ishikawa M, Morita M (2003) A proton conductor based on a polymeric complex of poly(ethylene oxide)-modified poly(methacrylate) with anhydrous H3PO4. Chem Mater 15:2005–2010
99. Peng XB, Chen XL, Zheng HB, Cheng JG, Xia CR (2007) Electrical performance of NH4PO3-SiO2 composite electrolytes prepared by a sol-gel method. Chin J Inorg Chem 23:1567–1571
100. Li YM, Noda K, Hinokuma K (2007) Nano-composite of phosphoric acid derivative of fullerene with SiO2 and its proton conductivity. Electrochem 75:243–247
101. Krishnan P, Park JS, Yang TH, Lee WY, Kim CS (2006) Sulfonated poly(ether etherketone)-based composite membrane for polymer electrolyte membrane fuel cells. J Power Sources 163:2–8
102. Sel O, Kim LTT, Debiemme-Chouvy C, Gabrielli C, Laberty-Robert C, Perrot H, Sanchez C (2010) Proton insertion properties in a hybrid membrane/conducting polymer bilayer investigated by ac electrogravimetry. J Electrochem Soc 157:F69–F76
103. Zoppi RA, Nunes SP (1998) Electrochemical impedance studies of hybrids of perfluorosulfonic acid ionomer and silicon oxide by sol-gel reaction from solution. J Electroanalyt Chem 445:39–45
104. Mioc UB, Milonjic SK, Stamenkovic V, Radojevic M, Colomban Ph, Mitrovic MM, Dimitrijevic R (1999) Structure, properties and proton conductivity of the 12-tungstophosphoric acid doped aluminosilicate gels. Solid State Ionics 125(1–4):417–424
105. Lavrencic-Stangar U, Orel B, Groselj N, Colomban Ph, Stathatos E, Lianos P (2002) In situ resonance Raman microspectrometry of a solid-state dye-sensitized photoelectrochemical cell. J Electrochem Soc 149(11):E413–E423
106. Surca A, Orel B, Spreizer H, Colomban Ph (2003) In situ resonance micro Raman and UV-visible spectroelectrochemical studies of an electrochemical device with an I3-/I- redox sol-gel electrolyte. Solid State Ionics 165(1–4):247–255

Chapter 5
Proton Exchange Membranes for H_2 Fuel Cell Applications

Kenneth A. Mauritz, Amol Nalawade and Mohammad K. Hassan

Abstract This chapter presents studies of sol-gel modifications of perfluorinated and hydrocarbon H_2 fuel cell membranes using metal alkoxides, organoalkoxy-silanes and combinations of these monomers. The impacts of these modifications on proton conductivity at high temperature and low relative humidity, fuel cell performance, chemical and mechanical durability and H_2 and O_2 crossover are discussed. Methods of tailoring the energetic environment, or polarity, within the polar cluster domains by insertion of inorganic oxide or organically modified silicate nanostructures by different chemistry routes are presented.

Keywords Proton exchange membrane fuel cells · Hybrid membranes · Sol-gel process · Mechanical and thermal properties · Nafion® · Nafion®/ORMOSIL membranes · Nafion®/Inorganic oxide membranes · Non-perflurosulfonated ionomer templates · Proton conductivity

A fuel cell (FC) is an electrochemical device that converts chemical energy of a fuel directly into electrical energy for practical use. Among the various types of FCs under development, those using proton exchange membranes (PEMFCs) are

---

K. A. Mauritz (✉) · A. Nalawade · M. K. Hassan
School of Polymers and High Performance Materials,
University of Southern Mississippi, Hattiesburg, MS 39406, USA
e-mail: Kenneth.mauritz@usm.edu

A. Nalawade
e-mail: nalawade.amol@gmail.com

M. K. Hassan
e-mail: Mohamed.Hassan@usm.edu

M. K. Hassan
Department of Chemistry, Faculty of Science,
Bani Suef University, Bani Suef, Egypt

M. Aparicio et al. (eds.), *Sol-Gel Processing for Conventional and Alternative Energy*,
Advances in Sol-Gel Derived Materials and Technologies,
DOI: 10.1007/978-1-4614-1957-0_5,

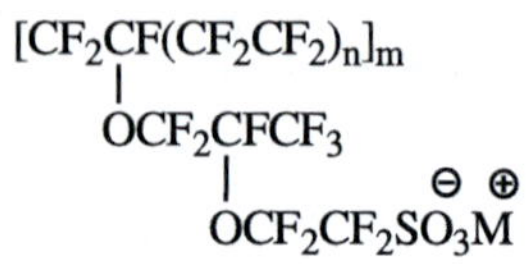

Fig. 5.1 Chemical structure of Nafion®

the leading technology for applications that largely include transportation vehicles, although stationary applications, such as back-up power systems, are also being considered. PEMFCs use solid polymer electrolyte membranes coated with platinum-based electrodes that serve as a proton conducting medium while also being an effective gas separation barrier. The required features of a PEMFC are high power density and high energy efficiencies at moderate operational temperatures, long-term durability, all solid construction and low/zero emission of pollutants [1, 2].

There are two electrodes on either surface of a proton conducting membrane. Electrochemical reactions taking place on catalyst surfaces at respective electrodes are shown below. At the anode, supplied hydrogen is split into protons and electrons. The protons migrate through the hydrated membrane towards the cathode while the electrons arrive at the cathode side having traversed an external circuit so as to supply useful electric current to a device. At the cathode, supplied oxygen is reduced on a catalyst surface to combine with protons and electrons to form water as a benign byproduct.

At anode:

$$H_2 \longrightarrow 2H^+ + 2e^-$$

Cathode reaction:

$$\tfrac{1}{2}\,O_2 + 2H^+ + 2e^- \longrightarrow H_2O$$

Other than PEMFCs, polymer electrolyte membranes are used in direct methanol fuel cells (DMFC) where hydrogen is replaced by liquid methanol as the fuel, although our discussion will be limited to H_2 fuel cells [3, 4] as DMFCs are discussed elsewhere in this volume.

Membranes in FCs must possess high proton conductivity, high dimensional stability, excellent chemical and mechanical durability, good thermo-mechanical properties, zero fuel permeability, good water management, ease of fabrication into membrane-electrode assemblies (MEAs) and low cost. While the perfluorosulfonic acid membrane Nafion® (E.I. DuPont Co.), in Fig. 5.1, is considered the benchmark fuel cell membrane for historical reasons, it has serious problems.

M is either a metal cation in the neutralized form or, of interest here, a proton. Different grades of Nafion® are classified on the basis of equivalent weight (EW), which is the weight of the polymer neutralized by one equivalent of base. The ion exchange capacity (IEC) is related to EW by IEC = 1,000/EW. A comprehensive review of the structure and properties of Nafion® is given by Mauritz and Moore [5]. Nafion® derives its usefulness in part because of its oxidative and chemical stability. Hydrated Nafion® morphology consists of hydrophilic regions, containing sulfonic acid groups and water, distributed throughout a hydrophobic tetrafluoroethylene matrix. The hydrophilic domains are interconnected above a water percolation threshold.

There are also rigid chain aromatic hydrocarbon ionomer FC membranes that are being investigated from the standpoint of avoiding problems associated with Nafion®, namely operating temperatures confined to under 100°C and poor mechanical durability. We will discuss this class of materials in this review.

Absorbed water in Nafion® lowers its T_g causing mechanical property deterioration because macromolecular rearrangements due to stress perturbations occur more easily. Water is produced on the cathode side and also migrates from anode to cathode due to electro-osmotic drag of water molecules. Some water back-diffuses from cathode to anode under a concentration gradient. The net accumulation of water results in flooding at the cathode causing membrane drying away from the cathode resulting in conductivity decrease while excessive membrane swelling toward the cathode side causes limiting proton current. Poor water management in Nafion® results in membrane swelling/de-swelling (shrinking) to different degrees at different locations causing irreversible macromolecular rearrangements and mechanical hysteresis. Also, shrinking-swelling might compromise contact between the membrane and the deposited electrode layer. During FC operation in a vehicle, relative humidity (RH) cycling due to start-stop affects mechanical property deterioration leading to cracks. It is observed during accelerated FC tests that membrane durability is mostly governed by the mechanical strength of the MEA [6, 7]. Formation of cracks, tears and pinholes accelerates fuel crossover which, in turn, causes additional chemical degradation [8]. Physical (mechanical) degradation associated with RH and temperature variations create new fuel crossover pathways and contribute synergistically with chemical degradation [9–11].

Given the poor durability of Nafion® and its inability to retain water above 100°C, FC research shifted to the development of alternative low cost proton conducting membranes [12] that are able to withstand high operating temperatures. These high-T_g materials include poly(perflurosulfonic acid) membranes other than Nafion® [5, 13], poly(arylene ether)s [14–16], polysulfones [17], poly(etheretherketone) (PEEK) [18–20], PEMs based on poly(imide)s [21, 22], polybenzimidazole (PBI) and polyphosphazenes.

Efforts are underway to develop membranes in which water is replaced by non-volatile proton ‘solvents’, e.g., phosphoric acid (PA). PA—doped PBI membranes appear promising at high temperatures. There is a need to improve water retention at high temperatures while resisting attack on the polymer by radicals that are decomposition products generated from peroxide. The issue of morphological stability against RH changes is also important. While high T_g aromatic polymers address these problems it appears that the only route for improving Nafion® is to intimately incorporate robust nanoparticles that retain hydration at high temperatures while maintaining high proton conductivity and provide reinforcement that resists structural hysteresis caused by RH cycling. These particles must be well-dispersed rather than aggregated, so that the mixing-in of preformed particles is not an optimal route. A method that is capable of affecting this modification is membrane—in situ sol-gel processes for inorganic alkoxide monomers.

The sol-gel route is attractive because of high metal alkoxide purity, solution homogeneity and lower processing temperatures. However, silicate structures can degrade in PEMFCs. Nonetheless, the high level of control over sol-gel processes for silicon alkoxides affords opportunities for model studies of factors not involving chemical durability. Sol-gel reactions of other alkoxides, e.g., Ti, Al and Zr, occur too rapidly for monomers to permeate deep into preformed membranes so that rate-limiting catalysts must be employed. Nanoscopic particles resulting from sol-gel reactions possess high surface area-to-volume that favors intimate incorporation and many interfacial interactions. There is much literature dealing with synthesis of bulk inorganic oxide gels synthesized from alkoxides of silicon, organo-alkoxysilanes, titanium, aluminum, zirconium, and boron [23–25]. Reviews of inorganic–organic polymer membranes for FCs by sol-gel processes [26–31] are available.

Sol-gel nanoparticle synthesis in Nafion® (and similar strategies in hydrocarbon PEMs) is performed to improve the following membrane properties:

(1) Increase hydration due to numerous M–OH surface groups to which water molecules can be hydrogen bonded.
(2) Nanoparticles through interfacial interactions will elevate T_g and thermomechanical properties to improve durability, and resist pinhole formation and cracking due to humidification–dehumidification cycling.
(3) Well-dispersed nanoparticles introduce tortuous gas diffusion pathways to diminish fuel crossover, thereby improving FC efficiency while reducing chemical degradation.

Nafion® nanocomposite membranes can be prepared by the following method [26, 27] which involves formation of hydrophilic inorganic oxide nanoparticles in preformed ionomer films. The ionomer acts as a 'template' in that sol-gel reactions which preferentially occur in acid-containing domains to which hydrolyzed inorganic alkoxide monomers migrate on the basis of energetic affinity. Alkoxide monomers permeate ionomer films from external solutions that include a solvent and water and an acid on occasion. Films are prehydrated to facilitate in situ alkoxide hydrolysis and there can be a co-solvent to cause swelling to improve monomer uptake. Drying and heating of reacted films removed from solution after prescribed times results in a nanocomposite membrane that can then be rendered into a membrane-electrode assembly.

Approaches taken to create perfluorosulfonic acid-based nanocomposites via in situ sol-gel processes [32] are summarized below.

5.1 Perflurosulfonate Ionomers

5.1.1 Nafion®/[Silicate] Hybrid Membranes

The potential of these hybrid membranes in PEMFC technology has been discussed elsewhere as well as here [33, 34].

Small angle X-ray scattering (SAXS) analyses of Nafion®/silicate hybrids established that the phase separated morphology of unfilled Nafion® persists based on the fact that the ionomer peak remains and the Bragg spacing associated with average intercluster spacing at ~4 nm is unchanged despite invasion of this morphology by the sol-gel–derived silicon oxide phase [35, 36]. This reinforced the hypothesis of a three-dimensional interactive template in which the reactants (TEOS and organically modified silicates = ORMOSILs) are confined to polar clusters wherein sol-gel reactions initiate and the quasi-ordered nanoscale morphology controls the geometrical distribution of the resultant silicate phase of isolated particles. FTIR and ^{29}Si NMR spectroscopy probed molecular connectivity in silicate and other inorganic oxide and ORMOSIL nanophases [37, 38].

Mechanical tensile studies revealed progressive strengthening with decreasing elongation-to-break followed by a ductile-to-brittle transition, owing to silicate phase percolation that occurs with increasing silicate content [39]. TGA/FTIR showed an upward shift in degradation temperature for silicate-modified composites compared to that of unfilled Nafion® [40]. Incorporation of silicate structures improves mechanical and thermal stability which may improve membrane durability in FCs.

In some cases, Nafion®/silicate membranes showed higher relative water uptake than that of a corresponding unfilled membrane which was attributed to numerous SiOH groups to which H_2O molecules can be hydrogen-bonded as depicted in Fig. 5.2 [41, 42]. This might be beneficial in membranes operating at high temperatures in FCs.

Later, Jung et al. [43] showed that a silicate phase in Nafion® imparts greater hydration than unfilled acid form Nafion®. They prepared Nafion®/silicate membranes, with various uptakes, which were evaluated for DMFC performance. Methanol crossover rates showed decreased methanol permeability with increased silicate content. DMFC performance of the composite with 12.4 wt.% silicate was better than that of higher silicate-containing membranes at temperatures of 60, 90, 110 and 125°C. As silicate content increases above 12.4%, FC performance decreases. The decreased proton conductivity of composites at higher silicate uptake was responsible for inferior fuel cell performance.

Miyake et al. formulated Nafion®/silicate membranes within the context of FC membranes using a sol-gel method [44]. They concluded that the hybrid has higher water content than unmodified Nafion® at 25 and 120°C, but not at 150 and 170°C. Conductivity decreased with increasing silicate content which may be rationalized in terms of increased proton migration tortuosity posed by impenetrable silicate

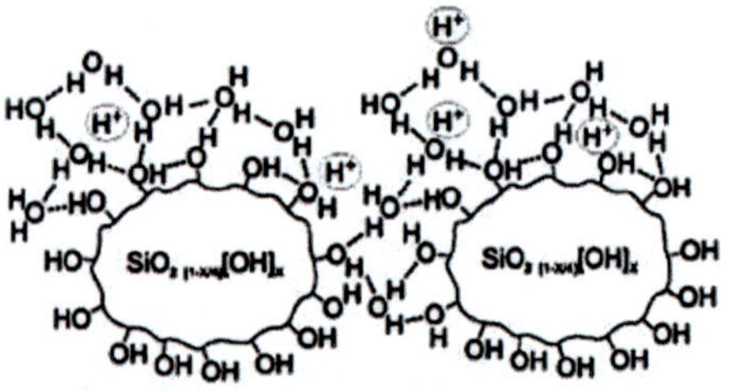

Fig. 5.2 Proton migration along hydrogen-bonded pathways structured around high surface/volume silicate nanoparticles [42]

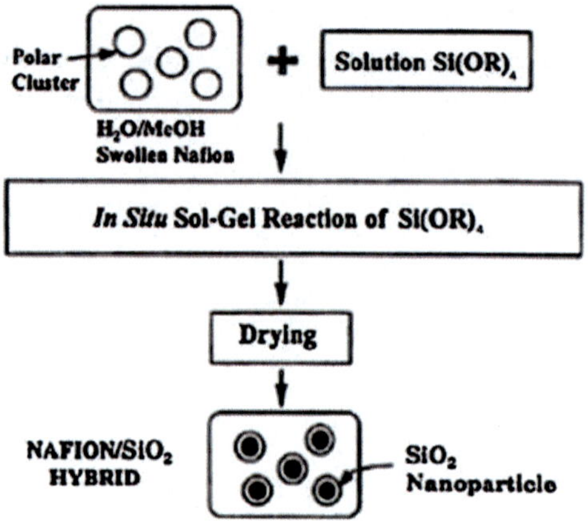

Fig. 5.3 Schematic depiction of formulating Nafion®/[silicon oxide] nanocomposites via in situ sol-gel reactions within nanometers in size polar clusters [51]

nanoparticles, or by sulfonic acid groups being shielded as proton hopping sites by encapsulating silicate structures, or a combination of both factors.

Bocarsly et al. performed fuel cell characterizations of perfluorosulfonic acid type/silicate membranes for PEM FCs [45, 46]. Silicate incorporation improved water retention which increased proton conductivity at elevated temperatures (in conflict with results of Miyake et al.). At operating conditions of 3 atm, a FC temperature of 130°C and constant potential of 0.4 V, Nafion®115/silicate showed four times the current density as that of unmodified Nafion® in a H_2/O_2 PEMFC (Nafion®115 has EW = 1100 g/mol and is 5 mil thick). These composites showed better stability at high temperature FC operation and thermal cycling. Improved mechanical properties and high temperature stability was attributed to perfluoro-organic/silicate interfacial interactions. Cases were reported where methanol uptake, permeation and diffusion were lowered [43, 47, 48]. It was concluded that membranes with high silicate content have potential for DMFCs operating on either liquid feed or vapor feed fuels.

Nafion®/silicate membrane preparation is illustrated in Fig. 5.3 [35, 37, 49–51]. First, a Nafion® film is preswollen in alcohol/water solution. At given pH and temperature, polar hydrolyzed TEOS molecules preferentially migrate to the hydrophilic clusters. Having removed membranes from TEOS source solutions, the samples were dried under heat and vacuum to remove volatiles and further drive condensation of SiOH groups.

The composites were probed by environmental scanning electron microscopy (ESEM) across membrane cross-sections to elucidate silicate structure distribution between membrane surfaces. These studies indicated that the greatest concentration occurs in the near-surface regions for relatively thick membranes.

Symmetric U-shaped profiles [35, 52] seen using ESEM were attributed to high TEOS overall sol-gel reaction rate compared to the slower rate of inward hydrolyzed TEOS diffusion which becomes progressively more difficult due to already-precipitated silicate structures in near-surface regions. Also, asymmetric silicate profiles were affected across the film thickness by allowing TEOS to permeate the film only from one direction [38]. FTIR and ^{29}Si NMR spectroscopies successfully demonstrated condensed silicate structures in the membrane interior.

Later, Baradie et al. created Nafion® 117 (1100 equivalent weight, 7 mil thick)/silicate membranes having non-uniform Si composition profiles across the thickness direction [53]. They created more silicate in the middle rather than near the edges by having TEOS (with no water) and deionized water (with no TEOS) separately permeate from opposite sides in a two-compartment cell. TEOS and water meet in the middle where sol-gel reactions ensue. A great benefit of this silicate distribution is that membrane-electrode interfaces would be similar to that for unmodified Nafion® after membrane electrode assembly formation.

Based on our early work and that of Baradie et al., different Nafion®/silicate membranes were formulated by Nalawade and Mauritz [54] via three routes for in situ sol-gel reactions of TEOS: (1) TEOS and water both permeate films from the same bulk solution across both surfaces; (2) the same, but for TEOS and water permeating only one membrane side, the other being in contact with air; (3) two-sided permeation, but with only TEOS entering one side (no water) and only water (no TEOS) from the opposite side. These three schemes are depicted in Fig. 5.4.

Composition profiles for similar silicate uptakes (~12.5 wt.%) across the thickness direction investigated by EDAX/ESEM showed silicate to be somewhat uniformly distributed for composites prepared by bulk TEOS permeation (Fig. 5.4a). Symmetric U-shaped profiles were not obtained as in earlier studies [52, 55] because the membranes used in the later studies were thinner (2 vs. 5 and 7 mil thick) so that hydrolyzed TEOS molecules do not require as much time to permeate to the middle. The composite prepared by one-side TEOS permeation (Fig. 5.4b) has the greatest silicate concentration toward the TEOS-permeated side, thereby generating asymmetric silicate composition profiles. The composite prepared by scheme 3 (Fig. 5.4c) has the greatest silicate concentration in the middle, which is reasonable.

TGA showed improved thermal degradative stability for composites over unmodified Nafion®. All composites showed water vapor pressure isotherms at 80°C to be close to and somewhat less than that of pure Nafion®. Proton conductivity versus RH curves for Nafion®/silicate membranes are rather close to each other and slightly lower than that of pure Nafion® at different RH from 20 to 100%. The slightly lower composite conductivities are attributed to slightly lower

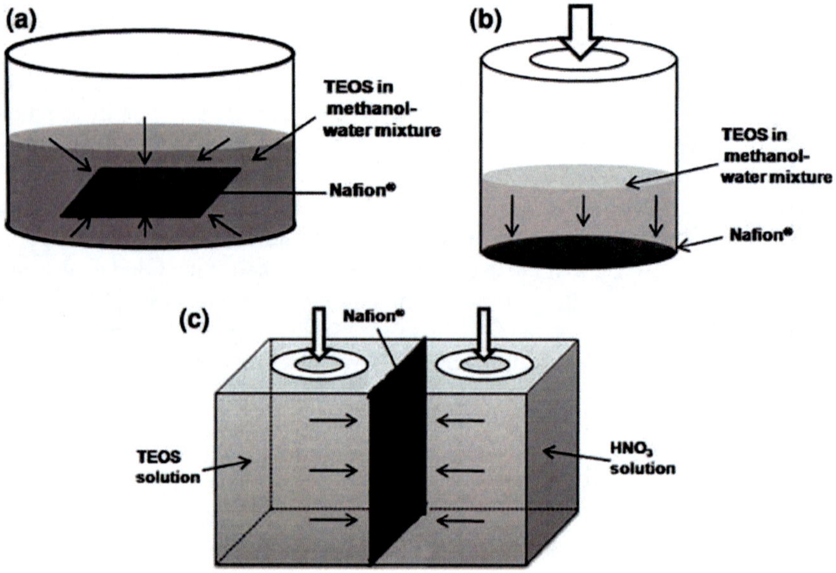

Fig. 5.4 Experimental configuration for the silicate sol-gel modification of Nafion® by (**a**) two-sided TEOS permeation from bulk solution; (**b**) one-sided TEOS permeation and (**c**) simultaneous TEOS|water—permeation from opposite membrane sides

water uptake and increased proton migration tortuosity posed by dispersed silicate particles.

Contractile stress versus dehydration time curves for Nafion® and sol-gel-derived Nafion®/silicate membranes are shown in Fig. 5.5. At the end of the test the stress generated by unmodified Nafion® was only 1 MPa while those for the composite membranes were four to five times greater. The curve for unmodified Nafion® reflects a material that offers little resistance to the effect of progressive dehydration as there is no pronounced upward curve motion but a downward trend at the end that reflects yield. All silicate modifications offer reinforcement. The series of peaks on such curves are due to a sequence of non-catastrophic crazes initiated at mechanically weak regions [56]. Each spike on the curve is an event initiated by yielding chain slippage through entanglements and craze formation with fiber-like structures spanning the craze. The silicate particles resist this process through strong particle–polymer interactions or perhaps by side chains trapped in silicate structures during particle growth [54].

Perhaps the stress-time profiles for membranes prepared by schemes 1 and 2 are similar because they have uniform, symmetric silicate composition profiles. On the other hand, the membrane with asymmetric silicate composition (scheme 3) displays a profile higher than these over most times because one membrane side has a large concentration of connected silicate structures that increases load bearing capacity.

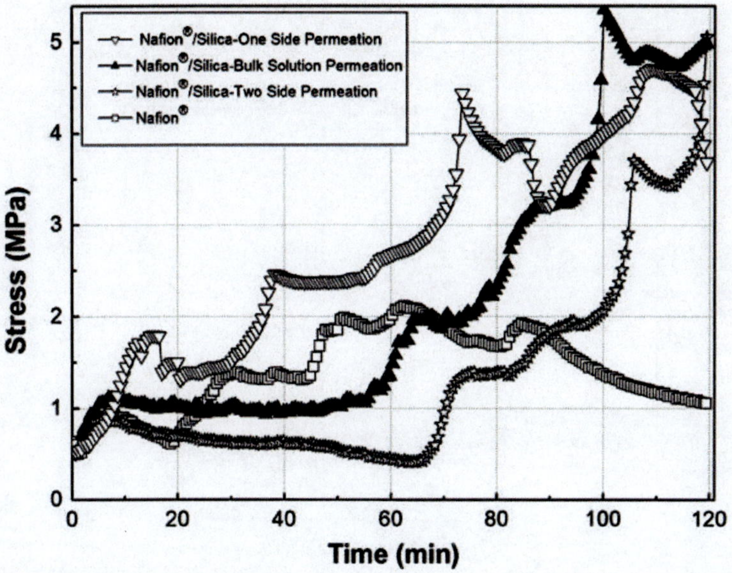

Fig. 5.5 Contractile stress versus time response to RH drop from 100 to 0% at 80°C for Nafion® and sol-gel-derived Nafion®/silicate membranes whose lengths were constrained during drying

Nafion® and Nafion®/silicate MEAs were prepared by spraying Pt/C catalyst onto the membrane surface at both electrodes and were evaluated for performance (voltage vs. current density curves) in H_2/O_2 FCs.

Linear sweep voltammetry (LSV) measurements were performed at 35°C and all composite MEAs showed lower fuel crossover flux than unmodified Nafion®. Silicate nanoparticles increase diffusion tortuosity of fuel gas molecules which reduce crossover rate.

Evaluation of membrane performance in a FC at a given temperature and degree of humidification involves determination of the voltage, V, across the cell terminals versus I. The goal is to have high V at a given I so that the power = IV is high. I–V curves for a Nafion® control and Nafion®/silicate membranes at 80°C and 75% RH are shown in Fig. 5.6. The performance of the scheme 3 composite is somewhat better than that of the control at high I because of reduced fuel crossover and better electrode kinetics. The composite prepared by bulk diffusion of hydrolyzed TEOS (scheme 1) has the worst performance because there is more silicate in the near-surface regions compared to the other composites. Thus, electrode kinetics and proton transport are diminished, reducing the performance [54].

In parentheses at the end of the membrane designations for the curves in Fig. 5.6 are corresponding open current voltages (OCVs). The MEA prepared by one-side TEOS permeation, having a greater silicate concentration on the side facing the anode, has the highest OCV owing to reduced fuel crossover, in

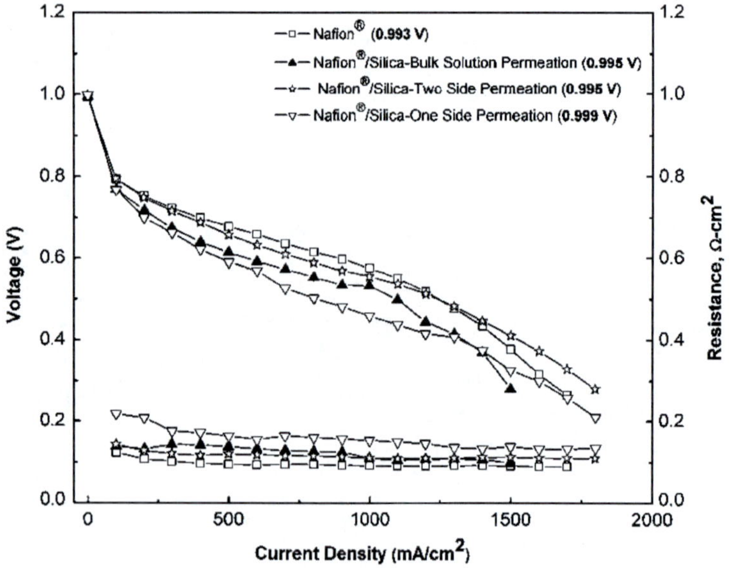

Fig. 5.6 FC performance (*top curves*) and membrane resistance (*bottom curves*) versus I for Nafion®and Nafion®/silicate membranes at 80°C and 75% RH. OCV values for each membrane are given in parentheses

harmony with the LSV experiments. Among all MEAs that which was made using pure Nafion® has the lowest OCV due to having highest fuel crossover. All composites showed slightly higher OCVs than unmodified Nafion®.

Membrane resistance calculated by the current interrupt technique is slightly higher for Nafion®/silicate composites than for unmodified Nafion® likely due to increased proton migration tortuosity. This can explain the inferior fuel cell performance of Nafion®/silicate membranes prepared by bulk solution (scheme 1) and one-side (scheme 2) TEOS permeation.

OCV versus time curves for Nafion® and Nafion®/silicate MEAs measured over 48 h at 90°C and 30% RH are shown in Fig. 5.7. The scheme 2 composite prepared has the least voltage loss, attributed to better gas barrier properties. Reinforcing this interpretation is the fact that this membrane showed least fuel crossover during LSV tests and highest OCV at no load on polarization curves (Fig. 5.6). The other two silicate-modified MEAs show greater voltage drops and their curves lie below that of unmodified Nafion® up to ~30 h after which the composite curves begin to level while for unmodified Nafion® OCV decay continues after 48 h due to poor gas barrier properties for unmodified Nafion®. These studies suggest that inorganic oxide quasi-ordered networks can enhance gas barrier properties by minimizing voltage loss associated with crossover without compromising proton conductivity.

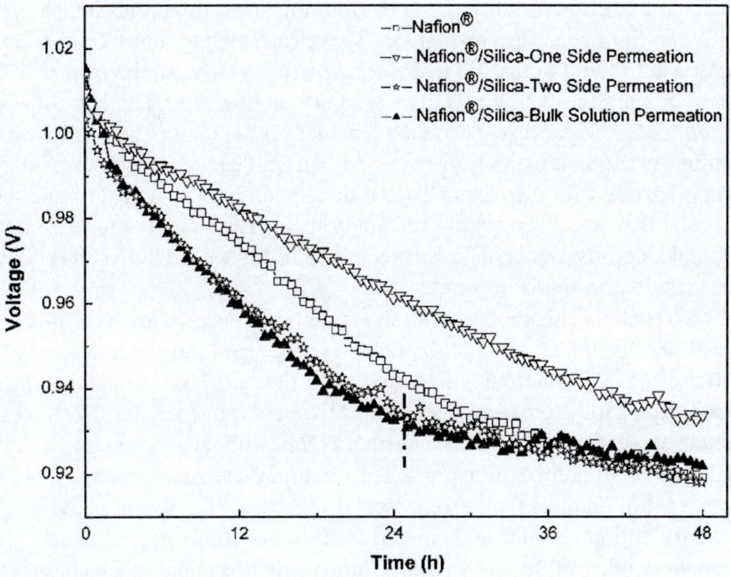

Fig. 5.7 OCV versus time *curves* for Nafion® and Nafion®/silicate membranes at 90°C and 30% RH

It is concluded from these model studies that in situ sol-gel-derived silicate reduces chemical and mechanical degradation and improves gas barrier properties without compromising fuel cell performance [54].

5.1.2 Hybrids of Nafion® with Other Inorganic Oxides

The following hybrid materials have been formulated although evaluation of their performance in FCs has not occurred at the time of this writing. These materials are discussed for the purpose of presenting similar simple sol-gel schemes for Nafion® and similar ionomers within the context of fuel cells. Composite membranes were formulated via simultaneous Nafion®—in situ sol-gel reactions for tetrabutyltitanate (TBT) + TEOS and aluminum tri-sec-butoxide (ATB) + TEOS that permeated the membrane from binary solutions in co-solvent 2-propanol [52, 55]. Ti and Al alkoxides undergo sol-gel reactions more rapid than Si alkoxides but co-reaction with TEOS renders the overall reaction more manageable. These rapid kinetics must be considered in formulating membranes with metal oxide composition gradients. Slow monomer diffusion in relation to rapid sol-gel reaction kinetics presents a challenge in affecting structures beyond the membrane surface. Environmental scanning electron microscopy studies showed U-shaped inorganic composition profiles across the film thickness directions indicating more Ti relative to Si in near-surface regions. The greatest TiO_2 concentration was near the surface, attributed to fast TBT reaction kinetics relative to the inward diffusion rate and comparatively slower TEOS sol-gel reaction rate. The large TiO_2 concentration near the surface accounts for the brittle nature of these materials. Elongation-to-break remains the same for all mixed compositions and tensile strength increases with increasing Ti/Si ratio as seen in mechanical tensile tests. In contrast, the $SiAl_2O_5$ phase within such hybrids is distributed homogeneously with mechanical ductility rather than brittleness. The high diffusivity of ATB within Nafion® resulted in uniform profiles.

Nafion®/zirconia membranes with asymmetric ZrO_2 profiles were obtained by one-sided permeation of tetrabutylzirconate (TBZ) [57] into hydrated membranes from dilute ethanol solutions contacting only one side for prescribed times. IR peaks reflecting ZrO_2 formation were detected near both sides. ESEM studies of Zr concentration across the thickness verified compositional asymmetry.

The tailoring of compositional gradients within these materials as PEMs would generate useful gradients in mechanical properties, T_g, water activity, proton conductivity and gas, water and solvent diffusion coefficients.

There were other efforts to generate Nafion® hybrid membranes via in situ sol-gel processes for other metal alkoxides within the context of FC applications.

Datta et al. synthesized Nafion®/MO_2 (M = Zr, Si, Ti) membranes for high temperature PEM FCs [58]. At 90 and 120°C, all hybrids showed higher water retention than unmodified Nafion® but only Nafion®/ZrO_2 exhibited better conductivity than Nafion®, attributed to increased acidity compared to membranes incorporating titania and silica. While all composites exhibited better thermomechanical properties than Nafion®, there was no significant improvement in FC performance, which was only comparable to that of unfilled Nafion®.

Rowshanzamir et al. synthesized Nafion®/TiO_2 membranes by both an in situ sol-gel procedure and a solution casting method in which a Nafion® membrane was dissolved in a DMF solution to which preformed TiO_2 powder was added [59]. SEM images showed uniform distribution of TiO_2 at low inorganic uptakes for the composites prepared by the sol-gel method. This uniform distribution is lost when the inorganic content increases above 3 wt.% due to TiO_2 particle agglomeration. For equal particle loadings, aggregates were larger in solution-cast composites compared to hybrids by the in situ sol-gel method. The composites prepared by both methods, compared to that of unmodified Nafion®, showed decrease in proton conductivity at 25°C and low RH conditions (~20%). The 3 wt.% TiO_2 containing membrane prepared by the sol-gel method showed higher water uptake. Reduced proton conductivity is attributed to increased proton pathway tortuosity. MEAs made from the composites (prepared by the sol-gel method having 3 and 5 wt.% of TiO_2) and unmodified Nafion® were evaluated for their fuel cell performance at 70 and 110°C. At 70°C, the Nafion® performance was better than that of the composites but at high temperature of 110°C, the composites showed superior performance. At the elevated temperatures increased water uptake is the major factor for improved performance which overrides the effect of decreased proton conduction with TiO_2 loadings.

Zhang et al. studied Nafion®/zirconia membranes, with fuel cell potential, produced via an in situ sol-gel process [60] and showed that these materials had greater water content compared to an unmodified recast Nafion® membrane at medium and high RH and at 100°C.

The model studies by Mauritz et al. demonstrated a mitigating influence of sol-gel inorganic modification on membrane durability [56, 61, 62]. In one study, Nafion® membranes were preswollen in methanol for 24 h and then removed for ~30 s during which time the sol-gel precursor monomer- titanium iso-propoxide and 2,4-pentanedione were added and stirred to form a homogeneous reaction mixture. Then, the membrane was reimmersed for 15 min and after removal from the reaction mixture, followed by a methanol wash, was dried at 100°C to remove water, methanol and further drive polycondensation reactions. 2,4-pentanedione acts as a catalyst to slow the otherwise rapid sol-gel reaction rate for titanium alkoxides that would result in titanium oxide precipitation in near-surface regions or on the surface itself. The resultant composite membranes did not show improvement in water uptake, proton conductivity and fuel cell performance. However, the Young's modulus increased, dimensional stability improved and mechanical creep lessened. These factors are important with regard to mechanical durability. As mentioned, the ability of membranes constrained under compression in fuel cells to withstand contractile stresses due to RH changes was greatly enhanced due to reinforcement by titania structures.

Contractile stress was monitored continuously for 2 h after an RH drop from 100 to 0% at 80°C in Nafion®/titania membranes [70]. The Nafion® control sample yielded after reaching critical stress. It was unable to withstand contractile stress and yields by chain slippage through entanglements. However, in situ grown titania structures in the composite provide reinforcement. It is clear that titania structures would be beneficial in FCs wherein clamped MEAs at the intersection of the lands and grooves of the gas distribution plate experience periodic swelling stresses due to RH changes during vehicle operation. Tensile stress–strain curves [70] at 80°C and 100% RH for Nafion® and sol-gel-derived Nafion®/titania membranes showed that the latter exhibits better dimensional stability due to titania reinforcements. These property enhancements are important for improving the mechanical durability.

Damage surface morphologies of the same clamped Nafion® and Nafion®/titania membranes subjected to this test were inspected using ESEM and micrographs are shown in Fig. 5.8. The arrows indicate the direction of contractile stress during drying. Nafion® shows significant signs of yielding when subjected to the humidity drop. The unmodified membrane shows cavities and stretching/orientation, but Nafion®/titania does not show such failure. These cavities could be nucleating loci for pinholes that result in membrane failure. These morphological results are supported by the contractile stress versus time curves.

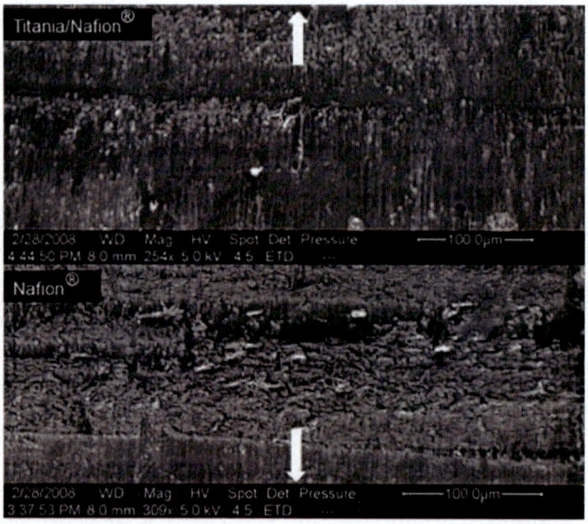

Fig. 5.8 ESEM micrographs of Nafion® (*bottom*) and Nafion®/titania (*top*) membrane surfaces after tests of contractile stress versus time at constant strain with drop in RH from 100 to 0%

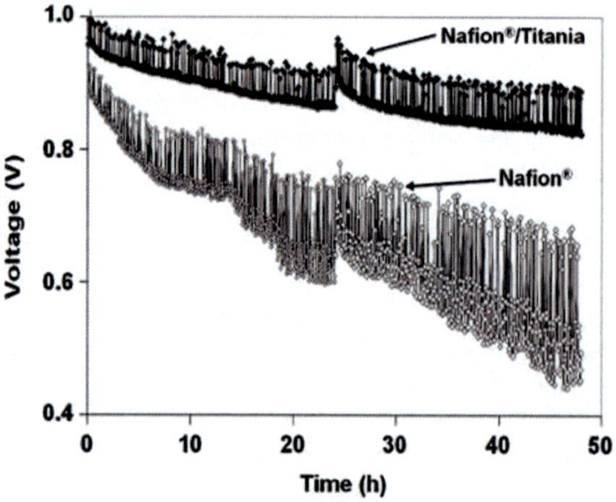

Fig. 5.9 OCV versus time for Nafion® and Nafion®/titania membranes at 90°C and 30% RH

The effect of this sol-gel modification on durability was evaluated by subjecting membranes to accelerated degradation in a FC under OCV conditions [62]. No load is connected so that protons and electrons formed at the anode do not reach the cathode. Since no reactants are consumed, fuel crossover is maximum.

H_2 and O_2 combine to generate H_2O_2 which decomposes to form $^{\bullet}OH$ and $^{\bullet}OOH$ radicals (and OH^- anions) which attack the polymer. OCV curves for Nafion® and Nafion®/titania membranes are in Fig. 5.9. The Nafion® curve is not continuous because the experiment was stopped every 8 h to check for fuel crossover. Nafion®/titania curves not only start high but also remain higher than that of Nafion®. This improvement is due to improved gas barrier properties owing

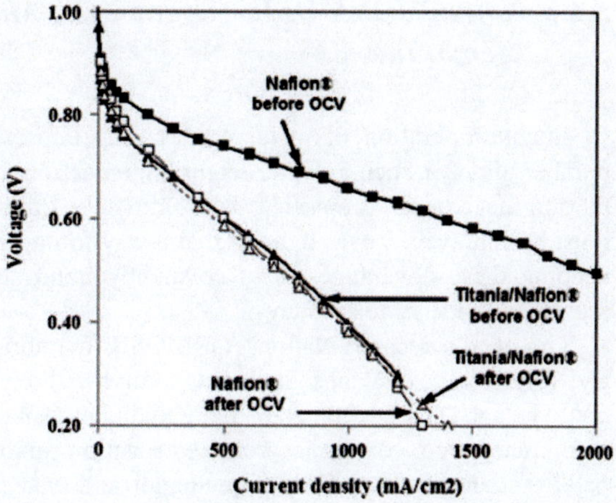

Fig. 5.10 FC performance *curves* for Nafion® and Nafion®/titania membranes before and after OCV treatment at 80°C and 75% RH

to improved dimensional stability and titania particles increase gas diffusion tortuosity. This claim is supported by a fluoride emission rate (FER) analysis. Peroxide radicals initiate Nafion® chemical degradation by attacking polymer end groups having H-containing end groups (–CF_2COOH). FER of both anode and cathode for the composite membrane were significantly lower than that of unfilled Nafion® due to better gas barrier and mechanical properties.

H_2 crossover current curves obtained by LSV after subjecting the membrane to an OCV test condition showed lower crossover current for Nafion®/titania compared to unmodified Nafion® which indicates that titania insertion improves the hydrogen barrier [56, 62]. Pinhole and crack formation due to physical-chemical degradation occurred more rapidly in unmodified Nafion®.

FC performance curves for Nafion® and Nafion®/titania before/after OCV degradation are shown in Fig. 5.10. Considering low water uptake and low proton conductivity, poor performance of titania-incorporated Nafion® was expected although it is stressed that these are model studies with focus on the issue of mechanical durability. Once durability problems are solved, in situ inorganic oxide structures will be modified/functionalized in order to affect an increase in proton conductivity.

It is significant that the performance of the Nafion®/titania membrane is essentially unchanged after OCV degradation compared to the large performance drop for unmodified Nafion®. The unmodified membrane, due to poor dimensional stability and mechanical properties, yields under periodic internal stresses resulting from humidity and temperature changes.

A weak region can be the focus of local failure thereby generating gas diffusion pathways and thickness reduction. Mechanically weak regions, after experiencing hysteresis owing to periodic swelling/de-swelling stresses and attack from hydroxyl radicals form pinholes which are prevented by this inorganic sol-gel modification.

5.1.3 Nafion®/ORMOSIL (Organically Modified Silicate) Membranes

In situ incorporation of metal oxides may improve membrane properties in a number of ways, such as water retention, reduction in fuel crossover and enhanced thermo-mechanical properties, but ultimately improvement in FC performance must be achieved. It was thought that a way to improve proton conductivity while keeping these advantages is to covalently bond acid groups, mainly SO_3H, to inorganic oxide nanostructures.

Kim et al. prepared Nafion®/ORMOSIL membranes for DMFCs using TEOS and organoalkoxysilanes including vinyltriethoxysilane (TEVS), diethoxydimethylsilane (DEDMS) and diethoxydiphenylsilane (DEDPS) [63]. Nafion® membranes were soaked in premixed reaction mixtures of TEOS and the organoalkoxysilane in a solution of methanol and water. Fracture surfaces of hybrid cross-sections prepared using the pure organoalkoxysilane (without TEOS) were observed using SEM. Images showed ball-shaped 'segregates' which reflected incompatibility between hydrophilic ionic clusters and ORMOSIL structures of lower hydrophilicity. Morphologies of Nafion®/ORMOSIL hybrids prepared with mixtures of TEOS and other organoalkoxysilanes (TEOS:ORMOSIL = 1:1 mol/mol) exhibited a greater compatibility between ORMOSIL and Nafion® irrespective of ORMOSIL type. Silicate structure was homogeneously distributed across fracture surfaces of hybrid membranes. These membranes were more flexible than the Nafion® composite prepared with pure TEOS which became brittle when the silicate phase exceeded 40 wt.%. As the fraction of organosilane (DEDMS or TEVS) increased, water uptake was reduced. The TEVS modification sorbed more water than that modified by DEDMS. Thus, the increased number of organic groups increased the hydrophobicity of composite. Membrane water uptake was found to be related to its hydrophilicity as controlled by number and type of organic groups. Methanol uptake was thought to be more controlled by 'connectivity' around SiO_4 units in the way as discussed by Young et al. [64] and Deng et al. [65] in the ORMOSIL phase as well as the compatibility of this phase with the polar clusters in Nafion®. It was showed that proton conductivity and methanol permeability could be controlled by varying organoalkoxysilane type and composition. Nafion®/ORMOSIL membranes showed improved DMFC performance which was attributed to reduced methanol crossover.

Li et al. synthesized a sulfonated diphenyldimethoxysilane (sDDS) monomer by adding chlorosulfonic acid to diphenyldimethoxysilane (DDS) at 50°C [66]. The sulfonated product was obtained by hydrolysis of sDDS by deionized water. Owing to the 2- rather than 3- or 4- alkoxy-functional nature of this monomer, network formation is not possible. The Nafion® composite with sDDS was prepared by adding the sulfonated sol to a Nafion® solution followed by film casting. Films were dried at 80°C and annealed at 145°C under vacuum. sDDS content in composites varied from 5 to 25 wt.%. The composite had lower methanol permeability than unmodified Nafion® but exhibited five times lower proton conductivity.

Liang et al. synthesized diphenylsilicate-incorporated Nafion® membranes for DMFCs. Preformed DPS particles were synthesized by adding DEDPS into a methanol–water mixture. Composites with 10 and 20 wt.% DPS were prepared by solution casting. SEM images showed layered crystalline microstructure of the synthesized DPS, with each layer ranging from 100 to 300 nm in thickness. DPS particles were found to be uniformly distributed in Nafion® membranes. The composites showed slightly lower proton conductivity, and lower rates of methanol crossover than a cast, unmodified Nafion® membrane which resulted in higher OCV and better DMFC performance for a high methanol concentration [67].

D'Epifanio et al. prepared Nafion®/ORMOSIL membranes for DMFCs. Nafion®/[sulfonated diphenylsilanediol] (sDPSD) composites were prepared by solvent casting [68]. sDPSD cannot form gel particles as this monomer has only two groups for condensation reactions. The composite membrane exhibited better thermal stability and proton conductivity than unmodified Nafion®, attributed to interactions between the ORMOSIL component and polar clusters in Nafion®. Higher proton conductivity and reduced methanol permeation resulted in better DMFC performance relative to unmodified Nafion®.

5.1.4 Nafion® with Proton Conducting Inorganic Fillers

Incorporation of heteropolyacids (HPA), zirconium phosphates (ZrP) and phosphonates is another way to modify PEMs to enhance proton conductivity through strong hydrogen bonds between water molecules and P-OH groups. Negami et al. reported good proton conduction (presence of water) silicophosphates, by the sol-gel method, for FC applications [69, 70]. Porous glasses were prepared by hydrolysis of $PO(OCH_3)_3$, $Zr(OC_3H_7)_4$ and $Si(OC_2H_5)_4$. P_2O_5–SiO_2 glasses of thickness 0.2–1.0 mm prepared by a sol-gel procedure resulted in a conductivity of 22 mS/cm at 50°C. This was the first reported performance of a H_2 fuel cell using a glass electrolyte. At 50°C and 90% RH, a P_2O_5–SiO_2 glass membrane resulted in power output of 6 mW/cm^2 [71].

Klein and co-workers prepared phosphosilicate gel-containing composites by 'infiltrating' Nafion® membranes with sols of SiO_2–P_2O_5–ZrO_2 (SPZ) and SiO_2–TiO_2–P_2O_5 (STP) using solutions of TEOS, tri-ethyl phosphate $P(OC_2H_5)_3$ (TEP) and zirconium-*n*-propoxide $Zr(OC_3H_7)_4$ (TPZ) [72–75]. To synthesize the SPZ-type composite, solutions of TEOS, TEP and TPZ was prepared. Also, separate solutions of silica doped with phosphoric acid (H_3PO_4) and with tungstophosphoric acid ($H_3PW_{12}O_{40}$) were prepared. Nafion® membranes were immersed separately in each of these solutions and composites obtained by in situ precipitation of the sols. The infiltrated membranes were dried and composites were further studied for dielectric and electrical properties versus dehydration time, RH and temperature. The composites exhibited only marginally higher conductivity than Nafion® at the high temperature of 130°C [72]. These composites, after sol-gel treatment, were generally heat treated at temperatures ranging from

80 to 150°C. Heating the infiltrated membrane at 150°C resulted in resistance increase of one order of magnitude compared to Nafion®. It was suggested that residues generated by decomposition of organic compounds at high temperatures (>100°C) could block proton conduction pathways. Implementation of a cleaning procedure for SPZ membranes using H_2O_2, H_2SO_4 and H_2O at 80°C improved performance by one order of magnitude over that of non-cleaned membranes with conductivity being close to that of unmodified Nafion® [73]. Nafion®/SPZ composite fuel cell performance at 130°C was comparable to that of the unmodified Nafion® fuel cell at 80°C. A 50°C increase in fuel cell operating temperature without compromising MEA performance was reported. The Nafion® membrane showed 20 times increase in resistance as temperature increased from 80 to 130°C, while for the SPZ composite, for the same rise in temperature, resistance increased by a factor of 5. Drop in Nafion® performance at high temperature was attributed to membrane water loss. Improved performance of Nafion®/SPZ at 130°C was affected by hygroscopic oxide particles containing hydroxyl groups that bind water molecules at high temperatures [75].

5.2 Non-Perflurosulfonated Ionomers

There has been considerable research on alternative hydrocarbon fuel cell membranes [12, 16, 76–78]. Aromatic polymers possess thermal and chemical stability, easy sulfonation, low cost [12] and availability. Examples include polyethersulfones (PES), poly(ether-ether-ketone) (PEEK), poly(arylene ether sulfone), polyesters and polyimides (PI). Proton conductivity depends on high sulfonation degrees although this can compromise mechanical properties and cause undesirable swelling and water solubility [20, 79]. Efforts were made to overcome these problems by sol-gel processes including embedding inorganic fillers in the polymer matrix or formation of covalently bounded inorganic networks within hydrophilic domains [12, 80–83]. Inorganic compounds may enhance proton conductivity, promote water retention and adjust swelling to mitigate fuel gas crossover and improve mechanical properties implicated in durability [30, 78].

The number of available sol-gel monomers provides opportunities for synthesizing hybrid membranes for high temperature FCs. Pezzin et al. used sol-gel processes with sulfonated poly(ether-ether-ketone) (SPEEK) and phosphonated polysilsesquioxanes (POSS) [84]. Copolysilsesquioxanes were obtained from a phosphonated silane and tetraethoxysilane by a modified sol-gel method and dispersed in SPEEK solutions. Flexible transparent membranes cast from these solutions were treated in different ways to hydrolyze diethylphosphonate to phosphonic acid groups and help form silicate semi-interpenetrating networks. Bifunctional diethylphosphatoethytriethoxysilane was used to introduce phosphonic acid groups into silicate networks or to form POSS by sol-gel methods. Phosphonic acid groups retain water better than sulfonic acid groups [85]. At 100% RH, the conductivities of composite membranes increase with increase in phosphonated POSS group concentration. Proton conductivity reached maximum (142 mS/cm) at 120°C with 40 wt.% of POSS with 80 mol% phosphonated units in SPEEK (SD = 50%) which was higher than that of Nafion® 117 for the same conditions [84]. SO_3H groups of SPEEK provided proton conductance at low temperatures while PO_3H_2 groups account for conductance at high temperature and low RH.

Smith and Zharov created membranes consisting of sintered self-assembled networks of nanoporous silica colloidal crystals modified with poly(3-sulfopropylmethacrylate) and poly(styrene sulfonic acid) brushes covalently attached to nanopore surfaces [86]. These membranes demonstrated high proton conductivity ($<2 \times 10^{-2}$ S cm^{-1} at 30°C, 94% RH; $<1 \times 10^{-2}$ S cm^{-1} at 85°C, 60% RH) and water uptake (~20 wt.%) at room temperature. They also prepared PEMs by self-assembly of silica nanospheres modified with poly(3-sulfo-propylmethacrylate) and poly(styrene-sulfonic acid) brushes of different thicknesses, which showed slightly higher conductivity and water uptake than the former, but had poor mechanical properties.

Niepceron et al. developed membranes based on fluorinated polymers such as polyvinylidene difluoride (PVDF) which is inert to various solvents and acids and with low permeability to gases and liquids [87]. Sodium 4-polystyrene sulfonate (PSSNa) chains were anchored onto nanoscopic fumed silica particle surfaces by '"grafting onto", "grafting through" or "grafting from" methods involving condensation reactions with surface silanol groups. Membranes had 40–60 wt.% silica particles with percolation threshold of 30%. Conductivity ranged from 30 to 100 mS.cm^{-1} at ambient temperature and 100% RH. Power densities above 1 W/cm^2 at 70°C were achieved using non-hydrated gas feeds in H_2/O_2 fuel cells.

Sel et al. produced networks of mesostructured hydrophilic materials in poly(vinylidene fluoride–*co*-hexafluoropropylene) (PVDF-HFP) by a sol-gel process [88]. Acid-functionalized, high surface area porous networks were created through co-condensation of TEOS with ORMOSIL monomers with sulfonic acid groups (2-(4-chlorosulfonylphenyl)) ethyltrimethoxysilane using a poly(ethylene oxide–propylene oxide–ethylene oxide) block terpolymer as structure-directing agent. The hydrophobic copolymer which surrounds mesostructured silicate networks restricts swelling in water and provides mechanical strength. Micron-sized cavities at the surface of the membrane limits interfacial proton transport resistance by increasing surface area at electrode/electrolyte interfaces [89], prevents flooding and MEA drying by controlling water management at the cathode interface and changes the interface through separation of hydrophilic and hydrophobic domains [88]. Also, crosslinkable poly(vinylidenefluoride-*ter*-perfluoro-(4-methyl-3,6-dioxaoct-7-*ene* sulfonyl fluoride)-*ter*-vinyltriethoxysilane) was incorporated, that caused conductivity to increase by an order of magnitude, regardless of temperature and RH [90]. This difference was attributed to polar (-OH and -SO_3H) domains at the hybrid interface when the terpolymer was added to the hybrid silicate -SO_3H/poly(VDF-*co*-HFP) membrane. The greater number of hydroxyl groups in this region might help solvated protons to migrate by formation of hydrogen bonded water network clusters. SO_3H groups located at these interfaces can be bridged via water molecules to form continuous pathways for proton hopping.

Kato et al. synthesized membranes from (trimethoxysilylmethyl)styrene (TMSMS) and phenylvinylphosphonic acid (PhVPA) at various monomer ratios using radical copolymerization followed by a sol-gel process [91]. Phosphonic acid groups were chemically bound in organosiloxane networks formed via alkoxide hydrolysis and polycondensation reactions between silanol groups of TMSMS in the side chains. Membrane conductivity increased with phosphonic acid content. The 1:4 TMSMS/PhVPA membrane had conductivity of 3.7×10^{-2} $S.cm^{-1}$ at 100% RH and 130°C, but was low (9.5×10^{-4} $S.cm^{-1}$) at 19.2% RH and 130°C.

Li et al. synthesized hybrid membranes using a sol-gel process for diethyl-4-(diethoxy(methyl)silyl)-1,1-difluorobutylphosphonate and diethoxyphosphorylethy ltriethoxysilane which were then hydrolyzed, in two schemes [92]. Proton conductivities increased with increasing $-PO_3H_2$ concentration and RH, reaching 62 $mS.cm^{-1}$ at 100°C and 100% RH, which is comparable to that of Nafion® under similar conditions. At 80 and 100°C and 20% RH, conductivity was ~2.3 $mS.cm^{-1}$, similar to that of Nafion® 117. Anhydrous membrane proton conductivity was enhanced by substitution of $-CH_2-PO_3H_2$ groups with $-CF_2-PO_3H_2$ groups because of the effect of the large electron withdrawing effect of C–F bonds on pKa. Durability needs to be assessed in actual fuel cells, considering instability of Si–O–Si linkages in the acidic membrane environment and susceptibility of aliphatic hydrogens to degradation by peroxide.

Protic ionic liquids (PILs), also known as ambient temperature molten salts, show promise to replace aqueous electrolytes in membranes for hot and dry FC operation [93]. PILs are a subset of ionic liquids formed by equimolar mixing of a Brønsted acid and a Brønsted base. In non-humidified FC conditions, the Brønsted base is typically an amine which plays the role of water (proton carrier) in sulfonic acid based membranes. An example of a PIL is diethylmethyl ammonium (dema) trifluoromethane sulfonate (TfO) Brønsted acid/base [93]. In some membranes, PILs can form well-connected nano-channels which promote proton conductivities up to ~100 $mS.cm^{-1}$ at 160°C without humidification; however, there was progressive PIL release in long-term operation and lack of stability toward water [94]. These problems were addressed by inorganic modifications of the PIL membranes [95].

Lin et al. prepared PIL/poly(styrene-*co*-acrylonitrile)/silica membranes by incorporating silica nanoparticles or mesoporous silica nanospheres in this copolymer to improve ionic liquid retention [95]. Monodispersed silica nanoparticles with diameters of 60, 130 and 220 nm were prepared by hydrolysis of TEOS in ethanol in the presence of ammonia and water. Mesoporous silica nanospheres were prepared from TEOS using a cationic surfactant and NaOH. For the former, the membranes had good thermal stability and mechanical properties. In anhydrous conditions, conductivity varied up to ~10 $mS.cm^{-1}$ at 160°C which was attributed to PIL continuous networks and ion transport channels. Proton conductivity increased with increased incorporation of silica nanoparticles although it decreased beyond a certain content. In PIL water retention tests the membranes lost more than 80% PLI, causing severe conductivity drop [95]. Compared with silica nanoparticles, mesoporous silica nanospheres having higher surface/volume were more effective in enhancing conductivity and preventing PLI release of PLI due to capillary forces in mesoporous silica nanospheres.

Another approach to improve PIL retention was by Lakshminarayana and Nogami through synthesis of phosphosilicate-based hybrid membranes by a sol–gel process using tetramethoxysilane/methyltrimethoxysilane/trimethylphosphate and 1-ethyl-3-methylimidazolium-bis (trifluoromethanesulfonyl) imide [EMI][TFSI] ionic liquid precursors [96]. FTIR and ^{31}P, ^{1}H and ^{13}C NMR spectroscopy showed good chemical stability and complexation of $PO(OCH_3)_3$ with [EMI][TFSI] ionic liquid in membranes. Highest conductivity was 5.4 mS.cm^{-1} at 150°C under non-humidifying conditions. For one composition, a maximum power density of 0.22 mW cm^{-2} at 0.47 mA cm^{-2} and current density of 0.76 mA cm^{-2} were obtained at 150°C under non-humidified conditions. Hydrogen permeability decreased as temperature increased from 20 to 150°C [96] which is good in terms of fuel crossover.

5.3 Future Trends for Sol-Gel-Derived Organic/Inorganic Fuel Cell Membranes

Although many sol-gel modifications have been conducted to date, there are still many possible variations for generating membranes with high proton conductivities at hot and dry conditions and having chemical and mechanical durability. Some of these ideas have been already implemented. For example, creative sol-gel chemistry schemes combined with polymer synthesis and processing using the electrospinning technique have been developed for the design of functionalized silica-based hybrid organic–inorganic membranes with highly structured conduction pathways and improved durability [78]. This approach could be enhanced by using other transition metal oxides, such as zirconia and titania-based materials, which are known to be more stable in real PEM fuel cells [11, 61, 62]. Also, porous inorganic networks functionalized with various acid groups such as $–SO_3H$ and $–PO_3H_2$ represent another interesting approach to construct fixed well-structured channels for proton hopping and improved conductivity. Organic/inorganic interfaces are important for catalysis, structured water for high temperature transport properties and mechanical integrity [78]. Thus, techniques and experimental methods to understand the nature of this interface are crucial for membrane performance. Those techniques include small angle X-ray and neutron scattering, microscopic techniques, FTIR and NMR and broadband dielectric spectroscopy. PVDF/immobilized ionic liquid membranes represent another interesting approach but need further improvements to prevent their leech-out during long-term fuel cell operation. And finally silica-immobilized imidazole water-free ionic conductors are an attractive approach for high temperature operation.

5.4 Conclusions

While limited studies of inorganic sol-gel modified fuel cell membranes are summarized here, the concept is largely unexploited and many opportunities exist. The important issues are whether inserted inorganic oxide nanostructures promote high temperature hydration while retaining proton conductivity (or whether water-free membranes can be constructed), prevent fuel crossover and improve mechanical durability. Exploring mesophase-separated, hydrocarbon-block copolymer templates with chemical stability, and resistance to chemical attack, would seem to be a logical route to pursue. However, to date, the list of organic polymer fuel cell membranes, whether sol-gel-modified or not, does not include any that are adequate for commercialization.

Acknowledgments The authors acknowledge the DOE Office of Energy Efficiency and Renewable Energy, contract # DE-FG36-08GO88106 for financial support.

References

1. Vielstich W, Gasteiger HA, Lamm A (2003) Handbook of fuel cells—fundamentals, technology and applications. Wiley, Chichester
2. Souzy R, Ameduri B (2005) Functional fluoropolymers for fuel cell membranes. Prog Polym Sci 30:644–687
3. Cho KY, Jung HY, Sung KA, Kim WK, Sung SJ, Park JK, Choi JH, Sung YE (2006) Preparation and characteristics of Nafion membrane coated with a PVdF copolymer/recast Nafion blend for direct methanol fuel cell. J Power Sources 159:524–528
4. Yamauchi A, Ito T, Yamaguchi T (2007) Low methanol crossover and high performance of DMFCs achieved with a pore-filling polymer electrolyte membrane. J Power Sources 174:170–175
5. Mauritz KA, Moore RB (2004) State of understanding of Nafion. Chem Rev 104:4535–4585
6. Kundu S, Simon LC, Fowler M, Grot S (2005) Mechanical properties of Nafion™ electrolyte membranes under hydrated conditions. Polymer 46:11707–11715
7. Huang X, Solasi R, Zou Y et al (2006) Mechanical endurance of polymer electrolyte membrane and PEM fuel cell durability. J Polym Sci Polym Phys 44:2346–2357
8. Schiraldi DA (2006) Perfluorinated polymer electrolyte membrane durability. J Macromol Sci, Polym Rev 46:315–327
9. Liu W, Crum M (2006) Effective testing matrix for studying membrane durability in PEM fuel cells: Part I. Chemical durability. ECS Trans 3:531–540
10. Crum M, Liu W (2006) Effective testing matrix for studying membrane durability in PEM fuel cells: Part II. Mechanical durability and combined mechanical and chemical durability. ECS Trans 3:541–550
11. Patil YP, Jarrett WL, Mauritz KA (2010) Deterioration of mechanical properties: a cause for fuel cell membrane failure. J Membr Sci 356:7–13
12. Hickner MA, Ghassemi H, Kim YS, Einsla BR, McGrath JE (2004) Alternative polymer systems for proton exchange membranes (PEMs). Chem Rev 104:4587–4612
13. Doyle M, Rajendran G (2003) In: Vielstich W, Gasteiger HA, Lamm A (eds) Handbook of fuel cells—fundamentals, technology and applications, vol 3. Wiley, Chichester, p 447
14. Wang S, McGrath JE (2003) Synthesis of poly(arylene ether)s. In: Synthetic methods in step growth polymers, Wiley, New York, p 327

15. Poppe D, Frey H, Kreuer KD, Heinzel A, Mulhaupt R (2002) Carboxylated and sulfonated poly(arylene-co-arylene sulfone)s: thermostable polyelectrolytes for fuel cell applications. Macromolecules 35:7936–7941
16. Wang F, Hickner M, Kim YS, Zawodzinski TA, McGrath JE (2002) Direct polymerization of sulfonated poly(arylene ether sulfone) random (statistical) copolymers: candidates for new proton exchange membranes. J Membr Sci 197:197–231
17. Nolte R, Ledjeff K, Bauer B, Mülhaupt R (1993) Partially sulfonated poly(arylene ether sulfone)—a versatile proton conducting membrane material for modern energy conversion technologies. J Membr Sci 83:211–220
18. Rozière J, Jones DJ (2003) Non-fluorinated polymer materials for proton echange membrane fuel cells. Annu Rev Mater Res 33:503–555
19. Alberti G, Casciola M, Massinelli L, Bauer B (2001) Polymeric proton conducting membranes for medium temperature fuel cells (110–160°C). J Membr Sci 185:73–81
20. Zaidi SM, Mikhailenko SD, Robertson GP, Guiver MD, Kaliaguine S (2000) Proton conducting composite membranes from polyether ether ketone and heteropolyacids for fuel cell applications. J Membr Sci 173:17–34
21. Fang J, Guo X, Kita H et al (2002) Novel sulfonated polyimides as polyelectrolytes for fuel cell application. 1. Synthesis, proton conductivity, and water stability of polyimides from 4, 4'-diaminodiphenyl ether-2, 2'-disulfonic acid. Macromolecules 35:9022–9028
22. Genies C, Mercier R, Sillion B, Cornet N, Gebel G, Pineri M (2001) Soluble sulfonated naphthalenic polyimides as materials for proton exchange membranes. Polymer 42:359–373
23. Hench LL, West JK (1990) The sol-gel process. Chem Rev 90:33–72
24. Hench LL, Ulrich DR (1984) Ultrastructure processing of ceramics, glasses and coposites. Wiley, New York
25. Brinker CJ, Scherer GW (1990) Sol-gel science: the physics and chemistry of sol-gel processing. Academic Press Inc, New York
26. Jones DJ, Rozière J (2003) In: Vielstich W, Gasteiger HA, Lamm A (eds) Handbook of fuel cells—fundamentals, technology and applications, vol 3. Wiley, Chichester, p 447
27. Alberti G, Casciola M (2003) Composite membranes for medium-temperature PEM fuel cells. Annu Rev Mater Res 33:129–154
28. Klein LC, Daiko Y, Aparicio M, Damay F (2005) Methods for modifying proton exchange membranes using the sol-gel process. Polymer 46:4504–4509
29. Klein LC (2009) Sol-gel process for proton exchange membranes. Key Eng Mater 391:159–168
30. Jones DJ, Rozière J (2008) Advances in the development of inorganic-organic membranes for fuel cell applications. Adv Polym Sci 215:219–264
31. Klein LC (2002) Opportunities for sol-gel materials in fuel cells. Mater Sci 20:81–93
32. Mauritz KA, Hassan MK (2007) Nanophase separated perfluorinated ionomers as sol-gel polymerization templates for functional inorganic oxide nanoparticles. Polym Rev 47: 543–565
33. Mauritz KA, Mountz DA, Reuschle DA, Storey DA (1999) Organic/inorganic nanocomposites as potential fuel cell membranes. In: Martin CW, Desmarteau DD (eds) Proc DOD workshop on advanced PEM fuel cell membranes and membrane-electrode assemblies for non-conventional fuels. Clemson University Press, Las Vegas, p 304
34. Mauritz KA (1999) Nanophase separated ionomers as sol-gel polymerization templates for inorganic alkoxides: applications as proton conducting membranes, conference on electroactive membranes: from fundamentals to applications, Monte Verita, Switzerland
35. Mauritz KA, Stefanithis ID, Davis SV, Scheetz RW, Pope RK, Wilkes GL, Huang HH (1995) Microstructural evolution of a silicon oxide phase in a perfluorosulfonic acid ionomer by an in situ sol-gel reaction. J Appl Polym Sci 55:181–190
36. Deng Q, Cable KM, Moore RB, Mauritz KA (1996) Small-angle X-ray scattering studies of Nafion®/[silicon oxide] and Nafion®/ORMOSIL nanocomposites. J Polym Sci Polym Phys 34:1917–1923

37. Mauritz KA, Warren RM (1989) Microstructural evolution of a silicon oxide phase in a perfluorosulfonic acid ionomer by an in situ sol-gel reaction. 1. Infrared spectroscopic studies. Macromolecules 22:1730–1734
38. Gummaraju RV, Moore RB, Mauritz KA (1996) Asymmetric [Nafion®]/[silicon oxide] hybrid membranes via in situ sol-gel reaction for tetraethoxysilane. J Polym Sci Polym Phys 34:2383–2392
39. Mauritz KA, Storey RF, Jones CK (1989) Perfluorinated-ionomer-membrane-based micro-composites. Silicon oxide filled membranes. In: Utracki LA, Weiss RA (eds) Multiphase polymer materials: blends, ionomers and interpenetrating networks. ACS Symp Ser 395, Washington, DC, p 401
40. Deng Q, Wilkie CA, Moore RB, Mauritz KA (1998) TGA-FTi r investigation of the thermal degradation of nafion® and nafion®/[silicon oxide]-based nanocomposites. Polymer 39: 5961–5972
41. Deng Q, Moore RB, Mauritz KA (1998) Nafion®/(SiO_2, ORMOSIL, and dimethylsiloxane) hybrids via in situ sol–gel reactions: characterization of fundamental properties. J Appl Polym Sci 68:747–763
42. Mauritz KA, Payne JT (2000) [Perfluorosulfonate ionomer]/silicate hybrid membranes via base-catalyzed in situ sol–gel processes for tetraethylorthosilicate. J Membr Sci 168:39–51
43. Jung DH, Cho SY, Peck DH, Shin DR, Kim JS (2002) Performance evaluation of a Nafion/ silicon oxide hybrid membrane for direct methanol fuel cell. J Power Sources 106:173–177
44. Miyake N, Wainright JS, Savinell RF (2001) Evaluation of a sol-gel derived Nafion/silica hybrid membrane for polymer electrolyte membrane fuel cell applications. II. methanol uptake and methanol permeability. J Electrochem Soc 148:A905–A909
45. Adjemian KT, Lee SJ, Srinivasan S, Benziger J, Bocarsly AB (2002) Silicon oxide Nafion composite membranes for proton-exchange membrane fuel cell operation at 80–140°C. J Electrochem Soc 149:A256–A261
46. Adjemian KT, Srinivasan S, Benziger J, Bocarsly AB (2002) Investigation of PEMFC operation above 100°C employing perfluorosulfonic acid silicon oxide composite membranes. J Power Sources 109:356–364
47. Jiang R, Kunz HR, Fenton JM (2006) Composite silica/Nafion® membranes prepared by tetraethylorthosilicate sol–gel reaction and solution casting for direct methanol fuel cells. J Membr Sci 272:116–124
48. Antonucci PL, Aricò AS, Cretì P, Ramunni E, Antonucci V (1999) Investigation of a direct methanol fuel cell based on a composite Nafion®-silica electrolyte for high temperature operation. Solid State Ion 125:431–437
49. Mauritz KA, Stefanithis ID (1990) Microstructural evolution of a silicon oxide phase in a perfluorosulfonic acid ionomer by an in situ sol-gel reaction. 2. Dielectric relaxation studies. Macromolecules 23:1380–1388
50. Stefanithis ID, Mauritz KA (1990) Microstructural evolution of a silicon oxide phase in a perfluorosulfonic acid ionomer by an in situ sol-gel reaction. 3. Thermal analysis studies. Macromolecules 23:2397–2402
51. Mauritz KA (1998) Organic-inorganic hybrid materials: perfluorinated ionomers as sol-gel polymerization templates for inorganic alkoxides. Mater Sci Eng C 6:121–133
52. Shao PL, Mauritz KA, Moore RB (1996) [Perfluorosulfonate ionomer]/[SiO2-TiO2] nanocomposites via polymer-in situ sol-gel chemistry: sequential alkoxide procedure. J Polym Sci Polym Phys 34:873–882
53. Baradie B, Dodelet JP, Guay D (2000) Hybrid Nafion®-inorganic membrane with potential applications for polymer electrolyte fuel cells. J Electroanal Chem 489:101–105
54. Nalawade AP (2011) Modification and evaluation of fuel cell membranes. Ph. D. Dissertation, University of Southern Mississippi
55. Shao PL, Mauritz KA, Moore RB (1995) [Perfluorosulfonate ionomer]/[mixed inorganic oxide] nanocomposites via polymer-in situ sol-gel chemistry. Chem Mater 7:192–200
56. Patil Y, Sambandam S, Ramani V, Mauritz KA (2009) Model studies of the durability of a titania-modified nafion fuel cell membrane. J Electrochem Soc 156:B1092–B1098

57. Apichatchutapan W, Moore RB, Mauritz KA (1996) Asymmetric Nafion/(zirconium oxide) hybrid membranes via in situ sil-gel chemistry. J Appl Polym Sci 62:417–426
58. Jalani NH, Dunn K, Datta R (2005) Synthesis and characterization of Nafion®-MO_2 (M = Zr, Si, Ti) nanocomposite membranes for higher temperature PEM fuel cells. Electrochim Acta 51:553–560
59. Amjadi M, Rowshanzamir S, Peighambardoust SJ, Hosseini MG, Eikani MH (2010) Investigation of physical properties and cell performance of Nafion/TiO_2 nanocomposite membranes for high temperature PEM fuel cells. Int J Hydrogen Energy 35:9252–9260
60. Pan J, Zhang H, Chen W, Pan M (2010) Nafion–zirconia nanocomposite membranes formed via in situ sol-gel process. Int J Hydrogen Energy 35:2796–2801
61. Patil Y, Mauritz KA (2009) Durability enhancement of Nafion® fuel cell membranes via in situ sol-gel-derived titanium dioxide reinforcement. J Appl Polym Sci 113:3269–3278
62. Patil Y, Kulkarni S, Mauritz KA (2011) In situ grown titania composition for optimal performance and durability of Nafion® fuel cell membranes. J Appl Polym Sci. doi: 10.1002/app.31500
63. Kim YJ, Choi WC, Woo SI, Hong WH (2004) Proton conductivityProton conductivity and methanol permeation in Nafion™/ORMOSIL prepared with various organic silanes. J Membr Sci 238:213–222
64. Young SK, Jarrett WL, Mauritz KA (2002) Nafion®/ORMOSIL nanocomposites via polymer-in situ sol–gel reactions. 1. Probe of ORMOSIL phase nanostructures by ^{29}Si solid-state NMR spectroscopy. Polymer 43:2311–2320
65. Deng Q, Moore RB, Mauritz KA (1995) Novel Nafion/ORMOSIL hybrids via in situ sol-gel reactions. 1. Probe of ORMOSIL phase nanostructures by infrared spectroscopy. Chem Mater 7:2259–2268
66. Li C, Sun G, Ren S, Liu J, Wang Q, Wu Z, Sun H, Jin W (2006) Casting Nafion-sulfonated organosilica nano-composite membranes used in direct methanol fuel cells. J Membr Sci 272:50–57
67. Liang ZX, Zhao TS, Prabhuram J (2006) Diphenylsilicate-incorporated Nafion® membranes for reduction of methanol crossover in direct methanol fuel cells. J Membr Sci 283:219–224
68. D'Epifanio A, Mecheri B, Fabbri E, Rainer A, Traversa E, Licoccia S (2007) Composite ormosil/Nafion membranes as electrolytes for direct methanol fuel cells. J Electrochem Soc 154:B1148–B1151
69. Nogami M, Miyamura K, Abe Y (1997) Fast protonic conductors of water-containing P_2O_5–ZrO_2–SiO_2 glasses. J Electrochem Soc 144:2175–2178
70. Nogami M, Nagao R, Cong W, Abe Y (1998) Role of water on fast proton conduction in sol-gel glasses. J Sol-Gel Sci Technol 13:933–936
71. Nogami M, Matsushita H, Goto Y, Kasuga T (2000) A Sol-gel-derived glass as a fuel cell electrolyte. Adv Mater 12:1370–1372
72. Damay F, Klein LC (2003) Transport properties of Nafion™ composite membranes for proton-exchange membranes fuel cells. Solid State Ion 162–163:261–267
73. Aparicio M, Damay F, Klein LC (2003) Characterization of SiO2–P2O5–ZrO2 sol-gel/Nafion™ composite membranes. J Sol-Gel Sci Technol 26:1055–1059
74. Daiko Y, Klein LC, Nogami M (2004) Modifying Nafion with nanostructured inorganic oxides for proton exchange membrane fuel cells. Mat Res Soc Symp Proc 822:S8.4.1–S8.4.6
75. Aparicio M, Klein LC (2005) Synthesis and characterization of Nafion/60SiO2–30P2O5–10ZrO2 sol-gel composite membranes for PEMFCs. J Electrochem Soc 152:A493–A496
76. Elabd YA, Hickner MA (2011) Block copolymers for fuel cells. Macromolecules 44:1–11
77. Nagarale RK, Shin W, Singh PK (2010) Progress in ionic organic-inorganic composite membranes for fuel cell applications. Polym Chem 1:388–408
78. Laberty RC, Vallé K, Pereira F, Sanchez C (2011) Design and properties of functional hybrid organic–inorganic membranes for fuel cells. Chem Soc Rev 40:961–1005
79. Xing P, Robertson GP, Guiver MD, Mikhailenko SD, Wang K, Kaliaguine S (2004) Synthesis and characterization of sulfonated poly(ether ether ketone) for proton exchange membranes. J Membr Sci 229:95–106

80. Kerres JA (2005) Blended and cross-linked ionomer membranes for application in membrane fuel cells. Fuel Cells 5:230–247
81. Li Q, He R, Jensen JO, Bjerrum NJ (2003) Approaches and recent development of polymer electrolyte membranes for fuel cells operating above 100°C. Chem Mater 15:4896–4915
82. Silva VS, Ruffmann B, Silva H et al (2005) Proton electrolyte membrane properties and direct methanol fuel cell performance: I. Characterization of hybrid sulfonated poly(ether ether ketone)/zirconium oxide. J Power Sources 140:34–40
83. Mecheri B, D'Epifanio A, Di Vona ML et al (2006) Sulfonated polyether ether ketone-based composite membranes doped with a tungsten-based inorganic proton conductor for fuel cell applications. J Electrochem Soc 153:A463–A467
84. Pezzin SH, Stock N, Shishatskiy S, Nunes SP (2008) Modification of proton conductive polymer membranes with phosphonated polysilsesquioxanes. J Membr Sci 325:559–569
85. Xu X, Cabasso I (1993) Preliminary study of phosphonate ion exchange membranes for PEM fuel cells. Polym Mater Sci Eng 68:120–121
86. Smith JJ, Zharov I (2009) Preparation and proton conductivity of sulfonated polymer-modified sintered and self-assembled silica colloidal crystals. Chem Mater 21:2013–2019
87. Niepceron F, Lafitte B, Galiano H et al (2009) Composite fuel cell membranes based on an inert polymer matrix and proton-conducting hybrid silica particles. J Membr Sci 338:100–110
88. Sel O, Laberty RC, Azais T, Sanchez C (2009) Designing meso- and macropore architectures in hybrid organic–inorganic membranes by combining surfactant and breath figure templating (BFT). Phys Chem Chem Phys 11:3733–3741
89. Zhang J, Xie Z, Zhang J et al (2006) High temperature PEM fuel cells. J Power Sources 160:872–891
90. Sel O, Soulès A, Améduri B, Boutevin B, Laberty RC, Gebel G, Sanchez C (2010) Original fuel-cell membranes from crosslinked terpolymers via a "sol-gel" strategy. Adv Funct Mater 20:1090–1098
91. Kato M, Sakamoto W, Yogo T (2007) Synthesis of proton-conductive sol-gel membranes from trimethoxysilylmethylstyrene and phenylvinylphosphonic acid. J Membr Sci 303:43–53
92. Li S, Zhou Z, Abernathy H et al (2006) Synthesis and properties of phosphonic acid-grafted hybrid inorganic–organic polymer membranes. J Mater Chem 16:858–864
93. Lee S-Y, Ogawa A, Kanno M et al (2010) Nonhumidified intermediate temperature fuel cells using protic ionic liquids. J Am Chem Soc 132:9764–9773
94. Yan F, Yu S, Zhang X et al (2009) Enhanced proton conduction in polymer electrolyte membranes as synthesized by polymerization of protic ionic liquid-based microemulsions. Chem Mater 21:1480–1484
95. Lin B, Cheng S, Qiu L et al (2010) Protic ionic liquid-based hybrid proton-conducting membranes for anhydrous proton exchange membrane application. Chem Mater 22:1807–1813
96. Lakshminarayana G, Nogami M (2010) Inorganic–organic hybrid membranes with anhydrous proton conduction prepared from tetramethoxysilane/methyltrimethoxysilane/trimethyl phosphate and 1-ethyl-3-methylimidazolium-bis(trifluoromethanesulfonyl) imide for H_2/O_2 fuel cells. Electrochim Acta 55:1160–1168

Chapter 6
Hybrid Materials for High Ionic Conductivity

Jadra Mosa and Mario Aparicio

Abstract Organic–inorganic hybrid materials are a blending of two components where at least one of them is at the nanoscale. The sol-gel process is an interesting method to synthesize these materials because it allows a wide variation in compositions and inorganic/organic ratios, together with an excellent control of porosity (volume, size and connectivity) and functional groups. These features enable the design of hybrid materials with high ion conductivity for different applications as electrolytes for proton exchange membranes fuel cells (PEMFC) and lithium ion batteries. Two of the main constraints of PEMFC are the operation conditions, limited to maximum temperatures of 80°C and relative humidity near 100%. Increasing of operation temperature above 100°C is a highly desirable goal because it increases the electrochemical kinetics, improves CO tolerance, facilitates heat rejection, and reduces the problems associated with water management. Although some of the properties have been achieved separately, no system has been able to gather all the necessary requirements. In the case of application of hybrid materials as solid electrolytes for lithium ion batteries, the ion conductivity is only slightly improved compared with poly(ethylene oxide)-based polymer electrolytes and still far from values of organic liquids. Besides, in the field of microbatteries, the electrolyte thickness can be as low as one micron, and, in this configuration, ionic conductivities of 10^{-5} S/cm or higher would be enough for practical applications.

J. Mosa · M. Aparicio (✉)
Instituto de Cerámica y Vidrio (CSIC), 28049 Madrid, Spain
e-mail: maparicio@icv.csic.es

J. Mosa
e-mail: jmosa@icv.csic.es

M. Aparicio et al. (eds.), *Sol-Gel Processing for Conventional and Alternative Energy*,
Advances in Sol-Gel Derived Materials and Technologies,
DOI: 10.1007/978-1-4614-1957-0_6,

Keywords High temperature PEMFC • Hybrid electrolytes • Hybrid structure • Ionic conductivity • Li-ion batteries • Lithium ion conductivity • Mesoporous hybrids • Nafion®/inorganic membranes • Organic–inorganic interaction • Poly(ethylene oxide)/inorganic electrolytes • Proton conductivity • Proton Exchange Membrane Fuel Cell (PEMFC) • Sol-Gel • Water-free electrolytes

6.1 Introduction

Currently, the word "hybrid" is seen very often in our daily world. Some examples are the hybrid car, hybrid version of a computer software, hybrid type mobile phone, etc. The word "hybrid" has also been used in materials science and engineering. The dictionary meaning of the word "hybrid" is: "a thing made by combining two different elements". Normally the terms "hybrid" and "hybrid materials" have been used to express materials produced by a combination of different materials. Therefore, an organic–inorganic hybrid is a combination of organic and inorganic materials. Also, the mixture of materials has been frequently called composites. In many cases, the difference between hybrid materials and composites is not so evident. It is widely accepted that conventionally known composites are mere mixtures of materials, and hybrid materials include at least one of the involved materials at either nanometer or molecular scale. Very interesting properties in these hybrid materials are expected that are not possible to find in organic polymer or inorganic material separately. For example, they can have features such as being flexible like plastic but have excellent mechanical strength and thermal stability. Inorganic–organic hybrids have received a great deal of attention from many different fields, and are a hot topic in materials science and engineering.

Yamada et al. [1] defined hybrid materials as mixtures of two or more materials with new properties created by new electron orbitals formed between each material, such as covalent bond between polymer and silanol. Makishima [2] categorized substances into three materials by their chemical-bond modes, i.e. metals, organic materials and their polymers, and ceramics. He also defined hybrid materials as mixtures of two or more materials with newly formed chemical bonds. His categorization of hybrid materials and their related materials were proposed as follows:

i. *Composites* Mixture of materials consisting of a matrix and a micron-level dispersion.
ii. *Nanocomposites* Sub-micron level mixture of similar kinds of materials.
iii. *Hybrids* Sub-micron level mixture of different kinds of materials with chemical bond between them.
iv. *Nanohybrids* Atomic or molecular level mixture of different materials with chemical bonds between their different materials.

Gómez-Romero and Sanchez [3] defined hybrid materials as organic–inorganic hybrid materials or inorganic biomaterials and classified them into two main kinds,

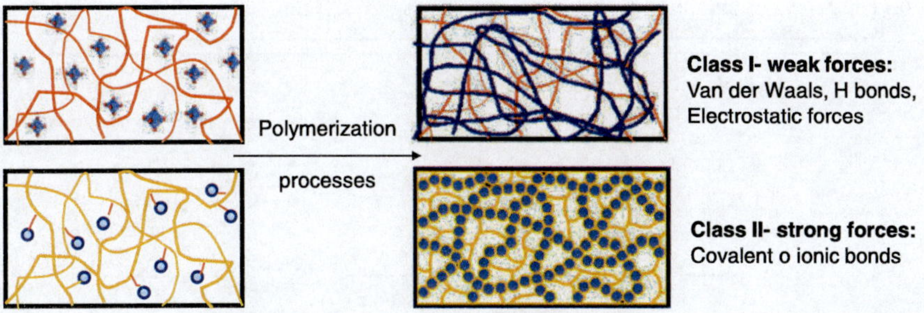

Fig. 6.1 Scheme of different kinds of hybrid materials (Class I—weak forces and Class II—strong forces) by Gómez-Romero and Sanchez [3]

depending on the nature of the links and interactions existing at the hybrid interface (Fig. 6.1). Class I hybrids include all systems where there are no covalent or ionic-covalent bonds between the organic and inorganic components. Thus, only Van der Waals, hydrogen bonding or electrostatic forces are present. On the contrary, in Class II hybrids, at least parts of the inorganic and organic components are linked through strong covalent or iono-covalent bonds. Hybrids can also be characterized by the type and size of the organic or the inorganic precursors. Precursors can be two separate monomers or polymers, or they can be covalently linked.

The above classifications were done from the viewpoint of chemical bonding between components and materials category. Their definitions of "hybrid materials" required an atomic or nanometer-level mixture of materials. Figure 6.2 shows a classification of hybrid materials and composites as a function of the scale level, as proposed by the Materials Science Society of Japan [4].

Suyama [5] proposed three inorganic–organic hybrids, categorizing them on the basis of structural differences of inorganic and organic materials. The first example of inorganic–organic hybrids is the organically modified silicates fabricated by sol-gel processing. The organic-modified silicates, as shown in Fig. 6.3a, have excellent mechanical properties due to strong covalent bonds between silica and organic molecules mixed in a molecular scale [5].

Hybrid materials fabricated in such a manner are characterized by the particular chemical bonds between silica and organic molecules, in contrast with traditional composites. Nowadays, there are many types of inorganic–organic hybrids. A clay/polymer hybrid is one example of another inorganic–organic hybrid material, as shown in Fig. 6.3b. Here, strong chemical bonds between silicate monolayers and polymer molecules offer much improved mechanical properties and lower gas permeability than those of polymer materials. On the other hand, the sol-gel approach is the preponderant one to obtain polymer-silica nanoparticles hybrid materials, as shown Fig. 6.3c. The method includes either the use of nanometer-size silica particles or their precipitation through hydrolysis and condensation reactions of silicon alkoxides. Silica is dispersed in the solution and then in situ polymerization of the monomer is carried out.

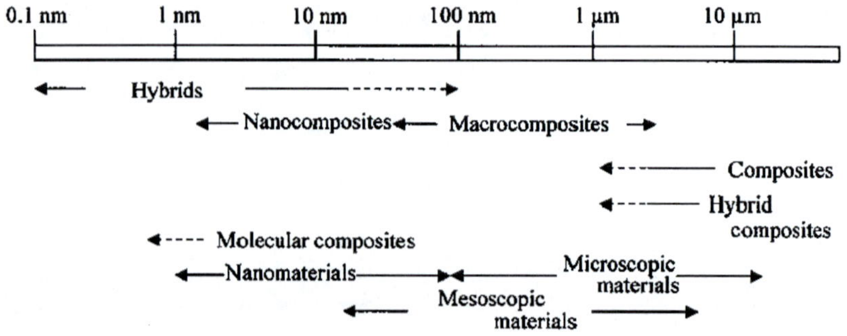

Fig. 6.2 Clasification of hybrid materials and composites at different scale levels proposed by Materials Science Society of Japan [4]

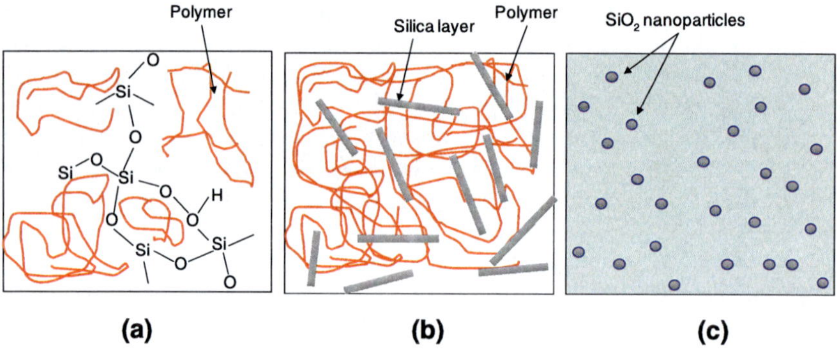

Fig. 6.3 Morphologies in inorganic–organic hybrid materials proposed by Suyama [5]: **a** polymer-modified silica, **b** clay-polymer layer and **c** silica particle—polymer matrix

The synthesis of hybrid materials using the sol-gel process allows the combination of different organic and inorganic precursors: alkoxides, organically modified alkoxides and monomers. The versatility of the method leads to a wide variation in compositions and inorganic–organic ratios, together with an excellent control of porosity (volume, size and connectivity) and functional groups. These characteristics allow the design of hybrid materials with high ion conductivity for different applications. The manufacturing of electrolytes for proton exchange membranes fuel cells (PEMFC) and lithium ion batteries are two of the most significant ones nowadays.

6.2 Proton Exchange Membrane Fuel Cells (PEMFC)

Proton exchange membrane (PEM) fuel cells are one of the most promising clean energy technologies currently under development [6–11]. The major advantages include: current prototype efficiency up to 60%, high energy densities (relative to batteries) and the ability to operate on clean fuels while emitting no pollutants. Despite these benefits, penetration of PEM fuel cells technology into the market place is being limited by cost and reliability issues. An immense worldwide effort to develop both catalyst and membrane and study the long-term behavior of fuel cells has so far met with moderate success. As a result, it is widely acknowledged that the goal of large scale fuel cell market penetration in areas including transport has moved from 2010 to 2015, and that there are still many technical and market issues to overcome. These challenges include: choosing the appropriate fuel source and infrastructure, industry regulation, safety and public acceptance.

Research into fuel cells has grown exponentially over the last 15 years [12–17]. In the case of the polymer fuel cell, the major breakthroughs in technology which have allowed a significant improvement in the overall performance of the PEM fuel cells, have been the modification of Nafion (Du Pont). In fact, Nafion is the benchmark by which all new materials are compared. A significant number of these modified derivatives of Nafion and other polymers (e.g. sulfonated polyetherketones or SPEK) are appearing in a wide range of the latest fuel cell prototypes. The history and current development of PEM fuel cells is linked to advantages and disadvantages of Nafion under different operating conditions.

Nafion and other perfluorinated PEMs have been widely used because of their excellent proton conductivity and electrochemical stability due to the PTFE backbone [18]. However, they are expensive, not as durable as desirable especially under cycling voltage, humidity, and freezing and thawing conditions, unstable at temperatures over 100°C, and effectively conduct protons only when they imbibe sufficient water, which limits operating temperatures of PEM fuel cells to around 80°C. A cell temperature above 100°C is a highly desirable goal. As the operation of fuel cells at higher temperature increases electrochemical kinetics, improves CO tolerance, facilitates heat rejection, and reduces the problems associated with water management [18–21], there have been extensive research efforts to find alternative membranes that are stable at higher temperatures. In fact, a desirable PEM must not only be highly proton conductive under hot and dry conditions, it should be thin for low resistance and high protonic conductivity, compliant to make a good contact with electrodes, but also rigid enough to provide support to the membrane electrode assembly (MEA), thermally and dimensionally stable, impervious to gaseous or liquid fuels, as well as to electrons, with a low electro-osmotic drag, and mechanically strong enough to last several years. This is a tall order indeed, and it is small wonder that success at finding alternatives to Nafion has been limited despite a very large-scale research effort.

It is important to understand that high temperature operation is being prevented by three main barriers [22, 23]:

1. Loss of hydration of the PEM and instantaneous increase in membrane resistance.
2. Polymer membrane degradation above 120°C.
3. Lack of intermediate proton conductors in the range of 100–400°C with a unique proton 'solvating' species supporting conduction in the regime.

The factor that has the highest influence on conductivity of proton conducting electrolytes is the degree of hydration. Hence, to understand the design parameters for high temperature PEMs, it is important to have a fundamental understanding of water and proton transport mechanisms. On the other hand, there are many other significant issues that will need to be resolved before alternative membranes are used in PEM fuel cells. These issues include standard operating considerations such as mechanical strength and durability, cyclability, synthesis, and integration in addition to design issues including catalyst compatibility and scale up (or down). Finally, it is important to recognize potentially that the greatest barrier faced will be start-up and shut-down operation, as typically the low temperature and high temperature conductivity mechanisms vary greatly.

6.2.1 New Polymer Electrolytes for High Temperature PEMFC

There have been numerous candidates developed for higher temperature operation of PEM fuel cells, both by modifying Nafion membranes and also by developing completely new systems of membrane [24–31]. Two main types of polymer membranes have dominated research efforts: sulfonated aromatic polymers (e.g. sulfonated polyetheretherketone, SPEEK and polyetherketone, SPEK) and perfluorosulfonic acid membranes such as Nafion which have been the industry benchmarks. These membranes both exhibit phase separated domains consisting of an extremely hydrophobic backbone which gives morphological stability and extremely hydrophilic functional groups. These functional groups aggregate to form hydrophilic nanodomains which act as water reservoirs. However, the conductivity of sulfonated aromatic polymers is also heavily dependant on the degree of sulfonation. At reduced levels of sulfonation the aromatic polymers have lower water contents and reduced conductivity, lower than 10^{-2} S/cm, which is not acceptable for use in fuel cell membranes. Nevertheless, if the degree of sulfonation is increased to improve conductivity, the mechanical properties of the membrane deteriorate.

With the aim of looking for alternatives, many efforts are being made to develop organic–inorganic hybrid membranes. The aim here is to review a variety of strategies that authors have developed within each class of hybrids and give the reader an idea of the enormous effort that has been carried out in the development of alternative hybrid electrolytes, alternatives to Nafion, which can operate at high temperatures. In the literature we have found different strategies on hybrid membranes and can be classified into two categories: Class I and Class II hybrids.

6.2.2 Class I Hybrid Electrolytes for High Temperature PEMFC

The incorporation of inorganic materials to improve chemical and mechanical properties, and thermal stability of polymer organic materials is a new procedure to get hybrid materials. Going from one extreme to another, organic components contribute to the formation of defect-free inorganic membranes and make it less brittle. The route followed in PEMFC field is based on the former type: modification of polymer matrix with an inorganic component. In the case of membranes for Direct Methanol Fuel Cells (DMFC), this kind of hybrid material is also of particular interest since it is a potential method for reducing methanol permeability and increasing proton conductivity. Examples of suitable polymer component are perfluorosulfonic matrix or alternative polymers that have been considered according to the following criteria: resistance to high temperatures, and presence along the backbone of an aromatic ring for sulfonation. The following polymers have therefore been checked as possible candidates for new proton conductive polymers: Polyamideimide (PAI), Polybenzimidazole (PBI), Polyetheretherketone (PEEK), Polyethersulfone (PES) and Polyimide (PI) [32–44].

There are several reviews in which different authors classified the state of the art in Class I hybrid membranes with different criteria: either taking into account the nature of the polymeric matrix (perfluorosulfonic or non-perfluorosulfonic), or the driving mechanism (water-based systems and alternative systems that use other solvents or heterocycles) [26–31, 45, 46].

The hydrophilic inorganic additive is added to retain water at higher temperatures and to increase acidity. However, there are two possibilities: adding the hydrophilic inorganic material into the perfluorinated membrane, or using the polymer as a binding medium for a large amount of inorganic proton conductors [47]. The inorganic component includes hygroscopic oxides such as SiO_2, TiO_2, SnO_2, ZrO_2, etc. These oxides do not have high intrinsic proton conductivity but increase the water retention of the membrane, especially at high temperature and/or low humidity conditions. Contents around 10 wt% are appropriate to improve these properties without limiting the conductivity properties of the polymer matrix. The method to incorporate the inorganic component has two main procedures: incorporation of oxide particles to the polymer before material consolidation, and precipitation from alkoxides through sol-gel reactions. The preparation of hybrid organic–inorganic materials using the sol-gel process has been a subject of growing interest because it leads to more homogeneous materials avoiding agglomeration and phase separation. Using this procedure, significant improvement in current–voltage response has been obtained at high temperature (130°C) when SiO_2-P_2O_5-ZrO_2 was in situ incorporated in a Nafion membrane [37, 48].

Examples of other inorganic additives are: (1) clays, formed with SiO_4 tetraedra and linked into two-dimensional networks [41], (2) zeolites, aluminosilicates enclosing cavities suitable for water molecules [49, 50], (3) hygroscopic layered-structures compounds as zirconium phosphate and other metal phosphates [42, 43], (4) heteropolyacids, clusters of tungsten, molybdenum, and others with high proton conductivity at low temperature because of the presence of protons and water molecules in the structure [39, 40] and (5) mesoporous inorganic micro or nanoparticles with pore sizes below 10 nm, sometimes ordered, that can be functionalized with protonic functions such as COOH, SO_3H or PO_3H_2 [51–53].

In the case of zeolites, exposing some inorganic oxides to water atmosphere at high temperature leads to oxygen vacancies in the oxides that are replaced by hydroxyl groups from the dissociation of water vapor, leaving protons to form covalent bonds with the lattice oxygen. The water uptake and the concentration of incorporated protons depend highly on the concentration of oxygen vacancies in oxides and the rate of water incorporation reactions. The diffusion of these incorporated protons in the lattice of oxides results in the high conductivity observed in these oxides. Among them, the highest proton conductivity was observed in oxides with perovskite-type structures (ABO_3) with cubic or slightly reduced symmetry [49, 50].

In regard to mesoporous hybrid systems, the very high and structured porosity of these new hybrid membranes obtained by sol-gel enhance the water storage, the intimate mixture of polymeric chains with inorganic phase, and mostly the exchange capacity (IEC) through a much larger surface area available for functionalization [51–53]. Indeed, this approach gathers together complementary essential properties: first, water uptake control due to the open mesoporosity; second, the enhancement of proton conductivity values when the membrane operates at high temperature and low relative humidity rate opening the possibility to graft sulfonic moieties in the silica phase; third, the mechanical property of the membrane, in which the formation of a finely dispersed mesostructured inorganic network leads to a flexible and non-swelling homogeneous material. Mesoporous silica/polymer hybrids can also be made through entrapping polymers in the nanochannels of mesoporous silica through the hydrogen bonding between the silanol groups on the silica surface and the groups in polymer chains. Hence, groups which can form hydrogen bonds with silanol groups are very necessary for a certain polymer if high loading of polymers has to be trapped. The hydrogen donating property of silanol moieties is so strong that organic polymers having hydrogen accepting groups such as amide moieties form hydrogen bonds with silanol groups. For polymers without hydrogen accepting groups, other methods should be used to graft hydrogen accepting groups onto polymer chain [54, 55].

Although this concept seems promising, the success so far seems limited and the interaction between the inorganic phase and proton conductor is not sufficiently understood. Further, the stability of the material in the polymer host is uncertain. Certain materials (e.g. heteropolyacids) dissolve in the water produced in the fuel cell leaching out if there is no chemical bond with the network.

Another approach is the replacement of water by a less volatile, non aqueous, and high proton conducting solvent developing the same function as water. Four relevant examples of such 'water replacements' are phosphoric acid, and systems that contain hetereocycles such as imidazole, solid acids as $CsHSO_4$ and CsH_2PO_4, and ionic liquids. The compounds should have the ability to act as a Brönsted base, and have a high dielectric constant and a low volatility. Proton conducting species are mainly responsible for transporting protons in short and long ranges. From the atomic level, only oxygen and nitrogen are suitable for carrying protons in the PEMFC due to chemical stability and interaction of oxygen and nitrogen with protons. Possible proton conducting species with oxygen include water, oxygen anion, hydroxyl anion, alcohol, ether, ester, carboxylic acid, inorganic oxoacid, and oxygen-containing heterocycles. For nitrogen, those include ammonia, amino groups, and nitrogen-containing heterocycles.

Phosphoric acid (H_3PO_4) is a highly viscous liquid and was first used in phosphoric Acid Fuel Cells (PAFCs), which were developed in the 1960s and also the first type of fuel cells to be commercialized. The operating temperature of PAFC is in the range of 150–220°C, indicating the excellent proton conductivity properties at high temperature. The estimated proton mobility of phosphoric acid at the melting point (42°C) is 2×10^{-5} cm^2/s with a conductivity of 7.7×10^{-2} S/cm [56]. In the case of PEMFC, the main application of phosphoric acid has been to dope polybenzimidazole (PBI)-based polymers. Wainright et al. [57] first proposed and developed PBI doped with H_3PO_4 for application as electrolyte in high temperature PEMFC. PBI provides enough mechanical strength, stability, and acid absorption ability to produce the composite membranes. Several methods have been used to incorporate phosphoric acid into the polymer: (1) the cast polymer membranes were immersed into the phosphoric acid solution, (2) PBI polymer and acid were directly cast from a solution of PBI and H_3PO_4 in trifluoroacetic acid solution, and (3) PBI was prepared in polyphosphoric acid (PPA) and directly cast to form membranes. Although the doping level is similar, the properties of membranes prepared by different methods are quite different. Comparing results, it seems that a higher content of phosphoric acid leads to an improved conductivity.

Among various nitrogen-containing species, some heterocycles are most likely to be the proton carriers applied for high temperature PEMFC because the high electron density on nitrogen is partially reduced by the conjugated aromatic system and the oxidative stability of nitrogen is improved. Kreuer et al. [58] first proposed heterocylic compounds as imidazole and pyrazole like proton conducting groups. The proton conducting ability of imidazole was determined by measuring the conductivity of sulfonated PEEK doped with imidazole and pyrazole at elevated temperatures, and also by studying the proton mobility. Imidazole also acts as a proton donating groups through the dissociation of the proton from the –NH group in imidazole. The conductivity of liquid imidazole was reported to be as high as 10^{-3} S/cm at the melting point of 90°C. Later work has been focused on the study of (i) imidazole doped acidic perflurosulfonic polymer such as Nafion and (ii) imidazole derivatives bonded to alternative polymers [56]. The imidazole grafted polymer shows noticeable proton conductivity at high temperatures without external humidification, which provides a potential way to design so-called "intrinsic proton-conductive" polymers for high temperature applications. By covalently tethering the proton carriers to the polymer backbones, the leaching or evaporation of these free-water proton conducting molecules can be avoided getting stabilized conductivity and performance of the membrane at high temperatures. However, the first fuel cell tests reported based on acidic polymers with imidazole doping or imidazole grafted polymers were not successful, possibly due to the poisoning effect of imidazole to the platinum catalysts [59]. Although the boiling point of imidazole is as high as 256°C, the slow evaporation or leaching of small imidazole molecules may still lower the conductivity of the imidazole doped membrane. For polymers with imidazole tethered, the difficulty is in either very low conductivity or insufficient mechanical strength. Besides the low conductivity, the thermooxidative degradation of imidazole and other heterocylic compounds may also be a difficulty for applying these materials for long-term high temperature-PEMFC applications.

Other typical inorganic proton conducting materials are acidic salts of oxo-acids such as $CsHSO_4$ [60]. At temperatures around 140°C, there exists a first-order phase transition of $CsHSO_4$, and above this point the solid shows proton conductivity as high as 10^{-2} S/cm, which results from the hydrogen transportation between sulfate groups. However, to apply these inorganic proton conducting materials to high temperature PEMFC, there are still some severe disadvantages: (1) the low conductivities at temperatures below 200°C, which may be even worse when it was doped into a polymer matrix due to the large interfacial resistance; (2) the instability of these oxides in liquid water; and (3) insufficient mechanical properties of these solid oxides and their composites with polymers.

Recently, a few groups [61, 62] reported that some kinds of ionic liquid, such as imidazolium bis-(trifluoromethanesulfonyl)imide (HImTFSI) and 2,3-dimethyl-1- octylimidazolium triflate, have proton conducting ability and may be applied in PEMFC at elevated temperature. However, only part of the high total conductivity is related to proton conduction in these systems. The stability, leaching, and electro-osmotic drag of ionic components in the membrane are disadvantages for these systems. Till now, there is no acceptable cell performance data reported for ionic liquid-based membranes. Unfortunately, all these water-free systems that use alternative solvents to transport protons are typically soluble in the water produced in the fuel cell and can eventually leach out.

6.2.3 Class II Hybrid Electrolytes for High Temperature PEMFC

The second approach involved to replace Nafion membrane is based on hybrid organic–inorganic materials with nano-sized phases and interfaces, where chemical bonds are created between the inorganic moieties and the polymer matrix. The synthesis of class II hybrid organic–inorganic materials using sol-gel technology has attracted the continuous attention of researchers from different fields. This method is part of the process of "soft chemistry" and allows the incorporation of organic molecules in inorganic matrices, an operation that cannot always be done by other methods without degradation of any component. This fact makes possible the incorporation of organic groups with a range of reactive functionalities on the inorganic network and may generate structures in which the two networks, one organic and the other inorganic, are mutually intertwined and interconnected. Such structures can be generated from different synthesis routes: (1) simultaneous synthesis of organic and inorganic networks, (2) sequential synthesis: inorganic polymerization followed by organic polymerization or organic polymerization and subsequent inorganic polymerization. Sol-gel processes are complex because during synthesis side effects and competitive reactions, including esterification and transesterification depolymerisation, can be produced. It is essential a strict control of the organic and inorganic polymerization kinetics, controlling all the parameters that govern them and avoiding phase separation processes, to achieve the appropriate conversion degrees and maintain the desired organic/inorganic ratio. The presence of an organic group covalently bound to the inorganic network (alkylalkoxide) affects the kinetics of sol-gel reactions, since the degree of condensation of trifuncion-tionalized silicon decreases when the size of alkyl organic group increases. Using a polymerizable alkylalkoxide unit as a precursor, organic polymerization shows strong steric hindrance, which affects the degree of polymerization of organic and inorganic, and the level of crosslinking. Therefore, it is necessary to control the kinetics of polymerization and the factors that govern them in order to develop both networks simultaneously for preparing homogeneous nanostructured membranes with no phase separation. The key factor to obtain hybrid structures with specific properties is the design of the synthesis, including the choice of precursors, the organic/inorganic ratio and the strategy of organic and inorganic synthesis. The main advantage of synthesized class II hybrid membranes by the combination of sol-gel technique and polymerization process is that the growth of inorganic and organic phases allows the improvement of the interface comparing with hybrids class I and enlarge the dispersion of the networks [3]. However, the success of the procedure depends on the molar ratio of organic and inorganic precursors, the kind of precursors used, and reaction kinetics [63, 64]. Moreover, if the process includes monomers, the possibilities of synthesis of organic–inorganic membranes with the desired properties grow exponentially. The incorporation of monomers also allows to define the organic/inorganic ratio in a large range of compositions [65, 66]. An adequate design of the hybrid structures should generate membranes with high thermal stability able to operate at temperatures up to 200°C, and enough chemical stability against water, hydrogen, oxygen, and methanol in this temperature range. The major problem associated with the incorporation of an inorganic network is the increase of brittleness, because this originates membranes with mechanical stability not enough to resist the preparation process of the Membrane-Electrode Assembly (MEA). For this reason, the precise design of the hybrid structure is very important to solve this problem, and an adequate organic–inorganic ratio, crosslinking and synthesis procedure is crucial. The membranes ought to have high proton conductivity to be applied in PEMFC. Two main alternatives are used to provide high proton conductivity in hybrid membranes: a high concentration of SO_3^- groups, as in Nafion-type membranes, and incorporation of high proton conducting compounds. In general, silica structure coming from silicon alkoxides and organically modified alkylalkoxides fulfills these requirements. On the other hand, inorganic components can also contribute to reach higher proton conductivities through both hydroxyl groups and nanosized porosity that attract and retain water molecules.

The class II hybrid membranes prepared by sol-gel with high proton conductivity for application in high temperature PEMFC can be classified considering the organically modified alkoxide uses to bond the organic and inorganic components. Glycidoxypropyl trimethoxysilane (GPTMS) is an organically modified alkoxide that has been widely used in the design of class II hybrid membranes.

Its three methoxy groups allow inorganic polymerization with the formation of siloxane chains, while the opening of epoxy group results in the formation of a bridge between components. Membranes obtained through hydrolysis and condensation reaction of this precursor without any proton donor shows low proton conductivity at high relative humidity: 1×10^{-4} to 4×10^{-3} mS/cm. Some of the hybrids produced use Poly (vinyl alcohol) (PVA) for the formation of the organic structure connecting with the inorganic component through of GPTMS. PVA is a cheap polymer widely used in the formation of hybrid membranes for fuel cells. Specifically, PVA-based membranes, doped with strong acids such as phosphomolybdic and phosphotungstic acids, have been tested as proton conductors [67, 68]. Other inorganic precursors such as tetraethyl orthosilicate (TEOS) and additives as chitosan, a linear polysaccharide composed of randomly distributed β-(1-4)-linked D-glucosamine (deacetylated unit) and N-acetyl-D-glucosamine (acetylated unit), have been incorporated to improve permeability and selectivity [69, 70].

However, the proton conductivity is not sufficient for use in PEMFC, and it is necessary to introduce other precursors such as phenyl triethoxysilane (SPS) and proton conductor additives as phosphoric acid to achieve higher conductivity values (2×10^{-3} S/cm at 120°C and 15% RH), but still insufficient [71, 72]. Other proton donors have been analyzed in these GPTMS-based hybrid systems. The introduction of heteropolyacids, as PWA y SiWA, and α-zirconium phosphate (ZrP) leads to membranes with high proton conductivities up to 20 mS/cm. The high proton conductivity of the composites is due to the proton conducting path through the GPTMS-derived 'pseudo-polyethylene oxide' [73]. One possibility to increase the proton conductivity is to incorporate Nafion on a covalently cross-linked network composed of 4,4-methylenedianiline (MDA) and GPTMS. The hybrid membranes present a high degree of bound water (87%), adequate proton conductivity (0.034 S/cm), low methanol permeability (1.1×10^{-8} cm^2/s) and an adequate oxidative stability [74].

Another organically modified alkoxide used in the design of class II hybrid membranes is the 3-aminopropyltriethoxysilane (APTS). Its combination with other precursors as 3-triethoxysilylpropyl (TSPHI) and p-vinylphenylmethyldiethoxysilane (VFMES), and doping with 4,5-dihydroimidazol allows the synthesis of hybrid membranes with conductivities up to $10^{-2} \times$ S/cm at 130°C [75].

Several hybrid systems were synthesized using polysiloxanes, showing a low glass transition temperature and high thermal and oxidative stability, but are highly permeable to oxygen, decomposed into water, and are sensitive to acidic and alkaline medium grade [76]. The main drawbacks of these types of membranes are their poor mechanical properties, making it difficult to obtain freestanding membranes, but have the advantage of providing proton conductivity due to its special structure.

Other kinds of class II hybrid membranes are based on PEEK that gave such good results in pure polymeric membranes. In this case, a new synthetic approach is developed to synthesize a hybrid membrane, PhSiSPEEK, which includes phenylsilanol groups covalently bound to the aromatic backbone of sulfonated PEEK. These membranes exhibit conductivity around 10^{-3} S/cm at 120°C and 20% RH [77].

An alternative organically modified alkoxide used in the design of class II hybrid membranes is the hybrid precursor metacriloxy propyltrimethoxysilane (MPS). The hybrid structure is usually formed by silica-methacrylate, although only a few works are devoted to this network [78]. A combination of methacrylate and epoxy structures has been also studied to reach strong hybrid membranes with improved proton conductivity . For example, systems based on styrene (STY)—2-hydroxiethyl methacrylate (HEMA)–TEOS–PWA and GPTMS–TEOS–2-allyl-phenol (AP) have reach high proton conductivities, similar to Nafion, at temperatures higher than 100°C and 100% RH. The combination of sulfonation process with heteropolyacids that have been immobilised within the membranes through electrostatic interaction with inorganic component overcoming the leakage problem from membranes allows proton conductivity values of 10^{-3} S/cm at 40°C and 65% RH. This result can be considered a significant improvement considering the measurement conditions at low temperature and relative humidity [79, 80]. Smith et al. [81] used poly(3-sulfopropyl-methacrylate) and poly(styrenesulfonic acid) as precursors that can be attached to the nanopore surface getting a robust membranes. Proton conductivity was dependent on temperature and humidity, reaching a value of 2×10^{-2} S/cm at 30°C and 94% RH. Sel et al. [82] studied several membranes that contain a termopolymer poly[vinylidenefluoride-*ter*-perfluoro(4-methyl-3,6-dioxaoct-7-ene sulfonyl fluoride)-*ter*-vinyltriethoxysilane] [poly(VDF-*ter*-PFSVE-*ter*-VTEOS)]. Vinyltriethoxysilane (VTEOS) units of the terpolymer help to bind and assemble the inorganic network into a continuous medium and the perfluoro (4-methyl-3,6-dioxaoct-7-ene) sulfonyl fluoride (PFSVE) chains are able to bridge the conduction paths in the hybrid membrane by way of their SO_3H-functionalized chain ends. The incorporation of terpolymer into the hybrid SiO_2–SO_3H/poly(VDF-*co*-HFP) copolymer membrane allows conductivity values of around 10 mS/cm up to 120°C.

So far, all described hybrid membranes are based on silica as inorganic precursor. However, other precursors were used in the synthesis of such hybrid membranes. For example, Kim et al. [83] studied several hybrid membranes using titanium alkoxides as precursors and polydimethylsiloxane (PDMS). The hybrid membranes showed proton conductivities around 2×10^{-3} S/cm at 150°C and maximum power values of 30 mW/cm^2 at 150°C and 100% RH.

The incorporation of mesopores through templates can be useful to increase the water retention capability of the hybrid membranes, especially at high temperature operation. An approach consists of creating a three-dimensional interconnected network of mesostructured hydrophilic materials in an inert/uncharged thermostable polymer, PVDF-HFP. Specifically, the hybrid membranes contain a mesostructured SiO_2 network with acidic functionality (–SO_3H) and the fluorinated polymer PVDF-HFP. The sulfonic acid functionality is introduced by co-condensation of TEOS with an organically modified silane containing sulfonic acid groups, 2-(4-chlorosulfonylphenyl) ethyltrimethoxysilane. The mesoporosity is generated by the use of structure-directing agents such as P123 [poly(ethylene oxide–propylene oxide–ethylene oxide)] block copolymers. Final hybrid membranes that containing hierarchical porosities, macro-, meso- and micro-scale exhibit a proton conductivity of 50 mS/cm at 70°C under 100% relative humidity [84]. Also in this research line, hybrid membranes were fabricated by electrospinning nanofibers composed of a proton conducting polymer with proton conducting, water-retaining inorganic nanoparticles. Nanofibers are obtained from sulfonated poly(arylene ether sulfone) (S-PAE) to which sulfonated polyhedral oligomeric silsesquioxane (S-POSS) nanoparticles were added to improve water retention and proton conductivity [85, 86].

6.3 Secondary Lithium Ion Batteries

A lithium ion battery cell consists of a positive and a negative electrode (both sources of chemical reactions) separated by an electrolyte which enable ion transfer between the two electrodes. Rechargeable lithium batteries have been used as efficient energy storage devices in lightweight electronic appliances, cellular phones and laptop computers because of the high energy densities. There have been many efforts to make microscale lithium batteries for application in fields related to microsystems, such as microsensors, micromechanics, and microelectronics [87]. The electrolyte is usually a flammable and explosive solution containing dissociated salts. The use of solid electrolytes should permit development of batteries which increase safety and provide greater flexibility of battery shape.

Besides a large electrolyte window, the electrolyte should have a stable electrode/electrolyte interface during cycling, a Li-ion conductivity higher than 10^{-4} S/cm, an electronic conductivity lower than 10^{-10} S/cm, chemical stability with respect to the electrodes, thermal stability over ambient temperature ranges and temperatures in the battery under high power, and made of safety (nonflammable and non-explosive) and non-toxic materials [88]. Solid electrolytes could meet all of these requirements, especially those, stable electrode/electrolyte interface and safety, which are more difficult to obtain with liquid electrolytes. An exemplary solid polymer electrolyte is the poly(ethylene oxide) (PEO) system containing a lithium salt. They are low-cost, nontoxic, and have a good chemical stability. It is widely accepted that ion conduction occurs in the amorphous phase above the glass-transition temperature (Tg) by a liquid-like motion of the Li ions associated with the segmental reorientations of the neighboring polymer strands. However, the polymers in such systems tend to crystallize, resulting in low room temperature ionic conductivity ($<10^{-7}$ S/cm at 25°C) and Li^+ transference number (0.1–0.3) [89].

6.3.1 Class I Hybrid Electrolytes for Li-Ion Batteries

Most approaches to increase the ionic conductivity of PEO systems are based on lowering the degree of PEO crystallization or reducing the glass transition temperature. The introduction of inorganic fillers to PEO-based electrolytes improves

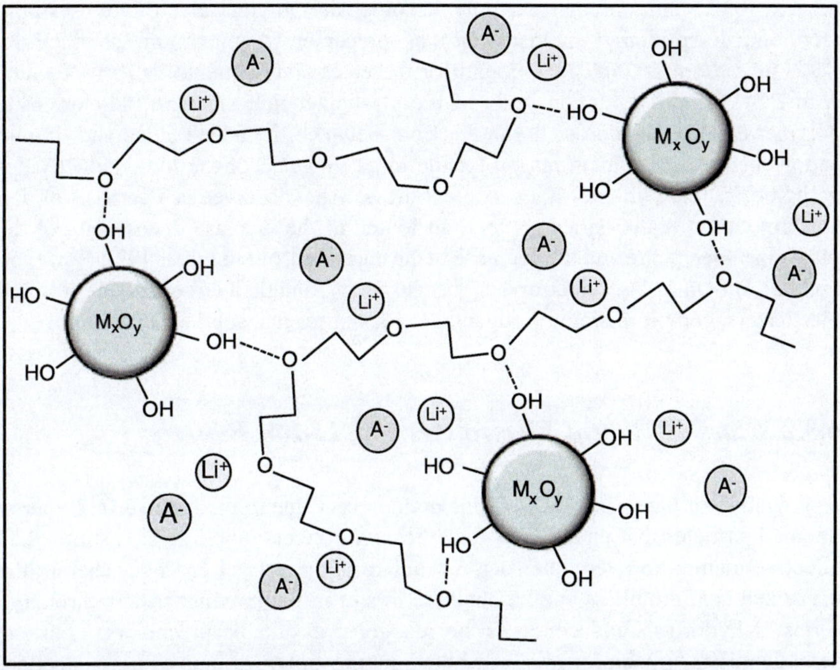

Fig. 6.4 Scheme of a Class I hybrid electrolyte for Li-ion batteries based on a PEO matrix and oxide particles (M_xO_y)

the conductivity and mechanical properties of the electrolytes. Although initial attempts sought to improve the conductivity of electrolytes through the use of conductive particles, the conductivity enhancement is commonly attributed to a decrease in polymer crystallinity, and an enlargement of the amorphous domains in the PEO matrix. Figure 6.4 displays a scheme of a polymer matrix electrolyte including oxide particles.

The surface hydroxyl groups of these particles interact with oxygen from polymer network. The inorganic component disrupts the organic structure opening new channels for the lithium ion diffusion. Reports show that smaller particles are the most effective, presumably because their high-surface area inhibits crystallization, and the result is an enhanced conductivity, but it is still not comparable to that of the carbonate electrolytes [90]. Almost all of these polymer-inorganic nanocomposites were prepared through mechanical blending of nanosize particles, polymer, and salt in a compatible solvent. Solid electrolytes of thickness around 100 μm were prepared by solution from acrylonitrile–methyl methacrylate–styrene (AMS) terpolymer, $LiClO_4$ in ethylene carbonate (EC)/dimethyl carbonate (DMC) and silica particles [91]. The capacity to retain electrolyte solution and the mechanical state of the polymer electrolyte film was found to be dependent on the molar composition. Ionic conductivity at room temperature was calculated to be 5 × 10^{-4} S/cm. Another example of composite polymer electrolyte uses poly (acrylonitrile-co-methyl methacrylate) as precursor. A maximum conductivity (2 × 10^{-3} S/cm at 25°C) was found for the electrolyte containing 10 wt% silica with EC–PC. The reduction in T_g of the plasticizer-rich phase in the composite polymer electrolytes due to the interaction between the added silica and lithium cation seems to be a main reason for the initial increase of the ion conductivities with the increase in the silica content. It can be observed a decrease of ion conductivity at higher silica content, attributed to the increase of viscosity of the plasticizer-rich phase and the decrease in the charge carrier numbers [92]. However, both are known as "gel electrolytes" because they contain a large amount of liquid plasticizers, compromising the advantages looked for in a solid electrolyte.

6.3.2 Class II Hybrid Electrolytes for Li-Ion Batteries

Aggregation of nanoparticles is an inevitable event due to the high surface energy of small particles that undermines the efficacy of the ceramic fillers. A simple and effective method to overcome such a problem is the sol-gel process, wherein the nanosized ceramic fillers can be obtained in situ in the polymer matrix through a series of hydrolysis and condensation reactions of suitable precursors. Thus the ceramic fillers are uniformly distributed in the polymer and exhibit excellent properties. Considerable advance was gained recently by designing new electrolytes based on these organic–inorganic hybrids or nanocomposite systems, among which silica-based materials hold a prominent place. They are called ORMOLYTEs (organically modified electrolytes) or ORMOCERs (organically modified ceramics) [89, 93, 94]. In contrast to hardly reproducible filling of polymer electrolytes with nanoparticles, the polycondensation of alkoxysilanes enables an extremely homogenised and reproducible distribution of the nano-sized oxidic units. On the other hand, the organic network is formed from reactive functional groups R′ of alkoxysilanes of the type $R'Si(OR)_3$, or by co-polymerizing reactive organic monomers with reactive functionalized alkoxysilanes. Figure 6.5 shows a scheme of a hybrid organic–inorganic electrolyte with chemical bonds between both components. Metal alkoxides and organically modified alkoxides produce the hybrid structure through inorganic condensation and polymerizable and non-polymerizable organic groups. The incorporation of reactive monomers allows a significant flexibility in the design of these materials.

Depending on the reactive organic functionalities and their thermal and uv-initiated organic crosslinking reactions, the materials can be adapted to meet all the electrolyte requirements. These inorganic–organic copolymers, synthesized by sol-gel processing, as host materials for ion conductors offer interesting properties: they are amorphous at room temperature, which causes comparatively high conductivities, and they provide higher thermal stability than pure organic matrices [95, 96]. Further, the inorganic–organic matrix enables mechanical and thermal stability of thin films/separators, even after the addition of plasticizers, fitting

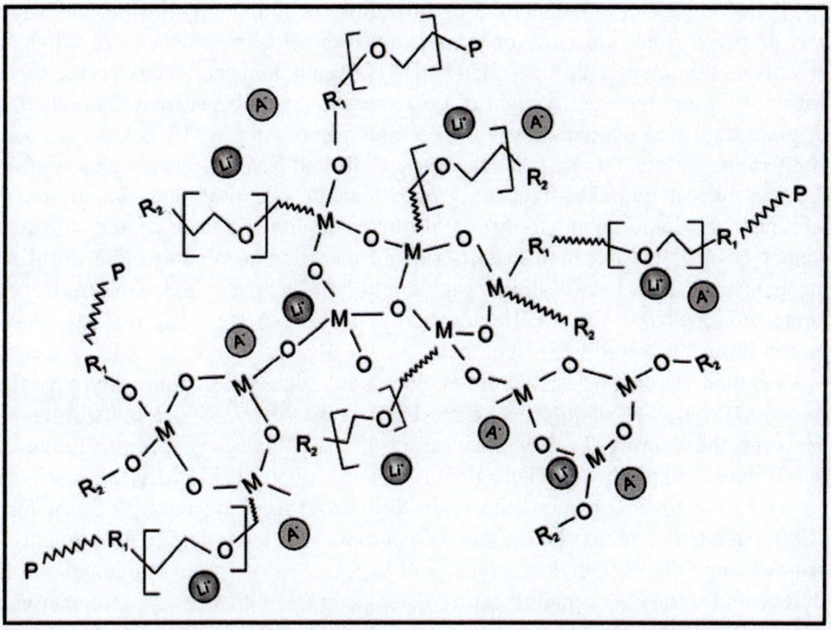

Fig. 6.5 Scheme of a Class II hybrid electrolyte for Li-ion batteries based on metal (M) alkoxides and organically modified alkoxides with polymerizable (R_1) and non-polymerizable (R_2) organic groups, and monomers (P)

perfectly for multi-layer technology and application in mini- and micro-batteries [97, 98].

Dahmouche et al. [99] have provided a comparative of organically modified sol-gel electrolytes producing chemical bonds between the organic (polymer) and inorganic (silica) phases, with others obtained from a mixture of tetraethoxysilane and poly(ethylene-propylene glycol) that are not chemically bonded. In the first case, lithium ion-conducting organic–inorganic nanocomposites based on poly(ethylene glycol) or poly-(propylene glycol) covalently attached to a silica network have been prepared using 3-isocyanatopropyltriethoxysilane, O,O′ Bis (2-aminopropyl)-polyethyleneglycol (or O,O′ Bis (2-aminopropyl)-polypropyleneglycol) and $LiClO_4$. The ionic conductivity is higher than 10^{-6} S/cm at room temperature. In the second family, prepared from tetraethoxysilane (TEOS), polyethyleneglycol (PEG) and lithium salt, the organic and inorganic phases are not chemically bonded. Ionic conductivity has been studied as a function of the polymer chain length and concentration. Values up to 10^{-4} S/cm at room temperature have been obtained, that are strongly related to the connectivity of the two phases and the mobility of both the structural network and the active species. After, the same authors [100] present a higher room-temperature maximum conductivity (6×10^{-2} S/cm) for optimized hybrid electrolytes without covalent organic–inorganic chemical bonds (type II). This value was obtained for

[O]/[Li] = 15 and PEG300/TEOS = 1.0 ratios. A similar hybrid ion conductor was prepared from a mixture of tetramethyl orthosilicate (TMOS), polyethylene glycol and lithium perchlorate ($LiClO_4$) [101]. Again, ionic conductivity increases with decreasing molecular weight of polyethylene glycol because of the reduction of glass transition temperature (T_g), reaching a maximum of 10^{-5} S/cm at room temperature. Structural characterization by FTIR and NMR measurements confirm the presence of hydroxyl-free end groups bonded to carbon and silicon atoms. Although the ionic conductivity is attributed to lithium ions, proton diffusion cannot be rejected and both can coexist. In the absence of water, the diffusion mechanism of both ions is similar and would be observed in the same frequency range of impedance spectra. However, it is to be expected that without water, proton diffusion would be insignificant.

Chemical stability is an important issue, and some amorphous poly(ethylene oxide) (PEO)–SiO_2 composites have been prepared by sol-gel transformation based on the hydrolysis and condensation of TEOS, concurrent with ultraviolet (UV) irradiation of poly(ethylene glycol) dimethacrylate (PEGDMA) and methoxy poly(ethylene glycol) monomethacrylate (PEGMA) to produce the PEO network [102]. An alternative non-hydrolytic sol-gel route uses citric acid (CA), tetraethyl-orthosilicate (TEOS), ethylene glycol and Li_2CO_3 [103]. The hybrid conductor is shown to be fully amorphous at room temperature with an ionic conductivity above 10^{-5} S/cm. The incorporation of SiO_2 nanoparticles (5 nm) to the matrix without the presence of Li has a conductivity value of 10^{-6} S/cm at room temperature. This effect can be explained by the features of SiO_2 nanoparticles surface, which generally is hydrated, indicating that proton conductivity could be also presented.

Hybrid with other oxides different from silica have also been synthesized. For example, poly(ethylene oxide) –$LiClO_4$–TiO_2–Al_2O_3 organic–inorganic hybrids were synthesized using titanium (IV) and aluminum isopropoxides as inorganic precursors [104]. Ionic conductivity increased with oxides content up to 10 wt% (3×10^{-5} S/cm at room temperature) and then decreased above this amount. Infrared spectra show the formation of molecular scale PEO-oxide hybrids, although also the presence of free hydroxyl groups allowing some proton conductivity interference. Other authors, claiming that TiO_2 is more surface active than SiO_2, have prepared PEO/TiO_2 nanocomposite polymer electrolytes with an in situ uniform dispersion of TiO_2 particles in the polymer matrix [89]. The sol-gel process involves the hydrolysis and condensation reaction of titanium (IV) ethoxide in the presence of high MW PEO and $LiBF_4$. The room-temperature ionic conductivity of these polymer electrolytes (10^{-6} S/cm) was an order of magnitude higher than that of the TiO_2-free sample. A high Li^+ transference number of 0.51 was recorded, and the nanocomposite electrolyte was found to be electrochemically stable up to 4.5 V versus Li^+/Li. The nanocomposites displayed enhanced ionic conductivity, in part because of the increase in the number of "active" charge carriers, as well as higher Li ion mobility. This behavior is attributed to the interaction between ceramic fillers, anions, and polymer chains.

The use of amino group is also an interesting approach to prepare hybrid electrolytes. Different chemically bonded hybrid electrolytes have been synthesized as an alternative to PEO from the reaction of diamino-terminated alkylene oxides with cyanuric chloride, followed by reaction with monoamino-terminated alkylene oxides and (3-aminopropyl) triethoxysilane [105]. The resulting films show improvements in both ionic conductivity and mechanical properties over PEO itself. The conductivity at 25°C for the optimum formulation is 4×10^{-5} S/cm, although lithium conductivity decreases and film strength increases with increasing amounts of cross-linking agents. Slightly higher conductivities may be obtained with polymers having less than 25% (3-aminopropyl) triethoxysilane cross-linker. However, dimensional stability is compromised as the precursor percent approaches zero. The incorporation of ionic liquids at ambient temperatures to these structures increases the conductivity up to 10^{-3} S/cm at room temperature [106]. However, the presence of a liquid reduces some of the advantages of solid electrolytes as the greater flexibility of battery shape.

An extension to improve the performance of this class of electrolytes is the incorporation of nanoporosity in the microstructure by a surfactant templating sol-gel route. An example is the composite PEO-based electrolyte containing micro-sized nanoporous zirconium-oxide-sulfate filler [107]. Cetyltrimethylammonium bromide (CTAB), zirconium propoxide, ammonium sulfate and $LiClO_4$ were used for the synthesis of membranes. The presence of nanoporous filler favors the ion transport, enhancing the ion conductivity almost one order of magnitude compared with ceramic-free electrolyte. Authors claim that the fillers promote lithium conducting pathways as a result of the Lewis acid–base interactions occurring between the ceramic nanoporous surface and both the lithium salt ClO_4^- anion and the PEO segments. Anyway, the maximum ionic conductivity at room temperature is around 10^{-5} S/cm, not enough for practical application. Only at temperatures higher than 70°C the conductivity value (10^{-3} S/cm) is interesting for application in batteries.

The increase of mechanical, thermal and chemical stability of solid electrolytes when an inorganic component is incorporated to produce hybrid organic–inorganic materials has been demonstrated. However, lithium ion conductivity is only slightly improved compared with poly(ethylene oxide)-based polymer electrolytes and still far from values of organic liquids. On the other hand, it is necessary to consider the electrolyte thickness which in the case of organic liquid is usually higher than a hundred microns. In the field of microbatteries, the electrolyte thickness can be as low as one micron. In this configuration, ionic conductivities of 10^{-5} S/cm o higher would be enough for practical applications.

Acknowledgments The authors acknowledge the Spanish Science and Innovation Ministry under National Program ACI-PLAN E (project PLE2009-0074) for financial support.

References

1. Yamada A, Sasabe H, Osada Y, Shiroda Y, Yamamoto I (1989) Concepts of hybrid materials, hybrid materials—concept and case studies. ASM International, Ohio
2. Makisima A (2004) Possibility of hybrids materials. Cers Japan 39:90–91
3. Gómez-Romero P, Sanchez C (2004) Functional hybrid materials. Wiley, New York
4. Materials Science Society of Japan (1993) Molecular hybridization and hybrid materials, composite system in materials. Shokabo Publishing, Tokyo
5. Suyama Y (2004) Research and development of organic–inorganic nanohybrids materials. Ceram Japan 39:92–93
6. Hoogers G (2003) Fuel cell technology handbook. CRC Press, USA
7. Ananthachar V, Duffy JJ (2005) Efficiencies of hydrogen storage systems onboard fuel cell vehicles. Sol Energy 78:687–694
8. Larminie J, Dicks A (2003) Fuel cell systems explained. Wiley, New York
9. Lassegues JC (1992) Proton conductors: solids, membranes and gels—materials and devices. Cambridge University Press, Cambridge
10. Srinivasan S (2001) Fuel cells: from fundamentals to applications. Springer, USA
11. Lee JH, Lalk TR (1998) Modeling fuel cell stack systems. J Power Sources 73:229–241
12. Gottesfeld S, Zawodzinski T (1997) Advance in electrochemical science and engineering. Wiley, USA
13. Cacciola G, Antonucci V, Freni S (2001) Technology up date and new strategies on fuel cells. J Power Sources 100:67–79
14. Barbir F, Gomez T (1996) Efficiency and economics of proton exchange membrane (PEM) fuel cells. Int J Hydrogen Energy 21:891–901
15. Wee JH (2007) Applications of proton exchange membrane fuel cell systems. Ren Sust Energy Rev 11:1720–1738
16. Costamagna P, Srinivasan S (2001) Quantum jumps in the PEMFC science and technology from the 1960s to the year 2000 Part II. Engineering, technology, development and application aspects. J Power Sources 102:253–269
17. Gamburzev S, Appleby AJ (2002) Recent progress in performance improvement of the proton exchange membrane fuel cell (PEMFC). J Power Sources 107:5–12
18. Heitner-Wirguin C (1996) Recent advances in perfluorinated ionomer membranes: structure, properties and applications. J Memb Sci 120:1–9
19. Pourcely G, Gavach C (1992) Proton conductors, solids, membranes and gels-materials and devices. Cambridge Univ Press, New York
20. Savadogo O (1998) Electrolyte membranes for fuel cell systems. J New Mater Electrochem Syst 1:47–55
21. Mauritz KA, Moore RB (2004) State of understanding of Nafion. Chem Rev 104:4535–4585
22. Noble RD, Stern SA (1995) Membrane separations technology principles and applications. Elsevier, Amsterdam
23. Thampan TM (2005) Systematic approach to design higher temperature composite PEMs. J Electrochem Soc 152:A316–A325
24. Hickner MA, Pivovar BS (2005) The chemical and structural nature of proton exchange membrane fuel cell properties. Fuel Cells 5:213–229
25. Tani J, Takagi T, Qiu J (1998) Intelligent material systems: application of functional materials. Appl Mech Rev 51:505–521
26. Kerres JA (2001) Development of ionomer membranes for fuel cells. J Memb Sci 185:3–27
27. Alberti G (2005) New preparation methods for composite membranes for medium temperature fuel cells based on precursor solutions of insoluble inorganic compounds. Fuel Cells 5: 366–374
28. Neburchilov V, Martin J, Wang H, Zhang J (2007) A review of polymer electrolyte membranes for direct methanol fuel cells. J Power Sources 169:221–238

29. Iojoiu C, Chabert F, Marechal M, Kissi NE, Guindet J, Sanchez JY (2006) From polymer chemistry to membrane elaboration. A global approach of fuel cell polymeric electrolytes. J Power Sources 153:198–209
30. Yang C, Costamagna P, Srinivasan S, Benziger J, Bocarsly AB (2001) Approaches and technical challenges to high temperature operation of proton exchange membrane fuel cells. J Power Sources 103:1–9
31. Curtin DE (2004) Advanced materials for improved PEMFC performance and life. J Power Sources 131:41–48
32. Nunes SP, Schultz J, Peinemann KV (1996) Silicone membranes with silica nanoparticles. J Mater Sci Lett 151:139–1141
33. Malhotra S, Datta R (1997) Membrane-supported nonvolatile acidic electrolytes allow higher temperature operation of proton-exchange membrane fuel cells. J Electrochem Soc 144:L23–L26
34. Antonucci PL, Aricò AS, Cretì P, Ramunni E, Antonucci V (1999) Investigation of a direct methanol fuel cell based on a composite Nafion-silica electrolyte for high temperature operation. Solid State Ion 125:431–437
35. Mauriz KA (1998) Organic-inorganic hybrid materials: perfluorinated ionomers as sol-gel polymerization templates for inorganic alkoxides. Mat Sci Eng C6:121–133
36. Adjemian KT, Srinivasan S, Benziger J, Bocarsly AB (2002) Investigation of PEMFC operation above 100°C employing perfluorosulfonic acid silicon oxide composite membranes. J Power Sources 109:356–364
37. Klein LC, Daiko Y, Aparicio M, Damay F (2005) Methods for modifying proton exchange membranes using the sol-gel process. Polymer 46:4504–4509
38. Aparicio M, Damay F, Klein LC (2003) Characterization of SiO_2–P_2O_5–ZrO_2 sol-gel/ Nafion^{TM} composite membranes. J Sol-Gel Sci Techn 26:1055–1059
39. Tazi B, Savadogo O (2001) Effect of various heteropolyacids (HPAs) on the characteristics of Nafion®–HPAS membranes and their H_2/O_2 polymer electrolyte fuel cell parameters. J New Mat Electrochem Syst 4:187–196
40. Staiti P, Freni S, Hocevar S (1999) Synthesis and characterization of proton-conducting materials containing dodecatungstophosphoric and dodecatungstosilic acid supported on silica. J Power Sources 79:250–255
41. Jung DH, Cho SY, Peck DH, Shin DR, Kim JS (2003) Preparation and performance of a Nafion®/montmorillonite nanocomposite membrane for direct methanol fuel cell. J Power Sources 118:205–211
42. Costamagna P, Yang C, Bocarsly AB, Srinivasan S (2002) Nafion®115/zirconium phosphate composite membranes for operation of PEMFCs above 100°C. Electrochim Acta 47:1023–1033
43. Yang C, Srinivasan S, Aricò AS, Cretì P, Baglio V, Antonucci V (2001) Composite Nafion/ zirconium phosphate membranes for direct methanol fuel cell operation at high temperature. Electrochem Solid-State Lett 4:A31–A34
44. Miyake N, Wainright JS, Savinell RF (2001) Evaluation of a Sol-Gel derived Nafion/Silica hybrid membrane for proton electrolyte membrane fuel cell applications: I. Proton conductivity and water content. J Electrochem Soc 148:A898–A904
45. Laberty-Rober C, Vallé TK, Pereira F, Sanchez C (2011) Design and properties of functional hybrid organic–inorganic membranes for fuel cells. Chem Soc Rev 40:961–1005
46. Pereira F, Vallé K, Belleville P, Morin A, Lambert S, Sanchez C (2008) Advanced mesostructured hybrid Silica–Nafion membranes for high-performance PEM fuel cell. Chem Mater 20:1710–1718
47. Dupuis AC (2011) Proton exchange membranes for fuel cells operated at medium temperatures: materials and experimental techniques. Prog Mater Sci 56:289–327
48. Aparicio M, Klein LC (2005) Synthesis and characterization of Nafion/60SiO_2–30P_2O_5–10ZrO_2 sol-gel composite membranes for PEMFCs. J Electrochem Soc 152:A493–A496
49. Holmberg BA, Hwang SJ, Davis ME, Yan YS (2005) Synthesis and proton conductivity of sulfonic acid functionalized zeolite BEA nanocrystals. Micro Mesoporous Mater 80:347–356

50. Alabi CA, Davis ME (2006) Proton-conducting solid electrolyte via ozonolysis of cationic ammonium organoalkoxysilane surfactant-templated MCM-41. Chem Mater 18:5634–5636
51. Fuller T, Hartnig C, Ramani V, Uchida H, Gasteiger H, Cleghorn S, Strasser P, Zawodzinski T, Jones D, Shirvanian P, Jarvi T, Zelenay P, Lamy C, Bele P (2009) Proton exchange membrane fuel cells 9. In: Hartnig C, Ramani V, Uchida H, Gasteiger H, Cleghorn S, Strasser P, Zawodzinski T, Jones D, Shirvanian P, Jarvi T, Zelenay P, Lamy C, Bele P (eds) The electrochemical society transactions. ECS, USA
52. Sahul AK, Pitchumani S, Sridhar P, Shukla AK (2009) Co-assembly of a Nafion–Mesoporous Zirconium Phosphate composite membrane for PEM fuel cells. Fuel Cells 9:139–147
53. Chai Z, Dong D, Wang C, Zhang H, Webley PA, Zhao D, Wang H (2010) Nanoporous niobium phosphate electrolyte membrane for low temperature fuel cell. J Memb Sci 356: 147–153
54. Ogoshi T, Kim KM, Chujo Y (2005) Synthesis of anionic polymer–silica hybrids by controlling pH in an aqueous solution. J Mater Chem 13:2202–2207
55. Posudievsky OY, Telbiz GM, Rossokhaty VK (2006) Effect of solvent nature on liquid-phase self-assembly of MEH-PPV/MCM-41 guest–host composites. J Mater Chem 16: 2485–2489
56. Kreuer KD, Paddison SJ, Spohr E, Schuster M (2004) Transport in proton conductors for fuel-cell applications: simulations, elementary reactions, and phenomenology. Chem Rev 104:4637–4678
57. Wainright JS, Wang JT, Weng D, Savinell RF, Litt MH (1995) Acid-doped polybenzimidazoles: a new polymer electrolyte. J Electrochem Soc 142:L121–L123
58. Kreuer KD, Fuchs A, Ise M, Spaeth M, Maier M (1998) Imidazole and pyrazole-based proton conducting polymers and liquids. Electrochim Acta 43:1281–1288
59. Yang C, Costamagna P, Srinivasan S, Benziger J, Bocarsly AB (2001) Water uptake and conductivity of composite membranes operating at reduced relative humidity. J Power Sources 103:1–9
60. Kreuer KD (1996) Proton conductivity: materials and applications. Chem Mater 8:610–641
61. Sekhon SS, Krishnan P, Singh B, Yamada K, Kim CS (2009) Morphology studies of high temperature proton conducting membranes containing hydrophilic/hydrophobic ionic liquids. Macromolecules 42:2054–2062
62. Nakamoto H, Akihiro N, Hayamizu K, Hayashi S, Hamaguchi H, Masayoshi W (2007) Proton-conducting properties of a bronsted acid-base ionic liquid and ionic melts consisting of Bis(Trifluoromethanesulfonyl)Imide and Benzimidazole for fuel cell electrolytes. J Phys Chem C 111:1541–1548
63. Sanchez C, Popall M, Julian B, Belleville P (2005) Applications of hybrid organic–inorganic nanocomposites. J Mater Chem 15:3559–3592
64. Sanchez C, Soler-Illia GJAA, Ribot F, Lalot T, Mayer CR, Cabuil V (2001) Designed hybrid organic–inorganic nanocomposites from functional nanobuilding blocks. Chem Mater 13: 3061–3085
65. Shea KJ, Loy DA, Webster O (1992) Arylsilsesquioxane gels and related materials new hybrids of organic and inorganic networks. J Am Chem Soc 114:6700–6710
66. Aparicio M, Mosa J, Durán A (2006) Hybrid organic–inorganic nanostructured membranes for high temperature proton exchange membranes fuel cells (PEMFC). J Sol-Gel Sci Tech 40:309–315
67. Arfat A, Banthia AK, Bandyopadhyay S (2008) Synthesis and characterization of polyvinyl alcohol copolymer/phosphomolybdic acid based crosslinked composite polymer electrolyte membranes. J Power Sources 179:69–80
68. Lin CW, Thangamuthu R, Yang CJ (2005) Proton-conducting membranes with high selectivity from phosphotungstic acid-doped poly(vinyl alcohol) for DMFC applications. J Membr Sci 253:23–31
69. Liu YL, Su YH, Lai JY (2004) In situ crosslinking of chitosan and formation of chitosan–silica hybrid membranes with using γ-glycidoxypropyltrimethoxysilane as a crosslinking agent. Polymer 45:6831–6837

70. Kulkarni SS, Kittur AA, Aralaguppi MI, Kariduraganavar MY (2004) Synthesis and characterization of hybrid membranes using poly(vinyl alcohol) and Tetraethylorthosilicate for the pervaporation separation of water–isopropanol mixtures. J Appl Pol Sci 94:1304–1315
71. Tezuka T, Tadanaga K, Matsuda A, Hayashi A, Tatsumisago M (2005) Utilization of glass paper as a support of protonconductive inorganic–organic hybrid membranes based on 3-glycidoxypropyltrimethoxysil. Electrochem Comm 7:245–248
72. Jacob S, Poinsignon C, Popall M (2005) Inorganic–organic hybrid protonic polymeric materials for fuel cells based on polycondensed and organically cross-linked sulfonyl- and styrene-functionalized alkoxysilagnes. Electrochim Acta 50:4022–4028
73. Park Y, Nagai M (2001) Proton exchange nanocomposite membranes based on 3-glycidoxypropyltrimethoxysilane, silicotungstic acid and α-zirconium phosphate hydrate. Solid State Ion. 145:149–160
74. Chen WF, Kuo PL (2007) Covalently cross-linked perfluorosulfonated membranes with polysiloxane framework. Macromolecules 40:1987–1994
75. Je-D Kim, Mori T, Honma I (2006) Organic–inorganic hybrid membranes for a PEMFC operation at intermediate temperatures. J Electrochem Soc 153:A508–A514
76. Schwab JJ, Lichtenhan JD (1998) Polyhedral oligomeric silsesquioxane (POSS)-based polymers. Appl Organometal Chem 12:707–713
77. Bonis CD, D'epifanio A, Di Vona ML, Mecheri B, Traversa E, Trombetta M, Licoccia S (2010) Proton conducting electrolytes based on silylated and sulfonated polyetheretherketone: synthesis and characterization. J Polym Sci 48:2178–2186 Part A: Polym Chem.
78. Aparicio M, Lecoq E, Castro Y, Durán A (2005) Proton conducting organic/inorganic sol-gel membranes produced from phenyltriethoxysilane and 3-methacryloxypropyl trimethoxysilane. J Sol-Gel Sci Tech 34:233–239
79. Aparicio M, Castro Y, Durán A (2005) Synthesis and characterisation of proton conducting styrene-co-methacrylate/silica sol-gel membranes containing tungstophosphoric acid. Solid State Ion 176:333–340
80. Mosa J, Durán A, Aparicio M (2010) Epoxy-polystyrene-silica sol-gel membranes with high proton conductivity at low relative humidity by combination of sulfonation and tungstophosphoric acid doping. J Memb Sci 361:135–142
81. Smith JJ, Zharov I (2009) Preparation and proton conductivity of sulfonated polymer-modified sintered and self-assembled silica colloidal crystals. Chem Mater 21:2013–2019
82. Sel O, Soulès A, Améduri B, Boutevin B, Laberty-Robert Ch, Gebel G, Sanchez C (2010) Original fuel-cell membranes from crosslinked terpolymers via a "sol–gel" strategy. Adv Funct Mater 20:1090–1098
83. Kim JD, Mori T, Honma I (2004) Proton conducting polydimethylsiloxane/metal oxide hybrid membranes added with phosphotungstic acid(II). Electrochim Acta 49:3429–3433
84. Sel O, Laberty-Robert Ch, Azais T, Sanchez C (2009) Designing meso- and macropore architectures in hybrid organic–inorganic membranes by combining surfactant and breath figure templating (BFT). Phys Chem Chem Phys 11:3733–3741
85. Choi J, Lee KM, Wycisk R, Pintauro PN, Mather PT (2008) Nanofiber network ion-exchange membranes. Macromolecules 41:4569–4572
86. Choi J, Lee KM, Wycisk R, Pintauro PN, Mather PT (2008) Membranes for PEM fuel cells. J Electrochem Soc 157:B914–B919
87. Dokko K, Sugaya J, Nakano H, Yasukawa T, Matsue T, Kanamura K (2007) Sol–gel fabrication of lithium-ion microarray battery. Electrochem Commun 9:857–862
88. Goodenough JB, Kim Y (2010) Challenges for rechargeable Li batteries. Chem Mater 22:587–603
89. Liu Y, Lee JY, Hong L (2003) Morphology, crystallinity, and electrochemical properties of in situ formed poly(ethylene oxide)/TiO_2 nanocomposite polymer electrolytes. J Appl Polym Sci 89:2815–2822
90. Walls HJ, Zhou J, Yerian JA, Fedkiw PS, Khan SA, Stowe MK, Baker GL (2000) Fumed silica-based composite polymer electrolytes: synthesis, rheology, and electrochemistry. J Power Sources 89:156–162

91. Kim DW (1998) Electrochemical characteristics of a carbon electrode with gel polymer electrolyte for lithium-ion polymer batteries. J Power Sources 76:175–179
92. Lee KH, Lee YG, Park JK, Seung DY (2000) Effect of silica on the electrochemical characteristics of the plasticized polymer electrolytes based on the P(AN-co-MMA) copolymer. Solid State Ion 133:257–263
93. Walcarius A (2001) Electrochemical applications of silica-based organic-inorganic hybrid materials. Chem Mater 13:3351–3372
94. Skaarup S, West K, Zachau-Christiansen B, Popall M, Kappel J, Kron J, Eichinger G, Semrau G (1998) Towards solid state lithium batteries based on ORMOCER electrolytes. Electrochim Acta 43:1589–1592
95. Popall M, Duband H (1992) Inorganic-organic copolymers as solid state Li^+ electrolytes. Electrochim Acta 37:1593–1597
96. Popall M, Andrei M, Kappel J, Kron J, Olma K, Olsowski B (1998) ORMOCERs as inorganic–organic electrolytes for new solid state lithium batteries and supercapacitors. Electrochim Acta 43:1155–1161
97. Popall M, Buestrich R, Semrau G, Eichinger G, Andrei M, Parker WO, Skaarup S, West K (2001) New polymer lithium secondary batteries based on ORMOCER® electrolytes inorganic–organic polymers. Electrochim Acta 46:1499–1508
98. Popall M, Du XM (1995) Inorganic–organic copolymers as solid state ionic conductors with grafted anions. Electrochim Acta 40:2305–2308
99. Dahmouche K, Atik M, Mello NC, Bonagamba TJ, Panepucci H, Aegerter MA, Judeinstein P (1997) Investigation of new ion-conducting ORMOLYTES: structure and properties. J Sol-Gel Sci Technol 8:711–715
100. Dahmouche K, Atik M, Mello NC, Bonagamba TJ, Panepucci H, Judeinstein P, Aegerter MA (1998) New Li^+ ion-conducting ORMOLYTEs. Sol Energy Mater Sol Cells 54:1–8
101. Nishio K, Okubo K, Watanabe Y, Tsuchiya T (2000) Structural analysis and properties of organic-inorganic hybrid ionic conductor prepared by sol-gel process. J Sol-Gel Sci Technol 19:187–191
102. Liu Y, Lee JY, Hong L (2004) In situ preparation of poly(ethylene oxide)–SiO_2 composite polymer electrolytes. J Power Sources 129:303–311
103. Souza FL, Bueno PR, Longo E, Leite ER (2004) Sol–gel nonhydrolytic synthesis of a hybrid organic–inorganic electrolyte for application in lithium-ion devices. Solid State Ion 166:83–88
104. Lee JK, Lee YJ, Chae WS, Sung YM (2006) Enhanced ionic conductivity in PEO-$LiClO_4$ hybrid electrolytes by structural modification. J Electroceram 17:941–944
105. Tigelaar DM, Meador MAB, Kinder JD, Bennett WR (2006) New APTES cross-linked polymers from poly(ethylene oxide)s and cyanuric chloride for lithium batteries. Macromolecules 39:120–127
106. Tigelaar DM, Meador MAB, Bennett WR (2007) Composite electrolytes for lithium batteries: ionic liquids in APTES cross-linked polymers. Macromolecules 40:4159–4164
107. Derrien G, Hassoun J, Sacchetti S, Panero S (2009) Nanocomposite PEO-based polymer electrolyte using a highly porous, super acid zirconia filler. Solid State Ion 180:1267–1271

Chapter 7
Giant Dielectric Constant Materials and Their Applications

Marcelo O. Orlandi, Miguel A. Ramirez, Cesar R. Foschini, Anderson A. Felix and José A. Varela

Abstract There is a constant need in the modern electronic industry for capacitors with high capacity per volume in order to use in many applications such as memories devices, energy storage, microwave filters, among others. The synthesis, characterization and study of materials with a very high or giant dielectric constant are in particular important and have been studied by many investigators. Then the search for high dielectric materials has been driven, which must requires some special properties such as to keep dielectric constant almost independent of temperature and frequency and low dielectric loss. In recent years many material systems have been searched, including perovskite and others, aiming to achieve a reliable material for practical applications. In this chapter we review many classes of materials including titanates, nickeletes, cuprates, multiferroics and composites such as cermets, ceramic/polymer and ceramic/ceramic. To the end of this work is presented the most important models in the literature to explain mechanism for giant dielectric constant. The main task is to understand the mechanism that controls materials properties in order to synthesize an optimized material that maximizes those properties by using sol-gel or other chemical-like synthesis for many applications in the electronic industries.

Keywords Capacitors · Giant dielectric materials · Perovskite-based materials · Ferroelectric · Multiferroics

M. O. Orlandi · M. A. Ramirez · C. R. Foschini · A. A. Felix · J. A. Varela (✉)
Chemistry Institute, University of São Paulo State—UNESP,
R. Prof. Francsico Degni 55, Araraquara, SP 14801-970, Brazil
e-mail: varela@iq.unesp.br

M. Aparicio et al. (eds.), *Sol-Gel Processing for Conventional and Alternative Energy*, Advances in Sol-Gel Derived Materials and Technologies,
DOI: 10.1007/978-1-4614-1957-0_7,

7.1 Introduction

The search for high dielectric constant materials has been driven by the continuing demand for electronic device miniaturization as a capacitor component once such a material enables the reduction of the circuit size and can realize tera-bit density static/dynamic random access memory [1]. However, in the most promising materials, e.g. ferroelectric oxides such as $BaTiO_3$ or relaxors such as $(Bi,Sr)TiO_3$ [2], a temperature dependence of the permittivity is found which can lead to failure of microelectronic devices based on these materials under a variety of conditions, i.e. the application is often limited.

Giant (or colossal) dielectric ceramics with good thermal stability and Ba/Pb-free have recently attracted much attention due to their potential applications in microelectronics. Oxides with the perovskite structure are well known for their ability to produce high dielectric constants. However, dielectric constants higher than 1,000 have always been associated with ferroelectric or relaxor properties. In both cases, the dielectric constant presents a peak as a function of temperature, and this resulting temperature dependence is undesirable for many applications.

The $ACu_3Ti_4O_{12}$ family of compounds has been known since 1967 [3]. This family was expanded and accurate structures were determined in 1979 [4]. Subramanian et al. [5] reported on a class of compounds with a perovskite-related structure which have dielectric properties very different from those of ferroelectrics or relaxors. One advantage of these compounds is that their high dielectric constants show only a small dependence on temperature. From all compounds studied by Subramanian et al. [5] the CCTO (A = Ca) shows the higher dielectric constant (10,000 for polycrystalline ceramic and over 100,000 for single crystals) becoming an interesting material to be deeply analyzed.

Several series of dielectric materials have been investigated looking for giant dielectric properties, including (M, N)-doped NiO (M = Li, Na, K and N = Ti, Al, Si, Ta) ($k \sim 10^4$–10^5), CuO ($k \sim 10^4$), $A_{2/3}Cu_3Ti_4O_{12}$ (A = trivalent rare-earth or Bi), Fe-containing complex perovskite $A(Fe_{1/2}B_{1/2})O_3$ (A = Ba, Sr, Ca and B = Nb, Ta, Sb) and $BaTi_{1-x}(Ni_{1/2}W_{1/2})_xO_3$. Other interesting materials are composite ceramic–metal (cermet). In particular, the singular electrical properties of these materials, especially at concentrations close to the percolation threshold, which can only be analyzed by percolation theory [6] have been widely studied in the past twenty years, but these properties are not yet totally understood. Moreover, at this concentration, the composite undergoes a metal–insulator transition, where the value of electrical conductivity changes by several orders of magnitude [7].

In this chapter we will review some materials which can present giant dielectric constant and the main advantages and drawbacks of each material are reported. Once applications of such materials are related to capacitive devices, concepts about capacitance and the desire of obtaining materials with high (or giant) dielectric constant is discussed. Because $CaCu_3Ti_4O_{12}$ (CCTO) material presents exceptional high dielectric constant, this material will be more explored and the existing models trying to explain their properties are presented.

7.2 Revisiting Capacitance

Before discussing all the advantages and disadvantages of using these high dielectric compounds in practical applications like capacitors and/or memories, it is important to define several capacitor parameters.

The definition of capacitor is an array of two isolated conductive surfaces (electrodes) which allows the energy storage in the electric field. The capacitance of a device depends on its geometry and the presence (or not) of a dielectric (with dielectric constant k or permittivity ε_m) between the electrodes.[1] The general way to calculate the capacitance of a device is:

$$C = \frac{dq}{dV}; \tag{7.1}$$

where dq is an infinitesimal element of charge and dV is an infinitesimal increment of voltage. Based on Eq. 7.1 the SI unit of capacitance is C/V, which is called Farad (F) in honor of Michael Faraday. So, high capacitance materials are those which allow high charge storage by applying a small voltage.

The dielectric constant of a material is related to their polarizability, mainly due to the existence of permanent electrical dipoles. From the physical point of view, it is the ability of electrical dipoles of the material to be orientated in such a way to create an electric field in opposition to an external applied electric field.

Capacitors have a lot of practical use in modern electronic, and all home electronic appliances must have capacitors inside. Besides, depending on their characteristics, capacitors geometry devices can also be used as random access memory (RAM) devices. The simplest arrange of a capacitor is two parallel electrodes of area A separated by a distance d. If a material with permittivity ε_m (ε_m = k ε_0) is inserted between the electrodes, the capacitance can be written as:

$$C = \frac{k\varepsilon_o A}{d} \tag{7.2}$$

where ε_0 is the free space permittivity; $\varepsilon_0 = 8.854 \times 10^{-12}\ C^2N^{-1}m^{-2}$.

Based on Eq. 7.2, it is possible to increase the capacitance of a device in several ways, such as:

- increasing the electrode area;
- decreasing the distance between the electrodes and
- inserting a dielectric material with high permittivity (ε_m = k ε_0).

---

[1] k and ε_m are different physical quantities but they are related by $k = \varepsilon_m/\varepsilon_0$. k is called dielectric constant and it is a dimensionless quantity while ε_m is called permittivity and has a dimension of $C^2N^{-1}m^{-2}$, which is also the dimension of ε_0. k and ε_m are directly proportional, and in this chapter we will use k and ε_m in similar connotation. However, it is important for the reader to understand the difference between both quantities, mainly quantitatively.

As demanded by electronic industry, from one generation to the next one, capacitors should keep its capacitance in a reduced geometry. So, increasing the electrode area, and consequently the device size, is not a good idea. Besides, decreasing the spacing between electrodes limits the maximum voltage which can be applied in the capacitor by the dielectric strength of vacuum (or material). Obviously, it limits the capacitor use. So, the best way to produce a small device with high capacitance is using materials with higher dielectric constant. In a simplest way, this is the reason why high (or giant) dielectric constant materials are so interesting from the technological point of view. Besides, most of them can also be used as memory devices, which allow a wide field of applications.

7.3 Giant Dielectric Constant (GDC) Materials

7.3.1 Composites

7.3.1.1 General View

In recent years, there has been an increasing interest on high dielectric constant flexible particulate composites (0–3 composites) made up of a ferroelectric ceramic and a polymer for high-density energy storage and capacitor applications [8]. However, invariably the dielectric constant of such polymer-based 0–3 composites is rather low (about 50) because of the lower dielectric constant of the matrix (usually below 10) [8]. Besides, increasing demand for high-density circuits in electronics has greatly accelerated the miniaturization and integration of chip electronic components. This requires the multifunctional components to work as both capacitors and inductors. These composite materials showing both high inductive and capacitive properties have been identified to be an enabling solution to fabricate miniaturized filter and antenna, and electromagnetic interference devices among other applications.

Some works [9–11] focused on ferromagnetic-ferroelectric ceramic composites, and studied the dielectric relaxation and the associated high values of dielectric constant. Yang et al. [12] presented a study on a novel ferroelectric-ferromagnetic ceramic composite with the effectively combined properties of giant dielectric constant and high permeability. The composites $Ba_{0.6}Sr_{0.4}TiO_3/Ni_{0.8}Zn_{0.2}Fe_2O_4$ (BST/NZO) were synthesized via the conventional solid-state reaction method. For the 10%BST/90%NZO composite, the dielectric constant and permeability in low frequency range are about 150,000 and 29, respectively.

Semi-crystalline polymer filled with conductive particles composite usually exhibit inimitable important insulator-conductor transitions, and significant temperature-activated switching feature (positive temperature coefficient effect, PTC; or a negative temperature coefficient effect, NTC) when the filler content is close to the critical volume fraction referred as the "percolation threshold" [13].

Research and development activities in such composites have brought successful industrial applications in self-regulating heaters, over-current and over-temperature protection devices and dielectric materials applied in capacitors [14, 15] by employing a semicrystalline polyvinylidene fluoride (PVDF) with high melting temperature ($T_m \sim 165°C$) as polymer host, and nickel (Ni) as the conductive filler. Prior to use, the surface of Ni particles was embellished with 1.0 wt% γ-aminopropyl triethoxysilane (KH550) coupling agent to improve the dispersion and affinity of Ni particles in the PVDF host. The dielectric constant of all composites decreases by increasing frequency and the dielectric constant increases by increasing the filler concentration. The dielectric constant for this composite is $\sim$90 at the lower frequencies [15].

Besides particles filled composites, the dielectric properties of carbon nanotubes (CNT)/polymer composites have also been investigated, and it was found that the percolation threshold in these materials can differ depending on the experimental conditions [16]. These results have not only been attributed to weak interactions between the CNTs and the polymer matrix but also they reflect poor CNT dispersal in the composites. To overcome these difficulties, several methods have been developed to improve CNT dispersions in host polymers. For example, CNTs can be dispersed in certain polymer solutions via ultrasonication or by using surfactants [17].

Recently, Maensiri et al. [18] reported high values of dielectric constant ($\sim 10^5$) observed in polycrystalline $CaCu_3Ti_4O_{12}$ ceramics reinforced with 10 mol % of $Li_{0.3}Ti_{0.02}Ni_{0.68}O$ (LTNO) nanoparticles. The dielectric behavior of these composites exhibits Debye-like relaxation which can be explained based on the Maxwell–Wagner model and the dielectric dispersions of the composites arise from the internal barrier layer capacitor (IBLC) mechanism (see Sect. 7.4).

7.3.1.2 Cermets

In general, formation of solid solutions, texture engineering and mixing rules have been used to obtain high ε_m materials. Polymer metal composites [19] and ceramic–metal composites [20] are the two extensively studied systems with high relative permittivity. In a ceramic–metal composite (cermet), if the volume fraction of the conductive powder is lower than a critical value it behaves as an insulator. As the volume fraction of the conductive powder reaches the critical value its conductivity and relative permittivity increases by several orders of magnitude [7]. This critical value is referred to as the percolation threshold at which a conductive network is formed to span the entire cross-section of the body of material. The conducting powders used in cermet composites are Pt, Pd, Ag, Al, Cu, Ni, etc. [21]. The Pt and Pd are too expensive and Al, Cu and Ni have oxidizing problem while sintering at higher temperatures. As a result, these materials are not suitable for preparing the cermets for practical applications. Nickel-loaded composites [20] provide high *k* materials and are consistent with the percolation law, but these must be sintered at high temperature in inert atmosphere. Silver-loaded composites do not have such drawbacks and show excellent permittivity near the percolation threshold. However, it is necessary to lower the sintering temperature of composites to prevent silver losses. In a multilayer structure, it is necessary to lower the sintering temperature of dielectrics below 950°C in order to co-fire with low melting and highly conducting electrodes. Recently, Pecharroman and Moya [20] reported that ferroelectric $BaTiO_3$ with high values of ε_m was obtained through the addition of Ni particles to utilize a percolative phenomenon whereby the electric field inside the matrix was enhanced microscopically [20]. This result suggests that a complex material composed of paraelectric ceramics and metal powder is useful for fabrication of paraelectric ceramics with high permittivity.

Cho et al. [22] reported dielectric properties of $SrTiO_3$ (ST)-Pt cermet composites as a function of Pt contents and compared the variation of ε_m in measurements with those predicted theoretically using the normalized percolation theory. Compositions with 5–30 vol% of Pt were prepared and pellets were sintered at 1,300°C for 2 h in air. The dielectric constant increased from 316 to 2,150 at 1 MHz by increasing the amount of Pt from 0 to 27 vol%. In particular, authors observed a considerable increase of ε_m in the narrow range from 26 to 27 vol% Pt. The increase in material permittivity was attributed to the effective electric field developed between the dispersed Pt particles in the matrix.

George et al. [23] prepared $Ba_{6\text{-}3x}Sm_{8+2x}Ti_{18}O_{54}$ (BST) ($x = 2/3$) + 10 wt% barium copper borate (BCB) + Ag composites by solid-state ceramic route using a sintering temperature less than the melting temperature of silver (961°C). The dielectric properties of these composites increased with silver addition in agreement with power law [23]. Addition of 0.14 volume fraction of silver in BST + 10 wt% BCB increases both the k value to 450 and the conductivity to the order of 10^{-5} S/cm. A 0.15 volume fraction of silver addition in BST + 10wt% BCB increases the conductivity to 10^{-1} S/cm along with k in the order of 10^5.

Kumar et al. [24] have prepared CCTO thin films with silver nanoparticles using the sol-gel method. Films of thickness ~0.6 μm were deposited on platinized Si (111) substrates by spin coating (25 coatings) at 5,000 rpm for 30 s. The dielectric constant is observed to increase by increasing Ag concentration and it was attributed to the polarization at particle-dielectric interfaces. The maximum k value (~1,070 at 100 kHz) for intermediate Ag concentration (Ag = 0.06) caused by dispersing the conducting particles in dielectric matrix is consistent with the percolation law.

The cermets are interesting materials as the critical concentration of metal particles reaches the percolation, as the dielectric constant is maximized. However, the dielectric loss is also high which limits its application in dielectric systems.

7.3.2 *Nickelates*

Nickel oxide (NiO) has been widely studied and used in many applications such as optical fibers, electrochromic materials and temperature sensors [25]. Pure phase

NiO is insulator (electrical conductivity less than 10^{-13} S/cm at room temperature), but the addition of Li atoms forms the compound $Li_xNi_{x-1}O$ and its conductivity can be increased to 1 S/cm [26]. The compounds $Li_xNi_{x-1}O$-type present two different structures depending on the Li cations fraction. $Li_xNi_{x-1}O$ with $x > 0.3$ is hexagonal and used as the electrode for Li batteries, whereas $Li_xNi_{x-1}O$ with $x < 0.3$ is cubic and their electrical transport properties have been studied to explicate the conduction mechanism in these compounds [27].

However, more recently researchers have reported that the dielectric properties of $Li_xNi_{x-1}O$ can be enhanced by addition of alkali or transition metal cations such as Ti, Al, Si and Zr [28–34]. These compounds have attracted much attention due to observation of a giant dielectric constant mainly because it presents a non-perovskite structure and non-ferroelectric properties. Nan et al. have reported a dielectric constant between 10^4 and 10^5 at low frequencies and room temperature in the compounds (Li, Ti)-doped NiO [28–31], (Li, Al)-doped NiO [32] and (Li, Si)-doped NiO [33] synthesized by sol-gel method. In these compounds, it was noted that by decreasing the frequency the tan δ peak is shifted to a lower temperature, which is related to a thermally activated relaxation process. At high temperatures, the increase in tan δ is mainly due to the dc conductivity contribution which becomes more apparent by increasing the temperature and decreasing the frequency. The dielectric loss in these materials at a fixed frequency also shifts to a lower temperature by increasing Li concentration or decreasing Ti concentration. The same behavior is observed for (Li, Al)-doped NiO, (Li, Si)-doped NiO and (Li, Zr)-doped NiO [34] compounds, i.e., the dielectric properties of these compounds are dependent on the adjusting of Li concentration and of the cations Ti, Al, Si and Zr on the NiO. These results have shown that these compounds have semiconducting grains (i.e., Li-doped NiO), while the grain boundaries are cation-rich forming insulating boundaries. This behavior has been explained by internal barrier layer capacitor (IBLC) model (See Sect. 7.4).

$Li_xTi_yNi_{1-x-y}O$ compounds can display high values of dielectric constant at low frequencies but it decreases drastically by increasing the frequency. This characteristic is not required for applications in electronic devices such as capacitors. Recently, giant dielectric responses at high frequency was observed in the $A_xSr_yNiO_4$ (A: La, Nd, Sm) ceramics [35–38]. The valuable characteristic of these ceramics is that high dielectric constant is stable over a wide temperature and frequency range. The dielectric constant is between 10^4 and 10^5 for frequencies up to 5 MHz in temperatures above 200 K. The dielectric constant in $Sm_xSr_yNiO_4$ (with $x > 1.5$ and $y < 0.5$) compounds increases by increasing strontium content while $A_{1.5}Sr_{0.5}NiO_4$ compounds (with A: La, Nd, Sm) have smaller dielectric constants due to larger ionic radius of rare-earth cations. The best dielectric properties are obtained in the $Sm_{1.5}Sr_{0.5}NiO_4$ ceramics with values of dielectric constant and loss of about 100.000 and 0.1, respectively, up to 5 MHz in temperatures above 200 K [39]. In comparison with other materials such as titanates, the main disadvantage of these ceramics is the high dielectric losses (tan $\delta \sim 10$) at low frequencies. On other hand, excellent dielectric properties were observed in $La_{1.875}Sr_{0.125}NiO_4$ single crystals [40]. The advantage of

$La_{1.875}Sr_{0.125}NiO_4$ single crystals is its high dielectric constant at room temperature (about 10^4) into the gigahertz range and it coupled with its low dielectric loss (tan $\delta \sim 0.1$) makes this material very attractive for applications in microwave devices, among others.

7.3.3 Cuprates

Copper oxide (CuO) has been intensely studied because of its relative low cost and because it is a material of easy processing in the several forms, such as nanostructures and single crystals, among others. Moreover, copper oxide is commercially available on a large scale. This material is of great interest due to its potential for applications in many fields such as high-critical-temperature superconductors, photovoltaic materials, gas sensing and catalysis [41].

Recently, some studies have reported giant dielectric properties on CuO polycrystalline ceramics. Sarkar et al. showed that CuO ceramics have dielectric constant of about 10^4 that is temperature-independent in the range between 230 and 400 K for frequencies up to kHz [42–44]. Some works have reported that the purity of starting CuO powder for fabrication of ceramics can also affect the dielectric response in the CuO materials. Thongbai et al. have reported that the dielectric constant in high purity (99.999%) CuO ceramics increase by increasing the grain size due to the heterogeneous electrical properties between the grains [45, 46]. This behavior was also observed in CCTO and LNTO ceramics. However, like for the case of the CCTO and LNTO polycrystalline ceramics, the high dielectric response observed in CuO ceramics was attributed to IBLC mechanism. More recently, Li et al. have reported that an extrinsic effect associated with non-ohmic electrode contacts can also influence the dielectric response in CuO ceramics [47].

Up to now, few studies have been reported on the huge dielectric properties in the CuO ceramics and the origin of this phenomenon. Thus, mechanisms responsible for dielectric properties in the CuO are not yet a closed question and it is an interesting and open field.

7.3.4 Multiferroics

By definition, a multiferroic material exhibits two or three of these properties: ferroelectricity, ferromagnetism and ferroelasticity simultaneously in the same phase [48]. The simultaneous existence of ferroelectric and ferromagnetic properties in the same material have attracted scientific and technological interest in the last decade due to the potential application in magnetic data storage, sensors, non-volatile memories, actuators and micromechanical applications [49, 50].

Ferroelectromagnetic materials with similar ordering temperatures offer a unique possibility to manipulate the magnetic and polar states by applying electric and magnetic fields, respectively, which makes them attractive candidates for different applications, especially when coupled magnetic and polar states exist at room temperature.

Many ferromagnetic-ferroelastic (magnetostrictive) and ferroelectric-ferroelastic (piezoelectric) materials are known, but very few multiferroic single-phase materials exist in nature [51]. This derives from the fact that the two properties are often chemically incompatible; e.g., in oxide perovskites the existence of transition metal *d*-electrons are essential for ferromagnetism, but work against the tendency for off-center displacement of cations which cause ferroelectricity [52].

Alternative mechanisms to obtain both electrical and magnetic ordering in a single-phase material have included the construction of artificial multiferroic thin film sandwiches combining a ferroelectric and a ferromagnetic material [53].

Smolenskii and Chupis [54] divided them into four major multiferroic groups. Among them, compounds with perovskite crystal structure ABO_3, in which the B position is fully or partially occupied by magnetic ions. This group is the largest family, and the most important member of this family is the bismuth ferrite $BiFeO_3$ (BFO), which is considered as one of the most promising multiferroic materials due to high ferroelectric and ferromagnetic Curie temperatures. This class also involves complex perovskite-like layered compounds with a general formula $A_{x+1}B_xO_{3x+3}$ where A = Bi, Ca, Sr, Ba, and B = Fe, Ti, Mo, Ni. Hexagonal-structure rare-earth manganites $RMnO_3$, where R = Ho, Er, Tm, Yb, Lu, as well as Y and Sc, which are good ferroelectrics with anti-ferromagnetic or weak ferromagnetic properties also belong to this group. However, the most studied member of this group is the $YMnO_3$.

The dielectric properties of BFO in various forms have also been widely studied in single crystalline [55], ceramic [56–58] and thin film [59] forms. It is a challenging task to synthesize highly resistive BFO, as the ceramic powders and thin film methods usually result in secondary phases such as Fe_2O_3 [56, 60], Bi_2O_3 [56] and $Bi_2Fe_4O_9$ [61]. Additionally, the presence of slight amounts of non-stoichiometric oxygen in BFO often leads to the presence of Fe^{2+} and lattice defects thereby resulting in an unwanted increase in the conductivity at room temperature due to a hopping mechanism from Fe^{2+} to Fe^{3+}, significant Maxwell–Wagner-type contributions in the dielectric constant, weak ferromagnetism, etc. [62]. The high leakage current in bulk BFO synthesized by solid solution methods also leads to serious dielectric breakdown at fields >100 kV/cm which makes them unsuitable for many practical devices [63]. To overcome this problem and to get higher resistive material, various approaches have been attempted, such as doping with various rare-earth and transition metal ions and a solid solution of BFO with other ABO_3 perovskite materials.

Orthorhombic crystals of manganites RMn_2O_5 are ferroelectromagnetics in which long-range magnetic and ferroelectric orders are realized at $T < 30$–40 K. These materials are attracting considerable attention from researchers because of the strong coupling between magnetic and ferroelectric states resulting from the nearly equal magnetic and ferroelectric ordering temperatures [64–67].

Golovenchits et al. [68] reported on magnetic and dielectric properties of single crystal $Eu_{0.8}Ce_{0.2}Mn_2O_5$ (ECMO) in which the state with high values of dielectric constant ($k \approx 10^4$) has been observed above 185 K. A strong coupling between magnetic and polar states in ferroelectromagnetics semiconductors can arise at sufficiently high temperatures if charge carriers participate in formation of both the magnetic and ferroelectric long-range orders. Double exchange between Mn^{3+} and Mn^{4+} ions that gives rise to a ferromagnetic orientation of spins of these ions plays an important role in ECMO at all temperatures. Together with the Jahn–Teller and Coulomb interactions, it forms barriers at the superlattices and the superstructure layer boundaries.

Complex manganites $La_{1-x}Sr_xMnO_3$ (LSM-x) have been intensively investigated in the last decade and the LSM-x crystals became the model object for investigations of the giant magnetoresistance [69–74]. The recent observation of large magnetoelectric and magnetocapacitive effects in pure manganites with the formula $RMnO_3$ (when R—rare-earth ion) arises new interest to these materials [75, 76]. The contact capacitance can lead to apparently large values of dielectric permittivity [77–80] and thus great care is required to distinguish the true response of the sample.

The temperature dependence of the dielectric and magnetic properties of the $La_{1-x}Sr_xMnO_3$ single crystals were investigated by Mamin et al. [81] in a wide range of temperature and external magnetic fields. The feature properties are explained by the nanoscale dynamic inhomogeneties appearing at the prepercolation regime of the charge ordering. The external magnetic field strongly affects the charge separation contrast and the giant magnetocapacitance effect is observed. These results give more evidences of the existence of the charge and phase separation in manganites.

Toward this, many systems such as $Pr_{(1-x)}Ca_xMnO_3$ [82] and manganese-based oxides, such as $NaMn_7O_{12}$ [83], $CaMn_7O_{12}$ [84], $La_{0.65}Sr_{0.35}MnO_3$ [85] and $Ba_4YMn_3O_{11.5}$ [86] have been studied. The reports pertaining to these above-mentioned materials highlight the grain boundary and the sample-electrode interfacial polarization contribution to their GDC behavior. There is a recent report in the literature on the structural, magnetic and polarization versus electric field behavior of a double perovskite, Sr_2TiMnO_6 (STMO) [87]. These are the two different transition metal cations in octahedral coordination situated on the crystallographically distinct sites. STMO is a centrosymmetric compound with the space group I4/m. The ferromagnetic to paramagnetic transition temperature (T_c) of STMO is reported to be around 40 K.

Preethi et al. [88] studied the STMO system and after X-ray photoelectron spectroscopy measurements, on sintered samples, the results demonstrated that manganese is present only in the Mn^{4+} state and similarly titanium was also found to be in the Ti^{4+} state. The authors ruled out the possibility of having any Jahn–Teller distortions due to the cations in the lattice. The room temperature (300 K) dielectric constant attains a value as high as 70,000 at 1 kHz and since the intrinsic dielectric constant of STMO was found to be ~98, the origin of abnormally high dielectric constant values (~10^5) was rationalized using the Maxwell–Wagner (M–W) relaxation mechanism involving grain boundary and electrode contributions. The dielectric constant was found to vary significantly with the thickness of the samples and the nature of the contacts at low frequencies owing to its extrinsic nature.

7.3.5 Titanates

7.3.5.1 Hexagonal-$BaTiO_3$, Ultrafine Grain Size $BaTiO_{3-x}$ and $Ba_{0.95}La_{0.05}TiO_{3-x}$

Considerable interest has been focused on several non-ferroelectric materials exhibiting GDCs such as $CaCu_3Ti_4O_{12}$ and hexagonal $BaTiO_3$ (h-$BaTiO_3$) [89]. The common features of such materials are a high value of dielectric constant (up to 10^5 at room temperature) and a weak temperature dependence of k over a wide temperature range (100–500 K). Recently, giant permittivity values were also reported in hexagonal barium titanate (h-$BaTiO_3$) single crystals [89]. A spheroid oxygen-deficient h-$BaTiO_3$ single crystal with a diameter of 2 mm was grown by containerless processing using a pressurized electrostatic levitation furnace. High values of permittivity of about 10^5 measured on the oxygen-deficient materials were explained by a Maxwell–Wagner interfacial polarization effect due to the presence of interfacial boundaries consisting of crystal defects such as screw dislocations. Nevertheless, the internal interfaces as well as the nature of the polarization carriers in the h-$BaTiO_3$ single crystals were not fully identified. M–W polarization can arise from the inhomogeneity of microstructure and composition due to the presence of internal boundary layers or the formation of external depletion layers at the interface of the electrode/sample surface caused by the Schottky barrier effect [89].

From the viewpoint of practical application, only the GDCs from the internal contribution of the boundary layers are desired. However, a certain external contribution from the depletion layers usually exists in the measurement results. Hence, to confirm whether the materials exhibiting GDCs are indeed adequate for commercial application, it is necessary to determine the exact internal contribution of the boundary layers. Yu et al. [89] studied a h-BaTiO3 single crystal in a frequency range from 100 Hz to 3 GHz in order to determine the internal and external contributions, since three parallel RC elements should be related to the responses of the interfacial layer, the depletion layer and the semiconducting bulk crystal. The fitting results revealed that the interfacial boundary layers and depletion layers exhibited the same order of capacitance, 1 nF at RT which is three orders of magnitude higher than that of the semiconducting bulk crystal.

Fritsch et al. [90] reported colossal permittivity values achieved in dense ceramics displaying ultrafine grain size ranging from 70 to 300 nm. Dielectric constant values of 10^6 at 1 kHz ($0.1 < \tan\delta < 0.7$) were obtained for $Ba_{0.95}La_{0.05}TiO_{3-x}$ ceramics and k values of about 10^5 were achieved in $BaTiO_3$ micrometric grain-sized material at room temperature in which preparation included incorporation of metallic layers in a complex multistep process [90].

The achievement of giant permittivity values in this material was ascribed to interfacial polarization phenomena. According to the brick layer model, the effective permittivity ε_{eff} of the microstructure can be expressed as $\varepsilon_{eff} = \varepsilon_{gb}\, t_g/t_{gb}$, where ε_{gb} is the grain boundary permittivity and t_g and t_{gb} are the thickness of the grain and grain boundary, respectively [6, 91]. Thus the IBLC mechanism is usually associated with enhanced grain size. In this context, a decrease of the average grain size should not favor IBLC effect since dielectric permittivity in ultrafine ceramics (<300 nm) such as $BaTiO_{3-x}$ and $Ba_{0.95}La_{0.05}TiO_{3-x}$ materials are relatively low. Dense dielectrics ceramics were prepared from nanoscale chemically homogeneous powders by Spark Plasma Sintering $BaTiO_{3-x}$ at 1,323 K in oxygen-deficient atmosphere. Electron probe microanalysis (EPMA) showed a surprisingly large x value ($x = 0.37$). A giant permittivity value $\sim$ 000000.8x10^6 was determined for this material at 300 K and 1 kHz with a dissipation factor, tanδ, of 0.56. Doping by La^{3+} cation was previously reported to increase permittivity values [92]. In comparison to the $BaTiO_3$ nanomaterial, complex impedance spectrum of the $Ba_{0.95}La_{0.05}TiO_{3-x}$ nanomaterial revealed a significant decrease in the bulk and grain boundary resistances that could correspond to either La^{3+} doping or to the higher sintering temperature.

Strontium titanate displays a series of interesting electrical properties, including insulating, semiconducting, metallic, and superconducting behavior. Ang and Zu [93] reported a doped $SrTiO_3$ polycrystalline ceramic exhibiting a giant dielectric constant (49,600 at 10 kHz and 24,700 at 1 MHz) and exceptional large capacitance-temperature sensitivity at very low temperatures. These ceramics were prepared by the solid-state reaction method, sintered in air, and annealed in a reduced atmosphere (2% H_2 + 98% argon) with the composition $(Sr_{1-1.5x}Bi_x)$-$TiO_{3-\delta}$ (x = 0.05).

7.3.5.2 $CaCu_3Ti_4O_{12}$ (CCTO)

Values of $k > 1{,}000$ are generally found for materials with ferroelectric or relaxor behavior which possesses a dipole moment in the absence of an applied external electric field or the ferroelectric response lower at high applied electric fields or low temperatures as in the case of relaxor materials. CCTO is not ferroelectric and its relaxor behavior was also discarded because relaxor materials usually present solid solutions (for example, PMN, Pb $(Mg_{1/3}Nb_{2/3})$ O_3) in which a sub-lattice is the disorder source that can be identified by XRD through peak dislocations, reduction of the intensities or a diffuse baseline. Neither XRD nor neutron diffraction results presented any structural change between 35 and 600 K in both CCTO polycrystalline ceramics and CCTO single crystals. It means CCTO cannot be classified as ferroelectric or relaxor.

Until 2000, the most extensively investigated dielectric materials in the literature were oxides based on $BaTiO_3$ presenting high dielectric constant (HDC). However, a detailed and careful processing must be made to be able to reach the desired properties. In 2000 Subramanian et al. [5] studied the dielectric properties of 13 composites of the perovskite type ($ACu_3Ti_4O_{12}$) where $CaCu_3Ti_4O_{12}$ presented a giant dielectric constant (GDC) of 10^4 for bulk ceramics and thin films, and 10^5 for single crystals, value that remained constant in a broad temperature interval (from 100 to 600 K) and broad frequency range (few Hz until MHz). For frequencies in the order of MHz at room temperature, the dielectric constant of the CCTO suffers an accented decay (Fig. 7.1a) for a CCTO sample subjected to superfast cooling (quenching from the sintering temperature) [94]. Other decay in the dielectric constant is evident for temperatures below 100 K, as shown in Fig. 7.1b for two different samples, extracted from the research carried out by Ramirez et al. [95].

CCTO belongs to the space group Im3 [96] and as experiments at low temperature (100 K) showed, ε_m diminished some orders of magnitude without presenting any structural changes as commonly observed in ferroelectric materials. Homes et al. [97, 98] demonstrated that at 100 K the relaxation time increases drastically due to the freezing of the domains. First researchers who reported the giant dielectric constant of CCTO [5, 95, 98] attributed this great value to various intrinsic effects, such as displacement of Ti from the octahedral center along the <111> direction, which is common in materials with TiO_6 units, displacement of Ca atoms and distortions of the TiO_6 octahedral. These hypotheses had controversial acceptance in the literature since same-structured composites (of the perovskite type) based on titanates did not present the GDC phenomenon. Table 7.1 presents some dielectric constants for some titanates materials.

Beyond the GDC of CCTO, Chung et al. [99] reported a pronounced nonlinear voltage-current characteristics, characterized by values of the coefficient of nonlinearity (α) close to 900 (this value must be read with care because it was calculated in the interval 5–100 mA, which is not usual for varistors; the usual range is from 1 to 10 mA) for samples sintered at 1,100°C for 3 and 20 h and quenched from the sintering temperature. An intrinsic electrostatic barrier (Schottky type) located mainly at the grain boundaries was characterized by atomic force microscopy in Kelvin mode, being responsible for the unusual non-ohmic behavior. Although CCTO presents a high α value, it also has a high leakage current, and this limits its application as varistor. Ramirez et al. [100–103] studied the dielectric and non-ohmic properties of the composition $CaCu_3Ti_4O_{12}/CaTiO_3$ and showed the reduction of the leakage current and a better non-ohmic property. Beyond the GDC and the pronounced non-ohmic characteristics, CCTO has potential applications as gas sensors due selective gas absorption at grain boundaries, resulting in changes in the apparent electric properties [104].

A summary of important events in the research history of CCTO is presented in Table 7.2.

7.4 Mechanisms

The main models proposed in the literature to explain the CCTO giant dielectric constant will be discussed in this session. Some models also serve to explain the GDC properties observed in other materials, as presented above. It is important to

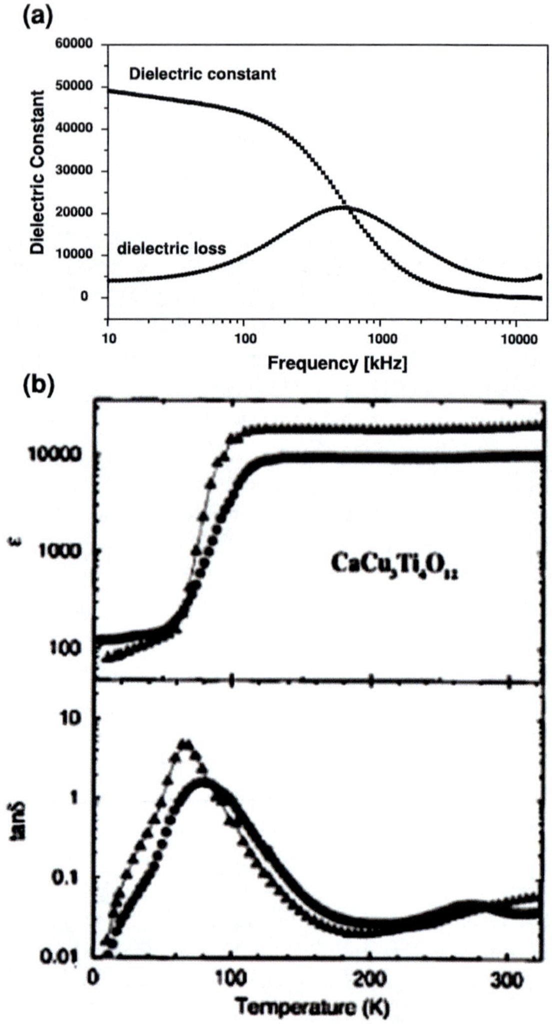

Fig. 7.1 **a** Dielectric constant as a function of frequency for CCTO samples with quenching [94], **b** Dielectric constant and dielectric loss as a function of the temperature for two samples of CCTO [95]

point out that there is no conclusive model accepted by the majority of investigators and this is an open field.

$CaCu_3Ti_4O_{12}$ is a unique material because of its giant dielectric constant. It has a perovskite structure, similar to the $CaTiO_4$, with the octahedrons distorted in the unit cell (Fig. 7.2). The giant dielectric constant presented by CCTO intrigues many investigators because other titanates perovskites does not present such property. Since the first report of this anomalous property in 2,000, many models have been constructed in order to explain this behavior.

Table 7.1 Dielectric constants for some titanates materials

Material composition	Dielectric constant (*k*)
$CaCu_3Ti_4O_{12}$	>10 286
$CdCu_3Ti_4O_{12}$	409
$La_{2/3}Cu_3Ti_4O_{12}$	418
$Sm_{2/3}Cu_3Ti_4O_{12}$	1,665
$Dy_{2/3}Cu_3Ti_4O_{12}$	1,633
$Y_{2/3}Cu_3Ti_4O_{12}$	1,743
$Bi_{2/3}Cu_3Ti_4O_{12}$	1,871
$SrCu_3Ti_4O_{12}$	200
$CaCu_3(Ti_{3.8}Cr_{0.2})O_{12}$	88
$CaCu_3(Ti_{3.8}Fe_{0.2})O_{12}$	87
$Ca_{0.9}Sr_{0.1}Cu_3Ti_4O_{12}$	~1,000
$Ca_{0.8}Sr_{0.2}Cu_3Ti_4O_{12}$	~200
$Ca_2Cu_2Ti_4O_{12}$	~1,500

Table 7.2 Chronology of CCTO research results published

Year	Event
1967	First studies regarding compositions of the family $ACu_3Ti_4O_{12}$ [3]
1979	Determination of the crystal structure of CCTO [4]
2000	First report of GDC of CCTO [5]
2002	ILBC model proposed to explain GDC of CCTO [105]
2004	First publication on non-ohmic properties of CCTO [99]
2005	Internal domains proposed for explain the GDC [106]
2006	Sensor applications [104] and synthesis of composition CCTO/CTO [100]
2008	Photoluminescent properties reported in CCTO thins films [107]
2009	Nanoscale barrier layer capacitance (NBLC) model proposed [108]

First models were proposed by Subramanian [5, 95] and Homes [98] which attributed the high values of permittivity to intrinsic process. Subramanian studied polycrystalline CCTO ceramics and proposed that both polarizability and dielectric constant are enhanced by tension on the Ti–O bonds caused by Ca^{2+} cations. Besides, they speculate that the ferroelectric transition is not observed due to the tilted octahedrons TiO_6 that can accommodate the Cu^{2+} cations in the correct planar coordination. Moreover, they also assumed that the giant dielectric constant can also be enhanced by the microstructure of CCTO ceramic due to the creation of several parallel arranged capacitors. Furthermore, Homes [98] observed that single crystals of CCTO grown by traveling-solvent floating-zone method also present the giant dielectric constant. He studied the dielectric properties of this sample using optical measurements and pointed out which the unit cell should present charge redistribution at low temperatures, becoming more ionic bond in CCTO, especially in the TiO_6 octahedrons, which may explain the low permittivity at low temperatures. Besides, they pointed out which physical processes in CCTO single crystals must be

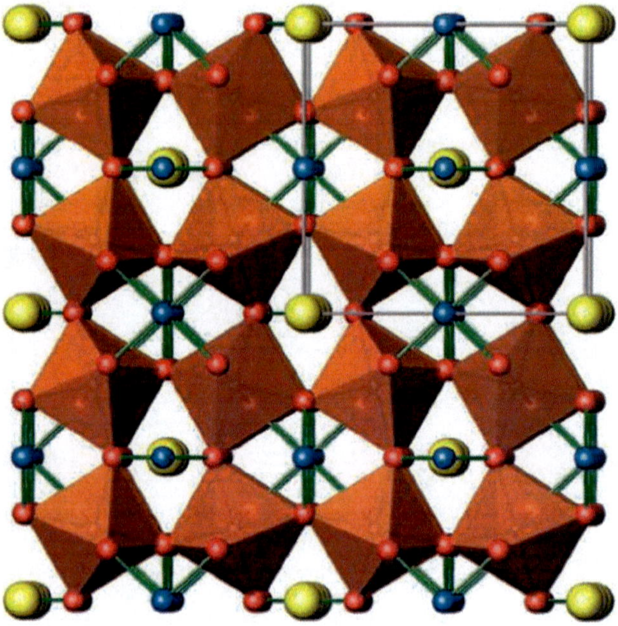

Fig. 7.2 Several unit cells of $CaCu_3Ti_4O_{12}$ shown distorted TiO_6 octahedra, Cu atoms (*brighter small atoms*) bonded to four oxygen atoms (*darker small atoms*), and Ca atoms (*larger ones*) without bonds. Adapted from Ref. [98]

dominated by relaxation effects, and assumed that the difference between the dielectric constant presented by CCTO (80 at higher frequencies and 10^5 at lower frequencies) is due to strong absorption in the low frequencies region.

However, applying density functional theory to CCTO and $CdCu_3Ti_4O_{12}$ (CdCTO) compounds, He et al. [109, 110] demonstrated that the lattice contribution to the static dielectric constant is similar in both materials, and it is not the determinant factor for the giant dielectric constant observed in CCTO. So, they suggest that the anomalous dielectric constant should have extrinsic origin.

Difficulties in understanding the phenomenon of the GDC based on the intrinsic origin and theoretical models [109, 110] took many researchers to study the extrinsic origin (linear and point defects, antiphase boundaries, twin boundaries, grain boundaries and general microstructure) as the most probable mechanism for explaining the dielectric phenomenon [105, 111]. Other extrinsic defects are related to interface electrode-sample phenomena, where a high resistance depletion layer would be the cause of the GDC in CCTO.

Using impedance spectroscopy (IS) measurements, which is a powerful technique to separate the *d.c.* contributions of material in the frequency spectra, it was verified that, at the same temperature, polycrystalline CCTO samples present low dielectric constant at high frequencies and giant permittivity at low frequencies. It is known that IS measurements at low frequencies take the whole contribution of material (fast and slow processes), while at high frequencies only fast processes are taken (usually bulk contribution). In this way, Adams and others [111] used this technique to show that at high frequencies CCTO present a modest value of permittivity of about 100. So, this value was attributed to the intrinsic grain permittivity and, according to authors, it is an experimental evidence that intrinsic model may not be correct to explain the GDC for this material.

Studying CCTO ceramics by impedance spectroscopy, Sinclair et al. [105] proposed that this material can be associated with an equivalent circuit composed by two parallel RC elements connected in series. They found that the grain and grain boundary resistances should be about 10^1 and 10^6 Ω cm, respectively, meaning that CCTO is electrically heterogeneous, consisting of semiconductor grains surrounded by isolating grain boundaries. Besides, they reported that post-thermal treatment in oxygen absent atmosphere significantly alters only the grain boundary resistance. Based on it, they proposed that insulating layers must be formed at grain boundaries during cooling to room atmosphere, and associated the high permittivity to this layer formation. So, they proposed that internal barrier layer capacitance (IBLC) model explains in an easier way the properties displayed by CCTO. Because of the thin insulating layer present at the grain boundaries the material is able to accumulate charge when exposed to external electric field.

To prove their model, they observed a dependence of the microstructure (basically the grain size) of CCTO on the electrical properties [111]. However, it was an intriguing result since sample with larger grain size (smaller grain boundary resistance) also presented larger permittivity. In this way a secondary phase at grain boundaries, as the main responsible for the giant dielectric constant, was not discharged. However, this secondary phase was not observed at grain boundaries by using electron energy loss spectroscopy (EELS) at a transmission electron microscope (TEM) [112].

The IBLC model is relatively accepted and explains very well the first results presented in the literature and also its application to samples with small grain size is reasonable. However, this model is not successful to explain the relation of grain growth and dielectric constant, as well as the dielectric property of CCTO single crystal.

In an attempt of conciliate models with the experimental results the presence of internal barriers was studied deeply. The existence of such barriers could explain the giant permittivity of CCTO single crystal and also the effect observed by Adams et al. [111]. Moreover, neutron and X-ray diffraction of CCTO samples indicated that CCTO crystals should be twined [110,113]. So, this or other planar defect could be the missing part of puzzle. Subramanian et al. [114] reported that such internal barriers should have activation energy of 0.24 eV, which is far higher than bulk energy of activation (0.087 eV). After, such internal domains were observed by Fang et al. [106] using transmission electron microscopy, a more detailed explanation of how planar defects resulting from a twinning parallel to the (100), (010) and (001) planes could became insulating barriers within grains in CCTO structure was reported [115].

Other controversy in the literature is the effect of the electrode-sample interface on the dielectric measurements. This was considered by Lunkenheimer et al. [80],

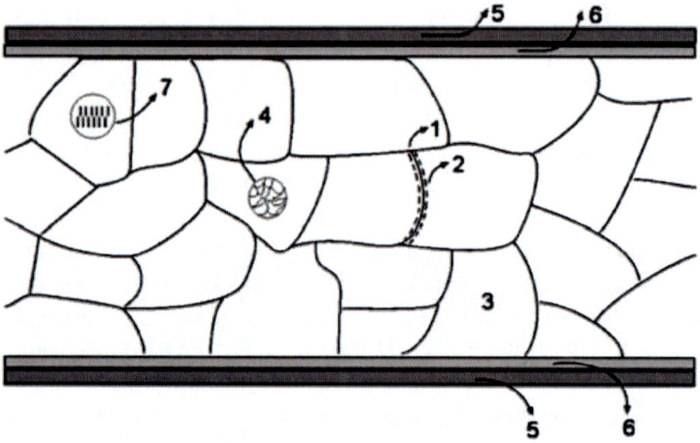

Fig. 7.3 The models most accepted in the literature to explain the anomalous permittivity in CCTO. *1, 2* IBLC; *3* intrinsic mechanism; *4* internal domains; *5* electrode-sample interface; *6* buffer layer and *7* NBLC (stacking fault) model

Ramirez et al. [95] and Homes et al. [97], which successfully separated the electrode contact interface contribution by means of introducing an Al_2O_3 thin film layer to form a "buffer layer" obtaining a dielectric constant for the bulk $\sim 10^4$, showing that the GDC is a material property. Bueno et al. [94, 108] separated the space charge contribution at the grain boundaries from the contribution of bulk, showing that in CCTO the latter is sufficiently significant to be attributed to internal domains.

Recently, a model was proposed which conciliates intrinsic and extrinsic mechanisms, which was called nanosized barrier layer capacitor (NBLC) [108]. In general, it is similar to IBLC model but it has different physical origin. In NBLC model, besides the grain boundaries barriers, there are nanosized barriers within grains which are based on planar defects (stacking faults) and the generation of polarons. These planar defects can increase the permittivity by up to three orders of magnitude, and can explain the giant dielectric constant presented by CCTO. So, processing parameters can influence the dielectric properties of materials due to creation of planar defects. The existence of such defects also explains the huge dielectric constant in single crystals. Besides, it evaluates the phenomena responsible for the high dielectric losses in CCTO.

Figure 7.3 summarizes the main models proposed in the literature to explain the anomalous permittivity in CCTO.

7.5 Summary and Outlook

In recent years giant dielectric constant (GDC) materials have shown great interest by investigators in both, science and engineering, due to its high applications such as memories devices, energy storage, microwave filters, among others. We have highlighted the increasing number of experimental reports focussing on colossal or giant dielectric constant.

We provide a general review of some materials class presenting giant dielectric constant and the main advantages and drawbacks of each material class are reported. Once applications of such materials class are related to capacitive devices, concepts about capacitance and the desire of obtaining materials with giant dielectric constant and very low dielectric loss is discussed.

Several classes of materials including nickelates, cuprates, multiferroics, titanates and composites such as cermets, ceramic/polymer and ceramic/ceramic were discussed in this chapter.

Finally critical discussions of the main proposed models in the literature to explain the giant dielectric constant were made. Models based on intrinsic mechanisms, internal barrier layer capacitor (IBLC), electrode-sample interface (including buffer layer) and nanosized barrier layer capacitor (including stacking fault) are critically discussed. The NBLC model was proposed to conciliate intrinsic and extrinsic mechanisms and predict in a satisfactory way the giant permittivity and dielectric loss of CCTO (108). It is similar to the IBLC model but it has different physical origin. Besides the grain boundaries barriers, there are nanosized barriers within grains that are based on planar defects (stacking faults) and the generation of polarons. Although NBLC model can predict in a satisfactory way the high permittivity and dielectric loss presented in ceramic and single crystal CCTO, it needs more experimental evidences about the amount of stacking faults in different kind of samples (single crystals and polycrystalline samples with different grain sizes).

References

1. Kingon AI, Maria JP, Streiffer SK (2000) Alternative dielectrics to silicon dioxide for memory and logic devices. Nature (London) 406:1032
2. Ang C, Scott JF, Yu Z, Ledbetter H, Baptista JL (1999) Dielectric and ultrasonic anomalies at 16, 37, and 65 K in $SrTiO_3$. Phys Rev B 59:6661
3. Deschanvres A, Raveau B, Tollemer F (1967) Substitution of copper for a bivalent metal in titanates of perovskite type. Bulletin de la Societe Chimique de France 11:4077
4. Bochu B, Deschizeaux MN, Joubert JC, Collomb A, Chenavas J, Marezio M (1979) Synthesis and characterization of series of isotype perovskite titanates of $CaCu_3Mn_4O_{12}$. J Solid State Chem 29:291
5. Subramanian MA, Li D, Duan N, Reisner BA, Sleight AW (2000) High dielectric constant in $ACu_{(3)}Ti_{(4)}O_{(12)}$ and $ACu_{(3)}Ti_{(3)}FeO_{(12)}$ phases. J Sol State Chem 151:323
6. Mauczok R, Wernicke R (1983) Ceramic boundary-layer capacitors. Philips Tech Rev 41:338
7. Kirkpatrick S (1973) Percolation and conduction. Rev Mod Phys 45:574
8. Bai Y, Cheng ZY, Bharti V, Xu HS, Zhang QM (2000) High-dielectric-constant ceramic-powder polymer composites. Appl Phys Lett 76:3804
9. Li YJ, Chen XM, Hou RZ, Tang YH (2006) Maxwell-Wagner characterization of dielectric relaxation in $Ni_{0.8}Zn_{0.2}Fe_2O_4/Sr_{0.5}Ba_{0.5}Nb_2O_6$ composite. Solid State Commun 137:120
10. Yu Z, Ang C (2002) Maxwell-Wagner polarization in ceramic composites $BaTiO_3$-$(Ni_{0.3}Zn_{0.7})Fe_{2.1}O_4$. J Appl Phys 91:794

11. Lin YQ, Chen XM (2009) Dielectric relaxations in $Sr_{(0.5)}Ba_{(0.5)}Nb_{(2)}O_{(6)}/CoFe_{(2)}O_{(4)}$ high-E magnetoelectric composite ceramics. Mater Chem Phys 117:125
12. Yang HB, Wang H, He L, Shui L, Yao X (2010) Polarization relaxation mechanism of $Ba_{(0.6)}Sr_{(0.4)}TiO_{(3)}/Ni_{(0.8)}Zn_{(0.2)}Fe_{(2)}O_{(4)}$ composite with giant dielectric constant and high permeability. J Appl Phys 108:074105
13. Xu HP, Dang ZM, Jiang MJ, Yao SH, Bai J (2008) Enhanced dielectric properties and positive temperature coefficient effect in the binary polymer composites with surface modified carbon black. J Mater Chem 18:229
14. Hindermann-Bischoff M, Ehrburger-Dolle F (2001) Electrical conductivity of carbon black-polyethylene composites—experimental evidence of the change of cluster connectivity in the PTC effect. Carbon 39:375
15. Xu HP, Dang ZM, Bing NC, Wu YH, Yang DD (2010) Temperature dependence of electric and dielectric behaviors of Ni/polyvinylidene fluoride composites. J Appl Phys 107:034105
16. Levi N, Czerw R, Xing SY, Iyer P, Carroll DL (2004) Properties of polyvinylidenedifluoride-carbon nanotube blends. Nano Lett 4:1267
17. Biercuk MJ, Llaguno MC, Radosavljevic M, Hyun JK, Johnson AT, Fischer JE (2002) Carbon nanotube composites for thermal management. Appl Phys Lett 80:2767
18. Maensiri S, Thongbai P, Yamwong T (2007) Giant dielectric permittivity observed in $CaCu_3Ti_4O_{12}$/(Li,Ti)-doped NiO composites. Appl Phys Lett 90:202908
19. Li YJ, Xu M, Feng JQ, Dang ZM (2006) Dielectric behavior of a metal-polymer composite with low percolation threshold. Appl Phys Lett 89:072902
20. Pecharromam C, Moya JS (2000) Experimental evidence of a giant capacitance in insulator–conductor composites at the percolation threshold. Adv Mater 12(4):294
21. Shih SJ, Tuan WH (2004) Solubility of silver and palladium in $BaTiO_3$. J Am Ceram Soc 87:401
22. Cho WW, Kagomiya I, Kakimoto KI, Ohsato H (2006) Paraelectric ceramics/metal dual composites $SrTiO_3$/Pt system with giant relative permittivity. Appl Phys Lett 89:15295
23. George S, Santha NI, Sebastian MT (2009) Percolation phenomenon in barium samarium titanate-silver composite. J Phys Chem Sol 70:107
24. Kumar P, Agrawal DC (2010) Dielectric and optical properties of $CaCu_{(3)}Ti_{(4)}O_{(12)}$ thin films containing Ag nanoparticles. Mat Lett 64:350
25. Lunkenheimer P, Loidl A, Ottermann CR, Bange K (1991) Correlated barrier hopping in NiO films. Phys Rev B 44:5927
26. Bosman AJ, Crevecoe C (1966) Mechanism of electrical conduction in Li-Doped NiO. Phys Rev 144:763
27. Vanelp J, Eskes H, Kuiper P, Sawatzky GA (1992) Electronic-structure of Li-doped NiO. Phys Rev B 45:1612
28. Wu JB, Nan CW, Lin YH, Deng Y (2002) Giant dielectric permittivity observed in Li and Ti doped NiO. Phys Rev Lett 89:217601
29. Wu JB, Nan J, Nan CW, Lin L, Deng Y, Zhao S (2003) Analysis of AC electrical properties of (Li and Ti)-doped NiO. Mater Sci Eng B 99:294
30. Dang ZM, Nan CW (2005) Dielectric properties of LTNO ceramics and LTNO/PVDF composites. Ceram Int 31:349
31. Lin YH, Zhao RJ, Wang JF, Cai JN, Nan CW, Wang YT, Wei L (2005) Polarization of high-permittivity dielectric NiO-based ceramics. J Am Ceram Soc 88:1808
32. Lin YH, Li M, Nan CW, Li JF, Wu JB, He JL (2006) Grain and grain boundary effects in high-permittivity dielectric NiO-based ceramics. Appl Phys Lett 89:032907
33. Lin YH, Wang JF, Jiang L, Chen Y, Nan CW (2004) High permittivity Li and Al doped NiO ceramics. Appl Phys Lett 85:5664
34. Lin YH, Jiang L, Zhao RJ, Nan CW (2005) High-permittivity core/shell structured NiO-based ceramics and their dielectric response mechanism. Phys Rev B 72:014103

35. Park T, Nussinov Z, Hazzard KRA, Sidorov VA, Balatsky AV, Sarrao JL, Cheong SW, Hundley MF, Lee JS, Jia QX, Thompson JD (2005) Novel dielectric anomaly in the hole-doped $La_2Cu_{1-x}Li_xO_4$ and $La_{2-x}Sr_xNiO_4$ insulators: signature of an electronic glassy state. Phys Rev Lett 94:017002
36. Rivas J, Rivas-Murias B, Fondado A, Mira J, Senaris-Rodriguez MA (2004) Dielectric response of the charge-ordered two-dimensional nickelate $La_{1.5}Sr_{0.5}NiO_4$. Appl Phys Lett 85:6224
37. Liu XQ, Wu SY, Chen XM, Zhu HY (2008) Giant dielectric response in two-dimensional charge-ordered nickelate ceramics. J Appl Phys 104:054114
38. Liu XQ, Wu SY, Chen XM (2010) Enhanced giant dielectric response in Al-substituted $La_{(1.75)}Sr_{(0.25)}NiO_{(4)}$ ceramics. J Alloy Compd 507:230
39. Liu XQ, Wu SJ, Chen XM, Zhu HY (2009) Giant dielectric response in two-dimensional charge-ordered nickelate ceramics. J Appl Phys 105:129902
40. Krohns S, Lunkehheimer P, Kant C, Pronin AV, Brom HB, Nugroho AA, Diantoro M, Loidl A (2009) Colossal dielectric constant up to gigahertz at room temperature. Appl Phys Lett 94:122903
41. Wang ZL (2003) Nanowires and nanobelts: materials, properties and devices, vols 1–2. Kluwer, Boston
42. Sarkar S, Jana PK, Chaudhuri BK (2006) Copper (II) oxide as a giant dielectric material. Appl Phys Lett 89:212905
43. Sarkar S, Jana PK, Chaudhuri BK (2008) Origin of electrically heterogeneous microstructure in CuO from scanning tunneling spectroscopy study. Appl Phys Lett 92:142901
44. Sarkar S, Jana PK, Chaudhuri BK (2008) Colossal internal barrier layer capacitance effect in polycrystalline copper (II) oxide. Appl Phys Lett 92:022905
45. Thongbai P, Yamwong T, Meansiri S (2008) Correlation between giant dielectric response and electrical conductivity of CuO ceramic. Solid State Comm 147:385
46. Thongbai P, Meansiri S, Yamwong T (2008) Effects of grain, grain boundary, and dc electric field on giant dielectric response in high purity CuO ceramics. J Appl Phys 104:036107
47. Li M, Feteira A, Sinclair DC (2009) Relaxor ferroelectric-like high effective permittivity in leaky dielectrics/oxide semiconductors induced by electrode effects: a case study of CuO ceramics. J Appl Phys 105:114109
48. Schmid H (1994) Multiferroic magnetoelectrics. Ferroelectr 162:317
49. Gajek M, Bibes M, Fusil S, Bouzehoune K, Fontcuberta J, Barthélémy A, Fert A (2007) Tunnel junctions with multiferroic barriers. Nat Mater 6:296
50. Bibes M, Barthélémy A (2008) Multiferroics: towards a magnetoelectric memory. Nat Mater 7:425
51. Dussan S, Kumar A, Scott JF, Katiyar RS (2010) Magnetic effects on dielectric and polarization behavior of multiferroic heterostructures. Appl Phys Lett 96:072904
52. Hill NA (2000) Why are there so few magnetic ferroelectrics? J Phys Chem B 104:6694
53. Ramesh R, Spaldin NA (2007) Multiferroics: progress and prospects in thin films. Nat Mater 6:21
54. Smolenskii GA, Chupis IE (1982) Segnetomagnetics. Usp Fiz Nauk 137:415
55. Redfern AT, Wang C, Hong JW, Catalan G, Scott JF (2008) 15th International Winterschoolon New Developmentsin Solid State Physics, Germany, 18–22 Feb
56. Jhih-Cyun C, Jenn-Ming W (2007) Dielectric properties and ac conductivities of dense single-phased $BiFeO_3$ ceramics. Appl Phys Lett 91(18):182903
57. Sitchai H, Prasit T, Teerapon Y, Rattikorn Y, Santi M (2009) Dielectric relaxations and dielectric response in multiferroic $BiFeO_3$ ceramics. Appl Phys 94:062904
58. Jun YK, Hong SH (2007) Dielectric and magnetic properties in Co- and Nb-substituted $BiFeO_3$ ceramics. Solid State Commun 144:329

59. Yun KY, Noda M, Okuyama M, Saeki H, Tabata H, Saito K (2004) Structural and multiferroic properties of $BiFeO_3$ thin films at room temperature. J Appl Phys 96:3399
60. Vijayanand S, Potdar HS, Joy PA (2009) Origin of high room temperature ferromagnetic moment of nanocrystalline multiferroic $BiFeO_3$. Appl Phys Lett 94:182507
61. Chen J, Xing XR, Watson A, Wang W, Yu RB, Deng JX, Yan L, Sun C, Chen XB (2007) Rapid synthesis of multiferroic $BiFeO_3$ single-crystalline nanostructures. Chem Mater 19:3598
62. Jaiswal A, Das R, Maity T, Vivekanand K, Adyanthaya S, Poddar P (2010) Temperature-dependent Raman and dielectric spectroscopy of $BiFeO_{(3)}$ nanoparticles: signatures of spin-phonon and magnetoelectric coupling. J Phys Chem C 114:12432
63. Pradhan AK, Kai Z, Hunter D, Dadson JB, Loiutts GB, Bhattacharya P, Katiyar R, Jun Z, Sellmyer DJ, Roy UN, Cui Y, Burger A (2005) Magnetic and electrical properties of single-phase multiferroic $BiFeO_3$. J Appl Phys 97:093903
64. Noda Y, Kimura H, Fukunaga M, Kobayashi S, Kagomiya I, Kohn K (2008) Magnetic and ferroelectric properties of multiferroic $RMn_{(2)}O_{(5)}$. J Phys Condens Matter 20:434206
65. Radaelli PG, Chapon LC (2008) A neutron diffraction study of $RMn_{(2)}O_{(5)}$ multiferroics. J Phys Condens Matter 20:434213
66. Harris AB, Kenzelmann M, Aharony A, Entin-Wohlman O (2008) Effect of inversion symmetry on the incommensurate order in multiferroic $RMn_{(2)}O_{(5)}$ (R = rare earth). Phys Rev B 78:014407
67. Kadomtseva AM, Krotov SS, Popov YuF, Vorob'ev GP (2006) Features of the magnetoelectric behavior of the family of multiferroics RMn_2O_5 at high magnetic fields. Low Temp Phys 32:709
68. Golovenchits EI, Sanina VA, Zalesskii VG, Scheglov MP (2010) Charge carrier self-organization in ferroelectromagnetic semiconductors $Eu_{(0.8)}Ce_{(0.2)}Mn_{(2)}O_{(5)}$. Low Temp Phys 36:522
69. Moritomo Y, Asamitsu A, Tokura Y (1997) Enhanced electron-lattice coupling in $La_{1-x}Sr_xMnO_3$ near the metal-insulator phase boundary. Phys Rev B 56:12190
70. Ivanov VYu, Travkin VD, Mukhin AA, Lebedev SP, Volkov AA, Pimenov A, Loidl A, Balbashov AM, Mozhaev AV (1998) Magnetic, dielectric, and transport properties of $La_{1-x}Sr_xMnO_3$ at submillimeter wavelengths. J Appl Phys 83:7180
71. Kawano H, Kajimoto R, Kubota M, Yoshizawa H (1996) Ferromagnetism-induced reentrant structural transition and phase diagram of the lightly doped insulator $La_{1-x}Sr_xMnO_3$ ($x \leq 0.17$). Phys Rev B 53:14709
72. Yamada Y, Hino O, Nohdo S, Kanao R, Inami T, Katano S (1996) Polaron ordering in low-doping $La_{1-x}Sr_xMnO_3$. Phys Rev Lett 77:904
73. Endoh Y, Hirota K, Ishihara S, Okamoto S, Murakami Y, Nishizawa A, Fukuda T, Kimura H, Nojiri H, Kaneko K, Maekawa S (1999) Transition between two ferromagnetic states driven by orbital ordering in $La_{0.88}Sr_{0.12}MnO_3$. Phys Rev Lett 82:4328
74. Mayr F, Hartinger C, Paraskevopoulos M, Pimenov A, Hemberger J, Loidl A, Mukhin AA, Balbashov AM (2000) High-frequency conductivity and phonon properties of $La_{7/8}Sr_{1/8}MnO_3$. Phys Rev B 62:15673
75. Kimura T, Goto T, Shintani H, Ishizaka K, Arima T, Tokura Y (2003) Magnetic control of ferroelectric polarization. Nature 426:55
76. Goto T, Kimura T, Lawes G, Ramirez AP, Tokura Y (2004) Ferroelectricity and giant magnetocapacitance in perovskite rare-earth manganites. Phys Rev Lett 92:257201
77. Biskup N, deAndres A, Martinez JL, Perca C (2005) Origin of the colossal dielectric response of $Pr_{0.6}Ca_{0.4}MnO_3$. Phys Rev B 72:024115
78. Gutierrez MP, Mira J, Rivas J (2004) Influence of charge-ordering on the dielectric response of $La_{1-x}Sr_xMnO_3$. Phys Lett A 323:473
79. Cohn JL, Peterca M, Neumeier JJ (2005) Giant dielectric permittivity of electron-doped manganite thin films, $Ca_{1-x}La_xMnO_3$ ($0 \leq x \leq 0.03$). J Appl Phys 97:034102
80. Lunkenheimer P, Bobnar V, Pronin AV, Ritus AI, Volkov AA, Loidl A (2002) Origin of apparent colossal dielectric constants. Phys Rev B 66:052105

81. Mamin RF, Egami T, Marton Z, Migachev SA (2007) Giant dielectric permittivity and colossal magnetocapacitance effect in complex manganites with high conductivity. Ferroelectr 348:7
82. Staruch M, Stan L, Lee JH, Wang H, Budnick JI, Jain M (2011) Magnetotransport properties of $Pr_{0.5}Ca_{0.5}MnO_3$ thin films grown by a solution route. J Appl Phys 110:013921
83. Cabassi R, Bolzoni F, Gauzzi A, Gilioli E, Prodi A, Licci F (2006) Dielectric properties of doping-free $NaMn_7O_{12}$: origin of the observed colossal dielectric constant. Phys Rev B 74:045212
84. Vilar SY, Castro-Couceiro A, Rivas-Murias B, Fondado A, Mira J, Rivas J, Señaris-Rodríguez MA (2005) High dielectric constant in the charge-ordered manganese oxide $CaMn_7O_{12}$. Z Anorg Allg Chem 631:2192
85. Hu J, Qin H (2001) Giant magnetoimpedance effect in $La_{0.65}Sr_{0.35}MnO_3$ under low dc magnetic fields. Mater Sci Eng B 79:186
86. Kuang XJ, Bridges C, Allix M, Claridge JB, Hughes H, Rosseinsky MJ (2006) Internal barrier layer capacitance effect in hexagonal perovskite $Ba_4YMn_3O_{11.5}$ ceramics. Chem Mater 18:5130
87. Roa-Rojas J, Salazar CM, Llamosa DP, León-Vanegas AA, Téllez DAL, Pureur P, Dias FT, Vieira VN (2008) Magnetoelectric response of new $Sr_{(2)}TiMnO_{(6)}$ manganite-like material. J Magn Magn Mater 320:104
88. Meher KRSP, Varma KBR (2009) Colossal dielectric behavior of semiconducting $Sr_{(2)}TiMnO_{(6)}$ ceramics. J Appl Phys 105:034113
89. Yu JD, Ishikawa T, Arai Y, Yoda S, Itoh M, Saita Y (2005) Extrinsic origin of giant permittivity in hexagonal $BaTiO_3$ single crystals: contributions of interfacial layer and depletion layer. App Phys Lett 87:252904
90. Guillemet-Fritsch S, Valdez-Nava Z, Tenailleau C, Lebey T, Durand B, Chane-Ching JY (2008) Colossal permittivity in ultrafine grain size $BaTiO_{3-x}$ and $Ba_{0.95}La_{0.05}TiO_{3-x}$ materials. Adv Mater 20:551
91. Adams TB, Sinclair DC, West AR (2006) Influence of processing conditions on the electrical properties of $CaCu_3Ti_4O_{12}$ceramics. J Am Ceram Soc 89:3129
92. Morrison FD, Sinclair DC, West AR (1999) Electrical and structural characteristics of lanthanum-doped barium titanate ceramics. J Appl Phys 86:6355
93. Ang C, Yu Z (2007) High capacitance-temperature sensitivity and "giant" dielectric constant in $SrTiO_3$. Appl Phys Lett 90:202903
94. Bueno PR, Ramírez MA, Varela JA, Longo E (2006) Dielectric spectroscopy analysis of $CaCu_3Ti_4O_{12}$ polycrystalline systems. App Phys Lett 89:191117
95. Ramirez AP, Subramanian MA, Gardel M, Blumberg G, Li D, Vogt T, Shapiro SM (2000) Giant dielectric constant response in a copper-titanate. Solid State Comm 115:217
96. Moussa MS, Kennedy BJ (2001) Structural studies of the distorted perovskite $Ca_{0.25}Cu_{0.75}TiO_3$. Mat Res Bull 36:2525
97. Homes CC, Vogt T, Shapiro SM, Wakimoto S, Subramanian MA, Ramirez AP (2003) Charge transfer in the high dielectric constant materials $CaCu_3Ti_4O_{12}$ and $CdCu_3Ti_4O_{12}$. Phys Rev B 67:092106
98. Homes CC, Vogt T, Shapiro SM, Wakimoto S, Ramirez AP (2001) Optical response of high-dielectric-constant perovskite-related oxide. Science 293:673
99. Chung SY, Kim ID, Kang SJL (2004) Strong nonlinear current-voltage behaviour in perovskite-derivative calcium copper titanate. Nat Mater 3:774
100. Ramírez MA, Bueno PR, Varela JA, Longo E (2006) Non-Ohmic and dielectric properties of a $Ca_2Cu_2Ti_4O_{12}$ polycrystalline system. Appl Phys Lett 89:212102
101. Ramírez MA, Bueno PR, Longo E, Varela JA (2008) Conventional and microwave sintering of $CaCu_{(3)}Ti_{(4)}O_{(12)}/CaTiO_{(3)}$ ceramic composites: non-ohmic and dielectric properties. J Phys D: Appl Phys 41:152004
102. Ramírez MA, Bueno PR, Tararam R, Cavalheiro AA, Longo E, Varela JA (2009) Evaluation of the effect of the stoichiometric ratio of Ca/Cu on the electrical and

microstructural properties of the $CaCu_{(3)}Ti_{(4)}O_{(12)}$ polycrystalline system. J Phys D: Appl Phys 42:185503
103. Ramajo L, Parra R, Varela JA, Reboredo MM, Ramírez MA, Castro MS (2010) Influence of vanadium on electrical and microstructural properties of $CaCu_{(3)}Ti_{(4)}O_{(12)}/CaTiO_{(3)}$. J Alloy Comp 497:349
104. Kim ID, Rothschild A, Hyodo T, Tuller HL (2006) Microsphere templating as means of enhancing surface activity and gas sensitivity of $CaCu_3Ti_4O_{12}$ films. Nano Lett 6:193
105. Sinclair D, Adams T, Morrison F, West A (2002) $CaCu_3Ti_4O_{12}$: one-step internal barrier layer capacitor. Appl Phys Lett 80:2153
106. Fang TT, Liu CP (2005) Evidence of the internal domains for inducing the anomalously high dielectric constant of $CaCu_4Ti_4O_{12}$. Chem Mater 17:5167
107. Parra R, Joanni E, Espinosa J, Tararam R, Cilense M, Bueno PR, Varela JA, Longo E (2008) Photoluminescent $CaCu_{(3)}Ti_{(4)}O_{(12)}$-based thin films synthesized by a sol-gel method. J Am Ceram Soc 91(12):4162
108. Bueno PR, Tararam R, Parra R, Joanni E, Ramírez MA, Ribeiro WC, Longo E, Varela JA (2009) A polaronic stacking fault defect model for $CaCu_{(3)}Ti_{(4)}O_{(12)}$ material: an approach for the origin of the huge dielectric constant and semiconducting coexistent features. J Phys D: Appl Phys 42:055404
109. He LX, Neaton JB, Cohen MH, Vanderbilt D, Homes CC (2002) First-principles study of the structure and lattice dielectric response of $CaCu_3Ti_4O_{12}$. Phys Rev B 65:214112
110. He LX, Neaton JB, Vanderbilt D, Cohen MH (2003) Lattice dielectric response of $CdCu_3Ti_4O_{12}$ and $CaCu_3Ti_4O_{12}$ from first principles. Phys Rev B 67:012103
111. Adams TB, Sinclair DC, West AR (2002) Giant barrier layer capacitance effects in $CaCU_3Ti_4O_{12}$ ceramics. Adv Mater 14(18):1321
112. Calvert CC, Rainforth WM, Sinclair DC, West AR (2006) EELS characterization of bulk $CaCu_{(3)}Ti_{(4)}O_{(12)}$ ceramics. Micron 37:412
113. Kim YJ, Wakimoto S, Shapiro SM, Gehring PM, Ramirez AP (2002) Neutron scattering study of antiferromagnetic order in $CaCu_3Ti_4O_{12}$. Solid State Commun 121:11
114. Li J, Sleight AW, Subramanian MA (2005) Evidence for internal resistive barriers in a crystal of the giant dielectric constant material: $CaCu_3Ti_4O_{12}$. Solid State Commun 135:260
115. Whangbo MH, Subramanian MA (2006) Structural model of planar defects in $CaCu_3Ti_4O_{12}$ exhibiting a giant dielectric constant. Chem Mater 18:3257

Chapter 8
Dye-Sensitized Solar Cells

Mateja Hočevar, Marko Berginc, Urša Opara Krašovec and Marko Topič

Abstract Dye-sensitized solar cells (DSSCs) are recognized as one of the world's leading innovation in nanosciences and photovoltaic technology. In contrast to the conventional silicon-based solar cells the demand on purity of materials for DSSC is lower and forecasted manufacturing costs are approximately halved which make the DSSCs attractive alternative. Nowadays the conversion efficiency of DSSCs exceeds 10% which makes it competitive with other thin film solar cells. In this chapter the structure and the fundamental operation of the DSSC are presented. Afterwards the properties and requirements for each individual component used in a DSSC are given. In addition advanced hybrid DSSC systems also are presented. In overall, the emphasis of this chapter is given to the involvement of sol-gel chemistry in the preparation of the individual DSSCs components, primarily of TiO_2 layers and electrolytes.

Keywords Hočevar · Absorption properties · Condensation · Conductivity · Conversion efficiency · Current density-voltage (J-V) · Dye · Dye-sensitized solar cells (DSSCs) · Electrolyte · Hybrid DSSC system · Hydrolysis · Iodide/tri-iodide (I^-/I_3^-) · Mesoporous TiO_2 · Ormosil · Photoactive electrode · Photoelectrochromic device · Photoelectrode · Photovoltaic device · Smart windows · Sol-gel · Thin film ·

M. Hočevar (✉) · M. Berginc · U. Opara Krašovec · M. Topič
Laboratory of Photovoltaics and Optoelectronics, Faculty of Electrical Engineering, University of Ljubljana, Tržaška 25, 1000 Ljubljana, Slovenia
e-mail: mateja.hocevar@fe.uni-lj.si

M. Berginc
e-mail: marko.berginc@fe.uni-lj.si

U. Opara Krašovec
e-mail: ursa.opara@fe.uni-lj.si

M. Topič
e-mail: marko.topic@fe.uni-lj.si

M. Aparicio et al. (eds.), *Sol-Gel Processing for Conventional and Alternative Energy*, Advances in Sol-Gel Derived Materials and Technologies,
DOI: 10.1007/978-1-4614-1957-0_8,

Thin film solar cell • TiO_2 • TiO_2 layers • TiO_2 nanoparticles • Titania • Transparent conductive oxide (TCO) • Tungsten oxide • WO_3

List of abbreviations

AM1.5	Air mass 1.5
ATO	Mixture of antimony and tin oxide; SbO_2/SnO_2
ITO	Indium-doped tin oxide; SnO_2:In
BIPV	Building integrated photovoltaics
CIGS	Copper indium gallium selenide
DSSC	Dye-sensitized solar cell
FTO	Fluorine-doped tin oxide; SnO_2:F
HOMO	Highest occupied molecular orbital
LUMO	Lowest unoccupied molecular orbital
N719	Commercial denotation of the dye: *cis*-diisothiocyanato-*bis* (2,2′-bipyridyl-4,4′-dicarboxylato) ruthenium(II) *bis*(tetrabutylammonium)
S^o	Ground state of the dye molecule
S^*	Excited state of the dye molecule
S^+	Positively charged dye molecule
STC	Standard test conditions
TCO	Transparent conductive oxide

Symbols

E_C (eV)	Conduction band
FF (%)	Fill factor
J_{MPP} (mA/cm^2)	Current density at maximum power point
J_{SC} (mA/cm^2)	Short-circuit current density
P_{Light} (mW)	Power density of incident light
P_{MPP} (mW)	Maximum power point
R_S (Ω)	Serial resistance of the cell
V_{MPP} (V)	Voltage at maximum power point
V_{OC} (V)	Open circuit voltage
η (%)	Conversion efficiency of solar cell
λ (nm)	Wavelength

8.1 Introduction

Photovoltaic devices are the renewable energy sources that directly convert solar into electrical energy and provide electricity without giving rise to any carbon dioxide emission during operation. World-wide forecasts estimate that in 2050 the contribution of photovoltaic will present 20–25% of global primary energy or even over 90% of the global production of electricity. Nowadays many different types of solar cells exist, but the most widely produced are crystalline silicon solar cells. However, their main drawback is relatively expensive production due to the high consumption of materials and energy. Other photovoltaic technologies are thin film solar cells based on silicon, copper indium gallium selenide (CIGS) or cadmium telluride (CdTe), and highly efficient concentrator solar cells in multi junction configuration [1].

In 1990s new solar cell technologies emerged such as organic solar cells [2] and dye-sensitized solar cells (DSSCs) [3–5]. The DSSC is recognized as one of the world's leading innovation in nanosciences and photovoltaic technology. The development of DSSCs has been driven by their many attractive features, e.g. low-cost potential (<0.7 €/Wp), high efficiency (up to 11%), and short energy payback time (<0.5 year). In the last decade the research in the field of DSSCs has been focusing on glass-based solar cells, but recently a strong emphasis has been given to the development of flexible DSSCs. Polymeric foil or flexible metal substrates offer a wider range of applications in DSSCs, and on the other hand, could foster these technologies toward their industrialization. The main steps for successful future commercialization of DSSCs are the improvement of stability and efficiency, simplification of the production processes, the use of flexible substrates, and scaling up of the technology, i.e. production of photovoltaic modules. Currently, the highest efficiency for a small area DSSC (0.219 cm^2) using electrolytes based on volatile organic solvents is 11.2%, while the officially confirmed efficiency for a 1.004 cm^2 DSSC is 10.4%. The greatest efficiency of a dye-sensitized submodule, which consists of eight parallel cells with a total active area of 17 cm^2 is 9.9% [1].

This chapter introduces the main properties of DSSC and describes the sol-gel processes for the preparation of their components. Sol-gel processes in DSSC are mostly used for the preparation of TiO_2 layers and electrolytes due to many advantages of the sol-gel methods. Among these are purity, homogeneity, and flexibility in introducing dopants in large concentrations, stoichiometry control, simplicity of processing, control over the composition, and relatively low energy budget as well as low manufacturing costs [6]. In addition, advanced hybrid DSSC systems are also presented.

8.2 Properties of DSSC

8.2.1 Structure and Operation

In the DSSC, a completely different type of thin film approach is used than in conventional thin film solar cells. It is based on photo-electrochemistry at the interference between a dye adsorbed onto a mesoporous titanium dioxide (TiO_2)

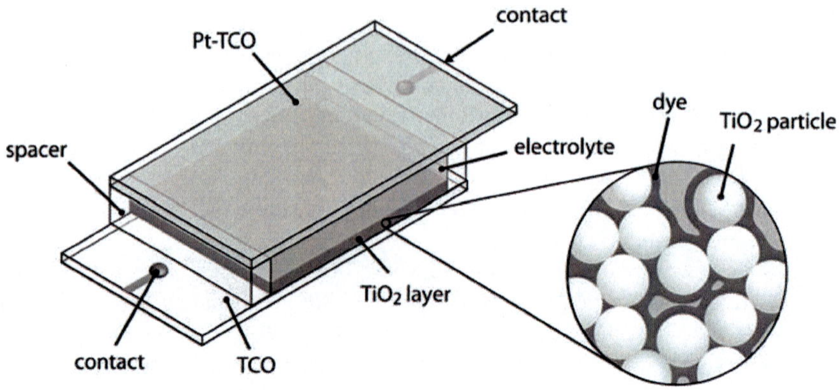

Fig. 8.1 Cross-section of a Dye-Sensitized solar cell

layer and an electrolyte [3]. According to electrolyte physical states, the DSSC can be divided into one of three types: liquid electrolyte-, quasi-solid-state electrolyte-, and solid-state electrolyte-based DSSC. Additionally, other types of DSSC exist, where the electrolyte is replaced by a solid-state hole conductor. A DSSC consists of a photoactive electrode, a counter electrode, and an electrolyte (Fig. 8.1).

The photoactive electrode is a transparent conductive oxide (TCO) on glass or flexible substrate, coated with mesoporous TiO_2 sensitized with a monolayer of a dye, while the counter electrode is a TCO on glass (or flexible substrate) coated with a thin catalytic layer. The gap between the two electrodes is filled with an electrolyte containing a redox couple. The most frequently used is an iodide/tri-iodide (I^-/I_3^-) redox couple that enables the highest cell performance.

In addition to the specific structure of DSSC its operation is also significantly different from other types of solar cells. The energy diagram and electron transfer paths during illumination of a DSSC are shown in Fig. 8.2. By absorption of a photon, the dye molecule is excited from ground state (S°) to the excited state (S*), which means that an electron in the highest occupied molecular orbital (HOMO) moves to the lowest unoccupied molecular orbital (LUMO). The exciton at LUMO orbital rapidly injects a free electron into the conduction band of the TiO_2 (E_C). Simultaneously, the dye molecule is transferred to the positively charged state (S^+). The conduction band electron is then transported to the front TCO (and further to the external load) through the nanostructured TiO_2 layer. Positively charged dye molecule returns to the ground state with the acceptance of an electron from the iodide (reaction 8.1).

Oxidation of I^- ion at the electrolyte/dye interface leads to the formation of I_3^- ion, while the opposite redox reaction, i.e. reduction of I_3^- ion to I^- ion takes place at the counter electrode (reaction 8.2) by afflux of electrons from the external circuit. For the later reaction a catalyst (usually platinum) is needed. A gradient in concentration of I^- and I_3^- in the electrolyte leads to the diffusion of the ions.

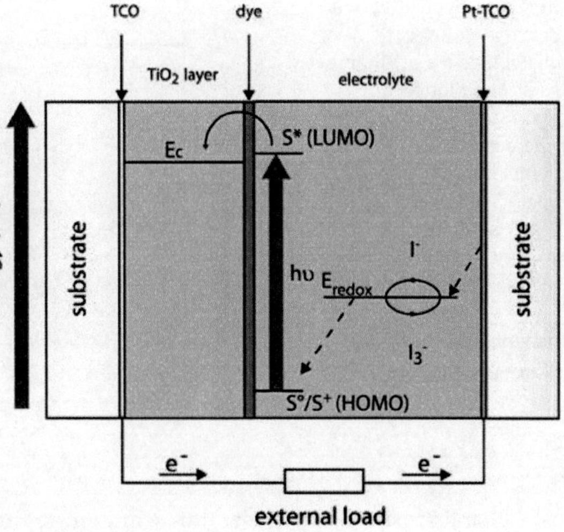

Fig. 8.2 Energy level diagram of Dye-Sensitized solar cell at open circuit conditions

With the diffusion of I^- ions to the front electrode and I_3^- ions to the counter electrode the electrical circuit of the DSSC is closed.

$$3I^- + 2S^+ \rightarrow I_3^- + 2S^o \tag{8.1}$$

$$I_3^- + 2e^- \rightarrow 3I^- \tag{8.2}$$

As in other solar cells also in DSSCs optical and electrical loss processes occur. The most significant electrical loss mechanism in the DSSC is the recombination of conduction band electrons in TiO_2 with the I_3^- ions in the electrolyte (reaction 8.2) [7]. Other recombination mechanisms are (i) an excited dye molecule may directly relax into its ground state and (ii) electrons from the conduction band of the TiO_2 may recombine with the oxidized dye molecule, before the dye is reduced by the redox couple in the electrolyte. In addition, a decrease in efficiency of DSSCs is also affected by its internal resistance (R_S), which consists of a series of resistances, e.g. resistance at the interface between the electrolyte and the platinum, resistance at the interface between the TCO and TiO_2 layer, and resistances of TCO and TiO_2 layer. Optical losses come from total reflection at the front side, absorption inactive layers and transmission through the cell in case of semitransparent cell type.

8.2.2 Characterization

The current density–voltage characteristic (*J*–*V*) is the most frequent method used for electrical characterization of the solar cell performance under different operation conditions (different light intensities, cell temperatures etc.). A typical

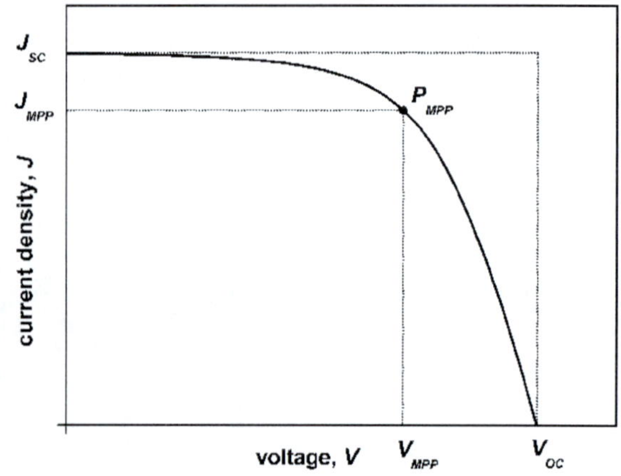

Fig. 8.3 Typical current density–voltage characteristics of illuminated solar cell plotted in generation operation regime. The denoted performance parameters are V_{OC}— open-circuit voltage, J_{SC}— short-circuit current density, V_{MPP}— voltage at the maximal power point, J_{MPP}— current density at the maximal power point and P_{MPP}— maximal power

J–V characteristic in the generator operation regime together with most common performance parameters is denoted in Fig. 8.3. A theoretically maximum value of short-circuit current density (J_{SC}) of DSSCs is determined by the absorption properties of dye as its absorption edge reflects the energy difference between the HOMO and LUMO energy levels of the dye molecule. Only the photons with energy exceeding the LUMO–HOMO difference can generate free electrons in the active layer. The sum of all electrons generated by the incident photons having different energies above the LUMO–HOMO difference defines the short-circuit current density (J_{SC}). In the generator operation regime, the solar cell generates maximum current density at short-circuit conditions while the voltage and thus the generated power equals zero. Similarly the generated power equals zero at open-circuit conditions where the cell provides the highest voltage i.e. open circuit voltage (V_{OC}) and on the other hand the current density equals zero. A maximum power point always exist somewhere in between the open-circuit and short-circuit conditions at which the product of current (J_{MPP}) and voltage (V_{MPP}) gives the highest generated power (P_{MPP}).

Another solar cell's performance parameter is defined at the maximum power point i.e. the fill factor (*FF*). It quantitatively describes ratio of actual power P_{MPP} to the hypothetical power of $V_{OC} \times J_{SC}$ product (Eq. 8.3) ranges between 25% for shunted solar cells and close to 90% for record solar cells.

$$FF = \frac{V_{\mathrm{MPP}} J_{\mathrm{MPP}}}{V_{\mathrm{OC}} J_{\mathrm{SC}}} \tag{8.3}$$

Another very important parameter is conversion efficiency of the solar cell (η) which is the ratio between the generated electric power at the maximum power point and the power density of the incident light (P_{Light}) (Eq. 8.4).

$$\eta = \frac{P_{MPP}}{P_{Light}} = \frac{V_{MPP} J_{MPP}}{P_{Light}} = \frac{V_{OC} J_{SC} FF}{P_{Light}} \tag{8.4}$$

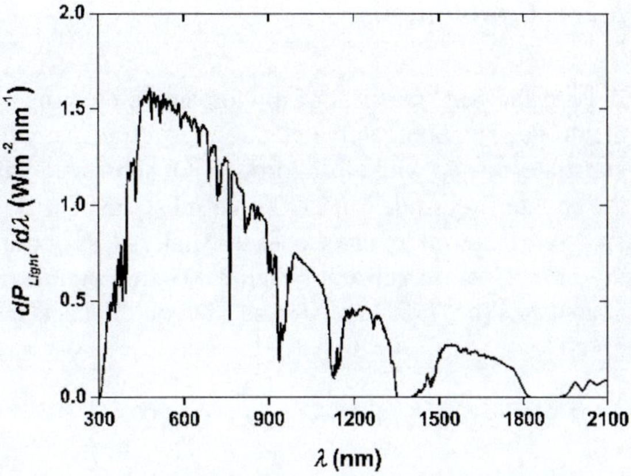

Fig. 8.4 The AM1.5 spectrum used for solar cell characterization defined by IEC 60904-3 standard

The current density–voltage characteristics can be measured under different controlled operating conditions (different cell temperatures, light intensities, etc.). However, for better comparison of the performance of different solar cell technologies and the influence of variation in materials the cells are usually measured under standard test condition (STC) at which the cell temperature and light intensity should be exactly 25°C and 100 mW/cm^2, respectively. Additionally the spectrum of the incident light should correspond to the solar spectrum AM1.5 specified in the international standard IEC 60904-3 [8]. In case of deviations from STC, calculation methods are standardized in the set of IEC standards (IEC 60904-X, X = 1–10) [9].

The prescribed spectrum AM1.5 consists of the UV, visible, and IR part of the solar spectrum (Fig. 8.4). It is almost impossible to perfectly reproduce the solar spectrum and only a few rather expensive solar simulators in the market can achieve adequate reproduction. Alternatively, the IEC 60904-4 standard [10] allows outdoor characterization of the solar cells if the light intensity ranges between 80 and 120 mW/cm^2. Afterwards the measured results should be scaled to 100 mW/cm^2. The outdoor method is not very convenient since it strongly depends on current weather conditions, season, or even the current air pollution, therefore a meteorological conditions should be determined during the characterization of the solar cells and absolute measurement of the light spectrum is needed [11, 12]. For practical reasons, the characterization of solar cells is most commonly performed under solar simulators based on xenon and/or tungsten lamps. In this case the spectrum usually matches the prescribed tolerances around AM1.5 solar spectrum, but a calibrated reference solar cell should be used to properly adjust the light intensity of the solar simulator. The spectral response of the calibrated reference solar cell should be as close as possible to the cells under test. In case of DSSCs, c-Si solar cell with a corrective optical filter should be used. Spectral mismatch factor needs to be calculated and applied in order to assure accurate measurement results [13].

8.2.3 Components

A typical DSSC consists of a photoactive electrode, an electrolyte, and a counter electrode. The photoactive electrode is a transparent conductive oxide (TCO) on a substrate, coated with mesoporous TiO_2 sensitized with a monolayer of a dye, while the counter electrode is a TCO on a substrate coated with a thin catalytic layer. The substrate material for electrodes is usually glass or transparent flexible foil, although a non-transparent substrate such as Ti foil, stainless steel, plain carbon steel, Zn coated steel or Ni foil etc. can be used on the back side of non-transparent DSSCs.

8.2.3.1 Transparent Conductive Oxide

For the photoactive electrode a transparent conductive oxide (TCO) can be deposited on different transparent substrates such as glass or plastic foils that support and encapsulate the solar cell's structure. The TCO should be as transparent as possible to enable maximization of light trapping in photoactive layer. Additionally, TCO collects photogenerated electrons from the active layer and transfers them to the external load; therefore the conductivity of the TCO layer has to be as high as possible. The TCO needs to form good ohmic contact to the TiO_2 layer to enable good electrical conductivity. An increased conductivity is achieved with thicker TCO layer, but on the other hand thicker layer introduces additional optical losses (lower transmission) which reduce the generation in the active layer. In addition to the highest conductivity and high transmission of the TCO layer on a certain substrate the TCO should also meet the following requirements: low reflectance to reduce optical losses (different antireflective coatings may be used to further reduce reflectance), and high scattering of light in transmission prolongs the optical path through the active layer and increases the light absorption. Nonetheless, TCO layer must not react with the electrolyte and it has to withstand the temperatures that are required for annealing of the TiO_2 layer. Only a few different types of the TCO layers satisfy all demands. The fluorine-doped tin oxide SnO_2:F (FTO) with the thickness around 500 nm is the most commonly used TCO for the DSSCs. The FTO crystal grains have different shapes and sizes ranging between 10 and 100 nm. They are firmly connected in order to assure good electrical conductivity. Alternatives to the FTO are: (i) an aluminum-doped zinc oxide ZnO:Al (ii) an indium-doped cadmium oxide CdO:In (iii) a zinc-doped indium oxide In_2O_3:Zn (iv) mixture of antimony and tin oxide SbO_2/SnO_2 (ATO), and (v) an indium-doped tin oxide SnO_2:In (ITO). The ITO is the most commonly used in flexible electronics [14].

8.2.3.2 Catalyst

The counter electrode contains a very thin layer or islands of a catalyst that is deposited on TCO or other conducting materials. The catalyst is used for reduction of the tri-iodide ions to the iodide ions (reaction 8.2). The most frequently used is platinum which shows excellent catalytic activity at thicknesses of only a few nm. The efficiency of the catalytic layer is defined by the charge transfer resistance at the interface between the catalytic layer and electrolyte and could be evaluated with the electrochemical methods [15]. The platinum layer could be deposited by sputtering [16, 17], electro deposition [18], electroplating [19], thermal decomposition of the H_2PtCl_6 solution [20, 21], or by screen printing and annealing of different Pt-based pastes [22]. The research also considers the use of other catalytic materials in order to replace relatively expensive platinum and further reduce the cost of DSSCs. The most commonly used alternatives are different carbon-based materials (carbon, single and multi wall carbon nanotubes, carbon black, carbon powder, activated carbon, carbon filament, nanohorn, graphite, graphene, and hard carbon spheres), conduction polymers [such as poly(3,4-ethylenedioxythiophene)] and recently also cobalt sulfide [23]. The catalytic materials, especially the carbon based ones, might detach from the counter electrode and could migrate toward the photoactive electrode during the long-term operation. This leads to reduction of the performance of the counter electrode and introduces very effective recombination centers at the photoactive electrode therefore the counter electrode has to be especially carefully designed.

8.2.3.3 TiO_2 Layer

The key breakthrough for DSSCs in 1991 was the use of a mesoporous TiO_2 electrode with a high internal surface area to support the monolayer of a sensitizer [3]. Since then many other metal oxide systems have been tested as an alternative electrode, such as ZnO, SnO_2, and Nb_2O_5 [5]. Besides these simple oxides, ternary oxides ($SrTiO_3$ and Zn_2SnO_4) have been investigated [5, 24, 25]. However, TiO_2 still gives the highest efficiencies and the majority of research in the field of DSSC has been focusing on TiO_2. Therefore, the main properties of TiO_2 and a short overview on the development of nanostructured TiO_2 layers for the active photoelectrode in DSSCs will be described in this section.

Titanium dioxide belongs to the family of transition-metal oxides. It has a high refractive index (2.4–2.5) and is widely used as a white pigment in paint, plastic, and paper, as an additive to toothpastes, sun cremes, self-cleaning materials, and food (E 171). There are three polymorphs of TiO_2 found in nature, i.e.: anatase, rutile, and brookite [26]. Nanocrystalline TiO_2 in anatase modification showing the highest diffusion coefficient and a band gap of $\sim$3.2 eV has been identified as the most appropriate material to be used as photoactive electrode in DSSCs, since its conduction band lies just beneath the LUMO level of the most studied ruthenium complex dyes. This leads to a higher V_{OC} in DSSCs for the same conduction band electron concentration [3].

In recent years, substantial progress has been made in the development of methods to synthesize new anatase nanostructures such as nanoparticles, nanorods, nanowires, nanobowls, nanosheets, and nanotubes as well as mesoporous materials such as aerogels, inverse opals, and photonic materials. These methods include sol-gel, micelle and inverse micelle, hydrothermal, solvothermal, sonochemical, microwave deposition techniques, direct oxidation, chemical vapor deposition, physical vapor deposition, and electrodeposition methods [5, 27]. For dye-sensitized solar cells, the most common technique for preparation of TiO_2 nanoparticles is the hydrolysis of a titanium precursor such as titanium(IV) alkoxide with excess water catalyzed by acid or base, followed by hydrothermal growth and crystallization. Acid or basic hydrolysis gives materials of different shapes and properties, and the rate of hydrolysis, temperature, and water content can be tuned to produce particles of different sizes [5]. So far, the spherical nanoparticles are the most efficient in the preparation of TiO_2 layer for DSSC. One commercially available product, widely used in DSSC is Aeroxide® P25 TiO_2, manufactured by Degussa through the high temperature (higher than 1200°C) flame hydrolysis of $TiCl_4$ in the presence of hydrogen and oxygen. The P25 product is nonporous 99.5% TiO_2 (anatase : rutile = 70 : 30), with surface area of $50 \pm 15\ m^2g^{-1}$ and an average particle diameter of 21 nm.

The nanoparticles are formulated in a paste with organic additives and deposited onto conducting glass/flexible substrates usually using doctor blading or screen printing techniques. After the deposition, the layer is usually annealed at about 400°C to remove organic components and to make good electrical interconnection between the nanoparticles [26].

In general, TiO_2 pastes suitable for screen printing usually contain a terpineol-based template [28]. A standard TiO_2 paste for DSSCs is composed of TiO_2 nanoparticles, a dispergant for TiO_2 nanoparticles (e.g. methoxy-benzoic acid in ethanol solution), a solvent (e.g. terpineol), and a binder (e.g. methyl cellulose) [28]. In recent years, many papers have been reported on the preparation of porous TiO_2 layers including the integration of inorganic polymeric binders (titanium isopropoxide, titanium butoxide, etc.) [29, 30] to the paste or applying sol-gel synthesis in the paste preparation processes [31–37].

The TiO_2 layer used in DSSCs has to meet the following requirements to enable high conversion efficiency: (i) good electron transport within TiO_2 layer (through TiO_2 grains) and TCO substrate (ii) good adsorbtance of dye with efficient injection of electrons (iii) good solar light harvesting, and (iv) diminished loss reactions of photogenerated electrons with the tri-iodide species present in the electrolyte. The TiO_2 layer has to meet the following characteristic: (i) high inner surface area available for dye molecules to be attached; (ii) appropriate porosity of the layer assuring electrolyte penetration through the layer; (iii) good connections between the TiO_2 grains; (iv) good adhesion to the TCO, and (v) the layer thickness between 7 and 14 μm [38].

One of the important characteristics of mesoporous TiO_2 layers for DSSCs is a good light trapping within the layer. To achieve this, pastes have been improved by integration of spherical voids or larger TiO_2 particles into the TiO_2 paste, which act as efficient light scatters in the visible light range. Light scattering increases the optical path of photons propagating through the layer, which results in an increased absorption and consequently an enhanced conversion efficiency of DSSCs. The scattering efficiency is dependent on the size, location and the refractive index of the perturbations within the TiO_2 layers [39, 40].

For the realization of high efficiency DSSC a three- or four-fold layer deposition of TiO_2 is reported. Wang et al. [41] introduced a complex fourfold TiO_2 layer. The TCO was first treated with an aqueous solution of $TiCl_4$ at 70°C for 30 min in order to make a thin compact TiO_2 layer assuring a good mechanical contact between the following printed TiO_2 layer and the TCO substrate. In the second step a transparent layer consisting of 20 nm size TiO_2 particles was screen printed on the $TiCl_4$-treated TCO and additionally coated with the third screen printed layer of 400 nm light scattering TiO_2 particles. Furthermore, the layers were gradually heated to 500°C in order to achieve nanostructured porosity of the TiO_2 layer with a high surface area. The fourth step was treating the triple layer with an ethanol solution of $TiCl_4$ to improve the connectivity of the grains present in the thick nanoporous layer. The preparation procedure of fourfold TiO_2 layer is complex, since different TiO_2 precursors as well as different deposition techniques are required. Therefore, in recent years the aim has been to simplify the preparation of TiO_2 layer in such a way that solely a deposition of a single TiO_2 paste followed by annealing of the layer would be sufficient to realize a highly efficient photoactive layer [30, 34–37].

However, the preparation of TiO_2 layers requires high-temperature treatment steps (above 400°C), which limit the choice of substrates to heat-resistant materials. Therefore, flexible materials such as plastic, textile, and paper cannot be used. Some efforts have been already made toward low-temperature preparation of porous TiO_2 layers TiO_2 layer [42–44]. In this case the porous TiO_2 layers were prepared by using a mixture of nanocrystalline TiO_2 powder and titania precursor with the UV irradiation treatment [43] or hydrothermal crystallization at low temperature in the solid/gas interface [44]. Lindström et al. [45, 46] introduced the press method (the pressing of oxide layers for production of porous thin layers) for low-temperature preparation of the nanostructured TiO_2 photoelectrodes. Using conductive plastic substrate photoelectrodes prepared by this method, they achieved the efficiency of 3% under 100 mW/cm^2 irradiation [45]. Dürr et al. [47] subsequently developed a lift-off process. In this process, the TiO_2 layer was first applied to a thin gold layer on a glass substrate. After sintering, the TiO_2 layer was removed from the glass by dissolving the gold layer. The TiO_2 layer was then transferred onto an indium tin oxide-coated polyethylene terephthalate layer by application of high pressure. By this process, Dürr et al. [47] achieved an efficiency of 5.8% under the illumination of 100 mW/cm^2. The most promising research results were published by Yamaguchi et al. [48]; they reported highly efficient (7.6%) DSSC based on plastic substrate using a press method. The high conversion efficiency was reached by optimization of the press conditions, the thickness of the layer and an additional UV-O_3 treatment of the TiO_2 layer and the plastic substrate.

Sol-Gel Processed TiO_2 Layer

The sol-gel process has been widely used for the synthesis of TiO_2 nanopowder (nanoparticles, nanowires, nanotubes, etc.) as well as for the preparation of pastes and layers [26]. Generally, sol-gel process is favored as appropriate process for preparing TiO_2 materials because the particle size, layer porosity and thickness can be controlled by adjusting parameters such as hydrothermal growth temperature, sintering condition, and sol concentration [6, 27].

Two types of the sol-gel process are known: alkoxide and non-alkoxide route [26]. The non-alkoxide route uses inorganic salts (such as nitrates, chlorides, acetates, carbonates, and acetylacetonates) [49, 50] and requires an additional removal of the inorganic anions. The more popular alkoxide route, on the other hand, uses metal alkoxides [27, 51–55] as the starting material. In a typical sol-gel process, a colloidal suspension, or a sol, is formed from the hydrolysis, condensation, and polymerization reactions of the precursors. The most commonly used precursors for TiO_2 are titanium(IV) isopropoxide ($Ti(OiPr)_4$) [27, 51, 52], titanium(IV) ethoxide ($Ti(OE)_4$) [53] and titanium(IV) butoxide ($Ti(OnBu)_4$) [54, 55]. However, hydrolysis and condensation can be exothermic and violent and usually lead to undesirable reactions. Therefore stabilizing agents have been added into the sols (acetic acid, ethyl acetoacetate…), which prevent the process of precipitation by decreasing the rate of the hydrolysis and condensation reactions [56–60].

In order to exhibit better control over the evolution of the microstructure, it is desirable to separate and temper the steps of hydrolysis and condensation [56]. One of possibilities is alkoxide modification by complexation with coordination agents such as carboxylates [57, 58], or diketonates [59, 60] that hydrolyze slower than alkoxide ligands [26]. Acid–base catalysis can also be used to enable separation of hydrolysis and condensation steps [56]. The acid catalysis increases hydrolysis rates and ultimately crystalline powders are formed from fully hydrolyzed precursors. Base catalysis is thought to promote condensation with the result that amorphous powders are obtained containing unhydrolyzed alkoxide ligands. On the other hand, acetic acid can be used in order to initiate hydrolysis via an esterification reaction, and alcoholic sols prepared from titanium alkoxide using amino alcohols have been shown to stabilize the sol, reducing or preventing the condensation and the precipitation of TiO_2 [26, 61].

Preparation of TiO_2 Nanoparticles via Sol-gel Process: The condensation/polymerization reactions are followed by an annealing treatment (above 400°C) to remove the organic components and to crystallize either anatase or rutile TiO_2 nanoparticles. However, a low-temperature preparation of TiO_2 nanopowder based on sol-gel method has been already prepared by Li et al. [26, 62]. They report on nanocrystalline TiO_2 powder, which has been synthesized at 100°C.

Cleaning of particles is usually achieved by washing the surface with a solvent, followed by centrifugation. The solvent can affect the chemical composition and crystallization. An alternative washing technique is to dialyze particles against double-distilled water, which could be an effective method of removing soluble impurities without introducing new species. It should be noted that nanosized particles tend to agglomerate. If separate nanosized particles are desired, often a deagglomeration step is necessary [26].

Preparation of TiO_2 Pastes/Layers via Sol-gel Process: The TiO_2 layer for DSSC can be prepared directly from the sol using dip-coating, spin-coating or ink-jet printing deposition technique or deposited by screen printing of the TiO_2 paste prepared by the integration of TiO_2 nanoparticles into the sol-gel matrix.

For nanostructured layers, the sols are often treated in an autoclave to allow controlled growth of the particles until they reach the desired size. Oswald ripening takes place during this process, leading to a homogeneous particle-size distribution. If a layer is made using these particles, different additives can be added to prevent cracking and agglomeration or increase the binding and viscosity after this ripening process. Thin layers (between 100 and 1 μm) can be made from the sol by dip-coating or ink-jet printing. The last step in preparation of the layers is annealing process, normally at air temperatures above 400°C. At this temperature, organic additives are removed from the layer and the particles are interconnected. Slow heating and cooling is important to prevent cracking of the layer [26, 27]. Since the optimal thickness for the effective operation of DSSCs ranges from 7 to 14 μm [38], multiple successive depositions of ink-jet printed or dip-coated layers would be required. However, the approach of multiple depositions is not economical, and it also introduces defects within the printed layers which hinder the performance of the DSSC.

Therefore, the integration of TiO_2 nanoparticles into the sol (or sol-gel matrix) to prepare TiO_2 paste is more suitable for production of TiO_2 layer. TiO_2 paste for DSSC is usually prepared by mixing TiO_2 nanoparticles and organic template (e.g. terpineol-based solvent). However, organic template can be replaced by the sol-gel matrix. Some attempts have been already made to introduce titanium alkoxide based sol into the TiO_2 paste in order to act as an inorganic binder [31–35, 63]. The results confirmed improved adhesion of the layers as well as improved connectivity of the crystalline grains. Ngamsinlapasathian et al. [31, 32] reported on the preparation of TiO_2 layers by using a mixture of TiO_2 powder and sol-gel matrix. The sol-gel matrix was synthesized by surfactant-assisted templating method using tetraisopropyl orthotitanate modified with acethylacetone and laurylamine hydrochloride. By this method, an efficiency of 8.4% for acetonitrile based DSSC has been reported [31, 32]. A similar approach has been reported by Chen et al. [33]. A commercial TiO_2 powder Degussa P25 has been successfully incorporated into a TiO_2 sol-gel matrix. The sol-gel network has been prepared by mixing titanium isopropoxide, acetylacetone and butanol. Additionally, a non-ionic surfactant Tween 20 was added, which acts as a binder precursor in order to immobilize commercial TiO_2 nanoparticles and enable dip coating deposition on a TCO glass. The results showed that DSSC with such a layer exhibits high conversion efficiency (about 6%), when a quasi-solid-state electrolyte was employed [33]. Other option is using a paste formulation with a sol-gel matrix prepared by Pechini type sol-gel method [30, 34–37]. The Pechini method is based on the preparation of a polymeric sol, which is achieved by mixing ethylene glycol with citric acid in which the metal alkoxide e.g. titanium

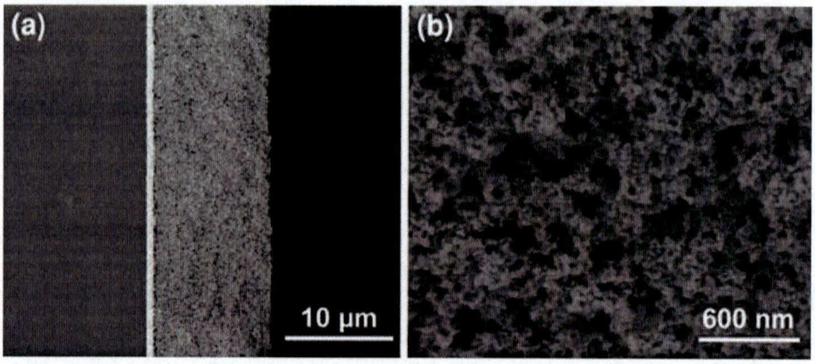

Fig. 8.5 The SEM micrographs of **a** TiO_2 layer deposited on the TCO-coated glass substrate (from *left* to *right*: glass, TCO and TiO_2) and **b** insight view of TiO_2 layer

isopropoxide is dissolved [64]. Titanium ions form mono- and bi-dentate bonds with side carboxylic groups of the polyester [65, 66]. The paste is prepared with the addition of commercial TiO_2 nanoparticles to the sol. The comparison study between the "standard" TiO_2 paste based on terpineoel and methylcellulose [28] and the paste based on Pechini sol-gel method has been reported for the first time by Hočevar et al. [30]. The results showed a two-fold increase in efficiency of DSSC if the Pechini-based TiO_2 paste was used for single mesoporouse TiO_2 layer. The main reasons are higher dye loading, improved electron transport through the TiO_2 layer, and improved adhesion to the TCO substrate. The SEM micrograph (Fig. 8.5a) shows the cross-section of the active electrode consisting of TiO_2 layer deposited on TCO-coated glass substrate (from left to right: glass, TCO, TiO_2 layer) revealing that the TiO_2 layer is homogenous and well attached to the TCO layer (Fig. 8.5a).

The spherical TiO_2 nanoparticles are homogenously distributed within the TiO_2 layer without agglomeration. The layer exhibits high porosity i.e. sponge like structure (Fig. 8.5b), because the sol-gel matrix exothermally decomposes during annealing of the layer. The highest reported efficiencies of DSSC determined under standard test conditions were 10.2% for acetonitrile and 7.3% for ionic liquid-based electrolytes [35].

The sol-gel processes have also been successfully applied for the preparation of low-temperature fabrication of TiO_2 layers for DSSCs. Murayama et al. [63] reported on the use of the sol-gel necking method with a titanium precursor solution added to a paste including TiO_2 nanoparticles. The DSSC showed an efficiency of almost 2%, even when layer was sintered just at 200°C [63].

8.2.3.4 Dye

The TiO_2 is a high band gap semiconductor, so only a small part of the light spectrum (below 350 nm) gets absorbed. In order to extend the absorption spectrum a dye monolayer is attached over entire inner surface of the TiO_2 layer using anchoring groups (–COOH, $-H_2PO_3$, $-SO_3H$, etc.). The dye molecules beside high extinction coefficient in a wide solar spectrum range have to fulfill many essential requirements: (i) anchoring groups have to facilitate the injection of the generated electrons into the conduction band of the semiconductor (ii) the dye monolayer has to cover entire inner surface since un-dyed part of the semiconductor introduce recombination centers (iii) the lowest unoccupied molecular orbital (LUMO) of the dye has to lie several hundreds meV above the conduction band edge of the semiconductor to secure efficient charge injection, and (iv) the highest occupied molecular orbital (HOMO) of the dye molecule has to lie below redox potential of the electrolyte for fast regeneration of the positively charged dye molecule.

The characteristic parameter of the dyes is the difference between LUMO and HOMO energy level, which is linked to the absorption band edge. Smaller difference would extend the absorption spectrum toward longer wavelengths (i.e. lower energy of the photons) which generally increase the short-circuit current of the solar cell. On the other hand, the open circuit voltage of the DSSCs could never exceed the difference between LUMO in HOMO layer, usually it is significantly smaller. Therefore there is always an optimal LUMO–HOMO difference which depends on (i) the incident solar spectrum and (ii) the energy losses for electron injection in the conduction band of the semiconductor and (iii) energy losses required for regeneration of the dye molecule [67]. Generally, the dye molecules should efficiently absorb in a wide solar spectrum range. Several different types of dye molecules exist today differing in molecular structure, their absorption spectrum and in visual appearance [5]. Variety of different dyes allows very creative design of the DSSCs especially for the building integrated photovoltaics (BIPV) and glass façade. The most commonly used dyes today are metal complex dyes based on ruthenium (Ru). Usually they have broad absorption spectrum, relatively long excited lifetime and good electrochemical stability [5]. The most widely used ruthenium complex in DSSC is a *cis*-diisothiocyanato-*bis*(2,2′-bipyridyl-4,4′-dicarboxylato) ruthenium(II) *bis*(tetrabutylammonium) known as N719 (Fig. 8.6a) which allows fabrication of the DSSCs with the efficiencies exceeding 10%. Ruthenium is rather expensive therefore many attempts have been reported to replace Ru with other metals like Os, Re, Fe, Pt, and Cu [67].

The metal complex-based dyes generally have very limited absorption in the near-IR part of the solar spectrum. In order to further increase the conversion efficiency of the DSSCs a broader part of the solar spectrum has to be harvest. Thus a porphyrin- and phthalocyanine-based dyes [5] have been developed which exhibit intense absorption in a near-IR while keeping most advantages of the metal complex-based dyes. In the recent years the research has been focusing also on different types of the organic dyes such as coumarin dyes, indoline dyes, tetrahydroquinoline dyes, triarylamine dyes, heteroanthracene dyes, carbazole dyes, *N,N*-dialkylaniline dyes, hemicyanines, merocyanines, squaraines, perylene dyes, anthraquinone dyes, boradiazaindacene dyes, oligothiophene dyes, polymeric dyes, and natural dyes, [5] . The organic dyes have several advantages compared to metal complex-based dyes. The molecular structures can be easily designed therefore many different organic dyes have been synthesized so far. They are

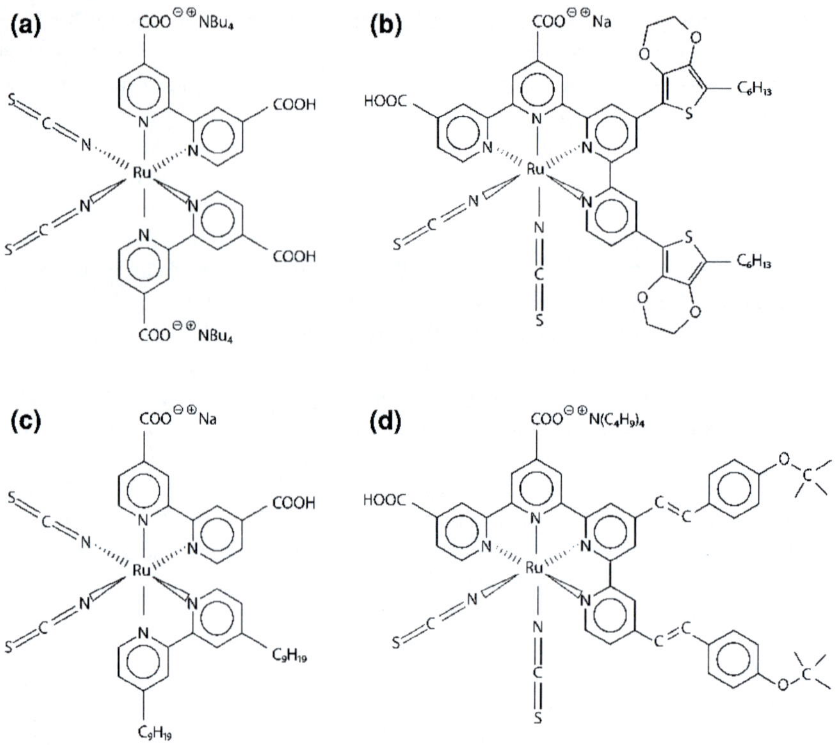

Fig. 8.6 The N719 (**a**), C103 (**b**), Z907 (**c**), and K77 (**d**) dye molecules

generally also more environmental friendly and cheaper. Their molar extinction coefficients are usually higher than those of Ru complexes, making them very attractive for thin film and solid-state DSSCs.

In addition to the wide absorption spectrum of the dye molecule, the dye desorption during long-term operation of the DSSCs is also a very important challenge and it has been recognized as the most crucial process limiting the long-term stability of the DSSCs [68–70]. The most frequently used N719 dye has been recognized as unstable [5, 71]. The most stable dyes so far have been heteroleptic polypyridyl ruthenium sensitizers such as C103 (Fig. 8.6b) [72], Z907 (Fig. 8.6c) [71], and K77 (Fig. 8.6d) [73, 74]. These dyes have successfully passed several aging test in a combination with acetonitrile, ionic liquid or polymer gel-based electrolytes.

8.2.3.5 Electrolyte

In DSSC, an iodide/triiodide redox couple is usually employed as an efficient electron mediator in an electrolyte. The function of the redox couple is to reduce

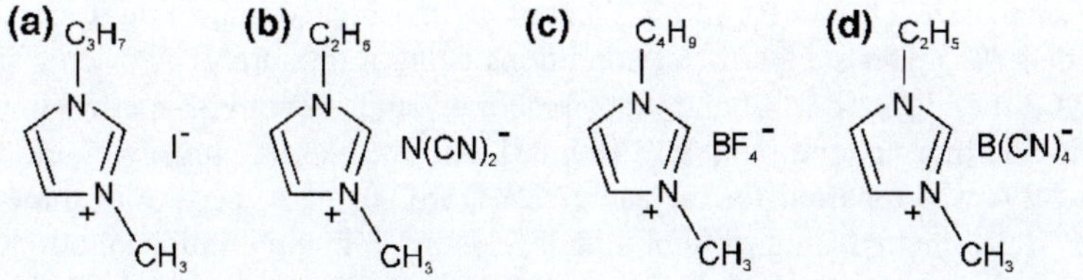

Fig. 8.7 The molecules of four different ionic liquids used in electrolytes; 1-propyl-3-methyl-imidazolium iodide (**a**), 1-ethyl-3methyl-imidazolium dicyanamide (**b**), 1-butil-3-methyl-imidazolium tetrafluoroborate (**c**), and 1-ethyl-3-methyl-imidazolium tetracyanoborate (**d**)

the dye cation (reaction 8.1), following electron injection, and carry the charge back and forth between the two electrodes. Generally, an electrolyte for DSSCs consists of a redox couple, and various additives (such as lithium salt, pyridine, ligands of the dyes, etc.), in a solvent. According to its physical states, an electrolyte can be divided into one of three types: liquid electrolyte, quasi-solid-state electrolyte, and solid-state electrolyte. Liquid electrolyte can be further divided into organic solvent-based electrolytes and ionic liquid-based electrolytes.

Organic solvent-based electrolytes have been widely used and investigated in DSSCs. The most commonly used solvent is acetonitrile since it has a large diffusion coefficient of I_3^- ions enabling the realization of high efficient DSSCs. The highest reported conversion efficiency of the DSSCs is reported for the cells using this type of the electrolyte. The efficiency of a small size DSSC ($\sim$0.2 cm^2) nowadays exceeds 11.3% when measured under standard test conditions [75]. The drawback of this electrolyte is the volatility of the acetonitrile thus a proper sealing must be used to maintain long-term operation. Alternatively, the acetonitrile could be replaced with non-volatile ionic liquids. They should be low cost, environmental friendly, and also electrochemically, thermally, and UV stable. Typical representatives of ionic liquids that fulfill the criteria are 1,3 alkyl imidazolium salts with different anions (iodide- Fig. 8.7a, dicyandiamide- Fig. 8.7b, tetrafluoroborate—Fig. 8.7c, tetracyanoborate- Fig. 8.7d, etc.) [76–81].

Ionic liquids based on 1,3 alkyl imidazolium iodide combine the properties of the solvent and the source of I^- ion. However, their viscosity is relatively high which is reflected in a low diffusion coefficient of the I_3^- ion in the electrolyte limiting the performance of DSSC [82]. Thus a concentration of the tri-iodide in the electrolyte should be higher compared to acetonitrile-based electrolyte in order to maintain sufficient transport of the ions. Higher tri-iodide concentration pronounces the recombinations i.e. the reactions between generated electrons in the semiconductor and the tri-iodide ions in the electrolyte and consequently reduces the conversion efficiency of the DSSCs. In order to keep adequate diffusion of tri-iodide ions in the electrolyte while maintaining lower tri-iodide concentration in the electrolyte a mixture of two or more ionic liquids have been used in electrolytes, i.e. one used as a source of iodide and other(s) with a significantly lower viscosity are used as a solvent [82]. The highest conversion efficiency published for a small size DSSC (0.158 cm^2) based on the binary ionic liquid electrolyte exceeded 7% under standard test conditions using a mixture of two ionic liquids 1-ethyl-3-methyl-imidazolium tetracyanoborate and 1-propyl-3-methyl-imidazolium iodide in a volume ratio 7:13 [80, 83]. An even larger conversion efficiency up to 8.2% was reported for the same size DSSC using a mixture of three ionic liquids 1,3-dimethyl-imidazolium iodide, 1-ethyl-3-methyl imidazolium iodide, and 1-ethyl-3-methyl imidazolium tetracyanoborate [84].

The iodide/triiodide (I^-/I_3^-) is the most common redox couple and performs better than all alternatives tested so far mainly due to the slow recombination reaction [85, 86]. However, the I^-/I_3^- based electrolytes are very corrosive, therefore other materials used in the DSSCs such as silver contacts have to be well protected and isolated in order to prevent deterioration of the solar cell performance. The corrosiveness of the I^-/I_3^-- based electrolytes is thus a strong driving force in the quests for the alternative redox couples. Two main directions can be identified nowadays, one toward molecular species (such as ferrocene-, hydroquinone-based redox couples [87], $LiBr/Br_2$ [88], Br^-/Br_3^- [89], $SCN^-/(SCN)_3^-$, $SeCN^-/(SeCN)_3$ [90, 91], stable organic radicals originating from 2,2,6,6-tetramethyl-1-piperidinyloxy or N,N′-di-m-tolyl-N,N′-diphenylbenzidine compound [92], and sulfur-containing species [93]) and another direction aiming at transition-metal-based systems (like Co(II)/(III) [94], Cu(I)/(II) [95], Ni(III)/Ni(IV) [96]). One of the most promising alternative redox couple is non-corrosive sulfur-containing redox couple for which the conversion efficiency up to 6.4% were measured at standard test conditions [93].

Another alternative, which furthermore overcomes the corrosive nature of the iodide/triiodide-based electrolytes, is to use a solid-state hole conductor. Among the possible inorganic p-type semiconductors, copper-based compounds were identified to be suitable alternatives. These hole-transporters can be cast from solution or vacuum deposition to fully interpenetrate the TiO_2 layer TiO_2 layer [14]. A solid-state device based on CuI was first demonstrated by Tennakone et al. [97], and a conversion efficiency as high as 4.7% has been reported [98]. Although CuI is a very promising material, the cells are generally unstable, which is attributed to a stoichiometric excess of iodine molecules adsorbed at the CuI surface, acting as hole-trapping sites [99]. An alternative is incorporated CuSCN as the hole-transporter, which improved stabilities. However, these DSSCs exhibit slightly lower efficiencies of around 2% under STC [100, 101]. Crystallization of the copper compounds in the pores is also thought to contribute to the cell degradation.

Another class of materials constituting effective hole conductors are organic molecular and polymer semiconductors. The charge mobility in organic hole-transporters is generally lower than in copper-based compounds, but the mobility is still reasonably high [14]. Moreover, the molecules are solution processable, rendering them well suited for mesopore infiltration, they are cheap, and allow creation of innumerable variations. For all these reasons, organic semiconductors are highly applicable [14]. The first solid-state DSSCs incorporating polymeric and molecular organic hole-transporters were realized by Murakoshi et al. [102] and

Bach et al. [103]. The conversion efficiency of best performing solid-state DSSCs incorporating organic hole transport materials exceeded 5% [104].

However, an ionic liquid-based electrolyte is still one of the most attractive charge mediators used in DSSCs for practical applications. One alternative to prevent the solvent leaking and to attain long-term outdoor operation of the DSSCs is the solidification of the liquid electrolyte. Thus many studies have been focusing on solidification of the electrolytes. Quasi- or solid-state electrolytes can be realized by mixing the liquid electrolytes with polymers, low molecular weight organic gelators [105–108], and inorganic nanoparticles [109, 110]. In the latter case the midazolium-based electrolyte forms organized structure with the nanoparticle in the center surrounded with the imidazolium cations adsorbed on its surface surrounded with a continuous chain of I^- and I_3^- ions. In addition to the diffusion of the ions the charge transport through organized structure is based also on electron exchange reaction along I^-/I_3^- chain which accelerates the charge transfer in the electrolyte [111] and improves the conversion efficiency of DSSCs up to 20% relatively [109, 110].

Sol-Gel-Based Electrolytes

The sol-gel processing has been successfully applied for the solidification/networking of the liquid electrolytes. Different approaches have been used. One of the promising approaches is to form the organically modified network (ormosil) through hydrolysable alkoxysilane derivatives in order to capture the volatile electrolyte [112–114]. Class I and Class II [6] ormosil materials have been tested, the difference between them is that in class II silica and poly(alkylene oxide) networks are covalently bonded. The Class II materials enable the incorporation of iodide/tri-iodide redox couple, allow the penetration of the electrolyte through the TiO_2 layer and give stable gels. The realization of N719 dye-based DSSC with the efficiencies up to 6% have been reported [114].

The second approach considers the modification of imidazolium iodide with alkoxysilane groups enabling the formation of gels through the solvolysis procedure. The highest reported efficiencies of DSSC assembled with this type of electrolyte exceeded 3.2% [115]. An attractive route for the preparation of efficient nanostructured silica-based electrolytes was reported by Cerneaux et al. [116], it includes the silica nanoparticles crosslinked with the alkoxysilane based precursors. Two different approaches were reported, one involves grafting of 3-aminopropyltriethoxysilane onto the activated surface of silica nanoparticles and further reacted with alkyl iodides. In the second approach synthesized treithoxysilane precursors bearing quaternary ammonium iodide moieties were grafted onto activated nanosized particles. The networks were added to the nitrile based electrolytes and tested in DSSCs, the highest reported efficiencies measured under STC exceeded 6% [116].

Sol-gel-based electrolytes are recognized as a good solution for preventing the leakage of the electrolyte in DSSCs, but additionally they could act as an adhesive that holds the two electrodes together. However, to achieve a long-term stability of DSSCs an edge encapsulating material is still needed. This is indispensible for the advanced hybrid system such as photoelectrochromic windows (see Sect. 8.3) for which also a class II sol-gel-based electrolyte has been successfully applied [117].

8.3 Advanced Hybrid DSSC Systems

Several types of DSSC systems have been reported that combine DSSC with other solar cells in tandem structures [118] or the ones with an additional charge storage layer [119–121]. In the later, the sol-gel chemistry in the preparation of their components is necessary; therefore the emphasis of this section is on hybrid DSSC systems with additional charge storage (Fig. 8.8).

Hybrid DSSC systems implement the device with chromogenic properties resulting in photoelectrochomic smart windows [119, 120] or allow the storage of the electrical energy produced by solar cell (solar-charging battery) [121]. In both systems tungsten oxide (WO_3) has been studied as a charge storage layer and lithium ions (Li^+) were incorporated into the electrolyte in order to maintain the charge balance of the systems during coloration/charging and bleaching/discharging processes (Eq. 8.5). The main difference between these systems is the thickness of the charge storage layer and the thickness of the photoactive electrode in the DSSCs, reflected consequently also in the transparency of the devices.

$$\begin{array}{c} \text{transparent/uncharged} \leftrightarrow \text{colored/charged} \\ WO_3 + xe^- + xLi^+ \leftrightarrow Li_xWO_3 \end{array} \tag{8.5}$$

The photoelectrochromic device (PE) could be described as a combination of DSSC and electrochromic layer [122]. In contrast to the electrochromic devices, the coloring is induced by the illumination and no external voltage source is needed. The transmittance can be decreased under illumination and in contrast to the photochromic devices the system is internally switchable. Applications of these devices include, for example, switchable sunroofs or mirrors in the cars or smart windows in buildings. In the first reported PE device the WO_3 layer was positioned on the TCO-glass substrate opposite to the photoactive electrode [120]. The coloration of the device was achieved during illumination by the injection of electrons to the WO_3 layer via external circuit, at the same time the lithium ions were intercalated in order to keep the charge balanced. The bleaching was possible solely in the dark by the electron transfer from the TiO_2 to the electrolyte, which is unwanted loss reaction in DSSC. Additionally, in this design of the PE system the coloration and bleaching processes are competing processes, which means that if the bleaching is fast, the coloring is weak and slow.

Better optimization of coloring and bleaching processes as well as bleaching during illumination and use of the device as a solar cell was enabled by changing the design of the PE system in such a way that the WO_3 layer was positioned

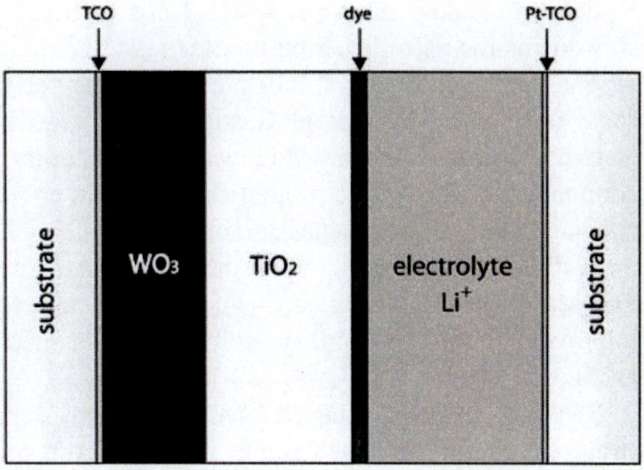

Fig. 8.8 Schematic drawing of hybrid DSSC system

between front TCO and TiO_2 layer of the DSSCs [119]. Solid-state PE devices in a size up to 10 × 10 cm^2 were developed enabling the change in photonic transmittance between 62% (bleached state) and 1.6% (colored state) in 15 min under 1 sun illumination [119].

The solar charged battery acts as a solar cell under illumination, but in addition the energy is stored during illumination and could be used in the absence of the light. The same design as for the PE system with the WO_3 layer positioned between front TCO and TiO_2 layer in DSSC was used for the realization of the solar charging battery. In this case the thickness of WO_3 layer was up to 50 μm enabling the charge storage up to 1.8 As/cm^2 after exposing the device for 1 h under 1 sun illumination. The application of this device could be foreseen in low-cost and low-power applications, examples include microelectronics, watches, smart cards, or transponders [121].

8.3.1 Sol-Gel Processed Components in Hybrid DSSC Devices

Nanostructured WO_3-TiO_2 layers for PE system were prepared by the sol-gel route [123, 124]. To obtain porous and crack-free layers with the thickness up to 0.8 μm with a single dipping cycle a templating strategy was successfully used. As a template three-dimensional network based on organically modified silane was introduced to the peroxopolytungstic acid and Ti-isopropoxide based sols. The study revealed that the presence of silica in the template was essential to avoid cracking of the layers [124]. First the WO_3 layers were dip-coated on the TCO-glass and annealed at 450°C, in the second step the TiO_2 layers were deposited on the top. The titanium sol is penetrating the WO_3 layer forming together with WO_3 and SiO_2 a 5–10 nm amorphous coating around the WO_3 spherical monoclinic grains with the 30 nm in diameter. The coating of the WO_3 grains with titanium oxide reduces loss reactions which leads to better coloration of the device and slowdowns the self-bleaching process [124, 125].

The synthesis of the solid-state electrolyte for the PE system by the sol-gel route is reported by Opara Krašovec et al. [125]. For the solidification of the electrolyte ormosilane network was used in order to form ormolyte. Ormolyte combines the electrolytic properties of organic polymers with the mechanical and chemical strength of the glassy inorganic backbone. This kind of electrolyte stays three-dimensionally stable upon the temperature changes that a window could be exposed to. The addition of a cationic surfactant to the ormolyte resulted in an improvement of the stability, coloration, and memory effect of the PE device [125].

The WO_3 layer in solar charged battery was deposited on the TCO-glass substrate by doctor blade technique from the corresponding paste. The paste was prepared by mixing the sol-gel made WO_3 nanopowder into the terpinoel-based matrix using classical procedure for the paste preparation [28]. The WO_3 nanopowder was achieved by heat treatment at 450°C of the WO_3 gel formed from peroxopolytungstic acid sol [126].

8.4 Future Outlook

Dye-Sensitized solar cells (DSSCs), due to the low-cost manufacturing processes based on screen printing technology on the one hand and due to broad range of color choices and optical semi transparency on the other hand, offer multiple opportunities for decorations and creative energy-active architectural solutions for facades in modern buildings. However, the DSSC technology is still at an early stage and several development steps are needed to be taken to increase its share in the photovoltaic market. After nearly two decades of intense research, the physical chemistry of several basic processes in the DSSCs is still far from fully understood. It is still not possible to accurately predict how the performance of DSSC could be affected by a small change of the system, e.g. replacing one component or changing the electrolyte composition. With time, the chemical complexity of DSSCs has become clear, and the main challenge for future research is to understand and master this complexity in particular at the oxide/dye/electrolyte interface. Thus, for future research, it will be important to carefully select several reference systems that emphasize different key aspects of the device and characterize these systems in-depth with all the different techniques that are available [5].

A challenging but realizable goal for the present DSSC technology is to achieve efficiencies above 15% by development of new dyes that would broaden the absorption in the longer wavelength range or new structure (even hybrid) that would combine several cells with different dyes for harvesting the entire solar spectrum. In the state-of-the-art DSSC device the main losses are the potential drop used for (i) the electron injection and (ii) dye regeneration process and (iii) the reduction of the recombination loss between electrons in the TiO_2 and acceptor species in the electrolyte. However, the challenge is to develop dye-electrolyte systems that give efficient regeneration of the oxidized dye. Such systems must be probably combined with the efficient blocking layers of the mesoporous oxide film and TCO substrate. To reach the aimed efficiency of DSSC components with better individual characteristics than the ones used in the state-of-the-art devices has to be developed and optimized to exploit their advantages in a complete DSSC device [5].

In addition, most laboratory research is conducted on cells up to 1 cm^2, while for commercial use the size of the devices should increase by at least a factor of 100. The shift to a larger surface area and production of photovoltaic modules is a complex challenge that itself requires a different methodological approach and careful optimization of the entire technological process. Each manufacturing step has to be simplified and automated for industrial production and at the same time optimized toward the low-cost and environmental friendly technologies. The ideal process for the preparation of the low-cost DSSC devices should involve solution processing of all layers on roll-to-roll substrates with as few coating and printing steps as possible. The process should be free from toxic solvents and chemicals and the final optoelectronic devices should have a low environmental impact and a high degree of recyclability [127]. Therefore, the development and optimization of optoelectronic devices is currently focused in several directions such as: (i) reduction of production cost (cheap materials, low-temperature processing, and printed technologies) (ii) deposition on flexible substrates, and (iii) improvement of the peak performance reproducibility and reliability [128]. The continuous development of the DSSCs, either through innovative design of new materials or by statistical trial-and-error approach, is needed to achieve efficiencies of DSSCs above 15%. In this respect the sol-gel processing can make a significant contribution in development of the efficient mesopourous oxide network and stable electrolytes.

Acknowledgments The authors are grateful for the financial support given by the Slovenian Research Agency (P2-0197).

References

1. Green MA, Emery K, Hishikawa Y, Warta W (2011) Solar cell efficiency tables (version 37). Prog Photovolt: Res Appl 19:84–92
2. Wöhrle D, Meissner D (1991) Organic solar cells. Adv Mater 3:129–138
3. O'Regan B, Grätzel M (1991) A low-cost, high efficiency solar cells based on dye-sensitized colloidal TiO_2 films. Nature 353:737–740
4. Peter LM (2007) Dye-Sensitized nanocrystalline solar cells. Phys Chem Chem Phys 9:2630–2642
5. Hagfeldt A, Boschloo G, Sun L, Kloo L, Pettersson H (2010) Dye-Sensitized solar cells. Chem Rev 110:6595–6663
6. Brinker CJ, Scherer GW (1990) Sol-Gel science. Academic Press, Boston
7. Haque SA, Tachibana Y, Klug DR, Durrant JR (1998) Charge recombination kinetics in dye sensitised nanocrystalline TiO_2 films under externally applied bias. J Phys Chem B 102: 1745–1749

8. Standard IEC 60904-3 (2007) Photovoltaic devices-Part 3: Measurement principles for terrestrial photovoltaic solar devices with reference spectral irradiance data. International Electrotechnical Commission, Geneva
9. Standard IEC 60904 (2006–2009) International standard—Photovoltaic devices. International Electrotechnical Commission, Geneva
10. Standard IEC 60904-4 (2009) Photovoltaic devices—Part 4: Reference solar devices - Procedures for establishing calibration traceability. International Electrotechnical Commission, Geneva
11. Pinazo JM, Cañada J, Boscá JV (1995) A new method to determine Ångström's turbidity coefficient: Its application for Valencia. Sol Energy 54:219–226
12. Suzuki M, Nose J, Nakano A, Imura Y, Igarashi M (1997) Outdoor measurement of multi-junction solar cells and modules based on the reference sunlight method. Sol Energy 60:63–70
13. Standard IEC 60904-7 (2009) Photovoltaic devices - Part 7: Computation of the spectral mismatch correction for measurements of photovoltaic devices International Electrotechnical Commission, Geneva
14. Kalyanasundaram K (2010) Dye-Sensitized solar cells. EPFL Press, Lausanne
15. Hauch A, Georg A (2001) Diffusion in the electrolyte and charge-transfer reaction at the platinum electrode in dye-sensitized solar cells. Electrochim Acta 46:3457–3466
16. Fang X, Ma T, Guan G, Akiyama M, Kida T, Abe E (2004) Effect of the thickness of the Pt film coated on a counter electrode on the performance of a dye-sensitized solar cell. J Electroanal Chem 570:257–263
17. Fang X, Ma T, Guan G, Akiyama M, Abe E (2004) Performances characteristics of dye-sensitized solar cells based on counter electrodes with Pt films of different thickness. J Photochem PhotobioL A: Chem 164:179–182
18. Tsekouras G, Mozer AJ, Wallace GG (2008) Enhanced performance of dye sensitized solar cells utilizing platinum electrodeposit counter electrodes. J Electrochem Soc 155:K124–K128
19. Li P, Wu J, Lin J, Huang M, Lan Z, Li Q (2008) Improvement of performance of dye-sensitized solar cells based on electrodeposited-platinum counter electrode. Electrochim Acta 53:4161–4166
20. Ito S, Nazeeruddin MK, Liska P, Comte P, Charvet R, Péchy P, Jirousek M, Kay A, Zakeeruddin SM, Grätzel M (2006) Photovoltaic characterization of dye-sensitized solar cells: effect of device masking on conversion efficiency. Prog Photovolt: Res Appl 14:589–601
21. Kroon JM, Bakker NJ, Smit HJP, Liska P, Thampi KR, Wang P, Zakeeruddin SM, Grätzel M, Hinsch A, Hore S, Würfel U, Sastrawan R, Durrant JR, Palomares E, Pettersson H, Gruszeck T, Walter J, Skupien K, Tulloch GE (2007) Nanocrystalline dye-sensitized solar cells having maximum performance. Prog Photovolt: Res Appl 15:1–18
22. Khelashvili G, Behrens S, Hinsch A, Habicht W, Schild D, Eichhöfer A, Sastrawan R, Skupien K, Dinjus E, Bönnemann H (2007) Preparation and characterization of low platinum loaded Pt:SnO_2 electrocatalytic films for screen printed dye solar cell counter electrode. Thin Solid Films 515:4074–4079
23. Li K, Yu Z, Luo Y, Li D, Meng Q (2007) Recent progress of counter electrodes in nanocrystalline dye-sensitized solar cells. J Mater Sci Technol 23:577–582
24. Hamann TW, Jensen RA, Martinson ABF, Ryswyk HV, Hupp JT (2008) Advancing beyond current generation dye-sensitized solar cells. Energy Environ Sci 1:66–78
25. Jose R, Thavasi V, Ramakrishna S (2009) Metal oxides for dye-sensitized solar cells. J Am Ceram Soc 92:289–301
26. Carp O, Huisman CL, Reller A (2004) Photoinduced reactivity of titanium dioxide. Prog Solid St Chem 32:33–177
27. Chen X, Mao SS (2007) Titanium dioxide nanomaterials: synthesis, properties, modifications, and applications. Chem Rev 107:2891–2959
28. Wienke J, Kroon JM, Sommeling PM, Kinderman R, Späth M, Roosmalen JAMV, Sinke WC, Baumgärtner S (1997) Effect of TiO_2 electrode properties on the efficiency of

nanocrystalline dye-sensitized solar cells (nc-DSC). 14th European photovoltaic solar energy conference and exhibition, Barcelona, Spain
29. Chittibabu KG, Gaudiana R, Beckenbaugh B, Li L, Lee M (2006) Low temperature interconnection of nanoparticles. US Patent 7 094 441
30. Hočevar M, Opara Krašovec U, Berginc M, Dražić G, Hauptman N, Topič M (2008) Development of TiO_2 pastes modified with Pechini sol-gel method for high efficiency dye-sensitized solar cell. J Sol-gel Sci Technol 48:156–162
31. Ngamsinlapasathian S, Sakulkhaemaruethai S, Pavasupree S, Kitiyanan A, Sreethawong T, Suzuki Y, Yoshikawa S (2004) Highly efficient dye-sensitized solar cell using nanocrystalline titania containing nanotube structure. J Photochem Photobiol A: Chem 164:145–151
32. Ngamsinlapasathian S, Sreethawong T, Suzuki Y, Yoshikawa S (2005) Single- and double-layered mesoporous TiO_2/P25 TiO_2 electrode for dye-sensitized solar cell. Sol Energ Mat Sol C 86:269–282
33. Chen Y, Stathatos E, Dionysiou DD (2009) Sol-gel modified TiO_2 powder films for high performance dye-sensitized solar cells. J Photochem Photobiol A: Chem 203:192–198
34. Hočevar M, Berginc M, Topič M, Opara Krašovec U (2010) Sponge-like TiO_2 layers for dye-sensitized solar cells. J Sol-Gel Sci Technol 53:647–654
35. Opara Krašovec U, Berginc M, Hočevar M, Topič M (2009) Unique TiO_2 paste for high efficiency dye-sensitized solar cells. Sol Energ Mat Sol Cells 93:379–381
36. Hočevar M, Opara Krašovec U, Berginc M, Topič M (2010) One step preparation of TiO_2 layer for high efficiency dye-sensitized solar cell. Acta Chim Slov 57:405–409
37. Drev M, Opara Krašovec U, Hočevar M, Berginc M, Maček MK, Topič M (2011) Pechini based titanium sol as a matrix in TiO_2 pastes for dye-sensitized solar cell application. J Sol-Gel Sci Technol 59:245–251
38. Ito S, Zakeeruddin SM, Humphry-Baker R, Liska P, Charvet R, Comte P, Nazeeruddin MK, Péchy P, Takata M, Miura H, Uchida S, Grätzel M (2006) High-efficiency organic-dye-sensitized solar cells controlled by nanocrystalline-TiO_2 electrode thickness. Adv Mater 18:1202–1205
39. Koo HJ, Park J, Yoo B, Yoo K, Kim K, Park NG (2008) Size-dependent scattering efficiency in dye-sensitized solar cell. Inorg Chim Acta 361:677–683
40. Hore S, Vetter C, Kern R, Smit H, Hinsch A (2006) Influence of scattering layers on efficiency of dye-sensitized solar cells. Sol Energ Mat Sol Cells 90:1176–1188
41. Wang P, Zakeeruddin SM, Comte P, Charvet R, Humphry-Baker R, Grätzel M (2003) Enhance the performance of dye-sensitized solar cells by co-grafting amphiphilic sensitizer and hexadecylmalonic acid on TiO_2 nanocrystals. J Phys Chem B 107:14336–14341
42. Bosc F, Ayral A, Albouy PA, Guizard C (2003) A simple route for low-temperature synthesis of mesoporous and nanocrystalline anatase thin films. Chem Mater 15:2463–2468
43. Gutiérrez-Tauste D, Zumeta I, Vigil E, Hernández-Fenollosa MA, Domènech X, Ayllón JA (2005) New low-temperature preparation method of the TiO_2 porous photoelectrode for dye-sensitized solar cells using UV irradiation. J Photochem Photobiol A: Chem 175:165–171
44. Zhang D, Yoshida T, Minoura H (2003) Low-temperature fabrication of efficient porous titania photoelectrodes by hydrothermal crystallization at the solid/gas interface. Adv Mater 15:814–817
45. Lindström H, Holmberg A, Magnusson E, Lindquist SE, Malmqvist L, Hagfeldt A (2001) A new method for manufacturing nanostructured electrodes on plastic substrates. Nano Lett 1:97–100
46. Lindström H, Holmberg A, Magnusson E, Malmqvist L, Hagfeldt A (2001) A new method to make dye-sensitized nanocrystalline solar cells at room temperature. J Photochem Photobiol A: Chem 145:107–112
47. Dürr M, Schmid A, Obermaier M, Rosselli S, Yasuda A, Nelles G (2005) Low-temperature fabrication of dye-sensitized solar cells by transfer of composite porous layers. Nat Mater 4:607–611
48. Yamaguchi T, Tobe N, Matsumoto D, Nagai T, Arakawa H (2010) Highly efficient plastic-substrate dye-sensitized solar cells with validated conversion efficiency of 7.6%. Sol Energ Mat Sol C 94:812–816

49. Iwasaki M, Hara M, Ito S (1998) Facile synthesis of nanocrystalline anatase particles from titanyl sulfate. J Mater Sci Lett 17:1769–1771
50. Bach U, Lupo D, Comte P, Moster JE, Weissortel F, Salbeck J, Spreitzer H, Grätzel M (1998) Solid-state dye-sensitized mesoporous TiO_2 solar cells with high photon-to-electron conversion efficiencies. Nature 395:583–585
51. Tonejc M, Djerdj I, Tonejc A (2001) Evidence from HRTEM image processing, XRD and EDS on nanocrystalline iron-doped titanium oxide powders. Mater Sci Eng B 85:55–63
52. Chen YF, Lee CY, Yeng MY, Chiu HT (2003) The effect of calcination temperature on the crystallinity of TiO_2 nanopowders. J Cry Growth 247:363–370
53. Okudera H, Yokogawa Y (2003) Fabrication of titania-coated silica fibers and effect of substrate shape on coating growth rate. Thin Solid Films 423:119–124
54. Yang P, Lu C, Hua N, Du Y (2002) Titanium dioxide nanoparticles co-doped with Fe^{3+} and Eu^{3+} ions for photocatalysis. Mater Lett 57:794–801
55. Arabatzis IM, Antonaraki S, Stergiopoulos T, Hiskia A, Papaconstantinou E, Bernard MC, Falaras P (2002) Preparation, characterization and photocatalytic activity of nanocrystalline thin film TiO2 catalysts towards 3, 5-dichlorophenol degradation. J Photochem PhotobioL A: Chem 149:237–245
56. Livage J, Henry M, Sanchez C (1988) Sol-gel chemistry of transition metal oxides. Prog Solid State Chem 18:259–341
57. Doeuff S, Henry M, Sanchez C (1990) Sol-gel synthesis and characterization of titanium oxo-acetate polymers Mater Res Bull 25:1519–1529
58. Campostrini R, Ischia M, Palmisano L (2003) Pyrolysis study of sol-gel derived TiO_2 powders: Part III. TiO_2-anatase prepared by reacting titanium(IV) isopropoxide with acetic acid. J Therm Anal Cal 71:1011–1021
59. Harizanov O, Harizanova A (2000) Development and investigation of sol-gel solutions for the formation of TiO_2 coatings. Sol Energy Mater Sol Cells 63:185–195
60. Guillard C, Beaugiraud B, Dutriez C, Herrmann JM, Jaffrezic H, Jaffrezic-Renault N, Lacroix M (2002) Physicochemical properties and photocatalytic activities of TiO_2-films prepared by sol-gel methods. Appl Catal B: Environ 39:331–342
61. Al-Salim NI, Bagshaw SA, Bittar A, Kemmett T, McQuilla AJ, Mills AM, ryan MJ (2000) Characterisation and activity of sol-gel-prepared TiO_2 photocatalysts modified with Ca, Sr or Ba ion additives. J Mater Chem 10:2358–2363
62. Li Y, White TJ, Lim SH (2004) Low-temperature synthesis and microstructural control of titania nano-particles. J Solid State Chem 177:1372–1381
63. Murayama M, Yamazaki E, Nishikawa N, Hashimoto N, Mori T (2006) Low-temperature fabrication of TiO_2 necking electrode by sol-gel method and its application to dye-sensitized solar cell. Jpn J Appl Phys 45:7917–7921
64. Pechini M (1967) Method of preparing lead and alkaline earth titanates and niobates and coating method using the same to form a capacitor. US Patent 3:330–697
65. Leite ER, Sousa CMG, Longo E, Varela JA (1995) Influence of polymerization on the synthesis of $SrTiO_3$: Part I. Characteristics of the polymeric precursors and their thermal decomposition. Ceramics Intern 21:143–152
66. Nakamoto K (1986) Infrared and raman spectra of inorganic and coordinate compounds. Wiley, New York
67. Snaith HJ (2010) Estimating the maximum attainable efficiency in dye-sensitized solar cells. Adv Funct Mater 20:13–19
68. Asghar MI, Miettunen K, Halme J, Vahermaa P, Toivola M, Aitola K, Lund P (2010) Review of stability for advanced dye solar cells. Energ Environ Sci 3:418–426
69. Sommeling PM, Späth M, Smit HJP, Bakker NJ, Kroon JM (2004) Long-term stability testing of dye-sensitized solar cells. J Photochem Photobiol A: Chem 164: 137–144
70. Ke L, Dolmanan SB, Shen L, Pallathadk PK, Zhang Z, Lai DMY, Liu H (2010) Degradation mechanism of ZnO-based dye-sensitized solar cells. Sol Energy Mater Sol Cells 94: 323–326

71. Wang P, Zakeeruddin SM, Moser JE, Nazeeruddin MK, Sekiguchi T, Grätzel M (2003) A stable quasi-solid-state dye-sensitized solar cell with an amphiphilic ruthenium sensitizer and polymer gel electrolyte. Nat Mater 2:402–407
72. Shi D, Pootrakulchote N, Li R, Guo J, Wang Y, Zakeeruddin SM, Grätzel M, Wang P (2008) New efficiency records for stable dye-sensitized solar cells with low-volatility and Ionic liquid electrolytes. J Phys Chem C 112:17046–17050
73. Kuang D, Klein C, Ito S, Moser J, Baker RH, Evans N, Duriaux F, Grätzel C, Zakeeruddin SM, Grätzel M (2007) High-efficiency and stable mesoscopic dye-sensitized solar cells based on a high molar extinction coefficient ruthenium sensitizer and nonvolatile electrolyte. Adv Mater 19:1133–1137
74. Kuang D, Klein C, Zhang ZP, Ito S, Moser JE, Zakeeruddin SM, Grätzel M (2007) Stable, high-efficiency ionic-liquid-based mesoscopic dye-sensitized solar cells. Small 3:2094–2102
75. Grätzel M (2006) The advent of mesoscopic injection solar cells. Prog Photovolt: Res Appl 14:429–442
76. Berginc M, Opara Krašovec U, Jankovec M, Topič M (2007) The effect of temperature on the performance of dye-sensitized solar cells based on a propyl-methyl-imidazolium iodide electrolyte. Sol Energ Mat Sol Cells 91:821–828
77. Matsumoto H, Matsuda T, Tsuda T, Hagiwara R, Ito Y, Miyazaki Y (2001) The application of room temperature molten salt with low viscosity to the electrolyte for dye-sensitized solar cell. Chem Lett 30:26–27
78. Wang P, Zakeeruddin SM, Exnar I, Grätzel M (2002) High efficiency dye-sensitized nanocrystalline solar cells based on ionic liquid polymer gel electrolyte. Chem Commun 24:2972–2973
79. Wang P, Zakeeruddin SM, Comte P, Exnar I, Grätzel M (2003) Gelation of ionic liquid-based electrolytes with silica nanoparticles for quasi-solid-state dye-sensitized solar cells. J Am Chem Soc 125:1166–1167
80. Kuang D, Wang P, Ito S, Zakeeruddin SM, Grätzel M (2006) Stable mesoscopic dye-sensitized solar cells based on tetracyanoborate Ionic liquid electrolyte. J Am Chem Soc 128:7732–7733
81. Wachter P, Zistler M, Schreiner C, Berginc M, Opara Krašovec U, Gerhard D, Wasserscheid P, Hinsch A, Gores HJ (2008) Characterisation of DSSC-electrolytes based on 1-ethyl-3-methylimidazolium dicyanamide: Measurement of triiodide diffusion coefficient, viscosity, and photovoltaic performance. J Photochem Photobiol A: Chem 197:25–33
82. Berginc M, Opara Krašovec U, Hočevar M, Topič M (2008) Performance of dye-sensitized solar cells based on Ionic liquids: Effect of temperature and iodine concentration. Thin Solid Films 516:7155–7159
83. Wang P, Zakeeruddin SM, Humphry-Baker R, Grätzel M (2004) A binary ionic liquid electrolyte to achieve $\geq$7% power conversion efficiencies in dye-sensitized solar cells. Chem Mater 16:2694–2696
84. Bai Y, Cao Y, Zhang J, Wang M, Li R, Wang P, Zakeeruddin SM, Grätzel M (2008) High-performance dye-sensitized solar cells based on solvent-free electrolytes produced from eutectic melts. Nat Mater 7:626–630
85. Hagfeldt A, Grätzel M (1995) Light-induced redox reactions in nanocrystalline systems. Chem Rev 95:49–68
86. Grätzel M (2005) Solar energy conversion by dye-sensitized photovoltaic cells. Inorg Chem 44:6841–6851
87. Pichot F, Gregg BA (2000) The photovoltage-determining mechanism in dye-sensitized solar cells. J Phys Chem B 104:6–10
88. Hara K, Horiguchi T, Kinoshita T, Sayama K, Arakawa H (2001) Influence of electrolytes on the photovoltaic performance of organic dye-sensitized nanocrystalline TiO_2 solar cells. Sol Energy Mater Sol Cells 70:151–161
89. Wang ZS, Sayama K, Sugihara HJ (2005) Efficient eosin Y dye-sensitized solar cell containing Br^-/Br_3^- electrolyte. J Phys Chem B 109:22449–22455

90. Oskam G, Bergeron BV, Meyer GJ, Searson PC (2001) Pseudohalogens for dye-sensitized TiO_2 photoelectrochemical cells. J Phys Chem B 105:6867–6873
91. Bergeron BV, Marton A, Oskam G, Meyer GJ (2005) Dye-sensitized SnO_2 electrodes with iodide and pseudohalide redox mediators. J Phys Chem B 109:937–943
92. Gregg BA, Pichot F, Ferrere S, Fields CL (2001) Interfacial recombination processes in dye-sensitized solar cells and methods to passivate the interfaces. J Phys Chem B 105: 1422–1429
93. Wang M, Chamberland N, Breau L, Moser J-E, Humphry-Baker R, Marsan B, Zakeeruddin SM, Grätzel M (2010) An organic redox electrolyte to rival triiodide/iodide in dye-sensitized solar cells. Nat Chem 2:85–389
94. Nelson JJ, Amick TJ, Elliott CM (2008) Transport of polypyridyl cobalt complexes in dye-sensitized solar cells with mesoporous TiO_2 photoanodes. J Phys Chem C 112:18255–18263
95. Hattori S, Wada Y, Yanagida S, Fukuzumi S (2005) Blue copper model complexes with distorted tetragonal geometry acting as effective electron-transfer mediators in dye-sensitized solar cells. J Am Chem Soc 127:9648–9654
96. Li TC, Spokoyny AM, She C, Fahra OK, Mirkin CA, Marks TJ, Hupp JT (2010) Ni(III)/(IV) bis(dicarbollide) as a fast, noncorrosive redox shuttle for dye-sensitized solar cells. J Am Chem Soc 132:4580–4582
97. Tennakone K, Kumara GRRA, Kumarasinghe AR, Wijayantha KGU, Sirimanne PM (1995) A Dye-sensitized nano-porous solid-state photovolatic cell. Semicond Sci Technol 10:1689–1693
98. Kumara GRRA, Okuya M, Murakami K, Kaneko S, Jayaweera VV, Tennakone K (2004) Dye-sensitized solid-state solar cells made from magnesiumoxide-coated nanocrystalline titanium dioxide films: enhancement of the efficiency. J Photochem PhotobioL A: Chem 164:183–185
99. Perera VPS, Tennakone K (2003) Recombination processes in dye-sensitized solid-state solar cells with CuI as the hole collector. Sol Energy Mater Sol Cells 79:249–255
100. O'Regan BC, Lenzmann F (2004) Charge transport and recombination in a nanoscale interpenetrating network of n-type and p-type semiconductors: transient photoc. J Phys Chem B 108:4342–4350
101. O'Regan BC, Scully S, Mayer AC, Palomares E, Durrant J (2005) The effect of Al_2O_3 barrier layers in TiO_2/Dye/CuSCN photovoltaic cells explored by recombinati. J Phys Chem B 109:4616–4623
102. Murakoshi K, Kogure R, Wada Y, Yanagida S (1997) Solid state dye-sensitized TiO_2 solar cell with polypyrrole as hole transport layer. Chem Lett 5:471–472
103. Bach U, Lupo D, Comte P, Moser JE, Weissortel F, Salbeck J, Spreitzer H, Gratzel M (1998) Solid-state dye-sensitized mesoporous TiO_2 solar cells with high photon-to-electron conversion efficiencies. Nature 395:583–585
104. Snaith HJ, Moule AJ, Klein C, Meerholz K, Friend RH, Grätzel M (2007) Efficiency enhancements in solid-state hybrid solar cells via reduced charge recombination and increased light capture. Nano Lett 7:3372–3376
105. Kubo W, Kitamura T, Hanabusa K, Wada Y, Yanagida S (2002) Quasi-solid-state dye-sensitized solar cells using room temperature molten salts and a low molecular weight gelator. Chem Commun 4:374–375
106. Kubo W, Kambe S, Nakade S, Kitamura T, Hanabusa K, Wada Y, Yanagida S (2003) Photocurrent-determining processes in quasi-solid-state dye-sensitized solar cells using ionic gel electrolytes. J Phys Chem B 107:4374–4381
107. Stathatos E, Lianos P, Zakeeruddin MS, Liska P, Grätzel M (2003) A quasi-solid-state dye-sensitized solar cell based on a sol-gel nanocomposite electrolyte containing ionic liquid. Chem Mater 15:1825–1829
108. Orel B, Vuk AŠ, Jovanovski V, Ješe R, Perše LS, Hočevar SB, Hutton EA, Ogorevc B, Jesih A (2005) Structural and electrical studies of a sol-gel derived nanocomposite ionic liquid based on positively charged polysilsesquioxane and iodide. Electrochem Commun 7:692–696

109. Yanagida S, Watanabe M, Matsui H, Okada K, Usui H, Ezure T, Tanabe N (2005) Dye-sensitized Solar Cells Using Nanocomposite lon-gel. Fujikura Tech Rev 34:59–65
110. Berginc M, Hočevar M, Opara Krašovec U, Hinsch A, Sastrawan R, Topič M (2008) Ionic liquid-based electrolyte solidified with SiO_2 nanoparticles for dye-sensitized solar cells. Thin Solid Films 516:4645–4650
111. Kawano R, Watanabe M (2003) Equilibrium potentials and charge transport of an I^-/I_3^- redox couple in an ionic liquid. Chem Commun 3:330–331
112. Stathatos E, Lianos P, Krontiras C (2001) Dye-sensitized photoelectrochemical cell using a nanocomposite SiO_2/poly(ethylene glycol) thin film as electrolyte support. Characterization by time-resolved luminescence and conductivity measurements. J Phys Chem B 105:3486–3492
113. Stathatos E, Lianos P, Stangar UL, Orel B (2002) A high performance solid state dye-sensitized photoelectrochemical cell employing a nanocomposite gel electrolyte made by the sol-gel route. Adv Mater 14:354–357
114. Stathatos E, Lianos P, Vuk AŠ, Orel B (2004) Optimization of a quasi-solid-state dye-sensitized photoelectrochemical solar cell employing a ureasil/sulfolane gel electrolyte. Adv Funct Mater 14:45–48
115. Jovanovski V, Stathatos E, Orel B, Lianos P (2006) Dye-sensitized solar cells with electrolyte based on a trimethoxysilane-derivatized ionic liquid. Thin Solid Films 511: 634–637
116. Cerneaux S, Zakeeruddin SM, Pringle JM, Cheng Y-B, Grätzel M, Spiccia L (2007) Novel nano-structured silica-based electrolytes containing quaternary ammonium iodide moieties. Adv Funct Mater 17:3200–3206
117. Opara Krašovec U, Georg A, Georg A, Wittwer V, Luther J, Topič M (2004) Performance of a solid-state photoelectrochromic device. Sol Energy Mater Sol Cells 84:369–380
118. Kubo W, Sakamoto A, Kitamura T, Wada Y, Yanagida S (2004) Dye-sensitized solar cells: improvement of spectral response by tandem structure. J Photochem PhotobioL A: Chem 164:33–39
119. Hauch A, Georg A, Baumgärtner S, Opara Krašovec U, Orel B (2001) New photoelectrochromic device. Electrochim Acta 46:2131–2136
120. Bechinger C, Ferrere S, Zaban A, Sprague J, Gregg BA (1996) Photoelectrochromic windows and displays. Nature 383:608–610
121. Hauch A, Georg A, Opara Krašovec U, Orel B (2002) Photovoltaically self-charging battery. J Electrochem Soc 149:A1208–A1211
122. Granqvist CG (1995) Handbook on inorganic electrochromic materials. Elsevier, Amsterdam
123. Hauch A, Georg A, Opara Krašovec U, Orel B (2002) Comparison of photoelectrochromic devices with different layer configurations. J Electrochem Soc 149:H159–H163
124. Opara Krašovec U, Topič M, Georg A, Georg A, Dražič G (2005) Preparation and characterisation of nano-structured WO_3-TiO_2 layers for photoelectrochromic devices. J Sol-Gel Sci Technol 36:45–52
125. Opara Krašovec U, Georg A, Georg A, Wittwer V, Luther J, Topič M (2004) Performance of a solid-state photoelectrochromic device. Sol Energy Mater Sol Cells 84:369–380
126. Opara Krašovec U, Ješe R, Orel B, Grdadolnik J, Dražič G (2002) Structural, vibrational, and gasochromic properties of porous WO_3 films templated with a sol-gel organic-inorganic hybrid. Monatsh Chem 133:1115–1133
127. Krebs FC (2009) Fabrication and processing of polymer solar cells: A review of printing and coating techniques. Sol Energ Mat Sol Cells 93:394–412
128. Grundmann M, Frenzel H, Lajn A, Lorenz M, Schein F, Wenckstern H (2010) Transparent semiconducting oxides: materials and devices. Phys Status Solidi A 207:1437–1449

Chapter 9
Sol-Gel Materials for Carbon Mineral Sequestration

V. Morales-Flórez, L. Esquivias and A. Santos

Abstract Sol-gel technology was used to design a new, energetically efficient and economically viable CO_2 sequestration technology. Calcium silicate grains were embedded into an inorganic silica matrix. These composites were used for the mineral sequestration of CO_2; that is, for permanent carbon fixation. It was found that the mineral grains presented better carbonation efficiency when embedded into an inorganic porous silica matrix. The reaction by-products were calcite particles of a few microns and silica particles of a few tenths of nanometres in size. They could be recycled as raw materials for commercial processes, such as coatings and mortar, for example, depending on their morphology, purity, and particle size and on calculations of the total cost scenarios. Finally, sol-gel composites can also be tailored to capture other atmospheric gases, not just CO_2, favouring the viability of this technology.

Keywords 3-Aminopropyltrimetoxisilane (APMES) · Aerogel · Alkoxide method · Calcium silicate · Carbon capture and storage (CCS) · Cavitation phenomenon · CO_2 emission · CO_2 sequestration · Composites · Condensation · Greenhouse effect · Microemulsion · Mineral carbonation · Mineral sequestration · Particle size ·

V. Morales-Flórez
Instituto de Ciencia de Materiales de Sevilla (CSIC-US), Av. Américo Vespucio 49, 41092, Sevilla, Spain
e-mail: victor.morales@icmse.csic.es

L. Esquivias (✉)
Dpto. Física de LA Materia Condensada, Universidad de Sevilla,
Av. Reina Mercedes s/n, 41012, Sevilla, Spain
e-mail: luisesquivias@us.es

A. Santos
Dpto. Ciencias de la Tierra, Universidad de Cádiz, Av. Republica Saharaui, s/n, 11510, Cádiz, Puerto Real, Spain
e-mail: Alberto.santos@uca.es

M. Aparicio et al. (eds.), *Sol-Gel Processing for Conventional and Alternative Energy*, Advances in Sol-Gel Derived Materials and Technologies,
DOI: 10.1007/978-1-4614-1957-0_9,

Polymerization · Porosity · Porous matrix · Silica matrix · Sol-gel · Sonocatalysis · Sorption degree of carbonation · Storing CO_2 · Supercritical drying · Tetraethylorthosilicate (TEOS) · Wollastonite ($CaSiO_3$) · Xerocomposites

9.1 Introduction

The environmental effects caused by emissions of CO_2 and other gases contributing to the "greenhouse effect" are stimulating the search for viable methods to remove these gases from the atmosphere or to reduce their emission without a severe impact on current industrial and social development. Carbon capture and storage (CCS) is presented as the main option for CO_2 emission mitigation [1]. However, it poses three major problems.

Since the common practice in this strategy is storing CO_2 in depleted oil and gas fields [2], the first problem is to develop environmentally secure CCS technologies, i.e., to focus efforts on the management of environmental safety and to ensure that the CO_2 captured and stored remains isolated. Nevertheless, this method bears a risk of the possible diffusion of CO_2 via cracks in rock and other zones of weakness, allowing portions of the injected CO_2 to re-enter the atmosphere and biosphere [3]. The second is to develop the various aspects of transport and storage infrastructure, i.e., how CO_2 can be captured from large point sources, compressed, transported over possibly longer distances and finally safely stored in suitable geological strata [4]. The third problem concerns CO_2 emitted by mobile or diffuse sources and the gas emitted in the past that cannot be captured or stored [5–7]. However, many research projects and tests are already being performed in this field (See for example, GRASP or CO2SINK projects [8, 9]).

On the one hand, these problems could be solved if the injected CO_2 is encapsulated in reactive geological reservoirs and likewise transformed in situ into a stable mineralogical phase. This strategy is an alternative to the long-term storage of CO_2 in geological reservoirs [10]. On the other hand, a procedure known as mineral sequestration [11, 12] is one of the most promising ways currently being studied for CO_2 sequestration. This can be understood as an artificial counterpart of the rock weathering process as it is similar to the natural mechanism that has regulated atmospheric CO_2 over geological time.

In natural systems carbonation takes place through the dissolution of minerals, mainly silicates, thus forming the corresponding carbonates and a solid by-product, namely amorphous silica. Thus, most of the processes currently under research are focusing on metal oxide-bearing materials containing divalent cations, usually alkaline-earth metals, such as wollastonite, $CaSiO_3$, olivine, Mg_2SiO_4, and multioxide silicates, such as serpentine $Mg_3Si_2O_5(OH)_4$. For example, carbonating wollastonite produces calcite ($CaCO_3$) and silica (SiO_2). This reaction accelerates if CO_2 and its subsequent carbonation reaction occur in aqueous solution according to:

(1) CO_2 (g) + H_2O (l) → HCO_3^-(aq) + H^+(aq)
(2) $CaSiO_3$(s) + 2 H + (aq) → Ca_2^+(aq) + SiO_2(s) + H_2O (l)
(3) Ca_2^+(aq) + HCO_3^-(aq) → $CaCO_3$(s) + H^+(aq)

Unfortunately, however, weathering is a very slow process that cannot keep up with the increasing levels of atmospheric CO_2. Therefore, the goal of mineral carbonation [13] has been, and still is, to increase the reaction rate of mineral carbonation in an environmentally sound way. And such a goal must also offer a useful process of massive carbonation for CO_2 sequestration. Among the research studies carried out to overcome this drawback, there have been studies on, for example, carbonation in aqueous media, the influence of several factors such as micro-emulsions and/or layered mineral phases previously activated to make these materials more reactive [14], mineral grain reactive surface studies [15–20], the use of acidic media for extraction of the divalent cation [21], and the carbonation reaction under supercritical CO_2 conditions [22, 23]. In aqueous systems the dissolution of the divalent cation from the precursor silicates seems to be the main rate-limiting step, and most research efforts have been devoted to finding ways to speed up the extraction of the cation from the precursor materials [15]. In addition, the inhibiting effect induced by the formation of a passivating layer on the reacted mineral surface is also relevant to carbonation efficiency [24].

In addition to the use of natural silicates, considerable effort is being devoted to the design and synthesis of new products and processes that can improve the reaction rates and optimise the efficacy of CO_2 sequestration. Laboratory tests under atmospheric conditions proved that aerogel [25] and the wollastonite encapsulated in a sol-gel porous matrix enabled very efficient carbonation reactions [26]. The composites were found to accelerate the carbonation kinetics enormously in comparison to their natural counterparts. Research under normal laboratory and atmospheric conditions indicated that grain size, gel porosity and the specific surface for the reaction performance are critical factors. The laboratory tests carried out on these samples confirmed a satisfactory performance of this CO_2 sequestering process.

Given the above considerations, sol-gel porous matrices with encapsulated mineral grains were prepared with the assistance of ultrasound [27]. Catalysis induced by the use of high power ultrasound (sonication) has been proven to be a useful tool for enhancing reactions, especially reactions in heterogeneous systems such as alkoxide/water mixtures. Cavitation is the phenomenon produced by acoustic waves through a solution. It consists of the creation of cavities within the liquid caused by the propagation of the pressure wave through the liquid. Cavitation serves as a means of concentrating the diffuse energy of sound into unique chemical nanoreactors to produce unusual materials from dissolved precursors.

As with the standard method, the most popular approach to forming the initial gel by sonocatalysis is the "alkoxide method". In the presence of any acid catalyst, alkoxides undergo hydrolysis inside the cavitation bubbles, producing alcohol as a by-product that serves as a solvent [28–30]. Therefore, the textural and structural properties of gels synthesised by the sonication of metal alkoxides and water mixtures differ from those prepared in a standard way in alcoholic solution. Supercritical extraction of the solvent to obtain aerogels [31], or *aerocomposites*, avoids collapse of the structure due to the drying process and allows the properties of the solid structure of sono-aerogels to be investigated. This solventless procedure and the cavitation phenomenon, responsible for homogenization of the alkoxide-water mixture, gives rise to a narrow pore size distribution, a high bulk density and a surface/volume ratio two or three times higher than gels prepared in alcohol solutions [32]. Sonogels are very homogeneous and have a continuous network on a scale of tens of nanometres. The larger the ultrasonic dose, the finer the porosity and the higher the homogeneity. These features make them appropriate for hosting small particles of the reactive phase. This paper describes the main characteristics and issues concerning these nanocomposites.

Recently, sol-gel technology has been used in CO_2 sequestration technologies to synthesise two main types of composites with two very different objectives. On the one hand, composites of calcium silicate grains embedded into an inorganic silica matrix have been used for mineral carbonation; that is, for permanent carbon fixation. We investigated the efficacy of incorporating wollastonite powders into sonogels [26, 33–35] because we found them attractive materials for encapsulating these fine particles since their structure prevents the powder from agglomerating and increases the area of the reactive surface. This type of processing allowed the synthesis of dicalcium silicate (larnite), which has a CO_2 fixation capacity 50% higher than that of wollastonite [36]. The reaction turned out to be highly efficient (100%) and fast.

On the other hand, the textural features of the synthetic samples (grain size, porosity and specific surface area) prevent the formation of passivating layers or films that inhibit the carbonation process. These properties make sonogels, both xero- and aerogels, attractive materials for use in the adsorption, capture and sequestration of pollutant gases, and they also have several general environmental applications [37–40]. Moreover, lithium silicate nanoparticles [41], membranes from aminosilicates [42] or mesoporous alumina gel (MA) and PEI (polyethylenimine)-impregnated MA [43] have also been used with great success to selectively separate and capture CO_2 from a mixture of gases and to reuse the captured carbon dioxide for other purposes, such as underground geological storage. These separation technologies are also being researched with calcium-rich minerals such as sequestering agents [44], but the capture efficiency is significantly lower after several carbonation/calcination cycles, probably due to sintering phenomena of the mineral grains. Currently, sol-gel technologies are being considered for embedding the grains of sequestering agents into a silica sol-gel matrix that might be able to inhibit the unwanted sintering and maintain a high level of carbonation efficiency for tens of carbonation/calcination cycles.

The first part of this section focuses on the aspects of the composite preparation, paying special attention to the characteristics of the resulting materials. The second part will discuss the efficiency of the carbon capture ability of the sol-gel samples. Finally, the last part of this section will focus on the technological viability of these technologies in order to design an industrial scale-up of the sequestration process.

9.2 Synthesis and Properties of Sequestration Agents Based on Silica Sol-gel Materials

Aerogel CO_2-sorbent materials were made by the sol-gel process in which a mixture of a metal alkoxide, for example tetraethylorthosilicate (TEOS, $Si(OC_2H_5)_4$) and calcium and magnesium salts ($Ca(NO_3)_2$ and $Mg(NO_3)_2$), was hydrolysed to form a sol. A catalyst was then added to start the condensation polymerization process to form a gel [25]. The gels were supercritically dried in an autoclave under supercritical conditions of the ethanol to extract it avoiding structure shrinking. Afterwards, the aerogels obtained were heat treated at 650–700°C to eliminate organic residues after autoclave drying. Following this procedure, opaque white samples containing Ca, Mg and Si oxides were obtained. Their density was approximately 0.12 g/cm^3, and the porosity was greater than 90%. The properties of the samples are listed in Table 9.1.

Later, lithium silicates were considered as sol-gel materials for carbon sequestration [41]. Lithium silicate nanoparticles were synthesised by coupling the sol-gel method and a reverse microemulsion. A reverse microemulsion was prepared using cyclohexane, Triton X-100 and n-hexane. A microemulsion containing lithium hydroxide and TEOS with a molar ratio of Li:Si of 2:1 was mixed with other microemulsions with ammonium hydroxide. The precipitated powders were recovered after 2 h of aging, centrifuged and then washed with acetone. They were then dried at 110°C for 12 h. Different heat treatments were considered from 600 to 1000°C. Samples with no heat treatment showed small diffraction peaks indicating very low crystallinity. The X-ray diffraction (XRD) pattern of the powder calcined at 800°C for 4 h shows the characteristic peaks of single-phase lithium metasilicate (Li_2SiO_3). Different Li:Si ratios and different heat treatments yielded different proportions of Li_2SiO_3 and $Li_2Si_2O_5$. The nanostructure of the lithium silicate powder was researched by transmission electron microscopy (TEM) and nitrogen physisorption. The particle size range observed was 4–12 nm, whereas a specific surface area of 7 m^2/g was measured (Table 9.1).

Calcium silicate mineral phases were also used for carbon sequestration by synthesizing silica-mineral composites by the sol-gel method. Moreover, different routes of synthesis have been considered for preparing the composites with the aim of reducing energetic and economic costs in order to develop an energetically and economically viable carbon capture process at an industrial scale. Therefore, different calcium silicates have been considered as sequestration agents, such as natural raw wollastonite and synthetic wollastonite and larnite, and different silica matrices have been employed as the porous embedding phase.

Synthetic wollastonite was synthesised using TEOS and calcium nitrate $Ca(NO_3)_2{\cdot}4H_2O$ [26]. These reagents were dissolved in an ethanol solution and then NaOH was added until the powder precipitated. The powder was extracted by centrifugation, washed several times and finally dried at 100°C. Once dried, the powder was heat treated in air at 1000°C. The chemical and structural properties of the powder are summarised in Table 9.1. X-Ray fluorescence (XRF) analysis

Table 9.1 Summary of the chemical composition and structural data of the different sequester agents considered

Sequester agents	Source	Heat treatment	CaO/SiO_2 ratio (wt.)[a]	Properties
Ca–Mg–Si (a)	$Ca(NO_3)_2 + Mg(NO_3)_2$ + TEOS	650–700°C	–	Density = 0.12 g/cm3 Porosity = 90%
Li–Si (b)	LiOH + TEOS + microemulsion	600–1000°C, 4 h	Li/Si = 1–2	$S_{BET} = 7\ m^2/g$ Particle size = 4–12 nm
WP (c)	TEOS + $Ca(NO_3)_2$ + APMES	100°C, 2 h	0.78	$S_{BET} = 0.92\ m^2/g$ $Vp = 2{\times}10^{-3} cm^3/g$
NWP (d)	Raw natural wollastonite sieved <345 microns + APMES	No	–	–
SWP900 (d)	$Na_2SiO_3 + CaCl_2$ + APMES	900°C, 1 h	0.44	$S_{BET} = 0.97\ m^2/g$ $Vp = 2{\times}10^{-3}\ cm^3/g$
Larnite2 (e)	Ludox + $Ca(NO_3)_2$ + Ethylene glycol + APTES	600°C, 1 h	2	$S_{BET} = 39\ m^2/g$
Larnite1 (e)	Ludox + $Ca(NO_3)_2$ + Ethylene glycol + APTES	600°C, 1 h	1,1	–
Natural W (f)	Raw natural wollastonite sieved <21 microns	No	0.30	Purity $CaSiO_3$ = 50% wt. SiO_2 = 40% wt. $CaCO_3$ = 5%

CaO/SiO_2 ratio (weight %) is given for all the sequester agents and composites, except for Lithium-bearing and Ca–Mg-bearing samples
(a) Ahmed and Attia [25] (b) Khomane et al. [41] (c) Santos et al. [26] (d) Santos et al. [33] (e) Santos et al. [36] (f) Morales-Flórez et al. [35]

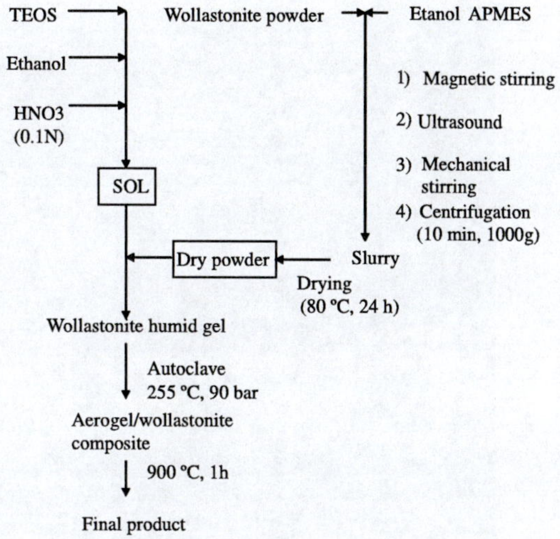

Fig. 9.1 Flowchart showing the routes for synthesis of the studied materials. (*Asterisks*) Ethanol and APMES volumes indicated were used for 1 g of precipitated powder. The different values of energy density submitted by ultrasound are described in J/cm^3 at the various steps in the whole process. A device delivering to the system 0.6 W/cm^3 of ultrasound power was employed. Reproduced with permission from [26]

showed a Ca/Si ratio similar to that of wollastonite, and X-ray diffraction (XRD) analysis confirmed the wollastonite nature of the precipitated powder. The crystals presented sizes of around several tenths of nanometres, with a bimodal pore size distribution of 10 and 100 nm pore radii. The specific surface area of the mineral grains was measured by nitrogen physisorption, obtaining a value of 0.92 m^2/g that can be associated with a particle size of around 1 μm (wollastonite density = 2.91 g/cm^3). In this case, the entire surface was reactive; that is, it was susceptible to reacting with CO_2.

Prior to its inclusion in the gel matrix, the surface of the powder was chemically modified with 3-aminopropyltrimetoxisilane (APMES) in order to facilitate and stabilise colloidal dispersion into the host matrix. Figure 9.1 shows a flowchart showing the route of synthesis. The APMES/powder mixture was diluted in ethanol with the help of ultrasound, and this slurry was added to a sol previously made by hydrolysis and polycondensation of TEOS, also ultrasonically assisted, ensuring a homogeneous distribution of the powders and avoiding decantation. The gels were mixed for about 5 min. Finally, the gel was supercritically dried, in an autoclave, by the supercritical extraction of ethanol, obtaining the aerogel composite. The resulting composite presented nanometric structural features (pores and particle sizes) corresponding to the silica matrix (~1000 m^2/g), and bulk values that depended on the content of wollastonite. Typically, the specific surface area was between 100–300 m^2/g. Figure 9.2 shows a micrograph of the wollastonite needles embedded in the silica matrix.

One of the strategies for lowering both the energetic and economic costs of this technology is the substitution of the TEOS reactant by sodium metasilicate and calcium chloride for wollastonite precipitation [33]. The use of low-cost reagents is an important step in improving the technological viability of sol-gel composites for CO_2 sequestration at an industrial scale. This capture agent was prepared from

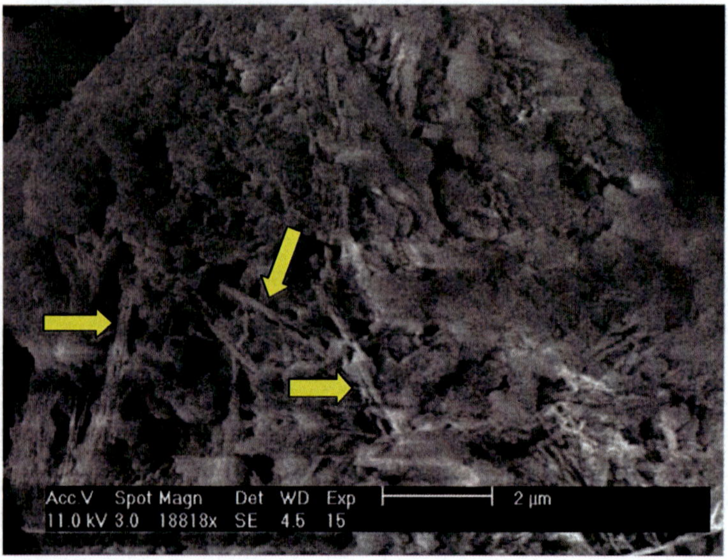

Fig. 9.2 SEM image of the simple SAWC40. *Arrows* point to embedded "needles" of wollastonite in the silica porous matrix. Reproduced with permission from [34]

a SiO_2–CaO gel with a Ca/Si molar ratio equal to 1. The gel was obtained from Na_2SiO_3 and $CaCl_2$ solutions. After powder precipitation, the gels were centrifuged and dried at 80°C. Later, the powder was ground and heat treated at 900°C for 1 h. The XRD phase identification revealed wollastonite precipitation, together with small amounts of silica. In addition, the structural features of the mineral grains were similar to those obtained using TEOS as the reagent (Table 9.1). The surface of the powder was chemically activated with APMES prior to its inclusion in the sol for composite preparation.

Composites were synthesised using this sequestering agent. A mixture of wollastonite powder and APMES was diluted in ethanol under vigorous magnetic stirring, followed by homogenization with the help of a high-speed rotary blender. The slurry was separated by centrifugation and dried at 80°C for 24 h. The powder obtained was added to a sol previously prepared with TEOS in a similar way as described before. The samples were again dried under supercritical conditions of ethanol. Some of the aerogel composites obtained were heat treated at 900°C and some at 500°C [34]. Despite the thermal treatment, both composites containing the same active phase, but processed differently, does not present large differences between their textures. The synthetic wollastonite powders were identified by XRD after heat treatment and after embedding it into the silica matrix. The texture and composition of these samples are summarised in Table 9.2.

Another different sequestering agent that was considered was raw natural wollastonite, obtained from mines of Aroche (Huelva, Spain). The sample was characterised by XRD, and wollastonite 1T was observed. This natural mineral

Table 9.2 Summary of the chemical composition and structural data of the different composites

Composites	Sequester agent	Gel type	Heat treatment	Properties
SiO_2-aerogel (a)	–	Aerogel	No	$CaO/SiO_2 = 0$ $S_{BET} = 1055$ m^2/g Vp = 2.18 cm^3/g
SAWC26 (a)	WP	Aerogel	500°C, 1 h	$CaO/SiO_2 = 0.38$ (wt.) $S_{BET} = 319$ m^2/g Vp = 0.611 cm^3/g
SAWC40 (a)	WP	Aerogel	500°C, 1 h	$CaO/SiO_2 = 0.73$ (wt.) $S_{BET} = 93.7$ m^2/g Vp = 0.262 cm^3/g
ASWC (b)	SWP900	Aerogel	No	$CaO/SiO_2 = 0.23$ (wt.) $S_{BET} = 122$ m^2/g Vp = 0.21 cm^3/g
ASWC500 (c)	SWP900	Aerogel	500°C, 2 h	$CaO/SiO_2 = 0.24$ (wt.) $S_{BET} = 75$ m2/g Vp = 0.370 cm^3/g
ASWC900 (b)	SWP900	Aerogel	900°C	$CaO/SiO_2 = 0.23$(wt.) $S_{BET} = 28.7$ m^2/g Vp = 0.04 cm^3/g
ANWC (b)	NWP	Aerogel	No	$CaO/SiO_2 = 0.41$ (wt.) $S_{BET} = 335.2$ m^2/g Vp = 0.52 cm^3/g
ANWC900 (b)	NWP	Aerogel	900°C, 1 h	$CaO/SiO_2 = 0.41$ (wt.) $S_{BET} = 67.7$ m^2/g Vp = 0.14 cm^3/g
Comp1A (d)	Larnite1	Aerogel	600°C, 1 h	$CaO/SiO_2 = 0.41$ (wt.) $S_{BET} = 258$ m^2/g
Comp1B (d)	Larnite1	Aerogel	600°C, 1 h	$CaO/SiO_2 = 0.69$ (wt.) $S_{BET} = 115$ m^2/g
Comp1C (d)	Larnite1	Aerogel	600°C, 1 h	$CaO/SiO_2 = 0.79$ (wt.) $S_{BET} = 53$ m^2/g
Comp2 (d)	Larnite2	Aerogel	600°C, 1 h	$CaO/SiO_2 = 1.60$ (wt.) $S_{BET} = 40$ m^2/g
Xero26 (e)	Natural W	Xerogel	No	$S_{BET} = 153$ m^2/g $<$ Rp $>=$1.75 nm Vp = 0.169 cm^3/g

(a) Santos et al. [26] (b) Santos et al. [33] (c) Santos et al. [34] (d) Santos et al. [36] (e) Morales-Flóreze al. [35]

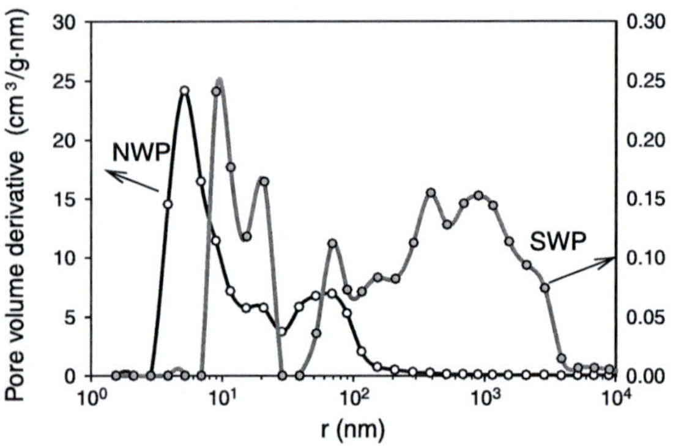

Fig. 9.3 Pore Volume Derivative of natural raw (*NWP, black line, left axis*) and synthetic (*SWP, grey line, right axis*) wollastonite powders before being submitted to CO_2 flow. The lines are eye-guides. The *black line* is represented with reference to the left axis, the *grey line* with reference to the right axis. Reproduced with permission from [33]

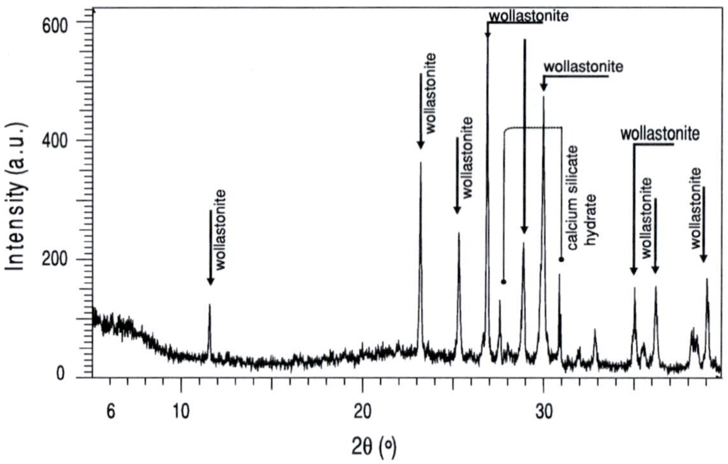

Fig. 9.4 Standard X-ray diffraction pattern of the aerogel/natural wollastonite powder composite. Reproduced with permission from [33]

was milled in an agate mortar and the resulting powder was sieved to <354 μm (−45 mesh). Mercury porosimetry revealed the presence of a bimodal pore size distribution curve (Fig. 9.3), with a first peak at ~5 nm that accounted for the irregular surface resulting from the acicular texture. The second part of the curve was formed by a maximum centred at ~60 nm, followed by a long tail. The surface of the powder was chemically modified with APMES in a similar procedure as before. Aerogel composites with raw natural wollastonite were also prepared and dried as usual. The composites were also observed by XRD. The preparation of the composites induced additional changes in the intensity of the reflections that were consistent with those of a hydrated calcium silicate (Fig. 9.4).

Synthetic larnite powders were also considered as a sequestering agent. In this case, a commercial colloidal silica (Ludox®) and calcium nitrate were used for

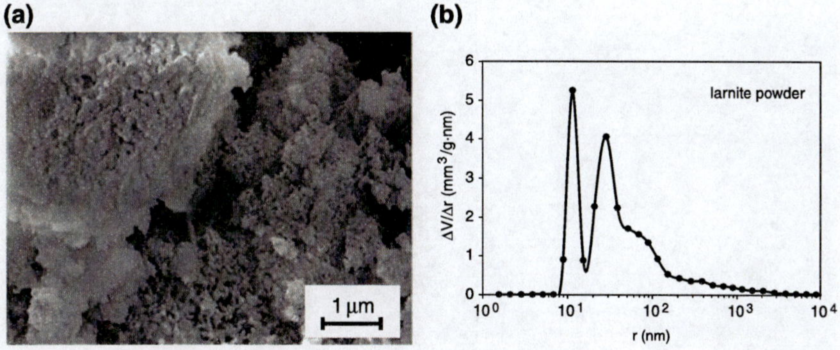

Fig. 9.5 Larnite2 samples, heat treated at 600°C. SEM micrograph (**a**) and pore size distribution from Hg porosimetry (**b**). *Line* is an eye-guide only. Reproduced with permission from [36]

preparation of the sequestering agent [36]. A mixture was prepared with colloidal silica, calcium nitrate and diluted (5%) ethylene glycol. The mixture was heated at 100°C with magnetic stirring until the solvent had completely evaporated. Two different Ca/Si molar ratios were fixed at 1 and 2, yielding two final Ca/Si ratios of the precipitated phase equal to 1.1 and 2.1, respectively (measured by XRF). This product was placed in an oven at 100°C for a further 48 h, then ground and the resulting powders were finally heat treated at 600°C for 1 h in one case, and at 1000°C in the other. The powders treated at 600°C were observed by scanning electron microscopy and not only dendritic morphologies but also particulate aggregates were observed (Fig. 9.5). Phase identification by XRD showed the presence of wide peaks in the sample heated at 600°C, corresponding to the early stage of crystal nucleation and growth. The identified phases were α-larnite, β-larnite, wollastonite and small amounts of amorphous phases, as well as tobermorite. With the heat treatment at 1000°C, the amorphous phases and tobermorite disappeared and the amount of α-larnite decreased; the major constituent was β-larnite. As a consequence of calcium silicate transformation, wollastonite appeared at 1000°C as well. The specific surface area of the larnite powders was measured and values close to 40 m^2/g were obtained (Table 9.2).

The processing of the composites was similar to that previously reported, but in this case the surfaces of the larnite powders were chemically modified with 3-aminopropyltriethoxysilane (APTES) in ethanol to facilitate and stabilise colloidal dispersion into the precursor sol of the host matrix. Magnetic stirring and ultrasound were used to homogenise the slurry. It was then centrifuged and dried at 80°C for 24 h. The composites were prepared with different loads of sequestering agents (2.5, 5, 10 and 15 g) on the previously prepared sol by hydrolysis and polycondensation of TEOS. The gels were supercritically dried as explained previously, and, finally, the composites were heat treated at 600°C for 1 h.

Phase identification by XRD showed broader peaks due to the amorphous contribution of the silica matrix. Nevertheless, the phases were the same as those observed in the pure non-encapsulated powder. Larnite grains appeared to be

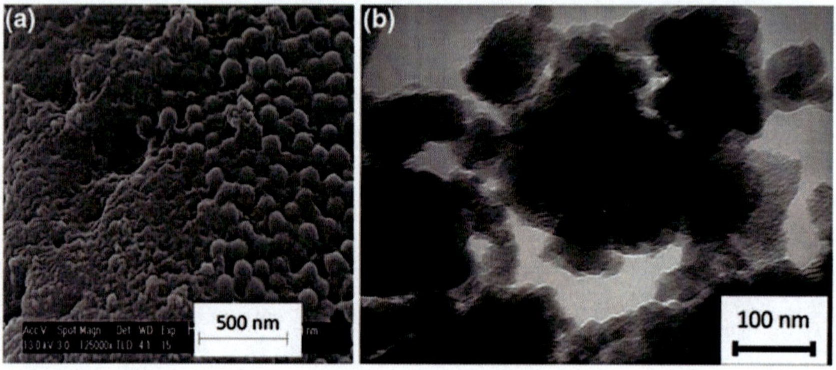

Fig. 9.6 Composite 1C samples, heat treated at 600°C, SEM (**a**) and TEM (**b**) micrographs. Reproduced with permission from [36]

coated by the silica matrix (Fig. 9.6). In these composites, the processing technique used made it possible to obtain textural features (particle size and specific surface area) appropriate for the carbonation process that are impossible to find in natural compounds, thanks to the contribution of the larnite particles and the porous matrix. The reported values for the specific surface area were 40 m^2/g for the composite bearing 15 g synthetic larnite and 258 m^2/g in the case of 2.5 g synthetic larnite. The structural properties of the larnite composites are summarised in Table 9.2.

The latest technological improvements concern drastic changes for reducing the energetic costs of the sample preparation, in turn reducing their carbon footprint. The final steps were given to avoid having to use surface chemical activators (APTES or APMES) and to substitute the supercritical drying by conventional drying, thus obtaining low-energy-costing xerogels [6]. The sequestering agent considered was the raw mineral wollastonite, as it is the simplest and easiest agent to deal with. This natural wollastonite was sieved to <20 μm and dispersed in ethanol with the help of high power ultrasound. Again, the powder dispersion was added to a previously prepared sol by acidic hydrolysis and the polycondensation of TEOS. No more than a theoretical 40% wt of CaO could be embedded into the composites without observing heterogeneous distribution and decantation of the mineral phase in the sol before the gelling occurred. Finally, the gels were dried in air, obtaining the final xerogel composites.

The purity of the raw natural wollastonite was studied by XRD and thermogravimetric analysis. This revealed the presence of wollastonite, quartz and a small amount of calcite. A detailed composition is shown in Table 9.1. The specific surface area of the xerogel matrix obtained was significantly lower than its aerogel counterpart, presenting around 250–300 m^2/g. Nevertheless, the porous matrix was similar, regardless of the amount of embedded wollastonite. The porous matrix coating of the mineral grains can be seen in the SEM figures (Fig. 9.7). It can be observed that the porous coating in these *xerocomposites* is partially broken.

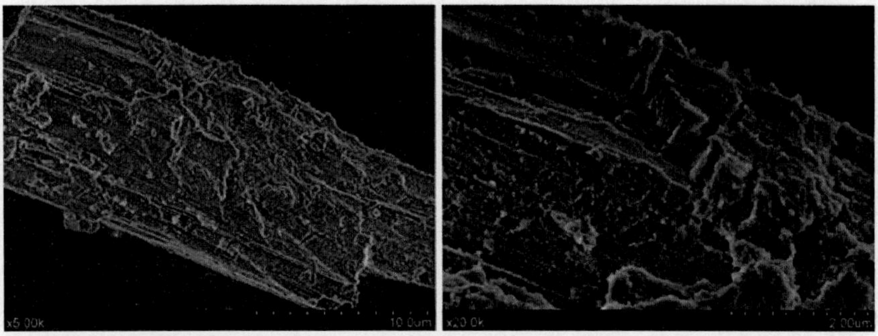

Fig. 9.7 *Left* SEM image of the broken coating of the mineral grains. Silica gel matrix does not completely cover the grains. *Right* detailed SEM image of the silica gel covering the mineral grain. Thickness of the coating can be estimated to be around 600 nm. Reproduced with permission from [35]

Moreover, the thickness of the porous coating was around 600 ± 100 nm. In this regard, it was confirmed that the coating of the *aerocomposites* was more homogeneous than that of the *xerocomposites*.

Currently, works are in progress to reduce economic costs by substituting the silica matrix precursor (TEOS) with low-cost reagents such as sodium silicate or commercial colloidal silica suspensions. We believe that this route is an important step for developing an energetically viable CO_2 sequestration technology based on calcium silicate sol-gel bearing materials.

9.3 Carbon Capture Efficiency

Carbon dioxide adsorption on the calcium-magnesium bearing silica aerogels [25] was measured by passing a mixture of CO_2, SO_2 and air (1000–2500 ppm SO_2, 6–15.5% CO_2) over 10–13 g of an aerogel sample enclosed in a stainless steel pressure filter vessel, and recording the change in the CO_2 concentration before and after the adsorption chamber. As the presence of water vapour enhances the adsorption process [45, 46], the aerogel was activated by introducing water vapour (steam) into the vessel for a short period (5 s). The gas was then allowed to flow at about 0.7 dm^3/min and 3 psi for selected periods of time. The precipitation of carbonates was ascertained by XRD. The adsorption of CO_2 ranged from 100% within the first 10 min of gas flow to 85% at up to 150 min gas flow. The role of the steam was to activate the surface of the gel by forming surface calcium and magnesium hydroxyl groups, thereby enhancing the kinetics of the chemisorption reactions on the gel's surface.

Regarding the lithium silicate nanoparticles [41], high-temperature sorption measurements for CO_2 were performed using a volumetric technique by pulse chemisorption up to 800°C. It was verified that the volume of adsorption of CO_2

Table 9.3 Details of the carbonation experiments and carbonation efficiencies of the samples

Sample name	CO_2 attack conditions	Carbonation degree AWL/MTWL[a] (%)
WP (a)	40 min	53
WP (a)	25 min	37
Larnite1 (d)	15 min	100
Larnite2 (d)	15 min	100
Natural W (e)	5 min, overnight	11
Natural W (e)	5 min, overnight, pH >7	52
SAWC26 (a)	25 min, overnight	100
SAWC40 (a)	40 min, overnight	83
ASWC500 (c)	15 min, overnight	100
ASWC900 (b)	30 min, ph >7	85
ANWC900 (b)	30 min, overnight	60
Comp1C (d)	15 min	100
Xero40 (e)	5 min, overnight, pH >7	51

[a] Actual weight loss and maximum theoretical weight loss
(a) Santos et al. [26] (b) Santos et al. [33] (c) Santos et al. [34] (d) Santos et al. [36] (e) Morales-Flórez et al. [35] (f) Ahmed et al. [25] (g) Khomane et al. [41]

increased with high pressure: up to 4 cm^3 under standard pressure and temperature per gram of sorbent. In addition, the effect of temperature was also tested. The aforementioned low rate of CO_2 sorption was observed at room temperature, and it slowly decreased at temperatures up to 450°C, where it then increased up to 610°C, where the maximum sorption was achieved (131 cm^3/g). It was further observed that the sorption started to decrease to small values again beyond 630°C up to almost 700°C (1.62 cm^3/g). In addition, nitrogen sorption experiments indicated that the material had no capacity for N_2, and could thus separate the CO_2/N_2 mixture (100% selectivity). This could be a useful attribute of the material for the separation and capture of CO_2 from flue gases under high temperatures. The higher sorption values on nanoparticles reported at high temperatures (>580°C) in this context than other values [47, 48] may have been due to the non-stoichiometric SiO_2 to SiO_x formation of the silicate molecule and strengthening of the Li_2O–SiO_x bond. If Li isomorphously replaces Si in the surface, the crystal field stabilization energy will tend to preserve a four-fold co-ordination of SiO_2.

The different calcium silicate composites synthesised were submitted to different fluxes of CO_2 in aqueous media to estimate their carbonation capacity, always working under ambient conditions. The experiments for monitoring the fixation of CO_2 were carried out in a reactor under agitation. The reactor consisted of a closed plastic container that was pierced and through which a pH electrode could be inserted. The container was connected to a CO_2 tank via two valves for entry and exit of the gas. For these CO_2 attacks, the samples were ground and dispersed in 25 ml of distilled water. Typically, the pH values dropped drastically when the CO_2 flux started, and then remained at values close to but higher than 6 for the rest of the experiment. The time of the CO_2 influx varied between 5 and 40 min, and its influence was studied.

The sequestering agent WP and composites containing WP as the sequestering agent: SAWC26 and SAWC40 (for sample specifications from now on, see Table 9.2) were subjected to a CO_2 influx for 25 and 40 min. Structural catalysis of the porous silica matrix was confirmed by measuring the degree of carbonation. The degree of carbonation was first estimated via mineral phase quantification by Rietveld analysis of the XRD patterns, and then also calculated by the weight loss due to CO_2 release in the thermogravimetric analysis with regard to the maximum theoretical weight loss. In the case of the NWP sample, for 25 and 40 min CO_2 subjected to a CO_2 influx for 25 and 40 min, the carbonation degrees obtained were 37 and 53%, respectively. In the case of the composites, sample SAWC26 was fully carbonated after 25 min of CO_2 influx, and sample SAWC40 was 83% carbonated after 40 min of CO_2 influx. Please see Table 9.3 for all experimental details of the CO_2 influx and degrees of carbonation.

Two important conclusions were drawn from these results. First, the performance of the composites in CO_2 capture was better than that of the powder. This fact shows that the dispersion of wollastonite powders in the aerogel matrix greatly enhanced the silicate dissolution and calcium carbonate precipitation kinetics; secondly, the efficiency of the carbonation process observed in the SAWC40 sample compared to the reports made by various authors working on samples of natural wollastonite [16, 19, 23] was reflected by a conversion rate of about 33% faster than that obtained for other natural silicates [20]. Moreover, these results were obtained under simple and gentle ambient conditions, so no high pressures or temperatures were employed. In fact, it is not possible to improve upon the simplicity of the experimental set-up.

The high efficiency obtained can be related to the high rate of wollastonite dissolution followed by the fast release of Ca^{2+} ions to solution. This speed of reaction produces a rapid increase of the solution pH [49] that is dependent on the specific surface area of the silicate [50]. This dissolution process can be considered as a consequence of the synergetic association of two main factors. First, the nanometric size of the wollastonite crystals that in turn increased the reactive surface, and, second, the homogeneous space distribution of the crystals provided by the aerogel support, avoiding agglomeration of the wollastonite grains and the production of retarding films [24].

The composites with different, less expensive forms of synthetic wollastonite (SWP900) were submitted to 15, 30 and 40 min of CO_2 influx in aqueous media, in a similar way as explained previously. In some cases, the pH was buffered at 7 with the help of some drops of 0.5 M $NaHCO_3$. Once the CO_2 flux had stopped, some of the samples were left to rest overnight before analysing their degree of carbonation, whereas others were immediately analysed. Generally, the XRD patterns showed the appearance of calcite peaks and a decrease in the wollastonite peaks in the carbonated samples compared to the samples that were not subjected to a CO_2 influx. In the case of the most favourable configuration for carbonation, namely, 40 min of CO_2 influx, pH buffered at 7 and rested overnight, it was concluded that sample ANWC900 achieved 60% carbonation whereas ASWC900 showed around 85% carbonation. Despite the reactivity of natural wollastonite

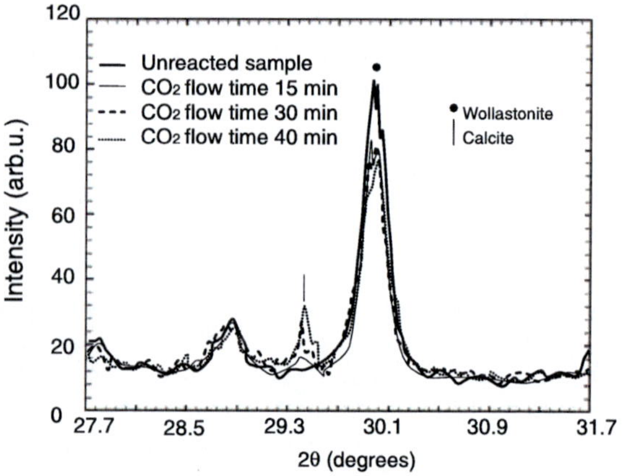

Fig. 9.8 Normalised diffraction patterns of the ASWC500 sample before and after being submitted to CO_2 flow. The mineral contents were checked immediately after completing every experiment. Reproduced with permission from [34]

(NWP) being increased before being embedded in the silica matrix, the synthetic wollastonite (SWP900) achieved a higher degree of carbonation thanks to its larger specific/reactive surface area and purity. The pH buffer seemed to help maintain the high carbonation degrees of these samples. Since the active phases considered in these samples were synthesised using inexpensive reagents, the economic cost of the process was reduced by 80% with respect to the previous sample. For this reason, realistic expectations of this method being an economically and energetically viable technology for CO_2 capture and elimination based on sol-gel composite materials were raised.

The different carbonation reaction times were tested using samples SAWC40 and ASWC500. The examination immediately after stopping the flow of CO_2 showed a similar behaviour of both, indicating that they had low degrees of carbonation. It was verified that the amount of reacted wollastonite increased with time (Fig. 9.8). When the samples were left to rest overnight in the CO_2-saturated water, the results were very different and high carbonation rates were obtained. The XRD patterns revealed that clear calcite peaks appeared whereas the wollastonite peaks almost disappeared (Fig. 9.9). Furthermore, in both cases, most of the carbonate precipitation occurred during the first 15 min. No relevant, additional carbonate precipitation was achieved by extending the reaction time period to 30 or 40 min. In this way, an important effect of the remaining dissolved CO_2 in the solution was confirmed. Moreover, it was also found that using a pH buffer such as $NaHCO_3$ favours the carbonation reaction.

Thermogravimetric analyses quantified the degree of carbonation, where the ASWC500 sample almost reached 100% carbonation. Thus, when comparing the ASWC500 and ASWC900 samples, it was found that the thermal treatment of these composites could be reduced from 900 to 500°C during the preparation procedure in order to reduce energy costs. Due to the carbonation reaction, microparticles of calcium carbonate were precipitated in the porous silica matrix, forming classic rhombohedral crystals, as could be seen by scanning electron

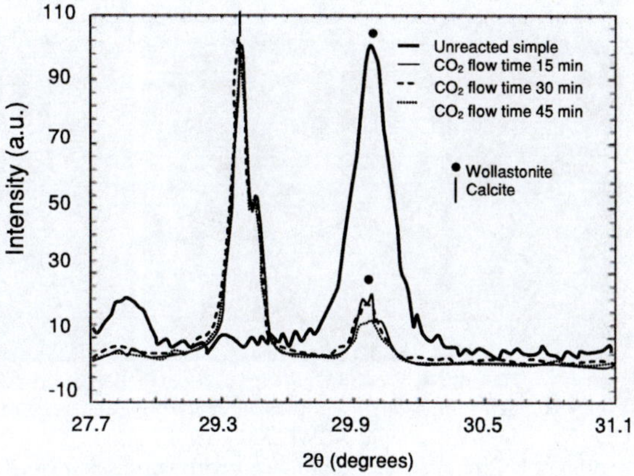

Fig. 9.9 Normalised diffraction patterns of the SAWC40 sample before and after being submitted to CO_2 flow. The samples were left to rest overnight in the reactor before checking the mineral contents. Reproduced with permission from [34]

microscopy (Fig. 9.10). In addition, no passivating layer was observed over the resting wollastonite grains (data not shown). The absence of a passivating layer significantly contributed to this high reaction rate. The reaction by-products were porous fine silica and calcite particles, highly appreciated raw materials for the construction industries, enhancing the technological and economic viability of the sequestration procedures being described.

Larnite-bearing composites were also tested as carbon dioxide sequestering agents in aqueous media and under conditions of room temerature and pressure, as explained before. The CO_2 influx time was set to 15 min. In this case, the samples were immediately analysed after the CO_2 influx was stopped. Two calcium carbonate polymorphs appeared: rhombohedral calcite crystals and spherulitic vaterite crystals. In 12 h, the metastable carbonate vaterite transformed into stable calcite, as monitored and verified by XRD. It is worth noting the absence of any trace of calcium silicates such as larnite or tobermorite, indicating complete carbonation or a carbonation degree of 100% after only 15 min of CO_2 treatment. After complete carbonation, these samples were heat treated again, by up to 900°C for 1 h. The resulting material was mainly composed of larnite; some parts of wollastonite were also observed, depending on the initial ratio of Ca/Si in the starting composite. Thus, an aerogel/larnite composite powder was made by starting from a raw material (larnite); when the composite was subjected to a flow of CO_2 for 15 min in aqueous media, it completely carbonated. Heat treatment of the resulting carbonation by-products yielded a final product, the majority of which was larnite. In this way, a closed cycle of carbonation/calcination with a high CO_2 storage capacity can be performed.

The process of carbonation begins with a high release rate of Ca^{2+} ions into the solution due to the dissolution rate of larnite, which depends upon the high specific surface area of the silicate [50]. This produces a rapid increase in the pH of the solution, by up to pH 10.5, and the subsequent production of monomers from the released silicic acid [49]. When the solution is submitted to CO_2 influx, a sharp decrease in pH followed, down to pH 6.5–7. In addition, the porous silica matrix

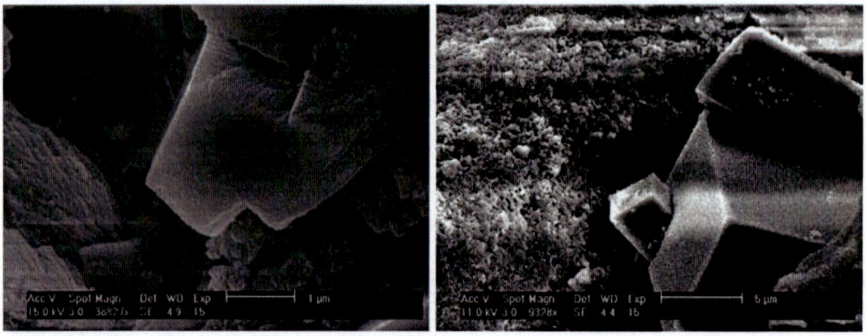

Fig. 9.10 SEM images of samples after carbonation reaction. Top, SAWC40 after 15 min of CO_2 flow. Bottom, sample ASWC500 after 15 min of CO_2 flow. Both samples were left to rest overnight before analysis. Reproduced with permission from [34]

also had a steric catalyst effect. The features of the nanoporous silica matrix prevented the possible agglomeration of the larnite particles and allowed the diffusion of CO_2 through the matrix, so all of the embedded calcium silicate particles could be accessed for the carbonation reaction. In summary, the intrinsic properties of these aerogel composites accelerated the carbonation kinetics under normal conditions of pressure and temperature.

The low-cost composites made without supercritical drying, i.e., the sequestering agent natural wollastonite, without surface chemical activation, were tested as a carbon sequestering agent under even simpler conditions of CO_2 influx. In this case, they were submitted to only 5 min of a pure CO_2 influx (20 cm^3/s, 1.5 bar), in the same aqueous media as used for previous samples. It was observed in these samples that this length of gas attack was not long enough to enhance the degree of carbonation of the embedded mineral phase, thanks to the catalysis of embedding in the porous silica matrix in comparison to the non-encapsulated mineral grains. Due to this result, the length of time of the CO_2 influx in aqueous media under room pressure and temperature was extended to 15 min. In addition, the broken coating obtained by conventional drying instead of the homogeneous one obtained by the more expensive supercritical drying was another drawback that lowered the carbonation efficiency; this should be resolved in order to develop a technologically viable low-cost composite synthesizing procedure.

The precipitation of calcium carbonate was also observed by SEM. The calcite by-product precipitated over the porous matrix (Fig. 9.11) and did not cover the mineral grain, and, consequently, did not hinder carbonation of the entire mineral phase. This fact has a major relevance because the gel coating of the mineral grains will continue, thereby allowing water to diffuse through the matrix despite the precipitation of $CaCO_3$. It would be possible to continue dissolving the mineral ions and diffusing the carbonic ions. This porous layer constitutes the phase that avoids the mineral grains becoming matted, and it also allows large degrees of carbonation without the formation of passivating layers. The carbonation procedures and efficiencies of the calcium-silicate-bearing sol-gel composites are listed in Table 9.3.

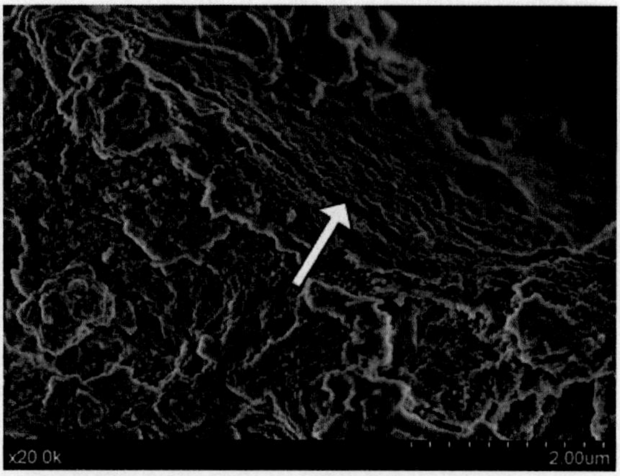

Fig. 9.11 Pointed by the *arrow*, carbonation layer over the silica gel coating of the grains. Reproduced with permission from [35]

9.4 Enhancing the Technological Viability of Calcium-Bearing Composites

Given the immediate industrial applications of the carbon sequestration procedures introduced herein, large research efforts were needed to develop a technologically and energetically viable industrial process based on the sol-gel calcium-bearing composites. Once the carbonation experiments were designed to be as simple as possible from an industrial scale-up point of view: just bubbling CO_2 through aqueous media where the samples are dispersed, research then focused on reducing the costs of the reagents and the carbon footprint, and on reducing energy consumption during sample synthesis. We consider this methodology to be economically viable on the basis of a reduction in the price of the starting materials and a reduction in energy consumption during sample preparation.

Thus, the calcium-bearing composites were synthesised using TEOS as a reagent for the porous silica matrix and as the sequestering agent. In addition, synthesis of the sample concerned several heat treatments at 900–1000°C for several hours and supercritical drying of the final aerogels. In this case, it was not possible to assess the energetic and economic balance in order to estimate technological viability, despite the fact that the sequestration efficiencies of the composites were excellent. Secondly, the sequestering agent was synthesised via low-cost reagents, and sodium metasilicate, calcium chloride or colloidal silica dispersions were employed to precipitate the calcium silicates. Even the carbonation bubbling time was reduced from 40 to 15 min. These new, active phases of NWP and SWP (see Table 9.1) are inexpensive compared to the most expensive samples (WP); therefore, the cost of the final composites (ANWC and ASWC) was reduced by 80% compared to samples SAWC26 and SAWC40. In this case, good carbonation efficiencies were again obtained, but the entire energetic cost of the process (and therefore its carbon footprint) was not low enough.

One way of making this carbonation process technically viable would be to valorise the carbonation by-products and to include the final market price of the by-products in the entire economic balance. In this case, the carbonation by-products are composed of fine silica and calcium carbonate (calcite) in the form of micro-crystals, raw materials that are very much appreciated in the construction industry as a component of stucco, for example.

In summary, raw wollastonite from Spanish mines was used as a CO_2 capturing agent and supercritical drying was avoided, working for the first time with *xero-composites* for carbon sequestration. In addition, the carbonation bubbling time was even reduced to 5 min, simplifying the experimental carbonation reaction set-up. And the heat treatments previously used were also avoided. In this case, the impurities found in the raw wollastonite mineral, the CO_2 bubbling time that was too short in length, and the worsening in the catalytic properties of the porous silica matrix xerogel mainly implied a decrease in the carbonation efficiency. However, on the other hand, these results are the first approximation of the synthesis of low-energy composites, essential for turning this technology into a viable process. Some of the steps required for improving the performance of this technology include working with purer sources of wollastonite, increasing the CO_2 bubbling time, and using basic pH buffers to maintain the pH close to 7.

As a general conclusion of this entire research, it must be noted that the major and unique source of support for turning this technology into a viable and affordable choice will come from the reuse of industrial mineral wastes as silica matrix reagents and sequestering agents. In this regard, there are several strategies that are being considered for industrial waste recycling for CO_2 mineral sequestration, such as portlandite waste from the acetylene industry [6], alkaline paper mill waste [51], coal combustion fly-ash [52], air pollution control residues [53], municipal solid waste incinerator bottom ash [54] and steel slag [55], for example. In these cases, the entire proposal will cover a completely enclosed cycle of waste and CO_2 management, including solid mineral wastes, aqueous basic wastes and greenhouse gases.

Acknowledgments The authors would like to thank the Ministerio de Ciencia e Investigación (Gobierno de España), who funded project CIT-440000-09-1 (Programa Nacional de Investigación Aplicada).

References

1. IPCC Intergovernmental Panel on Climate Change Special Report (2005) Carbon dioxide capture and storage. Cambridge University Press, Cambridge
2. Carey JW, Wigand M, Chipera SJ, Woldegabriel G, Pawar R, Lichtner PC, Wehner SC, Raines MA, Guthrie GD Jr (2007) Analysis and performance of oil well cement with 30 years of CO_2 exposure from the SACROC Unit, West Texas, USA. Int J Greenhouse Gas Control 1:75–85
3. Fabbri A, Corvisier J, Schubnel A, Brunet F, Goffé B, Barlet-Gouédard V, Rimmele G (2009) Effect of carbonation on the hydro-mechanical properties of Portland cements. Cem Concrete Res 39(12):1156–1163

4. Jaramillo P, Griffin WM, MCCoy ST (2009) Life cycle inventory of CO_2 in an enhanced oil recovery system. Environ Sci Technol 43:8027–8032
5. Lackner KS, Grimes P, Ziock HJ (2008) Capturing carbon dioxide froim air. 2nd US–China symposium on CO_2 emission control science and technology, 28–30 May. Hangzou, China
6. Stolaroff JK, Keith WD, Lowry GV (2008) Carbon dioxide capture from atmospheric air using sodium hydroxide spray. Environ Sci Technol 42:2728–2735
7. Morales-Flórez V, Santos A, Lemus A, Esquivias L (2011) Artificial weathering pools of calcium-rich industrial waste for CO_2 sequestration. Chem Eng J 166(1):132–137
8. CO2SINK Project. http://www.co2sink.org. Last visit: March, 2011
9. GRASP project (2011) http://www.grasp-co2.eu. Last visit: March
10. Hellevang H, Aagaard P, Oelkers EH, Kvamme B (2005) Can dawsonite permanently trap CO_2. Environ Sci Technol 39(21):8281–8287
11. Seifritz W (1990) CO_2 disposal by means of silicates. Nature 345:486
12. Lackner KS (2003) A guide to CO_2 sequestration. Science 300:1677
13. Sipila J, Teir S, Zevenhoven R (2008) Carbon dioxide sequestration by mineral carbonation. Literature review update 2005–2007, Heat Engineering Laboratory. Abo Akademi University, Turku
14. Inui T (1996) Highly effective conversion of carbon dioxide to valuable compounds on composite catalysts. Catal Today 29:329
15. Kojima T, Nagamine A, Ueno N, Uemiy S (1997) Absorption and fixation of carbon dioxide by rock weathering. Energy Convers Manag 38:461
16. Wu JCS, Sheen JD, Chen SY, Fan Ch Y (2001) Feasibility of CO_2 fixation via artificial rock weathering. Ind Eng Chem Res 40:3902
17. Mckelvy MJ, Chizmeshya AVG, Diefenbacher J, Bearat H, Wolf G (2004) Exploration of the role of heat activation in enhancing serpentine carbon sequestration reactions. Enciron Sci Technol 38:6897
18. Wolf GH, Chizmeshya AVG, Diefenbacher J, McKelvy MJ (2004) In situ observation of CO2 sequestration reactions using a novel microreaction system. Environ Sci Technol 38:932
19. Huijgen WJJ, Comans RNJ (2006) Mechanisms of aqueous wollastonite carbonation as a possible CO_2 sequestration process. Chem Eng Sci 61:4242–4251
20. Zevenhoven R, Eloneva S, Teir S (2006) Chemical fixation of CO_2 in carbonates: route to valuable products and long-term storage. Catal Today 115:73
21. Kakizawa M, Yamasaki A, Yanagisawa Y (2001) A new CO_2 disposal process via artificial weathering of calcium silicate accelerated by acetic acid. Energy 26:341–354
22. O'Connor WK, Dahlin DC, Rush GE, Dahlin CL, Collins WK (2002) Carbon dioxide sequestration by direct mineral carbonation: process mineralogy of feed and products. Miner Metall Process 19:95
23. Tai CY, Chen WR, Shih SM (2006) Factors affecting wollastonite carbonation under CO_2 supercritical conditions. AIChE J 52:292
24. Béarat H, Mckelvy MJ, Chizmeshya AVG, Gormley D, Nunez R, Carpenter RW, Squires K, Wolf GH (2006) Carbon sequestration via aqueous olivine mineral carbonation: role of passivating layer formation. Environ Sci Technol 40:4802–4808
25. Ahmed MS, Attia YA (1998) Multi-metal oxide aerogel for capture of pollution gases from air. App Thermal Eng 18:787–797
26. Santos A, Toledo-Fernández JA, Mendoza-Serna R, Gago-Duport L, De la Rosa-Fox N, Piñero M, Esquivias L (2007) Chemically active silica-wollastonite composites for CO_2 fixation by carbonation reactions. Ind Chem Eng 46:103–107
27. Esquivias L, Piñero M, Morales-Flórez V, De la Rosa-Fox N (2011) Aerogels synthesis by sonocatalysis–sonogels. In: Aegerter M, Leventis N (eds) Handbook of aerogels. Springer, London To be published
28. Tarasevich M (1984) Ultrasonic hydrolysis of a metal alkoxide without alcohol solvents. Cer Bull 63:500 (Abstract only)

29. Boudjouk P (1986) Electrochemical and sonochemical routes to organosilane. Precursors. In: Hench LL, Ulrich DR (eds) Science of ceramics chemical processing. Wiley, New York, p 363
30. Price GJ, Hearn MP, Wallace ME, Patel AM (1996) Ultrasonically assisted synthesis and degradation of poly(dimethyl siloxane). Polymer 37:2303–2308
31. Kistler SS (1932) Coherent and expanded aerogels. J Phys Chem 36(1):52
32. Blanco E, Esquivias L, Litrán R, Piñero M, Ramírez-del-Solar M, De la Rosa-Fox N (1999) Sonogels and derived materials. Appl Organomet Chem 13:399–418
33. Santos A, Ajbary M, Kherbeche A, Piñero M, De la Rosa-Fox N, Esquivias L (2008) Fast CO_2 sequestration by aerogel composites. J Sol-gel Sci Technol 45:291–297
34. Santos A, Ajbary M, Toledo-Fernández JA, Morales-Florez V, Kherbeche A, Esquivias L (2008) Reactivity of CO_2 traps in aerogel–wollastonite composites. J Sol-gel Sci Techol 48:224–230
35. Morales-Flórez V, Santos A, Esquivias L (2011b) Recent insights into xerogel and aerogel mineral composites for CO_2 mineral sequestration. J Sol-gel Sci Technol. Availabe doi: 10.1007/s10971-010-2276-8
36. Santos A, Ajbary M, Morales-Florez V, Kherbeche A, Piñero M, Esquivias L (2009) Larnite powders and larnite/silica aerogel composites as effective agents for CO_2 sequestration y carbonation. J Haz Mat 168:1397–1403
37. Leal O, Bolivar C, Ovalles C, Garcia JJ, Espidel Y (1995) Reversible adsorption of carbon dioxide on amine surface-bonded silica gel. Inorg Chim Acta 240(1–2):183–189
38. Song HK, Cho KW, Lee KH (1998) Adsorption of carbon dioxide on the chemically modified silica adsorbents. J Non-Cryst Solids 242:69–80
39. Graham UM, Thomas G (1999) Chemically active aerogel-fly ash composites for mercury clean-up, international ash utilization symposium center for applied energy research. University of Kentucky, Kentucky
40. Reynolds JG, Coronado PR, Urbes LW (2001) Hydrophobic aerogels for oil-spill clean up synthesis and characterization. J Non-Cryst Solids 292:127
41. Khomane RB, Sharma BK, Saha S, Kulkarni BD (2006) Reverse microemulsion mediated sol-gel synthesis of lithium silicate nanoparticles under ambient conditions: scope for CO_2 sequestration. Chem Eng Sci 61(10):3415–3418
42. Xomeritakis G, Tsai CY, Brinker CJ (2005) Microporous sol-gel derived aminosilicate membrane for enhanced carbon dioxide separation. Sep Purif Technol 42:249–257
43. Chen C, Ahn WS (2011) CO_2 capture using mesoporous alumina prepared by a sol-gel process. Chem Eng J 166:646–651
44. Gupta H, Fan LS (2002) Carbonation-calcination cycle using high reactivity calcium oxide for carbon dioxide separation from flue gas. Ind Eng Chem Res 41:4035–4042
45. Ahmed MS, Attia YA (1993) Sol-gel prepared aerogels for adsorption/capture of hazardous gases. First International Conference on Application Commercialization of Sol Gel Processing, Saarbrücken, Germany, October 10–13
46. Attia YA (1993) Capture of greenhouse gases by novel aerogel sorbents prepared by sol-gel technology. In: Parekh BK, Groppo JG (eds) Processing and utilization of high-sulfur coals V. Elsevier Science Publishers, B.V, pp 467–484
47. Nakagawa K, Ohashi T (1998) Novel method of CO_2 capture from high temperature gases. J Electrochem Soc 145:1344–1346
48. Yong Z, Mata V, Rodrigues AE (2002) Adsorption of carbon dioxide at high temperature—a review. Sep Purif Technol 26:195–205
49. Prabhalar S, Hanumantha Kao R, Forsling W (2005) Dissolution of wollastonite and its flotation and surface interactions with tallow-1, 3-diaminopropane (duomeen T). Miner Eng 18:691
50. Casey WH, Westrich HR, Banfield JF, Ferruzzi G, Arnold W (1993) Leaching and reconstruction at surfaces of dissolving chain silicate minerals. Nature 336:253
51. Perez-Lopez R, Castillo J, Quispe D, Nieto JM (2009) Mineral sequestration of CO_2 by aqueous carbonation of coal combustion fly-ash. J Hazard Mater 161:1347–1354

52. Montes-Hernandez R, Perez-Lopez R, Renard F, Nieto JM, Charlet L (2009) Mineral sequestration of CO_2 by aqueous carbonation of coal combustion fly-ash. J Hazard Mater 161:1347–1354
53. Priggiobe V, Polettini A, Baciocchi R (2009) Gas–solid carbonation kinetics of air pollution control residues for CO_2 storage. Chem Eng J 148:270–278
54. Rendek E, Ducom G, Germain P (2006) Carbon dioxide sequestration in municipal solid waste incinerator (MSWI) bottom ash. J Hazard Mater B 128:73–79
55. Huijgen WJ, Witkamp GJ, Comans RNJ (2005) Mineral CO_2 sequestration by steel slag carbonation. Environ Sci Technol 39:9676–9682

Chapter 10
Carbon Aerogels for Wastewater Treatment

Jonathan Fang, Justin Solis, Esther Lan and Bruce Dunn

Abstract Wastewater treatment technologies have become extremely important due to the dangers and health risks posed from water pollution. Adsorption is a common method for removing contaminants from wastewater with anthracite coal being commonly used. However, anthracite coal is ineffective at removing some contaminants, especially at low concentrations. We have investigated the use of carbon aerogels as effective adsorbents for wastewater treatment. Carbon aerogels are mesoporous materials with low density, high surface area, and a porous interconnected network, which are ideal properties for an effective adsorbent. The ma-terials were synthesized by supercritical drying and pyrolysis of organic gels. The resulting aerogels were tested using the contaminants hexavalent chromium and estrone in equilibrium adsorption experiments to determine their adsorption capa-bilities. Carbon aerogels significantly outperformed anthracite coal by removing nearly all of the contaminants from water, demonstrating that these materials are very promising adsorbents for wastewater treatment.

Keywords 17α-Ethinyloestradiol (EE2) · 17β-Estradiol (E2) · Adsorbate · Adsorbent · Adsorption · Adsorption capacity · Adsorption efficiency · Anthracite coal · Carbon aerogels · Condensation · Endocrine-disrupting compounds (EDC) ·

J. Fang · J. Solis · E. Lan · B. Dunn (✉)
Department of Materials Science and Engineering, University of California,
Los Angeles, CA 90095-1595, USA
e-mail: bdunn@ucla.edu

J. Fang
e-mail: jfang33@yahoo.com

J. Solis
e-mail: justinmsolis@gmail.com

E. Lan
e-mail: elan@ucla.edu

M. Aparicio et al. (eds.), *Sol-Gel Processing for Conventional and Alternative Energy*, Advances in Sol-Gel Derived Materials and Technologies,
DOI: 10.1007/978-1-4614-1957-0_10,

Estradiol · Estrogens · Formaldehyde (F) · Freundlich model · Hexavalent chromium (Cr^{6+}) · Mesoporous materials · Organic gels · Polymerization · Porosity · Pyrolysis · Resorcinol (R) · Resorcinol-formaldehyde (RF) · Supercritical drying · Surface area · Toxic contaminants · Wastewater · Wastewater treatment

10.1 Introduction

Water pollution, from industrial and natural sources, is a global problem that can have detrimental effects on the environment and many forms of life. Metallurgical industries, chemical and manufacturing plants, and mining have introduced heavy metals, halides, organics, and other contaminants into wastewater [1, 2]. In addition, endocrine-disrupting compounds (EDC) are also prevalent in wastewater [3, 4]. These contaminants pose a health risk to aquatic and human life and need to be removed to safe levels [3, 5, 6]. In this chapter, we will discuss the synthesis and application of carbon aerogels for the removal of toxic contaminants from wastewater. Aerogels are highly porous, solid materials which have low density, low thermal conductivity, and high surface area and porosity [7–9]. Applications for aerogels include thermal insulation, particle trapping, double layer capacitors, anti-reflective coatings, catalysis, and desalination [8–12]. Aerogels have been investigated as effective adsorbents for water treatment due to their high surface area and porosity [1, 2, 13, 14]. The open structure of the aerogel allows transport of molecules through the pores for binding to the internal surface. Therefore, adsorption of contaminants on aerogels can occur on both internal and external surfaces, thus enabling aerogels to have high adsorption capacities. For our studies, we will focus on the removal of the contaminants hexavalent chromium and the EDC estrone from water via adsorption onto carbon aerogels. The adsorption efficiency of carbon aerogels will be compared to that of a standard adsorbent, anthracite coal.

10.2 Hexavalent Chromium

The contamination of water sources from heavy metal processing presents a serious environmental problem. Many of these metals are ecologically harmful and can have toxic effects on marine life and humans. Therefore, it is imperative that these contaminants be removed from wastewater to levels significantly lower than the toxicity limit. Toxic metals, such as chromium, mercury, lead, and arsenic, have been observed to be at elevated levels in wastewater effluents [1]. Our investigation will focus on chromium, specifically hexavalent chromium (Cr^{6+}). Cr^{6+} is a stable form of the metal found in the environment and primarily comes from industrial processing [15]. In water, Cr^{6+} predominately exists as either dichromate ($Cr_2O_7^{2-}$) or chromate (CrO_4^{2-}) anions [2]. It is also a highly toxic carcinogen to many animals and humans via inhalation and ingestion [16–18]. Exposure to hexavalent chromium can lead to nausea, diarrhea, liver and kidney damage, lung cancer, and hemorrhaging [16, 19]. In Canada, Pawlisz et al. reported chromium concentrations in water between 1 and 165 μg/L [20]. In the United States, it has been reported that 18% of the population was exposed to chromium levels ranging from 2 to 60 μg/L, with a small fraction of the population being exposed to levels between 60 and 120 μg/L [21]. The Environmental Protection Agency (EPA) determined that lifetime exposure to 1 mg of chromium/kg of body mass per day will not cause severe health problems for many animals [18]. The EPA has also set the current standard of total chromium in drinking water to be 100 μg/L [15].

Various methods have been employed for the removal of chromium from water: precipitation, ultrafiltration, ion exchange, reverse osmosis, electrochemical treatment, and adsorption [19]. In comparison to other conventional methods, adsorption is low cost, simple to utilize, efficient, and adsorbents are readily available [17, 19]. Common adsorbents used for this method include activated carbon, polymers, biological materials, and carbon aerogels [2, 17, 19, 22]. The disadvantages of some of these treatment methods are high cost, undesired by-products (sludge), nonporous or low surface area materials, and ineffectiveness at low contaminant levels [19]. Recently, carbon aerogels have been investigated as an effective adsorbent for chromium, halides, and toluene [1, 2, 13, 14].

10.3 Endocrine-Disrupting Compounds

Another group of contaminants in wastewater that are of increasing concern are endocrine-disrupting compounds (EDCs) [3, 4]. EDCs, such as natural or synthetic estrogens, are substances that disturb the normal biological activity of an animal's endocrine system and can have damaging effects on marine life and humans [3, 5, 6]. Some of these effects include feminization of male fish, decreased fertility and egg production in female fish, sex reversal between genders, disruption of reproductive processes, and inducing cancer [23–28]. EDCs also have extremely high biological potency where their effects can be felt at concentrations as low as ng/L [23–28]. There are three prominent EDCs found in wastewater effluents (Fig. 10.1): estrone (E1), 17*β*-estradiol (E2), and 17α-ethinyloestradiol (EE2) [29–33]. E1 and E2 are synthesized in the human body while EE2 is an active ingredient in oral contraceptive pills [26, 29, 32, 34]. Thorpe et al. investigated the potencies of these estrogens in rainbow trout. The median concentrations for potency were 60 ng/L for E1, between 19 and 26 ng/L for E2, and between 0.95 and 1.8 ng/L for EE2 [31]. Typical estrogen concentrations observed in wastewater effluents across the United Kingdom ranged from 1 to 76 ng/L for E1, 4–48 ng/L for E2, and up to 7 ng/L for EE2 [30]. Vethaak et al. reported E1 and

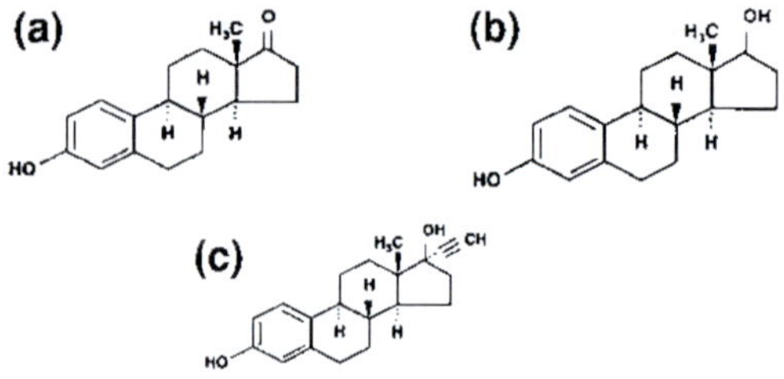

Fig. 10.1 Molecular structures of **a** estrone, **b** 17*β*-estradiol, and **c** 17α-ethinyloestradiol [33] (© Elsevier 1999), reprinted with permission

E2 concentration ranges that went above 100 ng/L for wastewater in the Netherlands [28]. The fact these EDC levels in wastewater are above the potency limit presents a significant health problem.

Current methods utilized for the removal of EDCs from wastewater include adsorption, filtration, lime softening, activated sludge treatment, and oxidative processes [35–39]. Unfortunately, current methods for EDC removal may be costly or ineffective. To the best of our knowledge, there has been no published literature on the use of carbon aerogels for the adsorption and removal of EDCs from water. Estrone is an EDC of particular interest due to its elevated levels in wastewater effluents [28–30, 40]. Oxidation of estradiol to estrone can occur during the treatment process, which increases the total estrone level in wastewater [40, 41]. In addition, European and Canadian wastewater treatment plants have shown that the fraction of estrone removed is significantly lower than that of estradiol [40, 41]. Estrone removal from wastewater varied from 61 to 98% for different wastewater treatment plants, with results being inconsistent [40, 41].

10.4 Water Treatment Plant

Figure 10.2 outlines a schematic for a typical water reclamation plant [42]. The primary treatment stage involves separation of lighter and heavier solids where the lighter solids, such as plastic and grease, are removed for further treatment while the remaining contaminants, such as organics and other dissolved species, continue onto secondary treatment [42]. The secondary stage introduces microorganisms into the tank to consume dissolved organic materials, after which the microorganisms settle to the bottom of the tank where they are removed and recycled back into the process [42]. Air is bubbled into the water as an oxygen supply to allow the organisms to breathe, which can lead to oxidation of estradiol to estrone as previously mentioned [42]. The tertiary stage involves replacing the ground with filters, containing an adsorbent, sand, and gravel, to remove any residual contaminants and disinfect the water [42]. The Sanitation Districts of Los Angeles County (SDLAC) uses anthracite coal as their adsorbent in the tertiary stage. Anthracite coal is a common adsorbent used for wastewater treatment that is abundant and inexpensive to produce. However, it is ineffective in removing some

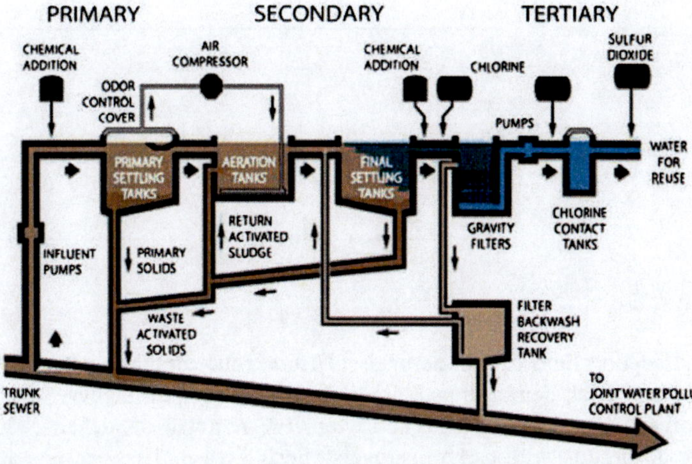

Fig. 10.2 Schematic of a typical water reclamation plant [42] (© LACSD), reprinted with permission

contaminants to safe levels. There is a need, therefore, to identify alternative adsorbents that are more efficient and could be mass produced. Carbon aerogels could fulfill this need and be incorporated into the filters used in the tertiary stage.

10.5 Carbon Aerogels

Alternative adsorbents to anthracite coal for wastewater treatment are carbon aerogels, which are mesoporous materials that have low density (0.05–0.8 g/cm^3), high surface area (400–1,200 m^2/g), high porosity (>80%), and tunable pore sizes (1–50 nm) [7, 9]. Carbon aerogels also have low thermal conductivity (0.029 Wm^{-1}K^{-1} in air), good electrical conductivity (0.6–20 Ω^{-1}cm^{-1}), a Young's modulus ranging from 100 to over 200 MPa, and a maximum strength of up to 50 MPa [43, 44]. Pekala et al. were the first to synthesize organic aerogels from the polycondensation of resorcinol (R) with formaldehyde (F) [45]. These resorcinol–formaldehyde (RF) organic gels can be further pyrolyzed to yield carbon aerogels [9, 46]. Researchers have also synthesized organic gels through the polycondensation of other functional monomers, such as melamine, cresol, and phloroglucinol, with formaldehyde [7, 47, 48]. It is hypothesized that the high surface area and porosity of carbon aerogels gives it more binding sites for adsorption than a non-porous, dense material, such as anthracite coal. This is supported by recent studies where carbon aerogels were used to remove heavy metals, halides, and organics from water [1, 2, 13, 14]. Table 10.1 summarizes some of the results obtained from the removal of various metals from water

Table 10.1 Removal of various metals from water using carbon aerogels [1]

Metal	Percent removal (%)
Hg	almost 100
Mn	93
Pb	92
Zn	91
Cu	79
Ni	76
Cd	70

through adsorption onto carbon aerogels [1]. For concentrations between 1 and 5 mg/L, Meena et al. demonstrated almost 100% removal of mercury, over 90% removal of manganese, zinc, lead, and over 70% removal of nickel, cadmium, copper for adsorption with carbon aerogels after 48 h [1]. These results demonstrate that carbon aerogels are promising adsorbents for the effective removal of many different metals.

For our studies, we chose to use the resorcinol–formaldehyde polymerization route to synthesize carbon aerogels. Resorcinol and formaldehyde undergo addition and condensation reactions to polymerize into a gel with a 3D cross-linked network structure (Fig. 10.3) [9, 46]. First, resorcinol, formaldehyde, and a catalyst, such as sodium carbonate, are mixed together. The sodium carbonate serves to create resorcinol anions that facilitate addition of formaldehyde to the aromatic ring, creating a hydroxymethyl derivative of resorcinol [9]. Then, these hydroxymethyl derivative monomers condense and form methylene and methylene ether linkages, forming polymeric clusters [9]. The polymeric clusters aggregate and bind to each other to yield an interconnected structure or an RF gel. The wet RF gel needs to be converted into an aerogel without significant shrinkage and collapse of the structure. Supercritical drying of these RF gels with carbon dioxide removes the solution phase of the gel while preserving the porous, interconnected structure to yield a robust aerogel. Finally, the organic aerogel can be pyrolyzed at 900°C to remove the organic groups and produce a relatively pure carbon aerogel [9].

The final microstructure and surface properties of carbon aerogels can be tailored by adjusting the amount of reactants or catalyst used for synthesis. For instance, increasing the amount of catalyst leads to faster reaction kinetics and the formation of smaller particles. Polymerization of smaller particles leads to gels that will have smaller pore sizes, higher density, and larger surface areas (Table 10.2) [7, 9]. The same effect can also be achieved by decreasing the amount of resorcinol, formaldehyde, or water. Conversely, increasing the amount of resorcinol, formaldehyde, or water leads to dilution of the solution and yields gels with larger pore sizes, lower density, and smaller surface areas [9]. The resorcinol:formaldehyde molar ratio typically used is 1:2 while the resorcinol:catalyst molar ratio normally ranges from 50 to 300.

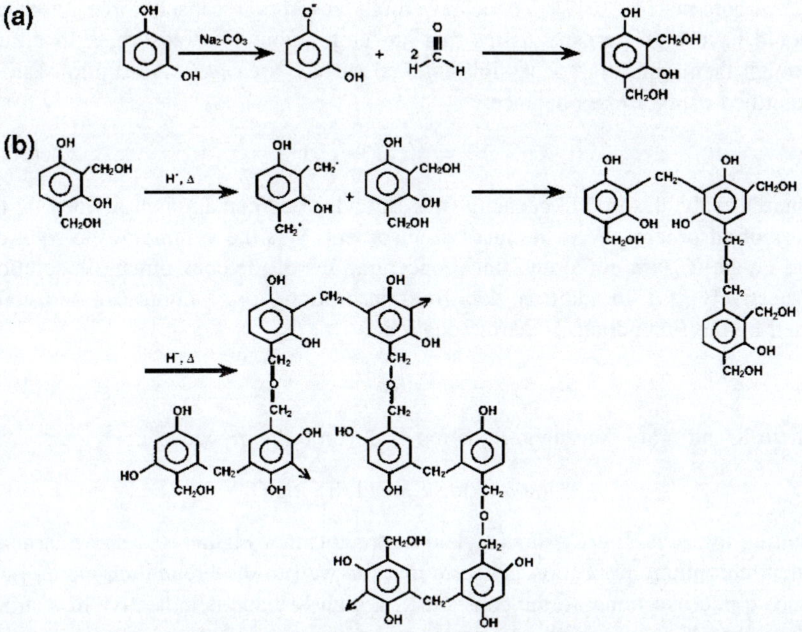

Fig. 10.3 Mechanism of polymerization of resorcinol with formaldehyde via a two step process: **a** addition and **b** condensation [9, 46] (© Wiley–VCH Verlag GmbH & Co. KGaA 2003, © Elsevier 1997), reproduced with permission

Table 10.2 Properties of RF aerogels for different R:C ratios [7]

	Low R:C (50)	High R:C (200)
Density (g/cm^3)	0.6	0.03
Surface area (m^2/g)	900	350
Particle size (nm)	3–5	11–14

10.6 Adsorption

A common and effective method for the removal of various contaminants from water is physical adsorption, which involves competing attractive and repulsive forces between the adsorbate (contaminant) and adsorbent molecules [49]. Attractive forces are attributed to dipole–dipole interactions between the adsorbate and adsorbent [49]. However, as the distance between the adsorbate molecule and adsorbent surface decreases, repulsive forces emerge due to electron cloud overlap between the two species [49]. These forces are relatively weak and are usually reversible [49]. Over the years, carbonaceous materials have been commonly used as adsorbent materials. Factors that affect physical adsorption include temperature, type of solvent used, pH of the solution, and the surface area and pore structure of the carbon material [49]. To achieve high adsorption capacity, the adsorbent should have high porosity, pores that are large enough for adsorbate to diffuse through them, and high accessible internal surface area [49]. Adsorption can be quantified using a mass balance:

$$qW = V(C_0 - C_e) \tag{10.1}$$

where q is the adsorption capacity (mass of adsorbed contaminant divided by the mass of adsorbent), W is the mass of adsorbent, V is the volume of the solution, and C_0 and C_e are initial and final concentrations of the contaminant in solution, respectively [49]. In addition, data from adsorption from a liquid can be usually fitted by the Freundlich isotherm equation:

$$q = KC_e^{1/n} \tag{10.2}$$

where K and n are constants. Equation 10.2 can be rearranged to:

$$\log q = \log K + (1/n)^* \log C_e \tag{10.3}$$

Plotting log q vs. log C_e should yield a straight line. Figure 10.4 shows a hexavalent chromium adsorption isotherm that fits well to the Freundlich model [22]. Since q is proportional to the constant K, a high K value is indicative of a strong adsorbent. K is a useful parameter in estimating and comparing adsorption capacities between different adsorbents. The Freundlich isotherm model does not assume that all binding sites for adsorption are equal, thus allowing for the possibility of multiple layers of coverage [49]. The Freundlich isotherm equation also implies that q can increase indefinitely as C_e increases, which is physically impossible so the equation would not fit data well at very high C_e [49]. Fortunately, most adsorption processes occur at relatively dilute concentrations so this does not become a problem [49].

10.7 Experimental Methods

Figure 10.5 shows a schematic of the process for synthesizing carbon aerogels. Resorcinol (R), formaldehyde (F), distilled water (W), and sodium carbonate (C) were mixed together such that the R:F, R:W, and R:C molar ratios were 1:2, 1:27, and 48, respectively. After stirring the solution for a couple of hours at room temperature, the sol was cast into glass tubes and cured at 50°C for two days to expedite polymerization and gelation. Next, the resorcinol–formaldehyde (RF) gels were cut into small pieces and immersed in acetone for solvent exchange twice a day for five days. The gels were then supercritically dried with carbon dioxide to create an aerogel. Finally, the aerogels were pyrolyzed at 900°C for four hours to remove any residual organic groups still present in the structure to yield a pure carbon aerogel.

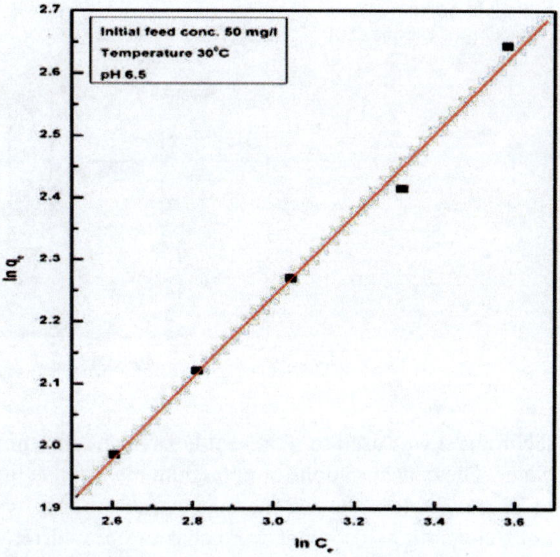

Fig. 10.4 Freundlich isotherm for hexavalent chromium adsorption onto activated carbon [22] (© Elsevier 2008), reprinted with permission

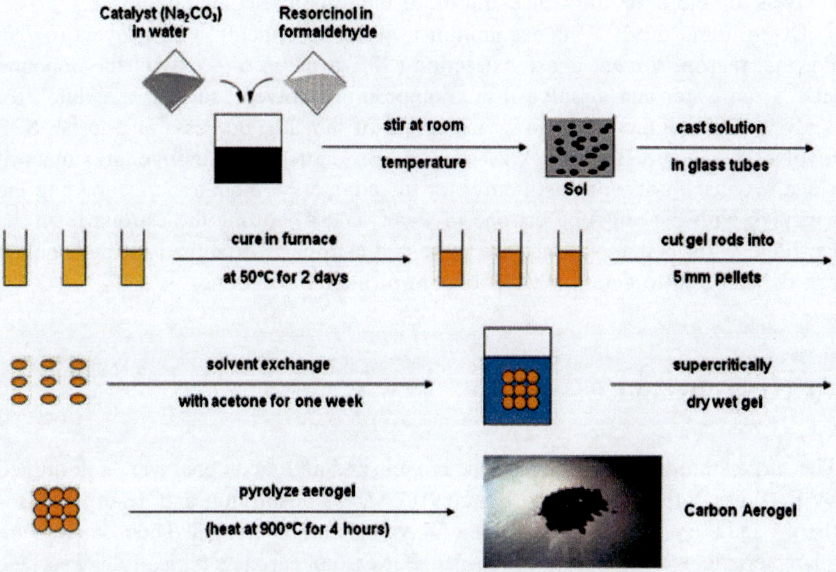

Fig. 10.5 Diagram outlining the synthesis of carbon aerogels

Adsorption experiments for various contaminants were performed by adding 100 mg of the adsorbent, either anthracite coal or carbon aerogel, to 20 mL of distilled water containing various concentrations of the contaminant. Potassium

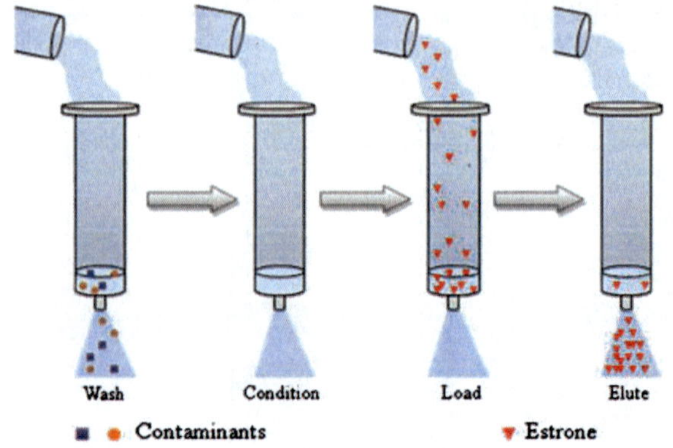

Fig. 10.6 Solvent phase extraction of estrone [50]

dichromate was used to represent hexavalent chromium contamination in distilled water. The concentrations of potassium dichromate in the solutions varied from 0.7 to 23 mg/L while the estrone concentration range was 0.1–50 mg/L. These solutions containing adsorbent were shaken on a stirrer for three days to ensure that equilibrium was reached. Finally, the solution was filtered to separate the unbound contaminant from the adsorbent. This filtered solution was sent to the SDLAC for analysis to determine the concentration of the unbound contaminant.

Determining the EDC concentration in water is difficult at low levels (ng/L). For that reason, solvent phase extraction (SPE) was used to extract the unbound EDC from water and dissolve it in an appropriate solvent, such as methanol, for analysis [50]. Figure 10.6 shows a diagram of the SPE process for estrone. SPE involves four major steps: 1) washing the SPE cartridge to remove any contaminants, 2) flushing it with distilled water to activate the membrane, 3) loading the cartridge with the unbound estrone in water, and 4) eluting the estrone from the cartridge with methanol. Once the extracted estrone is dissolved in methanol, it can be analyzed to determine the amount of unbound estrone.

10.8 Results and Discussion

The surface properties of the carbon aerogels and anthracite coal were determined by BET gas sorption analysis. Figure 10.7 shows adsorption and desorption isotherms and the pore size distribution of a typical carbon aerogel. The isotherms are characteristic of a mesoporous material. Our carbon aerogels typically had surface areas ranging from 800 to 1,000 m^2/g, pore diameters of 4–5 nm, and almost 90% porosity. In contrast, anthracite coal had a surface area of 3.2 m^2/g and no significant pore structure. Table 10.3 compares the properties of activated carbon (from wood), anthracite coal (from SDLAC), and one of our carbon aerogels. These results indicate that our carbon aerogels had high surface area and similar pore sizes to activated carbon. The carbon aerogel had significantly higher surface

Fig. 10.7 **a** Adsorption/desorption isotherms and **b** pore size distribution for one of our carbon aerogels

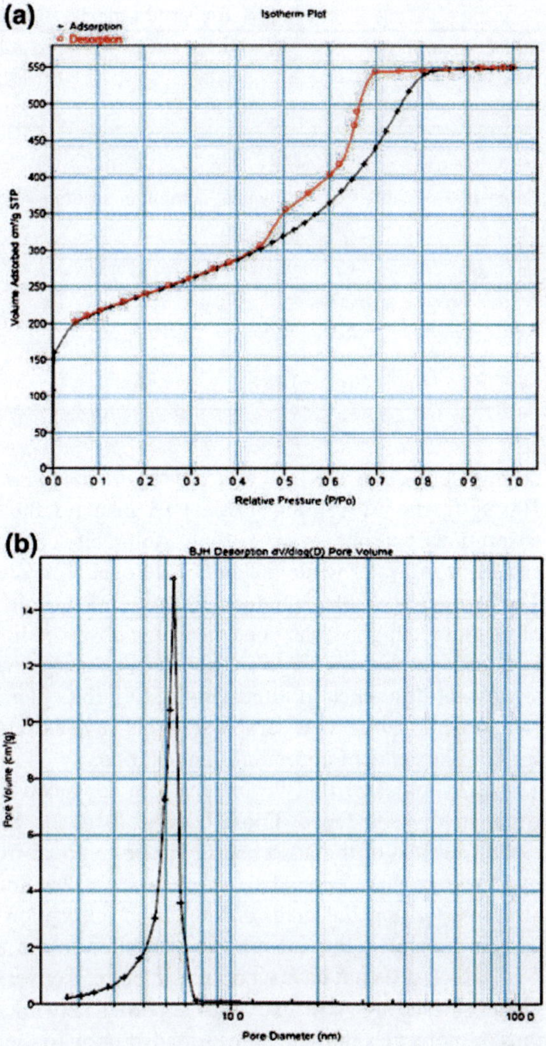

area and porosity than anthracite coal. In addition, the pore sizes of the aerogel were significantly larger than the molecular dimensions of trace metals and EDCs, which allows for diffusion of these species into the porous network. Therefore, the aerogel had accessible internal surface area and many more binding sites for adsorption than anthracite coal.

Table 10.4 shows the results from the equilibrium adsorption experiments for hexavalent chromium. Anthracite coal generally removed less than 20% of the Cr^{6+} from the distilled water while the carbon aerogel removed 95% or more of the Cr^{6+} from water. Even for an extremely high initial Cr^{6+} concentration (over

Table 10.3 Physical properties of various adsorbents [49]

	Activated carbon	Anthracite coal	Carbon aerogel
Surface area (m^2/g)	1,500	3.2	980
Pore volume (cm^3/g)	1.8	–	1.1
Average pore diameter (nm)	4.6	–	4.6

Table 10.4 Results from hexavalent chromium adsorption experiments

	Anthracite coal		Carbon aerogel	
C_0 (mg/L)	C_e (mg/L)	% removal	C_e (mg/L)	% removal
2.9	2.3	20	0.1	95
5.8	5.1	10.6	<0.1	>98.3
11.5	9.7	15.7	<0.1	>99
23	10	56.5	<0.1	>99.6

20 mg/L), carbon aerogel was able to reduce the Cr^{6+} concentration to under 100 μg/L, the EPA limit. Table 10.5 displays the results from the equilibrium adsorption experiments for estrone. Anthracite coal removed a minimal amount of estrone from water while the carbon aerogel removed nearly all of it (over 99.9%). The aerogel was able to reduce 100 μg/L of estrone in water (at least three orders of magnitude higher than concentrations observed in wastewater) to less than 1 ng/L. It appears that the large surface area of the porous carbon aerogel has made a significant difference in improving adsorption efficiency. These adsorption results clearly demonstrate that carbon aerogels have extremely high removal efficiencies for both hexavalent chromium and estrone.

We hypothesize that the mechanism for hexavalent chromium removal from water involves induced dipole bonding between the dichromate anions and the carbon surface of the adsorbent. Estrone removal from water can be attributed to hydrophobic interactions between estrone molecules and the carbon surface. We also suspect that the main difference in adsorption between anthracite coal and carbon aerogel is that adsorption is limited to the external surface of anthracite coal, which does not have a porous structure. Conversely, carbon aerogel has much higher surface area and an extensive porous network, which allows for transport of ions or molecules into the pores for adsorption to the internal surface. It seems that adsorption to the internal surface contributes a significant amount to the total adsorption capacity.

Adsorption of contaminants from solution was also characterized using the Freundlich isotherm model. Figure 10.8 shows a plot of log q vs. log C_e for estrone adsorption onto carbon aerogel. From the fitted line, K was calculated to be 18 μg of estrone/g of carbon. That is, the K value obtained for carbon aerogels indicates that one gram of carbon aerogel could reduce estrone levels from 100 ng/L to 1 for 180 Liters of water. As previously mentioned, estrone adsorption experiments with anthracite coal could not be performed over a wide range of estrone concentrations due to loading issues with the SPE cartridges. Therefore, there is insufficient data

Table 10.5 Results from estrone adsorption experiments

	Anthracite coal		Carbon aerogel	
C_0 (mg/L)	C_e (μg/L)	% removal	C_e (ng/L)	% removal
0.1	–	–	<1	>99.9
1	705	30	19	>99.9
10	–	–	430	>99.9
50	–	–	590	>99.9

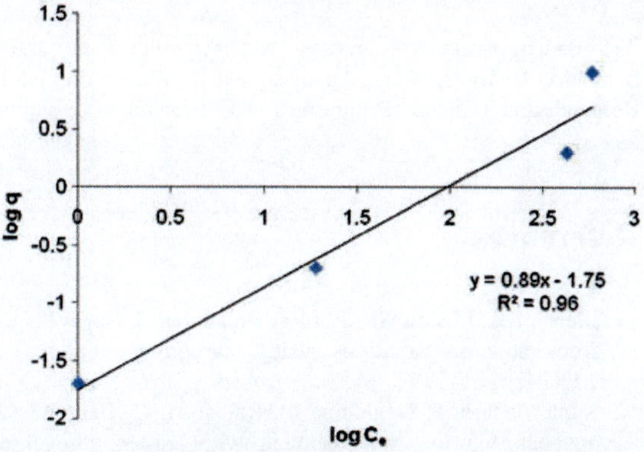

Fig. 10.8 Freundlich isotherm for estrone adsorption onto carbon aerogel (q has units of mg of estrone/g of carbon aerogel, C_e has units of ng/L)

to perform this analysis on estrone adsorption onto anthracite coal. However, the present results underscore the point that only a small amount of carbon aerogel is needed to treat a large volume of water with a high level of estrone.

10.9 Conclusion

Carbon aerogels have been investigated as an alternative adsorbent to anthracite coal for more efficient removal of contaminants from wastewater. The aerogels were successfully synthesized from organic gels using a resorcinol–formaldehyde polycondensation reaction. These mesoporous materials had high surface area with pore sizes that were significantly larger than the contaminant molecules, which allows transport of molecules through the porous network for adsorption onto the internal surface. Adsorption experiments were performed to compare the adsorption efficiency of carbon aerogels with anthracite coal. The contaminants of interest were hexavalent chromium and estrone due to their prevalence in wastewater and harmful effects on the environment. For high concentrations of contaminants in water, carbon aerogel removed 95% or more of the hexavalent chromium and virtually all of the estrone. In both cases, carbon aerogel significantly outperformed anthracite coal in the removal of these contaminants from water. Carbon aerogel adsorption of estrone was also characterized using the Freundlich isotherm model.

There is still much more to do before carbon aerogels can be widely employed for wastewater treatment. Additional studies are needed in order to quantify the adsorption capacity of carbon aerogels for hexavalent chromium, investigate adsorption kinetics, and to evaluate the adsorption efficiency of carbon aerogels for other important EDCs, such as E2 and EE2. Nonetheless, the present work demonstrates that there is a great opportunity for carbon aerogels in environmental applications because of the extraordinary adsorption properties derived from their chemistry and porous architecture.

Acknowledgments The authors greatly appreciate the assistance of Dr. John Strand and the Sanitation Districts of Los Angeles County. We would also like to thank the Toxic Substances Research and Teaching Program (TSRTP) for financial support.

References

1. Meena AK, Mishra GK, Rai PK, Rajagopal C, Nagar PN (2005) Removal of heavy metal ions from aqueous solutions using carbon aerogel as an adsorbent. J Hazard Mater 122(1–2):161–170
2. Rana-Madaria P, Nagarajan M, Rajagopal C, Garg BS (2005) Removal of chromium from aqueous solutions by treatment with carbon aerogel electrodes using response surface methodology. Ind Eng Chem Res 44:6549–6559
3. Diamanti-Kandarakis E, Bourguignon JP, Giudice LC, Hauser R, Prins GS, Soto AM, Zoeller RT, Gore AC (2009) Endocrine-disrupting chemicals: an endocrine society scientific statement. Endocr Rev 30(4):293–342
4. Falconer IR, Chapman HF, Moore MR, Ranmuthugala G (2006) Endocrine-disrupting compounds: a review of their challenge to sustainable and safe water supply and water reuse. Environ Toxicol 21:181–191
5. Meeker JD (2010) Exposure to environmental endocrine disrupting compounds and men's health. Matur 66:236–241
6. Tanabe S (2002) Contamination and toxic effects of persistent endocrine disrupters in marine mammals and birds. Mar Pollut Bull 45:69–77
7. Pekala RW, Alviso CT, Kong FM, Hulsey SS (1992) Aerogels derived from multifunctional organic monomers. J Non-Cryst Solids 145:90–98
8. Akimov YK (2003) Fields of application of aerogels (review). Instrum Exp Tech 46:287–299
9. Al-Muhtaseb SA, Ritter JA (2003) Preparation and properties of resorcinol-formaldehyde organic and carbon gels. Adv Mater 15:101–114
10. Fricke J, Emmerling A (1998) Aerogels–recent progress in production techniques and novel applications. J Sol-Gel Sci Technol 13:299–303
11. Choi J, Suh DJ (2007) Catalytic applications of aerogels. Catal Surv Asia 11:123–133
12. Jung HH, Hwang SW, Hyun SH, Lee KH, Kim GT (2007) Capacitive deionization characteristics of nanostructured carbon aerogel electrodes synthesized via ambient drying. Desalination 216:377–385
13. Sanchez-Polo M, Rivera-Utrilla J, Salhi E, von Gunten U (2007) Ag-doped carbon aerogels for removing halide ions in water treatment. Water Res 41:1031–1037
14. Maldonado-Hodar FJ, Moreno-Castilla C, Carrasco-Marin F, Perez-Cadenas AF (2007) Reversible toluene adsorption on monolithic carbon aerogels. J Hazard Mater 148:548–552

15. United States Environmental Protection Agency, Office of Water (2010) Chromium-6 in drinking water. EPA 815-F-10-005
16. Costa M, Klein CB (2006) Toxicity and carcinogenicity of chromium compounds in humans. Crit Rev Toxicol 36:155–163
17. Owlad M, Aroua MK, Daud WAW, Baroutian S (2009) Removal of hexavalent chromium-contaminated water and wastewater: a review. Water Air Soil Pollut 200:59–77
18. Zayed AM, Terry N (2003) Chromium in the environment: factors affecting biological remediation. Plant Soil 249:139–156
19. Mohan D, Pittman CU Jr (2006) Activated carbons and low cost adsorbents for remediation of tri- and hexavalent chromium from water. J Hazard Mater B137:762–811
20. Pawlisz AV, Kent RA, Schneider A, Jefferson C (1997) Canadian water quality guidelines for chromium. Environ Toxicol Water Qual 12:123–183
21. World Health Organization (1996) Guidelines for drinking-water quality. International Programme on Chemical Safety, Health Criteria, vol 2. WHO, Geneva
22. Acharya J, Sahu JN, Sahoo BK, Mohanty CR, Meikap BC (2009) Removal of chromium(VI) from wastewater by activated carbon developed from *Tamarind wood* activated with zinc chloride. Chem Eng J 150:25–39
23. Arcand-Hoy LD, Benson WH (1998) Fish reproduction: an ecologically relevant indicator of endocrine disruption. Environ Toxicol Chem 17:49–57
24. Jobling S, Nolan M, Tyler CR, Brighty G, Sumpter JP (1998) Widespread sexual disruption in wild fish. Environ Sci Technol 32:2498–2506
25. Nimrod AC, Benson WH (1998) Reproduction and development of Japanese medaka following an early life stage exposure to xenoestrogens. Aquat Toxicol 44:141–156
26. Panter GH, Thompson RS, Sumpter JP (1998) Adverse reproductive effects in male fathead minnows (*Pimephales promelas*) exposed to environmentally relevant concentrations of the natural oestrogens, oestradiol and oestrone. Aquat Toxicol 42:243–253
27. Purdom CE, Hardiman PA, Bye VJ, Eno NC, Tyler CR, Sumpter JP (1994) Estrogenic effects of effluents from sewage treatment works. Chem Ecol 8(4):275–285
28. Vethaak AD, Lahr J, Schrap SM, Belfroid AC, Rijs GBJ, Gerritsen A, de Boer J, Bulder AS, Grinwis GCM, Kuiper RV, Legler J, Murk TAJ, Peijnenburg W, Verhaar HJM, de Voogt P (2005) An integrated assessment of estrogenic contamination and biological effects in the aquatic environment of The Netherlands. Chemosphere 59:511–524
29. Desbrow C, Routledge EJ, Brighty GC, Sumpter JP, Waldock M (1998) Identification of estrogenic chemicals in STW effluent. 1. Chemical fractionation and in vitro biological screening. Environ Sci Technol 32:1549–1558
30. Routledge EJ, Sheahan D, Desbrow C, Brighty GC, Waldock M, Sumpter JP (1998) Identification of estrogenic chemicals in STW effluent. 2. In vivo responses in trout and roach. Environ Sci Technol 32:1559–1565
31. Thorpe KL, Cummings RI, Hutchinson TH, Scholze M, Brighty G, Sumpter JP, Tyler CR (2003) Relative potencies and combination effects of steroidal estrogens in fish. Environ Sci Technol 37:1142–1149
32. Runnalls TJ, Margiotta-Casaluci L, Kugathas S, Sumpter JP (2010) Pharmaceuticals in the aquatic environment: steroids and anti-steroids as high priorities for research. Hum Ecol Risk Assess 16:1318–1338
33. Hutchinson TH, Pounds NA, Hampel M, Williams TD (1999) Impact of natural and synthetic steroids on the survival, development and reproduction of marine copepods (*Tisbe battagliai*). Sci Total Environ 233:167–179
34. Harries JE, Janbakhsh A, Jobling S, Matthiessen P, Sumpter JP, Tyler CR (1999) Estrogenic potency of effluent from two sewage treatment works in the United Kingdom. Environ Toxicol Chem 18:932–937
35. Chang HS, Choo KH, Lee B, Choi SJ (2009) The methods of identification, analysis, and removal of endocrine disrupting compounds (EDCs) in water. J Hazard Mater 172:1–12

36. Fernandez-Alvarez P, Le Noir M, Guieysse B (2009) Removal and destruction of endocrine disrupting contaminants by adsorption with molecularly imprinted polymers followed by simultaneous extraction and phototreatment. J Hazard Mater 163:1107–1112
37. Fukuhara T, Iwasaki S, Kawashima M, Shinohara O, Abe I (2006) Adsorbability of estrone and 17β-estradiol in water onto activated carbon. Water Res 40:241–248
38. Yoon Y, Westerhoff P, Snyder SA, Wert EC (2006) Nanofiltration and ultrafiltration of endocrine disrupting compounds, pharmaceuticals and personal care products. J Membr Sci 270:88–100
39. Yu Z, Peldszus S, Huck PM (2008) AdsorptionAdsorption characteristics of selected pharmaceuticals and an endocrine disrupting compound–Naproxen, carbamazepine and nonylphenol—on activated carbon. Water Res 42:2873–2882
40. Johnson AC, Williams RJ (2004) A model to estimate influent and effluent concentrations of estradiol, estrone, and ethinylestradiol at sewage treatment works. Environ Sci Technol 38:3649–3658
41. Servos MR, Bennie DT, Burnison BK, Jurkovic A, McInnis R, Neheli T, Schnell A, Seto P, Smyth SA, Ternes TA (2005) Distribution of estrogens, 17[beta]-estradiol and estrone, in Canadian municipal wastewater treatment plants. Sci Total Environ 336(1–3):155–170
42. Wastewater Treatment and Water Reclamation. Available from: http//www.lacsd.org/about/wastewater_facilities/moresanj/default.asp
43. Lu X, Nilsson O, Fricke J, Pekala RW (1993) Thermal and electrical conductivity of monolithic carbon aerogels. J Appl Phys 73(2):581–584
44. Morales-Torres S, Maldonado-Hodar FJ, Perez-Cadenas AF, Carrasco-Marin F (2010) Textural and mechanical characteristics of carbon aerogels synthesized by polymerization of resorcinol and formaldehyde using alkali carbonates as basification agents. Phys Chem Chem Phys 12:10365–10372
45. Pekala RW (1989) Organic aerogels from the polycondensation of resorcinol with formaldehyde. J Mater Sci 24:3221–3227
46. Lin C, Ritter JA (1997) Effect of synthesis pH on the structure of carbon xerogels. Carbon 35(9):1271–1278
47. Zhu Y, Hu H, Li W, Zhao H (2006) Preparation of cresol-formaldehyde carbon aerogels via drying aquagel at ambient pressure. J Non-Cryst Solids 352:3358–3362
48. Kim HJ, Kim JH, Kim WI, Suh DJ (2005) Nanoporous phloroglucinol-formaldehyde carbon aerogels for electrochemical use. Korean J Chem Eng 22:740–744
49. Cooney DO (1999) Adsorption design for wastewater treatment. Lewis Publishers, Boca Raton
50. Solis JM (2010) Carbon aerogels for the removal of estrone from water. Thesis, University of California

Chapter 11
Sol-Gel Processed Oxide Photocatalysts

Jusang Lee and P. I. Gouma

Abstract The utilization of photochemical reaction on semiconductor materials has been investigated extensively. Photocatalysis is a reaction that uses light to induce chemical transformation on organic or inorganic substrates without being altered or consumed. Use of renewable solar light as a reagent under ambient conditions promotes photocatalysis as an economic and clean way to treat organic contaminants in polluted air and waste water. Among the several photocatalyst synthesis methods, the sol-gel technique has received attention due to the high chemical homogeneity achieved, low processing temperature required, and the possibility of controlling the various structural and morphological features of the synthesized product, such as the surface morphology, the surface area, nanocrystal size, the crystallinity, and the phase structure. Mechanism and a description of the attempts and possibilities to improve the reactivity along with summary of recent sol-gel synthesis of photocatalyst are discussed in detail in this chapter.

Keywords Activated oxygen · Advanced oxidation processes (AOPs) · Aggregation · Anatase · Band gap · Conduction band · Degussa P-25 · Diffusion time · Efficiency · Electron · Electron hole · Environmental pollution · Mineralizers · n-type semiconductor · Organic contaminants · Photocatalysis · Quantum size effect (QSE) · Quantum yield · Redox potential · Redox reactions · Rutile · Semiconductor · Sol-gel · TiO_2 · TiO_2 particles · Titania · Valence band · Waste water

J. Lee (✉) · P. I. Gouma
Center for Nanomaterials and Sensor Development,
State University of New York at Stony Brook,
New York, NY 11794, USA
e-mail: jusalee@ic.sunysb.edu

P. I. Gouma
e-mail: pgouma@notes.cc.sunysb.edu

M. Aparicio et al. (eds.), *Sol-Gel Processing for Conventional and Alternative Energy*, Advances in Sol-Gel Derived Materials and Technologies,
DOI: 10.1007/978-1-4614-1957-0_11,

11.1 Introduction

Over the past several decades, environmental pollution has been increasing as a result of a rapidly growing population and global industrialization. Consequently, finding novel and eco-friendly methods of cleaning the environment becomes a necessity [1]. Research efforts worldwide now focus on advanced oxidation processes (AOPs) for removing environmental contaminants [2]. AOPs utilize highly reactive radical species, mainly OH•, generating from solar, chemical and other forms of energy [3].

Among several AOPs, the photochemical reaction on semiconductor materials has been investigated extensively since the discovery of the water splitting effect on TiO_2 surface by Fujishima and Honda [4]. Photocatalysis is a reaction that uses light to induce chemical transformation on organic or inorganic substrates without being altered or consumed. Use of renewable solar light as a reagent under ambient conditions promotes photocatalysis as an economic and clean way to remove organic contaminants in polluted air and wastewater [5, 6].

Heterogeneous photocatalysis refers to semiconductor photocatalysis or semiconductor sensitized photoreactions [6]. Heterogeneous photocatalysis systems may be classified into two types, depending on where the initial reactions take place. When the initial photo-excitation occurs in an absorbed molecule which then interacts with the ground state catalyst substrate, the process is referred to as a "catalyzed photoreaction". If, on the other hand, the initial photo-excitation takes place on the catalyst substrate and the photo-excited catalyst then interacts with ground state molecule, the process is referred to as a "sensitized photoreaction" [5]. Since the semiconductor titanium dioxide (TiO_2) is predominantly featured in this field due to its cost effectiveness and inert nature and photostability, TiO_2 is used here as a model photocatalytic material.

11.2 Basic Principles of Photocatalysis

A semiconductor photocatalyst absorbs light with photon energies that are the same or greater than the band gap, $hv \geq E_g$, causing electron excitation from the valence band (VB) to the conduction band (CB) of the photocatalyst leaving behind holes (see Fig. 11.1) [7]. The activation equation can be written as:

$$TiO_2 + hv \rightarrow h^+ + e^-(\text{charge separation}).$$

Photogenerated h^+ and e^- are capable of oxidizing or reducing a surface-adsorbed substrate, forming a singly oxidized electron donor and singly reduced electron acceptor, super oxide $O_2\bullet^-$ and hydroxyl radical OH•:

$$h^+ + OH^- \rightarrow OH\bullet$$

$$e^- + O_{2ads} \rightarrow O_2\bullet^-.$$

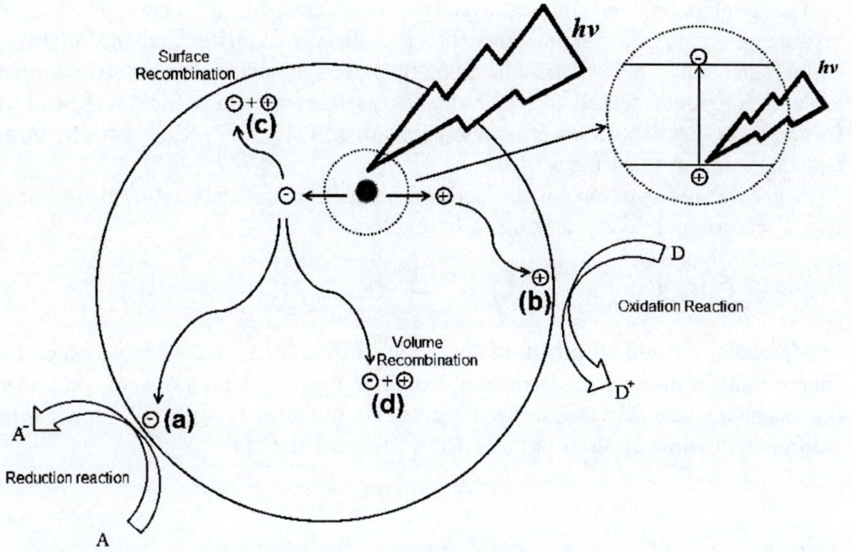

Fig. 11.1 Schematic of photo-excitation and photogenerated electron and hole pathways over a semiconductor particle (Adapted from Linsebigler et al. [5])

The relation between energy of photon (E) and threshold wavelength (λ_g), which is the greatest wavelength of radiation for a specified surface for the emission of electrons, can be described by Planck's relation

$$E = h\nu$$

where, h is the Planck's constant, $\nu(= c/\lambda_g)$ is the frequency and c is speed of light. Therefore, the threshold wavelength (λ_g) also can be expressed as:

$$\lambda_g = \frac{hc}{E_g}$$

As an example, the two phases of TiO_2 that is the polymorphs anatase and rutile have different band gaps E_g (anatase) = 3.20 eV and E_g (rutile) = 3.02 eV and as such have absorption thresholds of 380 and 410 nm, respectively [8].

Once the charge separation occurs, the photogenerated electron and hole can follow several pathways. Figure 11.1 illustrates the photo-excitation in a semiconductor particle followed by de-excitation events. At the surface of particle, the excited electron can reduce an electron acceptor (a) and a hole can combine with the electron donor species (b). The photogenerated electrons and holes recombination also occurs on the surface (c) and in the volume of particle (d). (The electron transferring process from the adsorbate to the semiconductor is referred as the oxidation mechanism, whereas electron transferring process from the semiconductor to the adsorbate is referred as the reduction mechanism).

The semiconductors photocatalysis has been used for the decomposition of organic and inorganic contaminants such as "alkanes, aliphatic alcohols, aliphatic carboxylic acids, alkenes, phenols, aromatic carboxylic acids, dyes, PCB's, simple aromatics, halogenated alkanes and alkenes, surfactants, and pesticides as well as for the reductive deposition of heavy metals (e.g., Pt^{4+}, Au^{3+}, Rh^{3+}, Cr(VI)) from aqueous solution to surfaces" [9].

A general net reaction for the heterogeneously photocatalyzed oxidation of a typical chlorinated hydrocarbon can be expressed as:

$$C_xH_yCl_z + \left(x + \frac{y+g}{4}\right)O_2 \xrightarrow{h\nu \geq E_g} xCO_2 + zH^+ + zCl^- + H_2O$$

Molecular oxygen adsorbed on the surface of the photocatalyst is activated by charge transfer of electrons generated from band gap excitation of semiconductor photocatalyst. The activated oxygen species are the primary agents and react with chlorinated hydrocarbon to produce CO_2, HCl and H_2O [9].

11.3 Mechanism of Photocatalysis

11.3.1 Trapping of Charge Carrier

Charge carrier trapping is associated with trapping the photogenerated electron, hole or both. It reduces electron hole recombination and increases the lifetime of electron–hole separation for efficient photocatalytic activity in a semiconductor. During the preparation of colloidal and bulk semiconductor photocatalysts, surface and bulk irregularities naturally occur. The irregularities affect the surface electron states which differ in their energy from the bands present in the bulk semiconductor, and these electron states work as charge carrier traps and help to reduce the electron and hole recombination rate [5].

Another way of hole trapping can be achieved by using degradable adsorbates or a sacrificial reagent. Shiragami et al. [10] used triethylamine as sacrificial electron donors in CdS mediated photoreductions, functioning as hole traps, so that the CB electron can be transferred more slowly without significant electron hole recombination. In a similar manner, electron trapping is associated with adsorbed oxygen in inert solvent that has almost same energy level of reduction potential with CB of TiO_2 [11].

The charge trapping phenomenon has been observed on the surface of an irradiated CdS single crystal electrodes using photoreflectance spectroscopy [12]. The frequency of charge tapping is dependent on the energy difference between the trap and the bottom of the CB (for electrons) or the top of the VB (for holes), and the decrease in entropy [13].

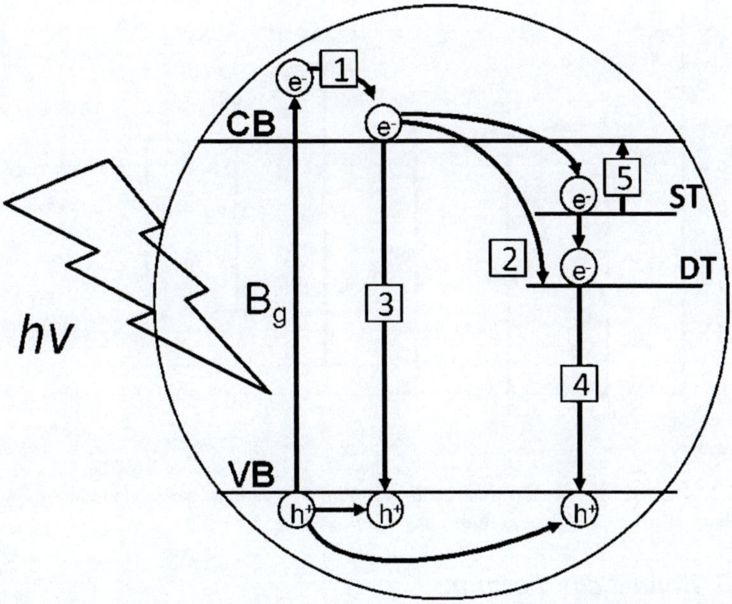

Fig. 11.2 Schematic illustration of charge carrier dynamics in a semiconductor nanoparticle. (1) electrons quenched to the bottom of the conduction band, (2) charge carriers captured by shallow trap (ST) and deep trap (DT) states and further trapping from ST to DP, (3) band-edge electron–hole recombination, (4) trapped electron–hole recombination, and (5) thermal excitation from traps (Adapted from Rothenberger et al. [14])

11.3.2 Diffusion of Charge Carrier

The mobility of charge carrier in a semiconductor is also an important factor for photocatalysis. In TiO_2 particles, the photogenerated electrons are trapped at Ti^{4+} sites located on the surface, the relaxation into the trap is preceded by diffusion of CB electrons to the particle's surface [14]. The average diffusion time (τ_d) of charge carriers from the interior to the surface of colloidal semiconductor is given by:

$$\tau_d = r_0^2/\pi^2 D_{e^-}$$

where r is the particle radius and D_{e^-} is the diffusion coefficient of the carrier. The diffusion coefficient of CB electrons in TiO_2 ($D_{e^-} = 1.2 \times 10^{-2}$ cm/s) is relatively small owing to their heavy effective electron mass ($m_{eff} = 30 \times m_{e^-}^{14}$) [15].

The average transition time could be further reduced with presence of electric field in the semiconductor [14]. Since the residence time for diffusion in colloidal particle is in the order of picoseconds and recombination time is nanoseconds, the diffusion of charge carriers from the interior to the particle surface can occur more rapidly than their recombination in TiO_2 [5]. A Schematic illustration of charge dynamic is shown Fig. 11.2.

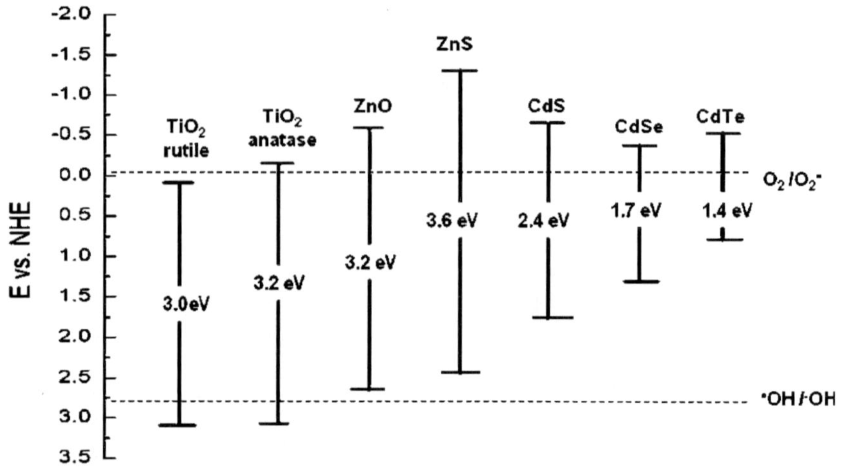

Fig. 11.3 Valence and conduction band position for various semiconductors in aqueous electrolytes at pH = 1 (Adapted from Langenhove et al. [55])

11.3.3 Band-Edge Position

Another important fact for the photoinduced electron excitation is band gap edge position and the redox potentials of the adsorbates. The produced electrons and holes may induce redox reactions with adsorbates having suitable redox potential. From the thermodynamic point of view, the relevant potential level of the acceptor species is required to be lower (more negative) than the CB potential level of the photocatalyst whereas the potential level of the donor needs to be higher (more positive) than the VB of the semiconductor in order to donate an electron to the vacant hole [5]. Figure 11.3 illustrates valence and CB position of various semiconductors. The energy scale is given versus normal hydrogen electrode (NHE) at pH = 1.

11.3.4 Space Charge Layers

At the contact between a solid semiconductor and another compound in a different physical state or of different type (e.g., liquid, gas, or metal), redistribution of electric charges generally occurs and the formation of a double charge layer. A space charge layer is generated when the mobile charge carriers transfer between the semiconductor and the contact phase, or the charge carrier trap at surface states of the interface. When TiO_2, an *n*-type semiconductor, contacts with gas, it will form a surface state for electron trapping. The surface of TiO_2 becomes negative and a positive space charge layer will produce to preserve electrical neutrality just within the semiconductor [5]. This phenomenon changes the

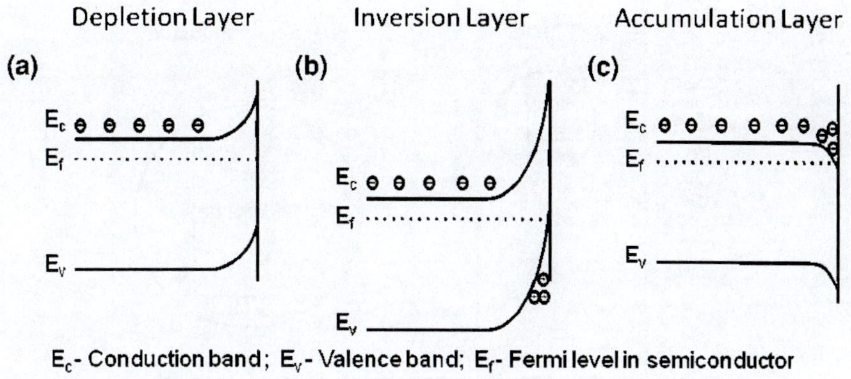

Fig. 11.4 Types of charged layers in n-type semiconductors. **a** Depletion layer. **b** Inversion layer. **c** Accumulation layer (Adapted from Linsebigler et al. [5])

electrostatic potential and moves the bands' edge position upward toward the surface. Three kinds of space charge layers may result from the mobility of charge across for an *n*-type semiconductor, as presented in Fig. 11.4.

When two different electrically neutral metal and semiconductor make contact, electron migration occurs from the semiconductor to the metal until the two Fermi levels are aligned. The electron migration results in excess negative charge of metal surface while excess positive charge of semiconductor forms a space charge layer. This depletion layer formed between metal-semiconductor interfaces is called the Schottky barrier [5].

From the electrochemical point of view, it is crucial to know the effective length of space charge layer over which the charge separation occurs and the band bending. Since the effect of band bending may cause a shift in the position of the band edge of the semiconductor that introduce thermodynamic limitations for the onset of photo initiated reactions that can be carried out with charge carriers [16].

11.3.5 Quantum Size Effect

The small clusters of atoms of semiconductor (Q-particles) on the order of 10–100 Å, having ability of showing quantum size effect (QSE), is an important subject relevant to the electron excitation in photocatalyst [17]. The significant enhancement in photocatalytic reactivity may show when the semiconductor's particle size is comparable to the De Broglie wavelength of the charge carriers in the semiconductor. The Q-particle does not experience electronic delocalization, instead, the photoinduced electrons and holes are confined in potential wells of a small geometric region. The confinement produces quantization of discrete electronic states and increases the effective band gap of the semiconductor. Such effects can change the color of the material and make a major increase on the band

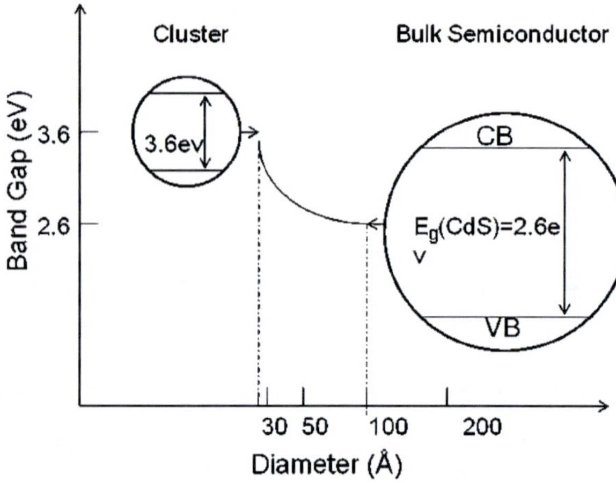

Fig. 11.5 Quantum size effect on semiconductor band gap (Adapted from Linsebigler et al. [5])

gap, which can reach up to six times that of the bulk band gap [5]. The increase in the effective band gap for CdS with decreasing particle size is illustrated in Figs. 11.4 and 11.5.

11.3.6 The Photocatalytic Efficiency

The efficiency of photocatalytic activity can be measured as quantum yield (Ø) which is defined as number of reactions from the absorption of one photon. However, the measurement of actual light absorption in heterogeneous system is difficult due to the light scattering. Therefore, it is usually assumed that all the light is absorbed and the efficiency is quoted as an apparent quantum yield. If there are measurable products made during the activity, the efficiency can be expressed as the yield of products [5]. The quantum yield (Ø) is defined as follows:

$$\varnothing = \frac{\mathrm{dN}_r/\mathrm{dt}}{\mathrm{dN}_{h\nu(\mathrm{inc})}/\mathrm{dt}} \propto \frac{K_{\mathrm{CT}}}{(K_{\mathrm{CT}} + K_R)}.$$

Ø is the ratio between the rate of molecules formed or degraded in the system, $\mathrm{dN}_r/\mathrm{dt}$ (molecules/s) and the rate of photons absorbed by the system, $\mathrm{dN}_{h\nu(\mathrm{inc})}/\mathrm{dt}$ (photons/s) at a given wavelength. Where K_{CT} is the rate of charge transfer and K_R is the electron–hole recombination rate. In ideal case, the quantum yield would be 1 with no recombination of electrons–holes and the rate of charge transfer would be dependent on the diffusion of charge carriers to the surface in the absence of excess surface charge. However, in real condition, recombination does occur and the concentration of electrons (n_s) and hole (p_s) at the surface of photocatalyst is not same in real system. As an example, the concentration of electrons and holes on TiO_2 during photo-oxidation process shows, $n_s > p_s$, owing to the electron process to molecular oxygen trapped at defect sites is relatively slow [18]. In addition, it is not necessary to all the incident photon will act upon the heterogeneous system and initiate the chemical transformation [19].

Therefore, an alternative for comparing efficiencies was introduced by defining relative photonic efficiency, ζ_r. A quantum yield can subsequently be determined from ζ_r, as $\emptyset = \zeta_r\ \emptyset_{phenol}$, where $\emptyset_{phenol}$ is quantum yield for photocatalyzed oxidation of phenol using Degussa P-25 TiO_2 as the standard catalyst with $\emptyset_{phenol} = 0.14 \pm 0.02$ at $\lambda = 365$ nm [20].

11.4 TiO_2 Semiconductors for Photocatalyst

Semiconductors are the most commonly used photocatalysts because of their relatively narrow gap ($E_g < 3.0$ eV) whereas metals possess no band gap and insulator ($E_g > 3.0$ eV) has wide gap. Metal oxides (e.g., TiO_2, WO_3, ZnO, etc.) and chalcogenides (e.g., CDS, ZnS, CdSe, ZnSe and CdTe) are most commonly studied semiconductor photocatalysts [21]. Among the possible semiconductors, TiO_2 is the most extensively used due to its chemical stability, non-toxicity and relatively low cost. Rutile and anatase are commonly used TiO_2 crystal structure in photocatalysis application and it is known that anatase shows higher photocatalytic activity than that of rutile [22]. As previously illustrated (Fig. 11.3), the VB redox potential of both TiO_2 structures are more positive than that of the (·OH/−OH) redox couple, whereas only CB of anatase is at a more negative redox potential than that of O_2/O_2·redox. It means adsorbed water and hydroxyl groups can be oxidized to highly reactive hydroxyl radicals on both TiO_2 structure, but oxygen molecules can be reduced to superoxide radical on anatase only, resulting more competitive than rutile for reduction reaction [15].

For bulk TiO_2, the rutile phase is considered more stable phase than the anatase phase at room temperature. However, Zhang et al. suggested and Gouma has confirmed that when the particle size is smaller than the critical size, *ca* 14 nm, the total free energy of rutile is higher than anatase so anatase becomes more stable phase [23, 24]. Recent studies have shown that rutile and anatase mixture exhibit higher photocatalytic activity than of pure anatase. The high charge separation in anatase/rutile mixture is result of electron transfer from anatase CB to rutile CB [25, 26]. Now the anatase (70–80%)/rutile (20–30%) mixture, Degussa P25, is most widely used heterogeneous photocatalysis for environmental application [13].

The photocatalytic activity on TiO_2 is following: absorb light having higher energy than its band gap and produce negatively charged electron and positively charged hole pare. The produced holes then oxidize with water resulting in hydroxyl radicals which are powerful oxidizers that can easily oxidize organic contaminants. The processes are summarized below (Fig. 11.6).

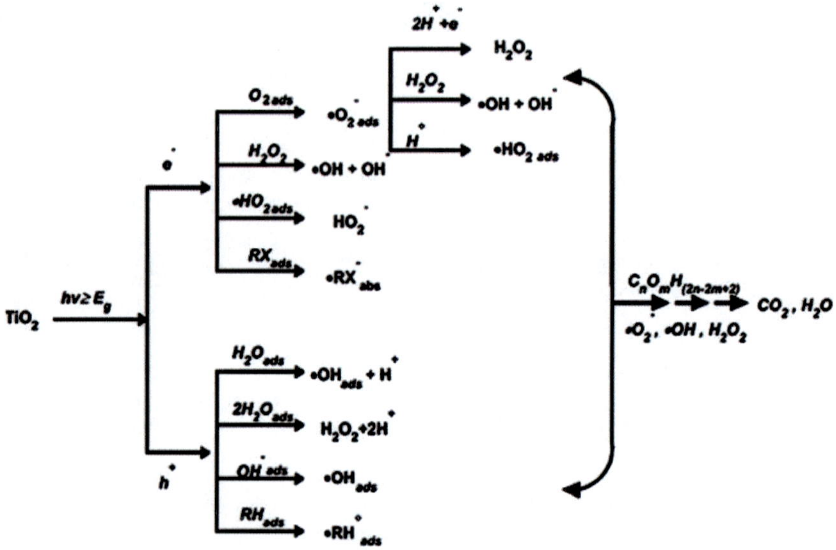

Fig. 11.6 Chain of reactions involved in the production activated oxygen species in the photoelectrochemical mechanism (Adapted from Langenhove et al. [55])

In the presence of absorbed water and oxygen couple (H_2O/O_2), absorbed water gets oxidized splitting into $\cdot OH$ and H^+ and absorbed organic pollutants (RH) can be directly oxidized producing cationic radicals $RH\cdot^+$ by holes in VB. Absorbed oxygen gets reduced by the electron results in generation of superoxide radical anions ($\cdot O_2^-$), which in turn reacts with H^+ to generate hydrogen dioxide radical ($\cdot HO_2$) in CB. On subsequent reaction with an electron and a hydrogen ion, a molecule of H_2O_2 is eventually produced. Electron also can directly react with adsorbed halogenated organic pollutant (RX) generating anionic radicals ($\cdot RX^-$). These chain reaction result in reactive oxygen species (ROS) such as H_2O_2, O_2– and hydroxyl radical $\cdot OH$ [23].

It is generally considered that photocatalytic reduction of organic compounds is less important than the oxidation, since the reduction potential of a $TiO_2(e^-)$ is lower than the oxidation potentials of a $TiO_2(h^+)$. In addition, O_2 shows much higher kinetics as an electron scavenges than most of the reducible substrates [13].

11.5 Modification of Photocatalyst

In order to determine the right purpose of particular semiconductor photocatalysis system needs to consider several factors such as the efficiency of photocatalytic process, stability under light, the selectivity of the product and activation range of wavelength [5]. For example, TiO_2, most extensively studied semiconductor

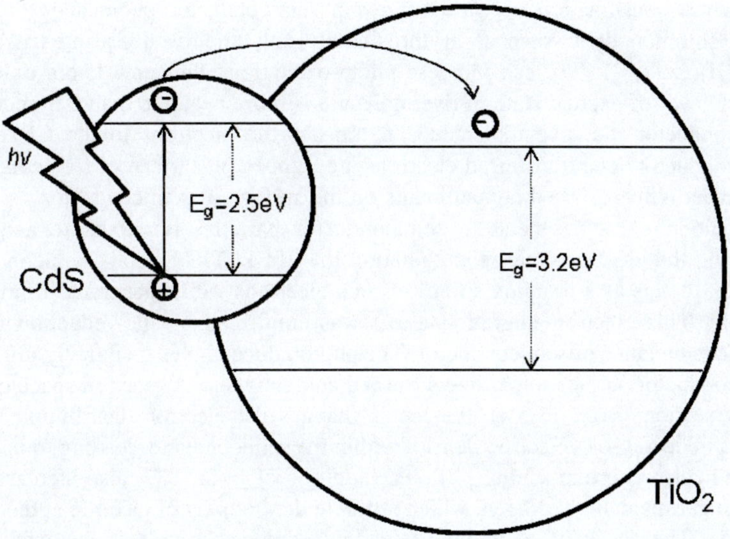

Fig. 11.7 Photo-excitation in composite semiconductor–semiconductor photocatalyst (Adapted from Gaya and Abdullah [2])

photocatalysis, has strong photostability. However, it only absorb UV light (absorption band at 390–400 nm) which is only about 3% of overall solar spectrum [27]. The limitation of certain semiconductors and necessity of optimum efficiency to meet specific application has increased research interest on new materials and the modification of materials. In order to enhance the photocatalytic efficiency, three main aspects should be considered. First, increase of the charge separation and recombination lifetimes of charge carriers. Second, increase of the solar spectrum response range. Third, change the selectivity or yield of a particular product [5].

One way to increasing charge separation and extending the solar spectrum response range is coupling semiconductors. The study of CdS–TiO_2 photocatalyst shows that a photon with wavelength less than 495 nm excites an electron from the valence band across the band gap of CdS ($E_g = 2.5$ eV) then those photogenerated electrons transfer to the TiO_2 CB [28]. The photogenerated holes in the CB of CdS particle migrate to the surface and participate in the oxidation of adsorbed organics. The electrons that are transferred to the CB of TiO_2 participate in reduction reactions increasing charge separation and photocatalytic efficiency. Figure 11.7 is the illustration of the CdS–TiO_2 coupled semiconductor photocatalyst.

The recent studies of nonmetal doping on TiO_2, such as carbon [29], nitrogen [30, 31] sulfur [32], phosphate [30, 33] show significant improvement utilizing visible spectrum. These anion dopants substitute the oxygen lattice on TiO_2 and lead to a band gap narrowing resulting in high visible light absorption [30]. Many doping processes employ a simple sol-gel method adding acid doping source in the precursor [29, 30, 31, 33]. Moreover, codoping of double nonmetal elements, such as N–F [34], C–N [35] and S–N [36], also shows high activity toward visible light.

Surface sensitization of wide band gap semiconductor photocatalyst (TiO_2) using common dyes, such as erythrosine B [37], thionine [38] and tris ruthenium(II)(RuL_3^{4-}) [39], can increase photo-excitation efficiency. If the oxidative energy level of excited state of dye molecule is more negative than CB energy of semiconductor, the dye molecule can transfer the electrons to the CB of the semiconductor. The transferred electrons are trapped on surface of semiconductor then react with adsorbed contaminants on the surface of semiconductor.

Addition of noble metal to semiconductor particles is another example to lowering the electron–hole recombination rate. In Pt-TiO_2 metal- semiconductor system, Pt acts as a trapping site of excited electrons which generated from TiO_2 particle. This electron migration effect was confirmed by the reduction in the photoconductance of semiconductor. The photoinduced holes then freely diffuse to semiconductor surface to oxidize adsorbed contaminants. As seen in space charge layers section (Sect. 11.3.4), the metal changes the electron distribution which results in decrease in electron density within the semiconductor leading to increase of the hydroxyl group acidity [40]. The addition of Pt on TiO_2 also increases gas evolving, especially hydrogen, which attribute the trapping of electron at the metal site [5]. The study of optimum Pt loading content shows that the loading of 0.6 wt% effect higher photocatalytic activity for H_2 evolution to achieve a maximum photocatalytic rate [41].

The relation between photocatalytic property and morphology including shape, size and surface area of particles is important aspect. TiO_2, particularly, shows different photocatalytic activity toward the same substrate in a liquid or a gas phase [42]. The advantage of using nanostructured semiconductor is flowing: (1) high absorption cross section (2) fast carrier diffusion on the interface that, in principle, can reduce recombination, (3) high surface area-to-volume ratios and (4) blue shift in the light absorption [27]. For example, the relation between the particle size and high reactive and selective of catalysts for the photocatalytic oxidation of organic compounds for nanostructure TiO_2 has been studied. Reaction studies conducted using gas-phase photo-oxidation of toluene as a probe reaction shows that the activity and selectivity of the TiO_2 catalyst were sensitive to the particle size. Higher conversions for toluene photo-oxidation were obtained by decreasing the TiO_2 particle size [43].

From the above it is inferred that there are many factors controlling the photocatalytic activity of a semiconductor and it is difficult to determine individual contributions. Below, the effect of sol-gel processing on oxide photocatalysts is reviewed, as reported in the literature, focusing on highlighting trends to achieve optimum performance.

11.6 Photocatalysts Prepared by the Sol-Gel Process

Nanocrystalline TiO_2 has been synthesized by chemical vapor deposition (CVD), sputtering, plasma spraying, electrospinning [44], coprecipitation, microemulsion, hydrothermal and sol-gel [45]. Among the various methods, the sol-gel technique has received attention due to the high chemical homogeneity achieved, low processing temperature required, and the possibility of controlling the various structural and morphological features of the synthesized product, such as the surface morphology, the surface area, nanocrystal size, the crystallinity and the phase structure. Especially for the titania photocatalysts, a simple sol-gel method allows control over the specific surface area, the crystallite size, the anatase/rutile content and the band gap energy of the product [30].

The typical sol-gel process of photocatalyst involves the hydrolysis and condensation of metal alkoxides (M(OR)n) as starting/precursor materials. The reaction process proceeds as follows:

1. Hydrolysis

$$M(OR)_n + H_2O \rightarrow M(OR)_{n-1}(OH) + ROH$$

2. Condensation
   Dehydration

$$M(OR)_n + M(OR)_{n-1}(OH) \rightarrow M_2O(OR)_{2n-2} + ROH$$

Dealcoholation

$$2M(OR)_{n-1}(OH) \rightarrow M_2O(OR)_{2n-2} + H_2O$$

The net reaction is

$$M(OR)_n + \frac{n}{2}H_2O \rightarrow M(OR)_{n/2} + nROH$$

Metal alkoxides are hydrolyzed and poly-condensed to form a metal oxide gel, whereas M = Si, Ti, Zr, Al, etc., and R = alkyl group.

The alkoxide groups are removed by acid or base catalyzed hydrolysis reactions and hydroxyl groups. Subsequent condensation and reactions involving the hydroxyl groups yield networks composed of oxo(M–O–M) linkages. Low volume fraction particles in colloidal solution may go through sedimentation or centrifugation methods to obtain gel-like properties. The remaining solvent can be removed by a drying process that causes shrinkage and densification. The solvent removal rate affects the porosity distribution in the gel. For further poly-condensation and enhancing the structural stability, a subsequent thermal treatment is necessary to acquire the final product. Even a small amount of dopants, such as organic dyes or rare earth metals, can be introduced in the sol and obtain uniformly dispersed in the final product [46].

11.7 Factors Affecting Sol-Gel Process for Photocatalyst

11.7.1 pH

The different pH levels in sol-gel process strongly influences particle size and phase composition of final products. Solid oxides in aqueous suspension tend to have electrical charge due to the amphoteric dissociation of surface hydroxyl groups, the adsorption of H^+/OH^- ions, or metal hydroxo compounds from the hydrolysis of solid material [23]. Bahnemann et al. [47] reported that the surface acidic/basic properties of TiO_2 can be determined by chemisorption of

$$H^+\left(TiO_2 + nH^+ \leftrightarrow TiO_2H_n^{n+}, \quad \text{for pH} < 3.5\right)$$

or

$$OH^-\left(TiO_2 + nOH^- \leftrightarrow TiO_2(OH)_n^{n-}, \quad \text{for pH} > 3.5\right)$$

Since the resultant surface charge is pH dependent, the strong repulsive force among charged particles reduces the probability of coalescence and thus more stable sols can be formed in acidic or alkaline media [23]. Wu et al. also suggested that since the isoelectric point of TiO_2 is 6.7, surface modification of colloids could be done by adding inorganic acid which results in preventing aggregation and the formation of a homogeneous colloidal solution [48].

Studies on the oxide crystallinity following sol-gel hydrolysis in different pH levels revealed that the anatase phase content in the products increases with increasing pH [49]. It was also shown that brookite was produced only in a certain range of acidity and disappeared prior to anatase onset with increasing acidity; and in high acid concentration it was favored compared to rutile crystallization. The study of hydrothermally processed product (Ti^{4+} concentration 0.44 mol dm^{-3}) showed mainly anatase and it contained some brookite and some rutile at pH = 1.0. The rutile phase disappeared in pH = 3.4, and pure anatase phase with a granular morphology was obtained in pH = 7.1. When the pH > 8, an amorphous phase began to appear [49].

Particle growth during hydrothermal aging in different pH values typically follows coarsening and aggregation. Isley and Penn [50] have studied hydrothermal aging with increasing pH in the range of −0.5–3.0 The synthesized TiO_2 showed that a sol-gel pH of −0.5 results in particle coarsening, and sol-gel pH of 3 results in particle growth involving both oriented aggregation and coarsening. The smaller particle size is expected to lead to higher TiO_2 solubility, which is consistent with the increase in the growth rate by coarsening; and the greater surface charge is expected to lead to higher electrostatic repulsions between particles, which is consistent with the decrease in the rate of growth by oriented aggregation [50]. The test results indicate that the products of hydrothermal aging are influenced by pH since initial phase composition and average particle size are also dependent on the sol-gel's pH.

11.7.2 Additives

It has been reported that additives (or mineralizers) can affect the morphology, phase, particle size and photocatalytic activity of TiO_2 particles [48, 49]. Cheng et al. [49] examined the effects of three mineralizers, NH_4C1, NaC1, and $SnC1_4$, on the TiO_2 formed in hydrothermal reactions with constant total Cl- concentration. The results showed that all three mineralizers ($SnC1_4$, NaC1 and NH_4C1) are favoring the formation of rutile phase and are decreasing grain size, but NH_4C1 promotes agglomeration among grains. It was considered that the additives could be adsorbed on TiO_2 particles and act as an nuclear agents changing the surface state to prevent particle growth and phase change from anatase to rutile. As the concentration of NaCl was increased, the size of rutile particle was reduced and rutile formations were promoted [48]. It was found that NaCl serves a dual rule: as a chemical catalyst and as a means of physical confinement. NaCl can favor to rutile formation and crystal growth and it may serve as an electrolyte promoting electrostatic adsorption at high concentrations.

Al-Salim et al. [51] found that alkaline earth ion (20 mol%) additives not only affect the TiO_2 formation but also approximately double the photocatalytic activity of Degussa P-25 without additive. The increasing concentration of alkaline earth ion additives resulted in the inhibition of anatase crystallization (Ba > Sr > Ca) and increase of surface areas with constant porosity. They also suggested that the high photodecomposition rate of aqueous oxalic acid on TiO_2 with alkaline earth additives might be due to an increase in the surface area of the TiO_2 crystallites.

11.7.3 Reaction Temperatures and Time

Reaction temperature and time have important role of thermodynamic and kinetic factors for hydrothermal reaction [48]. It is known that increasing reaction temperature may cause aggregation and/or phase and structure change of product, whereas lowering reaction temperature may decrease grain size and/or agglomeration among particles [52, 49].

Wu et al. [48] showed that 0.5 M of Titanium n-butooxide (TNB) and HCl mixture at 100°C initially showed mainly anatase with trace of rutile and brookite. Overall increase of reaction temperature and time accelerate the phase transformation from anatase (metastable) to rutile (stable) phase and condenser phase formation. Increasing reaction temperature caused the gradual disappearance of brookite and anatase phases and eventually led to the formation of a pure rutile phase and by prolonging the reaction time, an increased average grain size.

Cheng et al. [49] suggested that increase of reaction temperature generated a large amount of Cl^- coming from hydrolysis of $[Ti(OH)_nCl_m]^{2-}$ complex ions and gradually substituted OH^- ligands, while the linking between $[TiO_6]$ was carried out by the dehydration in $[Ti(OH)_nCl_m]^{2-}$. The produced TiO_2 crystallite formed an aggregate easily due to its large surface area.

11.7.4 H_2O/Alkoxide Molar Ratio

The Ti/alkoxide molar ratio is an important parameter which has an impact on both the titania product mixture and the average particle sizes during hydrolysis reaction of sol-gel. Isley et al. found that decreasing the Ti/H_2O molar ratio (from 1:4 to 1:700) resulted in both decreasing brookite content and anatase particle size in sol-gel synthesized TiO_2 nanocrystallines [50]. Wang et al. also showed effect of water: titania ration during the sol-gel synthesis at 450°C and it was note that increasing water: alkoxide ratio led to reduced crystal size. As seen in hydrolysis equation above, high water: alkoxide ratios generally promote hydrolysis, favoring nucleation versus particle growth [53]. Li et al. pointed out that as water concentration increases the gelation time is significantly reduced during the hydrolysis and polycondensation reaction of $Ti(O\text{-}Bu)_4$ [54]. In high water concentration, macromolecular networks form rapidly through the reactions of hydrolysis and condensation, and have low interconnectivity. Whereas, in low water concentration, networks form slowly, leading to a denser structured.

11.8 Various Photocatalysts Prepared by the Sol-Gel Process

The specific uses of sol-gel processed photocatalyst are derived from various materials and forms in different structure. Table 11.1 is the list of photocatalyst obtained by sol-gel process including precursor, reactant molar ratio and tested contaminants.

11.9 Summary

In this book chapter, the sol-gel synthesis of photocatalyst was introduced along with a summary of photocatalysis mechanisms. Several aspects should be considered in order to obtain the optimum photocatalytic efficiency. Among the several synthesis methods, sol-gel methods appeal to be very attractive due to their high chemical homogeneity, low processing temperature and the possibility of controlling various parameters. Many studies have been done by mainly using Ti alkoxides and further development of fabrication will be required to produce photocatalysts which can be an economical and clean way to remove organic contaminants in polluted air and waste water.

Table 11.1 Different types of sol-gel processed oxide photocatalysts

Catalyst	Precursor	Reactant molar ratio	Tested contaminants	Reference
TiO_2 powder	Titanium (IV) *n*-butoxide $(Ti(OC_4H_9))$	$Ti(O\text{-}Bu)_4:H_2O:$ *i*-PropOH:acac = 1:100:2:0.01	Salicylic acid	[52]
Mesoporous TiO_2 films	Titanium(IV) isopropoxide $(Ti(OC_3H_7)_4)$	Tween 80:i-PrOH:acetic acid:TTIP = 1:45:6:1	Methylene blue and creatinine in water	[56]
TiO_2 hollow fibers	Titanium (IV) nbsp;*n*-butoxide $(Ti(OC_4H_9)_4)$	Hydrochloric acid : $(Ti(OC_4H_9^n)_4)$, Pluronic P123: ethanol	Methylene blue and gaseous formaldehyde	[57]
Cr(III)-doped, TiO_2-coated active carbon (Cr-TiO_2/AC)	Titanium (IV) *n*-butoxide $(Ti(OC_4H_9))$	Tetrabutylorthotitanate : diethanolamine: ethanol	EDTA	[58]
TiO_2, Pt-TiO_2	Titanium(IV) isopropoxide $(Ti(OC_3H_7)_4)$	Anhydrous ethanol: hydrochloric acid: $Ti[OCH(CH_3)_2)]$	Toluene	[59]
TiO_2 pellet	Titanium(IV) isopropoxide $(Ti(OC_3H_7)_4)$	HNO_3: H_2O: $Ti(OC_3H_7)_4$	Trichloroethylene (TCE)	[60]
TiO_2 and SiO_2–TiO_2	Titanium(IV) isopropoxide $(Ti(OC_3H_7)_4)$	SiO_2: TiO_2 = 5:5, 2:8 and 1:9	Formaldehyde	[61]
N-TiO_2	Titanium (IV) *n*-butoxide $(Ti(OC_4H_9)_4)$	Tetra-butyl titanate $Ti(OBu)_4$: NH_4 Cl	Methylene blue (MB)	[62]

(continued)

Table 11.1 (continued)

Catalyst	Precursor	Reactant molar ratio	Tested contaminants	Reference
Fe(III)-doped TiO_2	Titanium(IV) isopropoxide ($Ti(OC_3H_7)_4$)	HNO_3:H_2O: $Ti(OC_3H_7)_4$	Methanol, Ethanol	[63]
Fe^{3+}-TiO_2, Pb^{2+}-TiO_2	Titanium (IV) *n*-butoxide ($Ti(OC_4H_9)_4$)	$Ti(OC_4H_9)_4$:C_2H_5OH:H_2O = 3:24:1 adding $Fe(NO_3)_3$ or $Pb(NO_3)_2$	Trichloroethylene, chloroform, dichloromethane, toluene, benzene, carbon tetrachloride	[64]
P-TiO	Titanium(IV) isopropoxide ($Ti(OC_3H_7)_4$)	P-TiO_2 = 0.01, 0.05, 0.1, 0.2 and 0.3 Ethanol	Ethanol	[30]
TiO_2/$YFeO_3$	Titanium (IV) *n*-butoxide ($Ti(OC_4H_9)_4$)	Metal nitrates: citric acid: ethylene glycol = 1:2:1 adding $Y(NO_3)_3$ and $Fe(NO_3)_3$	Benzene	[65]
TiO_2/Al_2O_3–SiO_2	Titanium (IV) *n*-butoxide ($Ti(OC_4H_9)_4$)	$Ti(OC_4H_9)_4$ adding Al_2O_3 and SiO_2 = 3:2	Acetaldehyde	[66]
N-doped SiO_2/TiO_2	Titanium(IV) isopropoxide ($Ti(OC_3H_7)_4$)	SiO_2/TiO_2 = 0.05, 0.10, 0.15, 0.2, 0.3 HNO_3: H_2O: $Ti(OC_3H_7)_4$ Tetraethyl orthosilicate $Si(OC_2H_5)_4$: ethanol, water: HCl adding NH_4OH	Ethylene	[67]

Acknowledgments This work has been funded by the National Science Foundation, grant # DMR 1046599 with Dr. Lynnette Madsen as the program manager.

References

1. Gogate P (2004) A review of imperative technologies for wastewater treatment I: oxidation technologies at ambient conditions. Adv Environ Res 8(3–4):501–551. doi:10.1016/S1093-0191(03)00032-7
2. Gaya U, Abdullah A (2008) Heterogeneous photocatalytic degradation of organic contaminants over titanium dioxide: a review of fundamentals, progress and problems. J Photochem Photobiol C Photochem Rev 9(1):1–12. doi:10.1016/j.jphotochemrev.2007.12.003
3. Kudo T, Nakamura Y, Ruike A (2003) Res Chem Intermed 29:631
4. Fujishima A, Honda K (1972) Nature 238:37
5. Linsebigler A, Lu G, Yates J (1995) J Chem Rev 95:735–758
6. Mills A, Hunte SL (1997) J J Photochem Photobiol A Chem 108:1–3
7. Xu Y, Schoonen MAA (2000) The absolute energy positions of conduction and valence band of selected semiconducting minerals. Am Miner 85:543–556
8. Dvoranov D (2002) Investigations of metal-doped titanium dioxide photocatalysts. Appl Catal B Environ 37(2):91–105. doi:10.1016/S0926-3373(01)00335-6
9. Hoffmann MR, Martin ST, Choi W, Bahnemann DW (1995) Environmental applications of semiconductor photocatalysis. Chem Rev 95(1):69–96. doi:10.1021/cr00033a004
10. Shiragami T, Pac C, Yanagida S (1989) Nonmetallised CdS-catalysed photoreduction of aromatic ketones to alcohols and/or. Photochemistry, 0–1
11. Fox MA, Chen CC (1981) Mechanistic features of the semiconductor photocatalyzed olefin-to-carbonyl oxidative cleavage. J Am Chem Soc 103(22):6757–6759. doi:10.1021/ja00412a044
12. Nakabayashi S, Kim A (1991) Surface-trapped charge at CdS electrode in aqueous solution as observed by photoreflectance spectroscopy (16):9961–9965
13. Fox MA, Dulay MT (1993) Heterogeneous photocatalysis. Chem Rev 93(1):341–357
14. Rothenberger G, Moser J et al (1985) Charge carrier trapping and recombination dynamics in small semiconductor particles. J Am Chem Soc 107(26):8054–8059
15. Gratzel M, Frank AJ (1982) J Phys Chem 86:2964
16. Vinodgopal K, Hotchandani S et al (1993) Electrochemically assisted photocatalysis: titania particulate film electrodes for photocatalytic degradation of 4-chlorophenol. J Phys Chem 97(35):9040–9044
17. Alivisatos P (1996) Semiconductor clusters, nanocrystals, and quantum dots. Science 271: 933937
18. Ollis DF, Al-Ekabi H (1993) Photocatalytic purification and treatment of water and air. Elsevier, Amsterdam
19. Serpone N, Emeline AV (2005) Modelling heterogeneous photocatalysis by metal-oxide nanostructured semiconductor and insulator materials: factors that affect the activity and selectivity of photocatalysts. Res Chem Intermed 31(4):391–432. doi:10.1163/1568567053956789
20. Serpone N (1997) Relative photonic efficiencies and quantum yields in heterogeneous photocatalysis. J Photochem Photobiol A Chem 104(1–3):1–12. doi:10.1016/S1010-6030(96)04538-8
21. Prousek J (1996) Advanced oxidation processes for water treatment photochem process. Chem Listy 90:307–315
22. Augustynski J (1993) The role of the surface intermediates in the photoelectrochemical behaviour of anatase and rutile TiO_2. Electrochimica Acta 38(1):43–46
23. Zhang H, Banfield JF (1998) Thermodynamic analysis of phase stability of nanocrystalline titania. J Mater Chem 8(9):2073–2076. doi:10.1039/a802619j

24. Gouma PI (2003) Nanostructured polymorphic oxides for advanced chemosensors. Rev Adv Mater Sci 5:147–154
25. Hurum DC, Agrios AG et al (2003) Explaining the enhanced photocatalytic activity of degussa P25 mixed-phase TiO_2 using EPR. J Phys Chem B 107(19):4545–4549
26. Wang R, Hashimato K, Fujishima A et al (1997) Light induced amphiphilic surface. Nature 388:431–432
27. Hashimoto K, Irie H, Fujishima A (2005) TiO_2 photocatalysis: a historical overview and future prospects. Jpn J Appl Phys 44(12):8269
28. Gopidas KR, Bohorquez M et al (1990) Photophysical and photochemical aspects of coupled semiconductors: charge-transfer processes in colloidal cadmium sulfide-titania and cadmium sulfide-silver(I) iodide systems. J Phys Chem 94(16):6435–6440
29. Sakthivel S, Kisch H (2003) Daylight photocatalysis by carbon-modified titanium dioxide. Angewandte Chemie (International ed. in English) 42(40):4908–4911. doi:10.1002/anie.200351577
30. Asahi R, Morikawa T, Ohwaki T, Aoki K, Taga Y (2001) Visible-light photocatalysis in nitrogen-doped titanium oxides. Science 293:269–271
31. Yokosuka Y, Oki K, Nishikiori H, Tatsumi Y, Tanaka N, Fujii T (2009) Photocatalytic degradation of trichloroethylene using N-doped TiO_2 prepared by a simple sol-gel process. Res Chem Intermediat 35:43–53
32. Umebayashi T, Yamaki T, Itoh H, Asai K (2002) Band gap narrowing of titanium dioxide by sulfur doping. Appl Phys Lett 81(3):454. doi:10.1063/1.1493647
33. Lin L, Lin W, Xie JL, Zhu YX, Zhao BY, Xie YC (2007) Photocatalytic properties of phosphor-doped titania nanoparticles. Appl Catal B 75:52–58
34. Huang D-G, Liao S-J, Liu J-M, Dang Z, Petrik L (2006) Preparation of visible-light responsive N–F-codoped TiO_2 photocatalyst by a sol-gel-solvothermal method. J Photochem Photobiol A Chem 184(3):282–288. doi:10.1016/j.jphotochem.2006.04.041
35. Li Q, Xie R, Li YW, Mintz EA, Shang JK (2007) Enhanced visible-light-induced photocatalytic disinfection of *E. coli* by carbon-sensitized nitrogen-doped titanium oxide. Environ Sci Technol 41(14):5050–5056. Retrieved from http://www.ncbi.nlm.nih.gov/pubmed/17711222
36. Naik B, Parida KM, Gopinath CS (2010) Facile synthesis of N- and S-incorporated nanocrystalline TiO_2 and direct solar-light-driven photocatalytic activity. Society 114: 19473–19482
37. Kamat PV, Fox MA (1983) Chem Phys Lett 102:379
38. Patrick B, Kamat PV (1992) J Phys Chem 96:1423
39. Vlachopoulos N, Liska P, Augustynski J, Gratzel M (1988) J Am Chem Soc 110:1216
40. Jafiezic-Renault N, Pichat P, Foissy A, Mercier R (1986) J Phys Chem 90:2733
41. Sreethawong T, Yoshikawa S (2006, May) Enhanced photocatalytic hydrogen evolution over Pt supported on mesoporous TiO_2 prepared by single-step sol-gel process with surfactant template. Int J Hydrogen Energy. doi: 10.1016/j.ijhydene.2005.06.015
42. Maldotti A, Molinari A et al (2003) Photocatalysis with organized systems for the oxofunctionalization of hydrocarbons by O_2. ChemInform 34(4)
43. Maira AJ, Yeung KL et al (2001) Gas-phase photo-oxidation of toluene using nanometer-size TiO_2 catalysts. Appl Catal B Environ 29(4):327–336
44. Sawicka KM, Prasad AK, Gouma PI (2005) Sens Lett 3:31
45. Baiju KV, Sibu CP, Rajesh K, Mukundan P, Warrier KGK, Wunderlich L (2005) Mater Chem Phys 90:123
46. Klein LC (1993) Sol-gel optical materials. Ann Rev Mater Sci 23(1):437–452. doi:10.1146/annurev.ms.23.080193.002253
47. Bahnemann D, Henglein A, Spanhel L (1984) Faraday discuss. Chem Soc 78:151
48. Wu M, Lin G et al (2002) Sol-hydrothermal synthesis and hydrothermally structural evolution of nanocrystal titanium dioxide. Chem Mater 14(5):1974–1980
49. Cheng H, Ma J, Zhao Z, Qi L (1995) Hydrothermal preparation of uniform nanosize rutile and anatase particles. Chem Mater 7(4):663–671. doi:10.1021/cm00052a010

50. Isley SL, Penn RL (2008) Titanium dioxide nanoparticles: effect of sol-gel pH on phase composition, particle size, and particle growth mechanism. Society 4469–4474
51. Al-Salim NI, Bagshaw SA, Bittar A, Kemmitt T, McQuillan AJ, Mills AM et al (2000) Characterisation and activity of sol-gel-prepared TiO_2 photocatalysts modified with Ca, Sr or Ba ion additives. J Mater Chem 10(10):2358–2363. doi:10.1039/b004384m
52. Su C, Hong BY et al (2004) Sol-gel preparation and photocatalysis of titanium dioxide. Catal Today 96(3):119–126
53. Wang C-C, Ying JY (1999) Sol-gel synthesis and hydrothermal processing of anatase and rutile titania nanocrystals. Chem Mater 11(11):3113–3120
54. Li Y, White ÃTJ, Lim SH (2004) Low-temperature synthesis and microstructural control of titania nano-particles. J Solid State Chem 177:1372–1381. doi:10.1016/j.jssc.2003.11.016
55. van Langenhove H, Demeestere K, Dewulf J (2007) Heterogeneous photocatalysis as an advanced oxidation process for the abatement of chlorinated, monocyclic aromatic and sulfurous volatile organic compounds in air: state of the art. Crit Rev Environ Sci Technol 37(6):489–538
56. Choi H, Stathatos E, Dionysiou D (2006, March) Sol-gel preparation of mesoporous photocatalytic TiO_2 films and TiO_2/Al_2O_3 composite membranes for environmental applications. Applied Catalysis B: Environmental. doi:10.1016/j.apcatb.2005.09.012
57. Zhan S, Chen D, Jiao X, Tao C (2006) Long TiO_2 hollow fibers with mesoporous walls: sol-gel combined electrospun fabrication and photocatalytic properties. Society 11199–11204
58. Gambhire AB, Lande MK, Mandale AB, Patil KR, Arbad BR (2008) Photocatalytic activity and characterization of sol-gel-derived Cr(III)-doped TiO_2-coated active carbon composites. Philosophical Magazine 88(5):767–779. doi:10.1080/14786430801955261
59. Cao L (2000) Photocatalytic oxidation of toluene on nanoscale TiO_2 catalysts: studies of deactivation and regeneration. J Catal 196(2):253–261. doi:10.1006/jcat.2000.3050
60. Yamazaki S, Abe H, Tanimura T, Yamasaki Y, Kanaori K, Tajima K (2009) Effect of thermal treatment on the photocatalytic degradation of ethylene, trichloroethylene, and chloroform. Res Chem Intermed 35(1):91–101. doi:10.1007/s11164-008-0001-9
61. Lee B-Y, Kim S-W, Lee S-C, Lee H-H, Choung S-J (2003) Photocatalytic decomposition of gaseous formaldehyde using TiO_2, SiO_2–TiO_2 and Pt–TiO_2. Int J Photoenergy 5(1):21–25. doi:10.1155/S1110662X03000072
62. Huan YU, Zheng X, Zhongyi YIN, Feng TAO (2007) Preparation of nitrogen-doped TiO_2 nanoparticle catalyst and its catalytic activity under visible light. Chin J Chem Eng 15:802–807
63. Piera E (2003) Relationship concerning the nature and concentration of Fe(III) species on the surface of TiO_2 particles and photocatalytic activity of the catalyst. Appl Catal B Environ 46(4):671–685. doi:10.1016/S0926-3373(03)00288-1
64. Zuo G-M, Cheng Z-X, Chen H, Li G-W, Miao T (2006) Study on photocatalytic degradation of several volatile organic compounds. J Hazard Mater 128(2–3):158–163. doi:10.1016/j.jhazmat.2005.07.056
65. Wang W, Li S, Wen Y, Gong M, Zhang L, Yao Y et al (2008) Synthesis and characterization of $TiO_2/YFeO_3$ and its photocatalytic oxidation of gaseous benzene. Acta Physico Chimica Sinica 24(10):1761–1766. doi:10.1016/S1872-1508(08)60072-8
66. Yuan J, Hu H, Chen M, Shi J, Shangguan W (2008) Promotion effect of Al_2O_3–SiO_2 interlayer and Pt loading on TiO_2/nickel-foam photocatalyst for degrading gaseous acetaldehyde. Catal Today 139(1–2):140–145. doi:10.1016/j.cattod.2008.08.016
67. Hou YD, Wang XC, Wu L, Chen XF, Ding ZX, Wang XX et al (2008) N-doped SiO2/TiO2TiO2 mesoporous nanoparticles with enhanced photocatalytic activity under visible-light irradiation. Chemosphere 72(3):414–421. doi:10.1016/j.chemosphere.2008.02.035

Chapter 12
Sol-Gel Coatings For Electrochromic Devices

S. Heusing and M. A. Aegerter

Abstract This chapter describes first the principles of electrochromic (EC) devices and then reviews the most important *sol–gel* developments related to the preparation and characterisation of the different layers used for the realisation of such devices: transparent conducting coatings, electrochromic coatings, counter electrodes and electrolytes. Finally the review shows how these coatings have been used for the realisation of prototypes and devices such as windows and displays describing their electro-optical properties, their long-term behaviour as well as their advantages and drawbacks. This review is a shorter but updated version based on earlier reviews published by the authors in 1996, 2005 and 2006.

Keywords Electrochromic coatings · Electrolytes · Counter electrodes · Sol-Gel · Electrochromic devices · Electrochromic windows · Electrochromic displays

12.1 Introduction

Electrochromic (EC) materials change their optical properties (transmittance or reflection) reversibly by applying a voltage and a current flow due to a reduction or an oxidation process [1, 2]. These materials are of considerable technological and commercial interest for the development of smart windows, displays and

S. Heusing
Optical Materials Group, INM–Leibniz Institute for New Materials,
Campus D2 2, 66123, Saarbruecken, Germany
e-mail: Sabine.Heusing@inm-gmbh.de

M. A. Aegerter (✉)
Ch. des Placettes 6, 1141, Bottens, Switzerland
e-mail: Michel.aegerter@bluewin.ch

M. Aparicio et al. (eds.), *Sol-Gel Processing for Conventional and Alternative Energy*, Advances in Sol-Gel Derived Materials and Technologies,
DOI: 10.1007/978-1-4614-1957-0_12,

automotive rear-view mirrors [3, 4]. Large electrochromic glazing can be used for architectural and automotive applications in order to control the solar radiation entrance into buildings and automotives for saving air-conditioning costs in summer and heating costs in winter. According to Granqvist [5], the amount of energy that can be saved annually for a normal office room equipped with such windows with transmittance varying between 75 and 7% can be at least 170 kWh m^{-2} but higher values can be expected in other applications [6, 7]. This is equal to the energy which can be generated by today's solar cell modules of the same size and positioned at the same place and presenting an efficiency of about 17% [5–7]. In addition, comfort factors such as privacy, glare and fading are also driven interest in EC-windows development. Smaller devices such as automotive rear-view mirrors adjusting automatically their reflection are already on the international market [4, see also Chap. 13].

Most of the electrochromic materials are inorganic and essentially transition metal oxides of tungsten (WO_3), niobium (Nb_2O_5), molybdenum (MoO_3) and nickel (NiO). Coatings of these compositions have been fabricated by practically all available deposition techniques. Other materials, most of them not discussed here, are organic polymers (e.g. polyaniline), some inorganic complexes (e.g. $Fe_4[Fe(CN)_6]_3$, prussian blue) and organic molecules (e.g. viologens).

A wide range of EC coatings and devices have been developed by the Sol-Gel process. The advantages of this process are the possibility to easily fabricate sols of pure, mixed and doped oxides by mixing different salts in a solvent and use its many attractive deposition techniques such as dip-coating, spin-coating, spray coating and printing to get small to even large area coatings. The process also offers advantages in controlling the microstructure of the deposited coatings, parameters that can influence the kinetics, durability, colouring efficiency and charge storage of the EC electrodes. Another advantage is the reasonable cost of the deposition processes for large areas compared to conventional vacuum methods.

This chapter is partly based on extended reviews on the same subject written by the authors in 1996 [8], 2005 [9] and 2006 [10] in which the readers will find extensive lists of references. This Sol-Gel field is still rather active (about 15 reports per year) so that an emphasis is given in this review on the most important developments and realisations reported during these last 6 years.

The review describes first the principle of EC-devices, the most important materials and coatings obtained by the *sol–gel* process and the configuration and properties of some EC-windows and EC-display devices (mostly prototypes) developed with them.

12.2 Electrochromic Devices

There are three major device configurations: battery-like, solution phase and hybrid structures [11]. The *battery-like* configuration (Fig. 12.1) is the most used in the EC-devices reported in this chapter; it consists of 5 functional layers located

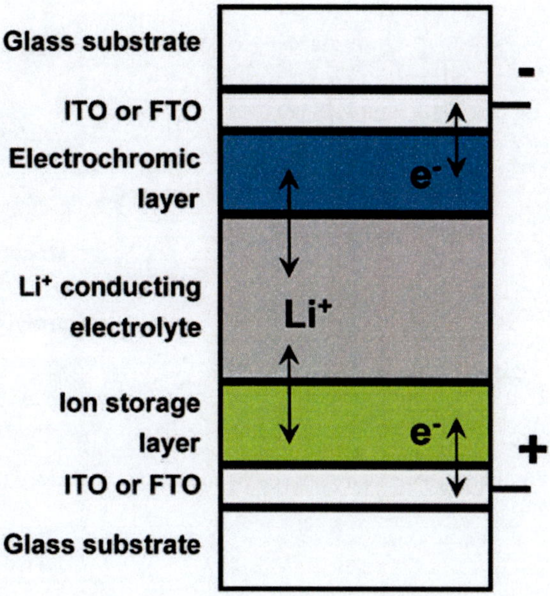

Fig. 12.1 Scheme of a battery-like EC-device, adapted from [9]

between two glass or plastic substrates. Each plate is coated with a thin transparent conductive electrode (TCE), to which the external electrical contacts are attached; they are usually a fluorine-doped tin oxide (FTO: SnO_2:F) or a tin-doped indium oxide (ITO: In_2O_3:Sn). One of the TCE is coated with an EC-layer, the other with a counter electrode (CE). Both layers are separated by an ionic conductive electrolyte either a liquid, a polymer, a gel or a coating having a very low electronic conductivity. Small ions such as protons (H^+) or lithium (Li^+) are preferred to assure fast optoelectrochemical kinetics. Polymer and gel electrolytes are preferred for large area EC-devices in order to avoid the buckling of the glass and the risk of leakage that can occur with liquid electrolyte. However their use requires the construction of laminated sandwich structures (Fig. 12.2); EC Sol-Gel devices with only one glass or plastic substrate can also be fabricated if a Sol-Gel film electrolyte (e.g. Ta_2O_5) is used [12, 13].

The operating mode of a *battery-like configuration* device is described below using as an example a *cathodic* electrochromic material colouring in the reduced state such as WO_3 (Fig. 12.1). The system is initially in a bleached transparent state. After applying a voltage with negative polarisation at the WO_3 side, this material is reduced and simultaneously the M^+ ions (M^+: H^+, Li^+) stored in the ion-storage layer diffuse through the electrolyte into the WO_3 layer to form a deep blue-coloured intercalation compound M_xWO_3. The reversible colouration and bleaching reactions are: $WO_3 + x\ M^+ + x\ e^- \longleftrightarrow M_xWO_3$. After reversing the polarisation, M_xWO_3 is oxidised and simultaneously the M^+-ions diffuse back to the counter electrode. The counter electrode, which was oxidised during the colouration step, is reduced and the EC-device is bleached. The counter electrode

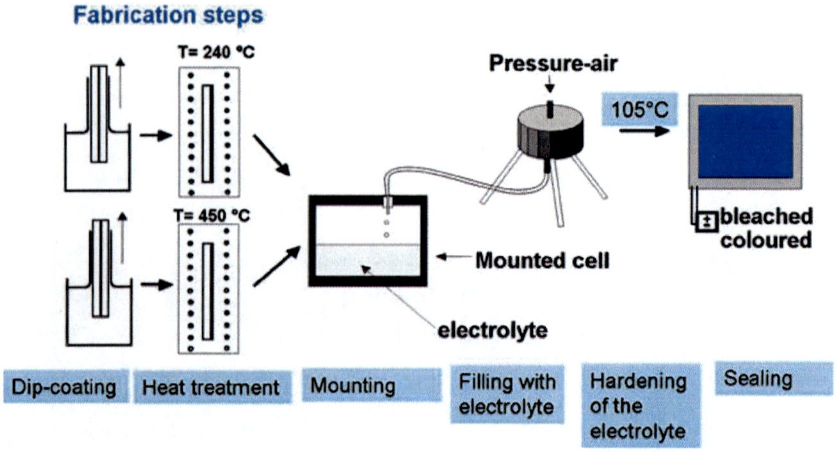

Fig. 12.2 Fabrication steps for the realisation of a battery-like EC-device using a liquid or a polymerizable electrolyte

(CE) can be either an ion-storage layer (IS-layer) which remains transparent and colourless during intercalation and deintercalation of the M^+-ions (e.g. cerium oxide-titanium oxide, CeO_2-TiO_2), or a complementary *anodic* electrochromic EC-layer colouring in the oxidised form (e.g. nickel oxide, NiO) so that the colouration of the EC-window can be intensified.

In *hybrid type EC-devices* [11], the counter electrode is missing and instead a redox material is dissolved in a liquid or gel electrolyte. This configuration was used in the development of automotive EC-mirrors by Donnelly [14] in which the WO_3 layer was coloured or bleached by intercalation and deintercalation of Li^+ cations, respectively. The back reaction (bleaching) without applying a voltage is due to a redox system in the electrolyte (ferrocene) (see Chap. 13).

In *solution EC-devices* both EC-materials are dissolved in a liquid or gel electrolyte, in which they can freely diffuse. This configuration glass/TCE/electrolyte containing 2 EC-redox materials/TCE/glass/mirror layer is used for EC-mirrors by Gentex [15–18]. After applying a voltage, one of the electrochromic dyes (bipyridine derivative) is reduced reversibly and coloured. The other anodic colouring dye is oxidised at the positive electrode and therefore also coloured. The back reaction of the two dyes to the bleached oxidation state happens without applying a voltage (see Chap. 13).

These last two types of EC-devices are of the so-called self-erasing type. They have been essentially used for the development of electrochromic mirrors but have also been applied in EC-windows. A continuous current is required to maintain the self-erasing device in the coloured state, while battery-like EC-devices have many hours open-circuit memory and need energy only for the change of the colouration, but not for maintaining it; this configuration is therefore advantageous for large

EC-glazing and -displays such as price labels and display panels which should show the same information during a long time.

The optoelectrochemical reactions in all EC-devices described above should be of course fully reversible to assure a long-term use.

12.3 Sol-Gel Coatings for Electrochromic Devices

This section reviews briefly the development of Sol-Gel coatings that can be used for the realisation of EC-devices. By lack of space, the list of references is mostly limited to reports emphasising device's development and to recent works bringing important new contributions. For more details, the readers should consult the earlier reviews on the same subjects [8–10].

12.3.1 Transparent Conductive Coatings (TCE)

TCE is a critical component in an EC-device as it should enable a rapid charge transport; hence its sheet resistance should be as small as possible and at least less than 50 $\Omega_{\square}$; this can only be obtained with In_2O_3:Sn (ITO) and SnO_2:F (FTO) coatings. The state of the art of TCEs fabricated by the Sol-Gel process was recently reviewed [19]. Unfortunately such coatings cannot be used for EC application as the lowest achievable sheet resistance is usually too high (>500 $\Omega_{\square}$) or requiring the deposition of about 10 coatings sintered at high temperature. Consequently TCEs for EC-devices are essentially made using physical techniques. A recent panoramic review of TCEs for EC-devices can be found in [5]. ITO-coated glasses are expensive and their sheet resistance increases during heat treatment required for the production of EC and IS Sol-Gel layers. On the other hand, FTO-coated glasses are cheap, their sheet resistance is stable up to 550°C and they are easily available in large sizes (Pilkington or Libbey Owens Ford); consequently they are preferred as conductive substrates for EC Sol-Gel layers. For plastic substrates, ITO-coated plastic foil fabricated by sputter deposition technique are available for flexible devices. For non-transparent devices, e.g. EC-mirrors, thin metallic films (e.g. Pt) can be used.

12.3.2 Inorganic Electrochromic and Ion-Storage Coatings

The most studied and used EC and IS layers are pure, mixed and doped oxides of transition metals (Fig. 12.3). Oxides of Ti, Nb, Mo, Ta and W are known as *cathodic* electrochromic materials i.e. they are coloured under reduction and charge insertion while those of Cr, Mn, Fe, Co, Ni and Rh, Ir are *anodic* materials

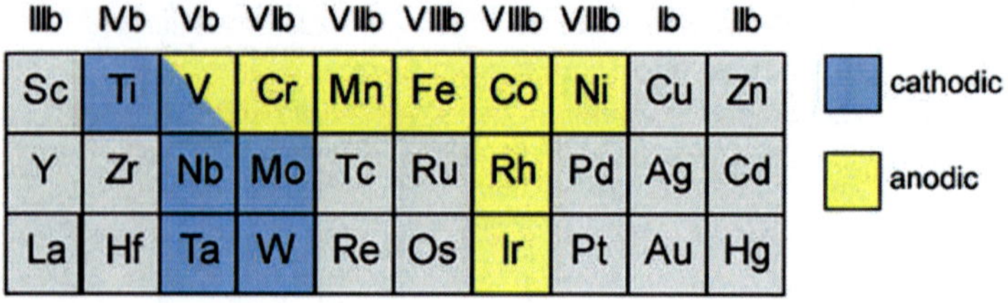

Fig. 12.3 Transition metals whose oxides lead to well-documented *cathodic* and *anodic* electrochromism (adapted from [9])

i.e. they are coloured under oxidation and charge extraction. The oxides of V are of intermediary nature and display both types of features in different wavelength regions.

The electrochromism is connected with the electronic structure of the materials; practically all of them have been prepared by the Sol-Gel process using deposition techniques such as spin and dip-coating [8, 9]. Their EC properties are usually tested using a special three-electrode cell equipped with optical windows with which all the electro-optical characteristics of the coatings can be measured under potentiostatic and/or galvanostatic control [20]. The used electrolyte is a well-adapted ionic liquid and consequently the properties obtained are usually better than those observed with complete EC-devices that are built with solid-type electrolyte. It is difficult to compare the results published in the literature as the obtained properties depend on many parameters such as thickness and heat treatment of the coatings, type and ions concentration of the electrolyte and voltage applied. An overview of the typical EC properties achieved with the most interesting Sol-Gel EC coatings is shown in Table 12.1 and described in more detail in the following section.

12.3.2.1 WO_3 and WO_3: X Coatings (X = Li, Ti)

Tungsten oxide changes its colour from transparent or slightly yellow to deep blue under proton (H^+) or lithium ions (Li^+) insertion; it is the most studied *cathodic* EC material because it has the largest colouration efficiency (CE), a value ranging from about 30 to 70 cm^2/C determined by the slope of the straight line obtained in the plot of the optical density change $\Delta OD = \log(T_{bleached}/T_{coloured})$ versus the charge consumed. The large spread in CE is due to the rate of the ion diffusion coefficient which depends on the microstructure of the coatings. Polycrystalline tungsten oxide, crystallizing between 300 and 400°C, modulates the optical transmittance by reflection (especially in the near-infrared region) whereas amorphous WO_3 does it by absorption [21], a structure better adapted for EC glazing. Four main Sol-Gel routes have been developed for the sols preparation. They are all described in detail in earlier reviews [9, 21] and only a brief summary is given below.

Table 12.1 Overview of typical properties obtained in a three-electrode cell of the most interesting EC Sol-Gel coatings (adapted from [9])

Material	Colouration	Colour a: amorphous, c: crystalline	$T_{bleached}$ [%]/$T_{coloured}$ [%] (wavelength)
WO_3	Cathodic	Blue	85/15
WO_3:Li	Cathodic	Blue	80/40
WO_3-TiO_2	Cathodic	Blue	80/36
Nb_2O_5	Cathodic	Brown	68/28 (550 nm)
		Blue	87/39 (550 nm)
Nb_2O_5 : Mo, Li or Ti	Cathodic	Grey, brown, blue	67/13 (550 nm, grey)
NiO, $Ni(OH)_2$	Anodic, H^+	Brown	80/50 (550 nm)
Li_xNiO_2	Anodic, Li^+	Brown	61/24 (633 nm)
$(NiO)_x(TiO_2)_{1-x}$	Anodic, Li^+	Brown	85/45 (550 nm)
TiO_2	Cathodic	Grey (a)	96/28 (550 nm)
		Blue (c)	96/72 (550 nm)
TiO_2 nanoporous	Cathodic	Blue (c)	52/22
TiO_2-viologen	Cathodic	Blue	Δ OD: 0.68
TiO_2-different viologen	Cathodic	Blue, red, green, black	45–50 /5 (600 nm)

- *Acidification of sodium tungstate*: thick coatings can be obtained with this earliest method but the sol stability is poor and consequently the method is not adequate
- *Hydrolysis of alkoxides*: it is the classical Sol-Gel route for any kind of oxides; however it is expensive and consequently not useful for industrial application. Large area coatings have been reported by Vroon et al. [22]
- *Reaction of tungsten chloride and oxychloride with alcohols*: such reactions with e.g. isopropanol are cheap methods leading to sols stable for several months. The coatings show better uniformity than those prepared from the colloidal route. Using this route, thin films have also been dip-coated on 60 $\Omega_\square$ flexible ITO/PET (polyethylene terephthalate) substrate and further heat treated at 80°C leading to films with the stoichiometry $WO_3{\cdot}18\ H_2O$. However the transmittance change was poor and the acidity of the solution degraded the ITO coating rapidly [23]
- *Use of peroxypolytungstic acid:* the use of hydrogen peroxide alone or together with an organic acid such as acetic or propionic acid at low temperatures (−10 to 12°C) is certainly the best method to obtain stable aqueous solutions of W(VI) precursors. W-peroxy acids are easily obtained by reacting tungsten or tungsten carbide powder. Such stable sols decompose into tungsten oxide during the heat treatment of the film at low temperature (100–200°C). The method can be in principle extended to other metals such as Mo, Mn, Cr, Rh, Ir and Ni. The addition of oxalic acid improves the colouration efficiency of the WO_3 coatings and the transmittance change with the switching cycles can be efficiently reduced. An interesting synthesis process was proposed to reduce the preparation time of the sols [24]. Addition of alkali metal salts (e.g. Li) into the sol converts W oxide into alkali metal tungstate without any adverse effect on the

electrochromic function or on the ability to process the layers and it accelerates the establishment of a stable operating state. Long-term switching experiments with almost cycling-independent colouration have been obtained with large area dip-coated WO_3 up to 50 cm × 80 cm^2 on FTO glass (Sect. 12.4.2). With addition of solvents having higher boiling point than ethanol and other additives, spray-coated WO_3 films of the size 35 cm × 35 cm^2 have been fabricated with homogeneous optical properties. The doping with Li (addition of $LiCF_3SO_3$ up to 15 mol% Li^+) resulted in improved bleaching kinetics. This peroxo route have been used to fabricate large area WO_3 Sol-Gel coatings [25–31].

The durability of WO_3 coatings can be enhanced by the addition of 10–15 mol% TiO_2 to the sol but the number of tungsten active sites and consequently the colouration efficiency is strongly reduced for high amount. Large area coatings with the composition WO_3–TiO_2 have been reported by Bell et al. [32, 33] (Sect. 12.4.2). On the contrary, the presence of adsorbed water and hydroxyl radicals was causing irreversible Li^+ trapping leading to an inferior cycling stability [34].

12.3.2.2 Nb_2O_5 and Nb_2O_5: X Coatings (X = Sn, Zr, Li, Ti, Mo)

Niobium oxide is also an interesting material for electrochromic and counter electrodes used in EC-devices. The first Sol-Gel attempt to fabricate such coating was reported in 1991 using a mixture of $NbCl_5$ dissolved in ethanol. However the durability of the EC response was only a few cycles. Several other precursors have then been reported such as niobiumethoxide ($Nb(OEt)_5$) or pentabutoxide of niobium ($Nb(OBu^n)_5$). Nb_2O_5 coatings present an interesting property as its cathodic colouration under Li^+ insertion depends on the heat treatment temperature of the films: amorphous ones ($T \leq 450°C$) colour in brown and crystalline ones ($T \sim 560°C$) colour deep blue.

Intensive studies leading to the development of devices have been reported in pure niobia layers prepared by dissolving, under a sonocatalytic process, $NbCl_5$ powder in ethanol or butanol and acetic acid leading to a niobiumchloroalkoxide sol [35–37]. Doping with Li, Ti and Mo at different concentration has been also realised using the same ultrasonic treatment. The incorporation of Mo led to a large increase in the optical density from $\Delta OD \sim 0.3$ (undoped) to 0.63 for a Mo:Nb = 0.2. An interesting grey colour was also obtained depending on the film crystallinity rather than on the structure (X= Ti, Mo, Li). With increasing crystallinity, the absorption band was found to shift to higher wavelength resulting, after Li^+-intercalation, to a brown colour for amorphous layers, a grey colour for layers with small crystallites (5–20 nm) and a blue colour for highly crystallized layers (crystallite size 35–100 nm) (Fig. 12.4). All coatings exhibited a stable behaviour with no change in the variation of the optical density up to 5,000 cycles and even up to more than 50,000 cycles when a small amount of water was added in the electrolyte [38], a process that led to the development of large area devices [39], Sect. 12.4.2).

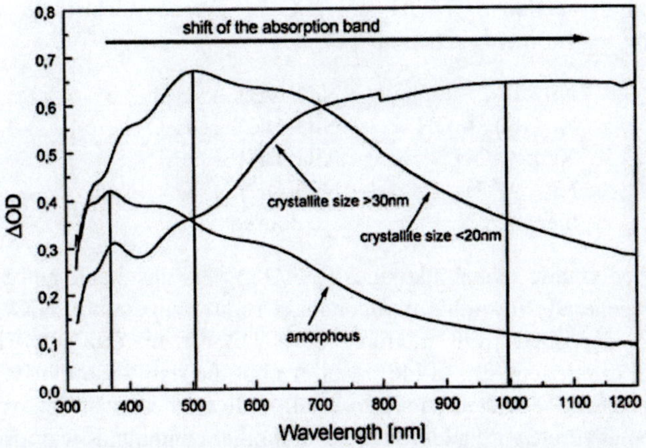

Fig. 12.4 Change in optical density (ΔOD) vs wavelength of pure and Ti-doped niobia EC-layers after Li^+ insertion: 120 s, −2.2, + 1 V. Layers with crystallite size >30 nm show a blue colour (Ti:Nb = 0, 600°C, hexagonal Nb_2O_5), those with crystallite size <20 nm exhibit a grey colour (Ti:Nb = 0.3, 600°C orthorhombic $Ti_2Nb_{10}O_{29}$) and amorphous layers colour in brown (Ti:Nb = 0.1 amorphous state, 450°C) [35–37]

Although the possibility to get different colours with Nb_2O_5:X layers is highly attractive for the technological point of view, their smaller colouration efficiency compared to WO_3 is a disadvantage as a larger amount of Li^+-ions (Q_{in}) have to be inserted to get a similar contrast. Consequently the fabrication of EC-devices requires the use of a counter electrode presenting a higher Li charge capacity (>30 mC/cm^2).

12.3.2.3 MoO_3

MoO_3 Sol-Gel layers have been deposited from alkoxides, chlorides, chloro-alkoxides or molybdic acid. Their electrochromic properties are not as good as those of WO_3 and Nb_2O_5 but mixed MoO_3:W [40], MoO_3:V [41] and MoO_3:Ce [42] coatings exhibit excellent electrochromic properties.

12.3.2.4 Nickel Oxide-Based Layers

Nickel oxide-based materials are well-known *anodic* electrochromic systems and have been extensively studied using the Sol-Gel process. They change from a colourless state to brown coloured oxyhydroxides upon electrochemical oxidation with usually high electrochromic efficiency (35–50 cm^2/C) related to the redox process Ni^{2+} (bleached)/Ni^{3+} (coloured). Most of the works on such electrodes discuss proton and/or hydroxyl ion intercalation/deintercalation but a few also describe Li^+ intercalation/deintercalation. Proposed mechanisms for the electrochromic reaction of NiOH and NiO in aqueous electrolyte are usually described by the following reactions [43, 44].

$$
\begin{array}{ll}
Ni(OH)_2 & \rightarrow NiO(OH) + H^+ + e^- \\
Ni(OH)_2 + OH^- & \rightarrow NiO(OH) + H_2O + e^- \\
NiO + OH^- & \rightarrow NiO(OH) + e^- \\
NiO + H_2O & \rightarrow NiO(OH) + H^+ + e^- \\
\text{bleached} & \quad \text{coloured}
\end{array}
$$

The use of polymeric nickel alkoxides $[(Ni(OR)_2]_n$ is unsuitable since these precursors are generally insoluble in alcohols at room temperature. NiO films were successfully deposited from mixtures of nickel nitrate, $[Ni(NO_3)_2(H_2O)_6]$, in alcohol [45] or ethylene glycol [46] solutions but the stability of $Ni(NO_3)_2$ during thermal processing restricted its use to small-scale applications. The use of nickel sulphate heptahydrate with addition of several other compounds was also reported, but the initially rather good EC properties of the coatings were rapidly degrading after the 10th cycle, indicating that the Ni^{3+} phase was progressively and irreversibly formed during cycling. Further studies showed that the thermal treatment of such layers was in fact crucial [47–50]. The same result was found for coatings obtained with sols of nickel chloride ($NiCl_2 \cdot 6\ H_2O$) in butanol and ethylene glycol for the deposition of NiO_xH_y thin films [51, 52]. However no real improvements have been obtained.

NiO_x films have been obtained using other interesting precursors such as nickel diacetate tetrahydrate $[Ni(ac)_2 \cdot 4\ H_2O]$ in methanol, nickel(II)acetylacetonate $[Ni(acac)_2]_3$, nickeldiacetate tetrahydrate or nickel diacetate dimethylaminoethanol $[Ni(ac)_2 \cdot (dmaeH)_2]$ in dry N, N-dimethylaminoethanol (dmaeH). The use of dmaeH as a solvent is however questionable for industrial application because of its low flash point (39°C). All the films studied in KOH electrolyte first showed an activation period, during which both the capacity and the electrochromic response increased, then a short steady-state period followed by a rapid degrading typically after less than 1,000 cycles.

Ni-based films with Si: The addition of tetraethyl orthosilicate (TEOS) to the alcoholic solutions of anhydrous halide proved to be a suitable method to prepare electrochromic Ni-based thin films [53], the coatings showing higher hardness, better adhesion and maintaining rather high and reversible transmittance at 550 nm. Mixture of nickel nitrate hexahydrate with ethylene glycol and TEOS was also proposed [46].

Ni-based films doped with Ti: These compositions present the greatest promise. The chemical deposition of such NiO_xH_y-based films was done using an alcoholic solution of $NiCl_2$ and Ti(IV)butoxide to which triethylamine was added. The higher the Ti/Ni ratio, the faster and more efficient was the electrochromic effect and high Ti content leads also to improved adherence and abrasive resistance and higher transparency of the films in their bleached state [54]. Using nickel acetate tetrahydrate and titanium n-propoxide deep brown coloured films with fast switching time (<10 s) have been obtained in a 1M KOH aqueous electrolyte [55]. The same overall cycling behaviour reported above for NiO films was also observed with however an extension of the degrading period up to about 10,000 cycles. An explanation of the rather complex electrochemical processes occurring during switching in a KOH electrolyte based on results obtained using an electrochemical quartz crystal microbalance was reported in [56]. 5 $\times$ 10 cm^2 prototypes exhibiting a stability up to 17,000 cycles have also been built [57] (Sect. 12.4.2).

Ni-based films doped with Li or Co: Lithiated Ni oxide thin films (Li_xNiO_2) have been prepared by dissolving a mixture of Ni(II)acetate [$Ni(CH_3COO)_2 \cdot 4\,H_2O$] and Li(I) acetate precursors with the slow addition of an aqueous solution (30 vol%) of H_2O_2 by Svegl et al. [58]. After heat treatment at 550° C the films showed an anodic electrochromic behaviour in 1 M $LiClO_4$ in propylene carbonate (PC) electrolyte with a transmittance change from 61 to 24% at 633 nm during CV cycling. These coatings, as well as lithiated Co-oxide (Li_xCoO_2) thin films, prepared with the same route using Co(II) acetate [58] have been used for the preparation of EC-devices (Sect. 12.4.2).

12.3.2.5 CeO_2 and CeO_2:X (X = Ti, Sn, Zr, Si, V, Fe)

Since their discovery by Baudry et al. [59] and a series of other papers (see [8, 9]), CeO_2 and mixed oxides, especially those with compositions of $(CeO_2)_x(TiO_2)_{1-x}$, have been intensively studied as electrochromic non-colouring (passive) Li^+ and H^+ *ion-storage electrode* and used in EC-devices in which the colouration is due to the other EC-electrodes, such as WO_3 or Nb_2O_5. The transmittance of mixed oxides of Ce/V or Ce/Fe is however not totally colourless in the bleached state, so that EC-devices fabricated with them remain slightly coloured in the bleached state.

Pure CeO_2 layers: The preparation and electrochromic properties of pure CeO_2 layers have been reported among others by Štangar et al. [60] and Orel et al. [61]. The reversibility of the lithium intercalation in pure CeO_2 was rather good, but the reaction kinetics was slow. To improve the intercalation kinetics of these electrodes several CeO_2-based mixed oxide films such as CeO_2–TiO_2 [59, 60], CeO_2–SnO_2 [61], CeO_2–ZrO_2 [62], CeO_2–SiO_2 [63] and CeO_2–V_2O_5 [64] have been proposed The intercalation/deintercalation reactions of such coatings are coupled with the redox reaction Ce^{3+}/Ce^{4+} in the layer but in those containing vanadium, the redox reaction of V^{5+}/V^{4+} takes place in addition.

CeO_2–TiO_2 layers: The CeO_2–TiO_2 sols have been prepared using different precursors such as cerium ammonium nitrate [$Ce(NH_4)_2(NO_3)_6$] and titanium isopropoxide or titanium ethoxide and a mixing of alkoxides $Ce(OBu^s)_4$ and $Ti(OBu^n)_4$. The use of ultrasonic irradiation has also been suggested [65]. These films consist of an amorphous matrix of titanium oxide-containing cerium oxide nanocrystallites. Both the lithium diffusion coefficient and the intercalated charge obtained e.g. for a ratio Ce/Ti=1 are much higher than in pure CeO_2, these parameters, recently studied in detail [66], remaining constant for several thousand of cycles. Using an aged sol, up to 200 nm thick crack free CeO_2–TiO_2 single

layers have been prepared, facilitating the fabrication of large area devices [28, 30], with stability up to 50,000 cycles [31, 38, 67] (Sect. 12.4.2). An interesting recent research by Sun et al. [68] reported on the mass changes of $CeO_{0.81}/TiO_2$ coatings during Li^+ ions exchange processes in a dry and wet $LiClO_4$-PC electrolyte using a quartz crystal microbalance. The composition of the coatings was found to change during the earlier cycles mainly because of an irreversible Li^+ intercalation and also involved adsorption/desorption or exchange of other ions contained in the electrolyte such as Li^+ and ClO_4^- in a dry electrolyte and Li^+, hydrated $Li(H_2O)_n$ and ClO_4^- in a wet electrolyte; such results allowed the authors to explain the drastic improvement of the reversibility of the exchange processes as well as the faster kinetics of the wet electrolyte/CeO_2–TiO_2 system. The electrochromic properties of CeO_2–TiO_2 and $CeTi_2O_6$ have also been recently reported [69–71].

CeO_2–ZrO_2 layers: ZrO_2 by itself is not able to intercalate lithium ions; therefore the electrochemical process in Ce-Zr mixed oxides is related to the oxidation/reduction of cerium (Ce^{4+}/Ce^{3+}). Several routes have been proposed using as precursors cerium nitrate [$Ce(NO_3)_3{\cdot}6H_2O$] and [$ZrOCl_2{\cdot}9H_2O$] in ethanol, $Ce(NO_3)_3{\cdot}6\ H_2O$ and zirconium propoxide in ethanol and addition of acetyl acetone, mixture of $Ce(NH_4)_2(NO_3)_6$ and zirconium alkoxide ($Zr(O^iPr)_4$) in isopropanol submitted to ultrasonic irradiation [62]. In all cases the charge capacity was rather small and was found to drop during cycling.

CeO_2–SnO_2 layers: Such mixed oxides (up to 50 mol% SnO_2) have been prepared by an aqueous-based process using $Ce(NH_4)_2(NO_3)_6$ and $SnCl_4$ [61, 72], by dissolving $Ce(NH_4)_2(NO_3)_6$ and stannous citrate monohydrate ($C_6H_6O_7Sn_2{\cdot}H_2O$) in ethylene glycol (EG) and adding citric acid. Unfortunately SnO_2 was found unstable under Li intercalation so that the long-term stability of EC-devices made with CeO_2–SnO_2 films is doubtful.

CeO_2–SiO_2 layers: CeO_2–SiO_2 thin films prepared using a mixture of cerium nitrate ($Ce(NO_3)_3{\cdot}6\ H_2O$) and silica sol as precursors present a very low charge density. Better results have been obtained using an aqueous solution of $(NH_4)_2Ce(NO_3)_6$ and a peptization process performed using nitric acid and water and GPTS ((3-Glycidyloxypropyl)trimethoxysilane) [63].

$CeVO_4$ and Ce/V-oxide: Such ion-storage films exhibit a high ion-storage capacity of 20 mC/cm^2 and have been prepared by adding V-oxoisopropoxide to ceria sols made with $CeCl_3{\cdot}7\ H_2O$ mixed with citric acid in a 1:2 mol proportion and dissolved in ethanol [64]. Higher lithium ion-storage capacity was achieved with Fe-containing $CeVO_4$ films [73]. However the cycling stability was poor in both cases.

12.3.2.6 *V_2O_5, Fe_2O_3, TiO_2, SnO_2, SnO_2:Sb, SnO_2:Mo and mixed oxides of V, Ti, Zr, Fe, Ce*

Besides CeO_2, other single oxide films such as SnO_2, TiO_2, Fe_2O_3 and V_2O_5 have been proposed as counter electrodes. All show a good transmittance in the visible range but, as for CeO_2, rather low charge capacity when cycled in safe potential ranges. Ion-storage electrodes with higher charge capacities and small optical modulation can be obtained with mixed oxides of V, Ti, Zr, Fe, Ce, W. An earlier overview of the photopic transmittance of 24 different counter-electrode films discussing their intercalated (T_{ins}) and deintercalated states (T_{ext}) and their maximum possible charge (typically 15–30 mC/cm^2) was reported by Opara Krašovec et al. [73]. Crystalline or amorphous Fe/V-oxide films (Fe:V=1:1 and 1:2) are other promising compounds with a high charge capacity [74] but with lower T_{ins} values (0.66–0.69) than $CeVO_4$. Sn/Mo-oxide films have also been proposed as optical passive counter electrodes [75]. Films with ratio Sn/Mo = 9:1 and 2:1 have charge capacities up to 20 mC/cm^2 and T_{ins} of 0.86. For Sn/Mo = 1:1 the charge capacity increases up to 52 mC/cm^2 but the photopic transmittance T_{ins} decreases to 0.68. The disadvantage of the Ce/V, Ti/V, Fe/V, and Sn/Mo-oxide films is their low cycling stability and for V containing oxides, the yellowish colour of the coatings in the bleached state. Recent work has been reported on W-doped V_2O_5 showing that the doping can remarkably increase the ion-storage capacity (to about 45 mC/cm^2) and transmission modulation of the films which display a reversible multichromism orange↔green↔blue↔brownish↔grey under Li^+ insertion/extraction [76]. Mo-doped V_2O_5 xerogel film prepared by cathodic electrodeposition from a sol also showed a multichromism orange↔green↔blue with a high charge capacity of 70 mC/cm^2 under Li^+ insertion/extraction depending however on the extent of the intercalation [77]. Probably the most promising recent result is the use of a V_2O_5–TiO_2 as ion-storage coating prepared by a Sol-Gel electrodeposition technique for EC-windows which exhibit high contrast (68%) and a minimum transmission at 580 nm of 1%, fast switching of about 5 s and cyclic durability tested up to 150,000 cycles [78]. A few earlier works report on the use of large area Sol-Gel V_2O_5–TiO_2 ion-storage coatings [22, 32, 79] as well as using Sol-Gel antimony-doped tin oxide (ATO) and mixed ATO–TiO_2, ATO–Al_2O_3 and ATO–SiO_2 powders. These later materials have been used for the development of printed displays by Coleman et al. [80] showing a reflectance change of up to 32%.

12.3.2.7 TiO_2

TiO_2 is also a widely studied electrochromic material. Its properties depend strongly on its preparation and structure, either amorphous or crystalline (anatase or rutile). TiO_2 coatings have been prepared initially from the classical alkoxy route [81, 82] as well as from other methods [83–85]. Pure TiO_2 colours grey; the charge capacity and colouration efficiency of amorphous films is higher (~10 cm^2/C at 630 nm) than those with the crystalline anatase structure (7.4 cm^2/C) and their transmittance change is rather small, typically between 90 and 70%. Materials doped with Al and Cr colour blue. The same colouration was obtained with ~4 μm thick highly porous films made with nanocrystalline TiO_2 anatase particles [86] heated at 450°C in air exhibiting a high and reversible intercalated charge of 110 mC/cm^2 [87] and high colouration efficiency, 20 cm^2/C at 780 nm. TiO_2

nanocrystalline films prepared at low temperatures are not adequate as a practical EC material. However, nanostructured TiO_2 modified by adsorbing molecular redox-chromophores such as viologens combine the advantages of an EC nanostructured metal oxide film with those of these chromophores i.e. a low redox potential, significant reversibility and large change of the photopic extinction coefficient in the reduced state [87, 88]. When possessing carboxylic groups ([N,N′-(bis-2-carboxyethyl or bis-2-phosphonoethyl)-4,4′-bipyridine]$^{2+}$) they can be directly attached to the TiO_2 surface. An absorbance change of 0.68 was achieved with thick films (3.5–4 μm thick) after inserting a charge of only 8 mC/cm^2 with a colouration efficiency of 85 cm^2/C [87, 89]. EC-window based on viologen-modified nanostructured TiO_2 films have been reported using the redox chromophore [N,N′-(bis-2-phosphonoethyl)-4, 4′- bipyridine]$^{2+}$, an electrolyte of $LiClO_4$ and ferrocene in γ-butyrolactone and a conducting glass as counter electrode [90, 91] (Sect. 12.4.2). Electrochromophores were used for transparent EC-displays [92] and for paper-like displays (trade mark NanoChromics) working in reflective mode [93] (Sect. 12.4.3). By using different organic redox-chromophores (viologens) adsorbed on mesoporous anatase TiO_2, EC-devices colouring in blue, green, red and black could be obtained [94] (Sect. 12.4.3).

We recall that TiO_2 was often used in mixed electrochromic oxides (see above) as WO_3–TiO_2, Nb_2O_5–TiO_2, CeO_2–TiO_2, V/Ti-oxide, Ni/Ti-oxide, etc. to improve the cycling stability (V/Ti-oxide), the adherence, the abrasive resistance or the transparency of the films in their bleached state (Ni/Ti-oxide) and the Li^+intercalation and deintercalation kinetics (Ce/Ti-oxide).

12.3.2.8 Hybrid Materials

This class of materials is rather new and a few recent researches have been reported for their application in electrochromism. The titania nano domains obtained in covalently bonded polyaniline (PANI)-TiO_2 were reported to act as electron acceptors, reducing the oxidation potential and band gap of PANI [95]. Silica-polyaniline core–shell composite nanoparticles have also been tested [96]. Both materials have been reported to improve the long-term EC stability.

12.3.2.9 Other Materials

The use of polyoxometalates (abbreviated POM), polyatomic three or more transition metal oxyanions of vanadium (V), niobium (V), tantalum (V), molybdenum (VI) and tungsten (VI) linked together by shared oxygen atoms to form a large, closed three-dimensional framework have been recently reported as potential materials for electrochromic layers prepared by the Sol-Gel process and other techniques [97]. A ternary system CeO_2–TiO_2–ZrO_2 has been proposed in which both the charge density and the ratio of the cathodic to anodic values were found to increase with the sintering temperature to reach a maximum at 500°C [98]. Electrochromism has also been found in CuO nanoparticles studied in a lithium-based electrolyte showing a transmission variation of about 50% but in a very narrow range of wavelength, typically 400–500 nm [99].

12.3.3 Electrolytes

The ionic conductor is an important element for all types of electrochromic systems (Fig. 12.1). The ions should move fast from the EC-layer into the electrolyte and back and for battery-type EC-devices to the second EC-layer or the ion-storage electrode and back.

The most important requirements for an electrolyte for EC-devices have been discussed in detail [1, 100–102]. They are:

(1) High ionic conductivity between 10^{-3} and 10^{-7} S/cm (depending on the application)
(2) Low electronic conductivity (smaller than 10^{-12} S/cm).
(3) Long cycling durability at operation temperature
(4) Good adhesion with the adjacent layers (no delamination for several years, even after temperature switching tests and more than 10^4 switching cycles)
(5) Optical transparency for most EC application (display may be an exception)
(6) Chemical compatibility with the functional layers
(7) Electrochemical stability in the voltage range used for switching the EC-device
(8) Long-term stability against UV-light if such light is not filtered by the functional layers for certain application (e.g. EC-devices for architectural or automotive glazing)

The literature on ionic conductors is vast and several reviews on inorganic ion conductors suitable for EC-devices [8, 9] and other applications have been reported (Chaps. 4 and 6). Most of the researches in the EC field were done on H^+- and Li^+-conductive coatings. They are briefly reviewed below.

Two main classes of materials can be used as a "Sol-Gel" electrolyte: the first one involves inorganic oxide materials (e.g. Ta_2O_5, $LiNbO_3$) and the other organic–inorganic hybrids which combine the better conductive properties of polymeric materials with the better mechanical strength of inorganic materials.

Sol-Gel proton conducting amorphous tantalum oxide (Ta_2O_5) films were deposited using a sol made with $Ta(OC_2H_5)_5$ as a precursor [103, 104] and exhibited a conductivity at room temperature of up to about 3.2×10^{-6} Scm^{-1} and low optical absorption. Crystalline Ta_2O_5 coatings had a slightly smaller value. The colouring and bleaching cycles of "half EC cells" consisting of Sol-Gel Ta_2O_5 layers (190 nm thick) coated onto WO_3 with H_2SO_4 as electrolyte were similar to those using WO_3 alone with H_2SO_4.

$LiNbO_3$ is a lithium conductor and the ionic conductivity at room temperature of Sol-Gel layers are in the range of 6–8 × 10^{-7} Scm^{-1} [105]. To our knowledge, no complete all Sol-Gel EC-device was reported with such electrolytes. Other metal oxides prepared by the Sol-Gel process show an adequate ionic conductivity but only at high temperatures (e.g. LiCl-ZnO-SiO_2 xerogels, with about 10^{-3} Scm^{-1} at 450°C) and consequently are not useful for EC-devices.

The most used electrolytes to build EC-devices are polymers, copolymers and gels [9, 106] and are not reported here; EC-devices made with liquid or polymeric gel electrolytes are often less durable than those made with inorganic solid-state electrolyte and those in a liquid form are not useful for large area EC-devices because of the bending of the glazing.

On the other hand lithium-doped organically modified Sol-Gel materials (called hybrids or ormolytes) are most promising low temperature Li^+ ion-conducting materials for their fabrication [107, 108]. The ionic conduction and the flexibility of the ormolytes are induced by the polymer phase and because of the presence of the inorganic network, their mechanical properties are improved in comparison to pure organic conducting polymers. As they present also a high transparency and high chemical stability, such materials are most adequate to build small and large area EC-devices. The hybrid materials are divided into two classes depending on the nature of the chemical bonding between the organic and inorganic constituents. In class I materials, covalent chemical bonds exist between the organic and inorganic components while in class II materials only weak physical bonds (van der Waals, hydrogen and ionic bonds) bind parts together. A typical class I ion-conducting ormolyte was prepared from a mixture of 3-isocyanatopropyltri-ethoxysilane, O,O′bis(2-aminopropyl)-polyethylene glycol and a lithium salt; its ionic conductivity was however rather low, about 10^{-6} Scm^{-1} [107]. Higher room temperature ionic conductivity was obtained for type II materials. An organic–inorganic composite electroyte based on the partial hydrolysis and condensation of a silane with addition of e.g. tetraethylene glycol and an alkali salt was patented [28]. For example, a nanocomposite electrolyte based on glycidyloxypropyl-trimethoxysilane (GPTS), tetraethoxysilane (TEOS), $LiClO_4$, $Zr(O^nPr)_4$ and tetraethylene glycol (TEG) was used for the lamination of many all Sol-Gel electrochromic devices up to a size of 50×80 cm^2 having the configuration glass/ITO (or FTO)/WO_3/electrolyte/CeO_2–TiO_2/ITO (or FTO)/glass [30, 31, 109–111]. The nanocomposite electrolyte was filled as a liquid between the functional layers and after hardening at 100°C for several hours (Fig. 12.2), it had an ionic conductivity of 1.6×10^{-6} Scm^{-1} at 298 K (Sect. 12.4.2). By adding small amounts of water to this composite electrolyte (1–4 wt%), the switching kinetics and the long-term stability of the EC-devices could be drastically improved up to 50,000 switching cycles [38, 67] (Sect. 12.4.2).

Orel et al. [112] developed a mixed class I-II ormolyte based on glycidyl-oxypropyl-trimethoxy-silane (GPTS), tetraethoxysilane (TEOS), tetraethylene glycol (TEG), Li salt and $Ti(O^iPr)_4$. The material presented an ionic conductivity of $(1–3) \times 10^{-4}$ Scm^{-1} and was also used for the construction of all-Sol-Gel EC-devices. The stability was high and no variation in the conductivity values at 70°C were observed during 12 h. The same group also developed a Li^+ ormolyte made of a non-hydrolysed hybrid silicon precursor, modified by polypropylene glycol (4,000) [113]. Another interesting development was reported using new

Sol-Gel redox I_3^-/I^- electrolytes; their application in hybrid electrochromic devices without counter electrode led however to a rapid electrochemical degradation [114]. A successful recent approach was developed based on a single-phase hybrid polyelectrolyte (SPHP) in which the ionic conductivity is not coupled to the segmental motion of the polymer chain yielding a Li^+ conductivity of 10^{-5} S/cm at room temperature; it was applied to build 5 × 10 cm^2 EC-devices comprising WO_3 and CeO_2–TiO_2 Sol-Gel layers withstanding at least 60,000 chronoamperometry cycles with an almost constant contrast [115]. Barbosa et al. [116] developed a hybrid cross-linked material based on poly(ethylene oxide (PEO)/siloxane (di-ureasil) with a wide range of concentration of lithium tetrafluoroborate ($LiBF_4$). Four-layer prototype EC-devices incorporating these electrolytes were claimed to have encouraging colouring/bleaching process.

Recently cheap electrolytes based on renewable primary products have been developed and tested in EC-devices. One is based on a Li^+ conducting amylopectin-rich corn starch and used in a 7 cm^2 EC-devices built with WO_3 and CeO_2–TiO_2 [117]; the device exhibited however a small ΔOD of about 0.27 but a rather fast kinetics (~10 s). Another one was a Li^+-conducting gelatine exhibiting a conductivity of 1.5 × 10^{-5} S/cm and tested in a grey colouring EC-device involving Sol-Gel electrodes of Nb_2O_5:Mo and $(CeO_2)_{0.81}$–TiO_2 [118]; the change in transmission at 550 nm was however rather low (60↔40%) but the stability was tested up to more than 25,000 cycles. A proton conducting gelatine plasticized with glycerol and containing acetic acid was also developed successfully and used in a 7 × 2 cm^2 EC-device made with WO_3 and CeO_2–TiO_2 Sol-Gel electrodes [119]. The conductivity was 5 × 10^{-5} S/cm at room temperature and, if the kinetics was fast (~10 s for the colouring and 2 s for the bleaching), the transmission variation was rather poor, typically 15% during more than 10,000 cycles. Finally a fourth approach was tested using a Li^+-conducting agar-based electrolyte [120]; its conductivity at room temperature was 1.1 × 10^{-4} S/cm and had a Vogel-Tammann-Fulcher dependence. This material was tested in an EC-device involving WO_3 and CeO_2–TiO_2 Sol-Gel electrodes up to 5,000 cycles; the charge which could be inserted reversibly was however rather low, ranging from about 5 to 7.5 mC/cm^2, so that the transmittance variation was also rather low, typically 24% at 550 nm.

12.4 Applications

12.4.1 Overview

The production of large area EC-devices has been proposed for architectural, automotive and aerospace glazing, for the realisation of automotive sunroofs, rear-view EC-mirrors (see Chap. 13) and for EC-displays [3] as well as EC-papers, EC-glasses and cathodic ray tubes [121].

The requirements are highly dependent on their use. Architectural EC glazing should have a long-term stability of at least 30 years, requiring a minimum stability of about 100,000 switching cycles and a switching time of some minutes is suitable. Further requirements are a high UV-stability, a high homogeneity of the colouration, a high homogeneity of a set of EC-windows on a facade as well a high stability of the properties between −30 and 85°C. Durability issues, tests and service lifetime prediction of electrochromic windows for buildings applications have been reported [3, 102, 122, 123].

For automotive glazing, a long-term stability of about 10 years with about 30,000 switching cycles is required, but the switching kinetics should be faster than above and the device should bleach automatically without current. Temperature stability range and UV-stability are more drastic and a good resistance to shock similar to toughened or laminated safety glass [124, 125]. Automotive rear-view mirrors adjusting automatically their reflection are already on the international market and are reported in this book (Chap. 13) and not reviewed here. The requirements for EC-displays depend on their location (inside or outside) and the rate at which the information should be changed. The lifetimes should be about 5 years, so that about 10^7 on/off cycles are required for most applications.

Unfortunately most of these requirements have not been tested *for sol–gel made devices* and therefore their commercial applications are very scarce.

12.4.2 EC-Windows

The first "all Sol-Gel" EC-device was reported by Judeinstein and Livage in 1988 [126] who laminated a device consisting of a SnO_2 counter electrode, a WO_3 layer and a TiO_2-gel ionic conductor. Unfortunately the SnO_2 counter electrode exhibited irreversible colouration but the device has been cycled for more than 4×10^5 times. Then Özer et al. [127] fabricated an EC-device with a Sol-Gel ITO counter electrode, an electrochromic TiO_2 layer and a Li^+-doped polymeric electrolyte. This device had a transmittance change of 40% (80–40%) but the kinetics was slower (50 s) than that of WO_3-based Sol-Gel EC-devices.

An "all Sol-Gel" EC-device consisting of a WO_3 layer, a CeO_2–TiO_2 counter electrode and a TiO_2 gel electrolyte was first described by Macêdo et al. [128]; it had an optical transmittance change from 60 to 20% during the first cycles and a response time of several seconds (Table 12.2) but a delamination was observed during extended cycling due to the degradation of the TiO_2 gel ion-conductor. The use of a hard silica-polyethyleneglycol ormolyte as electrolyte was also reported but the transmission change in the visible was rather poor (78–48%) [129].

Following the initial development of the $(CeO_2)_x$–$(TiO_2)_{1-x}$ counter electrode [59], an "all Sol-Gel" electrochromic system was patented by Schmidt et al. [28] consisting of a lithium-doped Sol-Gel WO_3 layer, a Sol-Gel $(CeO_2)_x$–$(TiO_2)_{1-x}$ counter electrode and an organic–inorganic nanocomposite electrolyte. Then many devices with size up to 50×80 cm^2 have been developed at the INM-Leibniz

Table 12.2 Some representative EC-devices made by the Sol-Gel process: Materials and configuration, size, colour, variation of transmittance (T) (or reflectance (R), diffuse reflectance (R_{diff}), contrast ratio (CR)) at a given wavelength (in nm), the potential range, number of colouration/bleaching (c/b) cycles tested and switching time (t_{sw}). G: glass, ITO: In_2O_3:Sn, FTO: SnO_2:F. (adaped from [9])

Device configuration	Size (cm^2)	Colour	T [%], R [%], R_{diff} [%], CR	c/b cycles	t_{sw} (s)	References
G/ITO/H_xWO_3/TiO_2 gel electrolyte/CeO_2-TiO_2/ITO/G	few cm^2	blue	$35 < T_{550} < 58$ (@360th cycle)	360	10	[128]
G/ITO/WO_3/nanocomposite electrolyte-Li^+/$(CeO_2\text{-}TiO_2)_{1-x}$/ITO/G	1,225	blue	$20 < T_{633} < 75$ (−2.5/+2 V)	10,000	100–300	[29, 30, 110]
G/FTO/WO_3/nanocomposite electrolyte-Li^+/$(CeO_2\text{-}TiO_2)_{1-x}$/FTO/G	4,000	blue	$25 < T_{550} < 70$ (−2.3/+2.3 V)	10,000	180	[31, 109]
G/FTO/WO_3/nanocomposite electrolyte-Li^+/$(CeO_2\text{-}TiO_2)_{1-x}$/FTO/G (bending radius 2 m)	1,200	blue	$25 < T_{550} < 67$ (−2.1/+1.9 V)	10,000	180	[111]
G/FTO/WO_3/nanocomposite electrolyte-Li^+ (+1–4 wt% H_2O)/ $(CeO_2\text{-}TiO_2)_{1-x}$/FTO/G	~150	blue	$28 < T_{550} < 70$ (−2/+2 V)	50,000	120	[38, 67]
G/FTO/Nb_2O_5:Mo/nanocomposite electrolyte-Li^+ (+1−4 wt% H_2O)/ $(CeO_2\text{-}TiO_2)_{1-x}$/FTO/G	~1,200	grey	$20 < T_{550} < 60$ (−2.5/+ 1.5 V)	50,000	120–180	[38, 67]
G/FTO/WO_3/ormolyte-Li^+/Nb/Fe-oxide/FTO/G	9	blue	$26 < T_{550} < 62$ (−2.6 / +1.6 V)	2,000	100	[112, 132]
G/FTO/WO_3/ormolyte-Li^+/V/Ti-oxide/FTO/G	9	blue	$30 < T_{550} < 63$ (−3/+2V)	3,000	120	[112, 132]
G/FTO/WO_3/ormolyte-Li^+/V/Ti-oxide (1:2)/FTO/G	9	blue	$4\% < T_{550} < 62$	–	–	[74]
G/FTO/WO_3/$(H^+)Li^+$ ormolyte/$Li_{0.94}Ni_{1.06}O_2$/FTO/G	9	blue	$2 < T_{550} < 68$	1,000	~180	[58]
G/ITO/WO_3-TiO_2/polymer electrolyte/V_2O_5:Ti/ITO/G	–	blue	CR: 2.6	< 2,000	–	[79]
G/Nb_2O_5:Li/nanocomposite electrolyte-Li^+/CeO_2-TiO_2/FTO/G	50	brown	$39 < T_{550} < 66$	40,000		[131]
G/FTO/NiO-TiO_2/KOH/CeO_2-TiO_2/FTO/G	50	brown	$30 < T_{550} < 65$	7,000		[10, 135]
G/FTO/TiO_2/Chrom 1/electrolyte: $LiClO_4$, ferrocene in γ-butyrolactone/FTO/G	6	blue	$10 < T_{608} < 65$ (−1/+1 V)	> 10,000	1	[90, 91]
G/FTO/TiO_2/Chrom 1/electrolyte: $LiClO_4$ in γ-butyrolactone)/ Chrom 2/SnO_2/FTO/G	6	blue–red	$23 < T < 64$ (@10000th cycle)	> 10,000	< 1	[92]
G/FTO/WO3/Ormolyte with I_3^-/I^- redox system/Pt/FTO/G	4	?	$30 < T_{550} < 68$	1,800	10 (c) 5 (b)	[114]

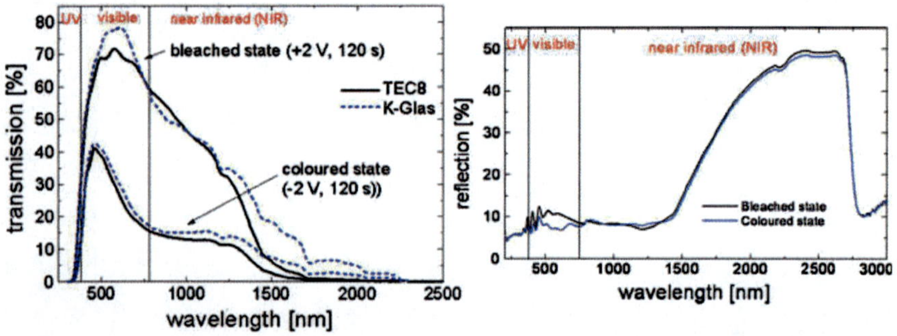

Fig. 12.5 Transmittance (*left*) and reflectance (*right*) spectra of a coloured (*blue*) and bleached EC-devices with size 50 × 80 cm^2 having the configuration glass/FTO/WO_3/nanocomposite electrolyte/$(CeO_2)_x$–$(TiO_2)_{1-x}$/FTO/glass; potentiostatic switching: −2.0 V, 2 min/+2.0 V, 2 min (from [9])

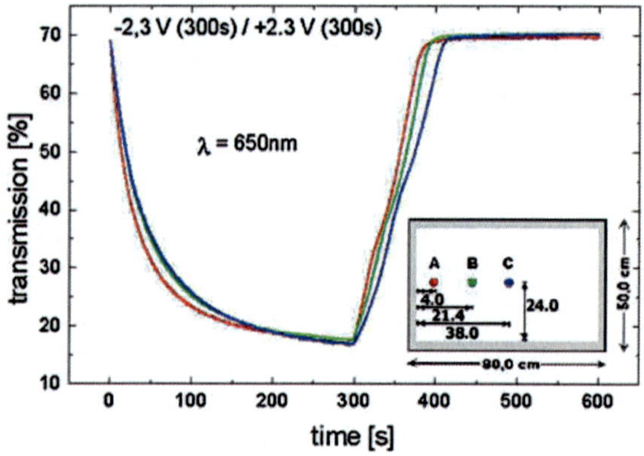

Fig. 12.6 Transmittance at 650 nm of the EC-device shown in Fig. 12.5 (size 50 × 80 cm^2) during colouration (−2.3 V) and bleaching (+2.3 V). The transmittance was measured at different points of the EC-device (points A, B, C, see inset), adapted from [9]

Institute for New Materials using these recipes with the configuration glass/ITO or FTO/WO_3/nanocomposite electrolyte/$(CeO_2)_x$–$(TiO_2)_{1-x}$/ITO or FTO/glass showing typical reversible transmittance change from 70 to 25% (−2.5 V/+2 V) in the visible range (Fig. 12.5), fast switching kinetics (as fast as 3 min), high cycling and UV-stability [29–31, 109–111, 130] (Table 12.2). The Sol-Gel films could be achieved using one dip-coating step, an advantage for industrial fabrication and the nanocomposite electrolyte was filled in the liquid form between the functional layers and then hardened by heat treatment at 100°C.

Large area EC-devices tend to colour faster at the edges than in the middle of the device; but using 8 $\Omega_\square$ TEC substrates (FTO glass), the difference between these transmittance could be reduced to about maximal 5% during colouration and maximal 14% during bleaching (Fig. 12.6) [31, 109]. The memory effect was quite good since without applying a voltage, the transmittance in the coloured state increased only 5% after 4 h at 25°C and that in the bleached state changed only about 1% after 24 h at 25 and 80°C.

Fig. 12.7 Prototype of an electrochromic profile glass module developed at INM with the configuration glass/FTO/WO_3/nanocomposite electrolyte/$(CeO_2)_x$–$(TiO_2)_{1-x}$/FTO/glass in the bleached (*left*) and coloured (*right*) state (Courtesy from INM - T Traulsen, A Rueff, P Zapp, I Buder, M Mennig)

Fig. 12.8 Bent EC-devices of the configuration glass/TEC8/WO_3/nanocomposite electrolyte/CeO_2–TiO_2/TEC8/glass; size: 420 × 310 × 7 mm, max. *height* 18 mm); *left*: in the coloured state (−2, 1 V, 180 s), right in the bleached state (+1, 9 V, 180 s), from [111]

The long-term stability of these devices was proved up to 10,000 switching cycles. The UV stability, tested during 1,000 h under illumination in an Heraeus suntest or under a xenon arc lamp, (800 W/m^2) was excellent and no delamination or degradation at the interfaces of the electrolyte have been observed even after storing the devices at 80°C for 24 h in the coloured state and then 24 h in the bleached state. A reversible photochromic colouration of the WO_3 layer was however observed but it could be bleached by applying the bleaching voltage for 2 min and did not affect their switching behaviour.

Figure 12.7 shows a prototype of a large EC-window described above and realised at INM.

Bent electrochromic windows of the size 42 × 31 cm^2 with a bending radius of 200 cm have also been developed at INM [111] (Fig. 12.8). The bending of the TEC8 glass substrates was performed in a metallic mould at 600°C and the homogeneous deposition of the WO_3 EC-layer and the CeO_2–TiO_2 counter electrode was then done by dip-coating, an advantage of the Sol-Gel technique. The realisation of the device was done as described in Fig. 12.2. After heat treatment at 105°C for 12 h the device was sealed. The transmittance change was

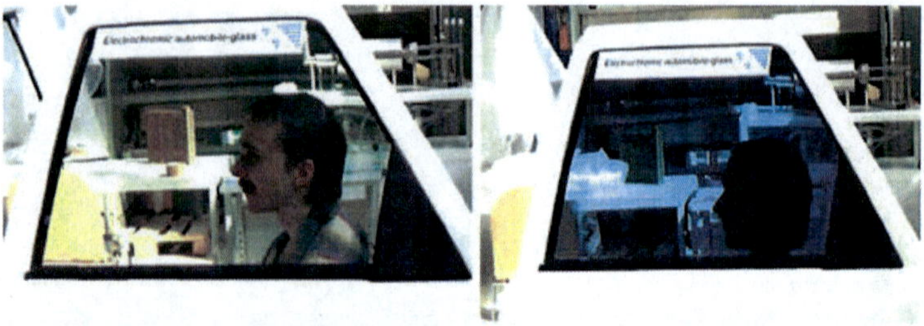

Fig. 12.9 Prototype of a bent EC- window for a car side window. *Left*: bleached state, *right*: coloured state, Courtesy from INM

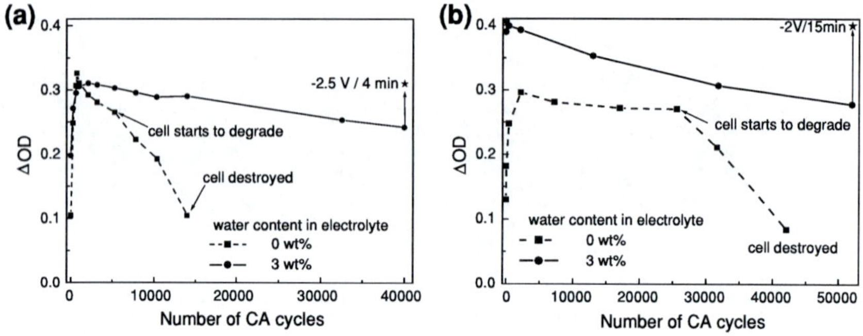

Fig. 12.10 Change of the optical density ΔOD at λ=550 nm of EC-devices with the configuration glass/FTO/EC-layer/composite electrolyte/$(CeO_2)_x(TiO_2)_{1-x}$/FTO/glass without and with addition of 3 wt% water in the electrolyte as a function of the CA cycle number, whereby the EC-layer is (**a**) Nb_2O_5:Mo (−2.5 V, 120 s/+1.5 V, 120 s) and (**b**) WO_3 (−2V, 120 s/+2 V, 120s) (from [38, 131])

homogeneous over the whole area from 67 to 25% with a switching time of 180 s. Such bent EC-devices could be used for automotive sunroofs and automotive glazing prototype (Fig. 12.9).

By adding 1–4 wt% of water to the nanocomposite electrolyte, the kinetics and the long-term switching behaviour of the above devices could be improved up to more than 50,000 switching cycles (Fig. 12.10b). However the memory effect of the devices made with a wet composite electrolyte was not as good but nevertheless still adequate [67, 68].

A similar improvement was also obtained with large area grey colouring EC-devices of the configuration K-glass/Nb_2O_5:Mo/nanocomposite electrolyte/$(CeO_2)_x$–$(TiO_2)_{1-x}$/K-glass (Fig. 12.10a) showing high transmittance change (60–20%) and long-term stability up to 40,000 switching cycles (−2.5, +1.5 V [38, 132]. The colouration efficiency of the Nb_2O_5:Mo Sol-Gel layers was 35 cm^2/C in EC-devices with dry electrolyte and 27 cm^2/C in EC-devices with wet electrolyte. A typical prototype colouring in grey is shown in Fig. 12.11.

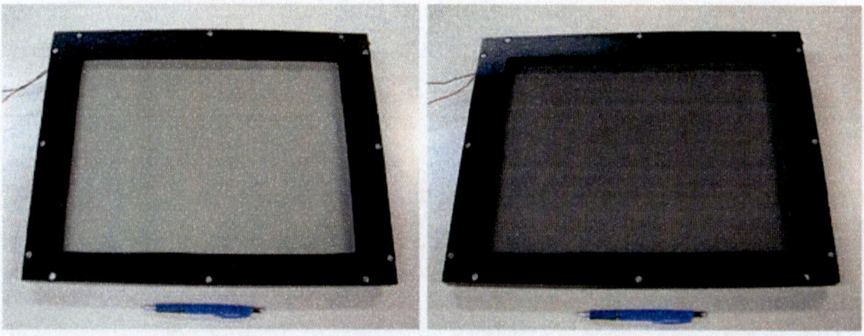

Fig. 12.11 30 × 40 cm^2 prototype of a grey colouring device with the configuration K-glass/Nb_2O_5:Mo/nanocomposite electrolyte/$(CeO_2)_x$–$(TiO_2)_{1-x}$/K-glass with a transmission change at 550 nm varying from 60% (*bleached state*) to 25% (*coloured state* [38, 131])

As stated in Sect. 12.3.3, cheap electrolytes based on starch, gelatine, gelatine plasticized with glycerol and agar have been developed and tested in EC-devices at INM. With the one based on a Li^+-conducting amylopectin-rich corn starch, a 7 cm^2 EC-device was built with WO_3 and CeO_2–TiO_2; it exhibited a small ΔOD of about 0.27 but a rather fast kinetics (~10 s) [117]. The Li^+-conducting gelatine was tested in a grey colouring EC-device involving Sol-Gel electrodes of Nb_2O_5:Mo and $(CeO_2)_{0.81}$–TiO_2 [118]; the change in transmission at 550 nm was also rather low (60↔40%) but it remains stable up to more than 25,000 cycles. A proton conducting gelatine plasticized with glycerol and containing acetic acid was also developed successfully and used in a 7 × 2 cm^2 EC-device involving WO_3 and CeO_2–TiO_2 Sol-Gel electrodes [119]. If the kinetics was fast (~10 s for the colouring and 2 s for the bleaching), the transmission variation was rather poor, typically 15% during more than 10,000 cycles. In the fourth approach using a Li^+-conducting agar-based electrolyte, EC-devices have been mounted with WO_3 and CeO_2–TiO_2 Sol-Gel electrodes; the charge which could be exchanged reversibly was however rather low, ranging from about 5 to 7.5 mC/cm^2, so that the transmittance variation was also rather low, typically 24% at 550 nm [120], a value however stable up to 5,000 cycles.

"All Sol-Gel" EC-devices of the configuration glass/WO_3 or Nb_2O_5)/ormolyte/counter electrode/glass have also been developed in Orel's group using different counter electrodes such as SnO_2(Mo,Sb) and LiCo-oxide, Nb/Fe-oxide, $CeVO_4$, Ce/V, V/Ti/Ce and V/Ti-oxide [74, 112, 132, 133]. A comparative study of "all Sol-Gel" EC-windows made with 24 different counter electrodes is given in [73]. The stability of WO_3/ormolyte/SnO_2(Mo,Sb) devices was tested up to only 500 cycles, whereby with high switching voltages (−4/+2.5 V) the ormolyte cracked and bubbles were observed in the final stage of degradation. With lower switching voltages or applying a small constant current (0.1 mA/cm^2) the ormolyte degradation could be avoided. The transmittance change, switching voltage and cycling stability of the different EC-devices are given in Table 12.2. EC-devices with Nb/Fe-oxide counter electrode were cycled up to 2,000 cycles, showing a transmittance change at 550 nm from 62 to 26% by switching between −2.6 and 1.6 V.

Applying more negative potentials also lead to a reduction of the cycling stability. Another problem is the low transmittance of the device below 500 nm, due to the low transmittance of the Nb/Fe counter electrode. Higher transmittance in the bleached state (72% at 550 nm) was obtained with Ce/V-oxide as counter electrode, however the transmittance of the coloured state is also higher (33%). With Sol-Gel V/Ti counter electrodes the transmittance in the bleached state at 550 nm is rather low (63%). In addition there is a weak absorption of V/Ti-oxide films below 550 nm, which changes the colour of the film from yellow to greenish-brown after charging.

Sol-Gel EC-devices with a WO_3 film deposited on FTO as electrochromic electrode, an $(H^+)Li^+$ ormolyte and a Sol-Gel counter electrode consisting of either $Li_{0.99}Co_{1.01}O_2$ or $Li_{0.94}Ni_{1.06}O_2$ deposited on FTO glass have been also reported [134] with dark blue colouration (T_{550} ~ 3%) after galvanostatic colouration. The bleached state had a higher transmittance for EC-devices with $Li_{0.94}Ni_{1.06}O_2$ (T_{550} ~68%) than with $Li_{0.99}Co_{1.01}O_2$ (T_{550} ~ 46%) as counter electrode. The devices showed good reversibility and stability after performing about 1,000 galvanostatic cycles. The disadvantage of $Li_{0.99}Co_{1.01}O_2$ films is the slight yellow colouring of the EC-devices in the bleached state.

$5 \times 10\ cm^2$ EC-devices comprising of WO_3 and CeO_2–TiO_2 Sol-Gel and a SPHP layers were successfully tested up to 60,000 chronoamperometry cycles with an almost constant contrast [115]. A four-layer prototype EC-device incorporating hybrid cross-linked materials based on poly(ethylene oxide (PEO)/siloxane (di-ureasil) with a wide range of concentration of lithium tetrafluoroborate ($LiBF_4$) was recently claimed to have encouraging colouring/bleaching behaviour [116].

An hybrid EC-device with a new Sol-Gel redox I_3^-/I^- electrolyte was tested up to 1,800 cycles and exhibiting good stability, high transmittance change from T_{vis} = 66 to 69% (bleached state) to 27–31% (coloured state) after 10 s (colouration) and 5 s (bleaching) switching time [114]. The Sol-Gel EC-device with the configuration glass/FTO/WO_3/α-PEO/NiO_xH_y/FTO/glass with a lithiated oxymethylene-linked poly(ethylene oxide) (α-PEO) ion conductor presented a short cycling lifetime and a small transmittance change (80–50% at 550 nm) but a fast optical response (50 s) [134]. Bell et al. [33, 34] constructed EC-devices using multiple dip-coating Sol-Gel titanium- doped WO_3 films as electrochromic layer and titanium-doped V_2O_5 films as counter electrode layer, both deposited on FTO- or ITO-coated glass and a polyether polyurethane copolymer containing a lithium salt and a cross-linking agent as electrolyte. The optical transmittance at 550 nm changed from a low value (44%) in the bleached state, due to the low transmittance of the V_2O_5:Ti layer, to 20% in the coloured state. The contrast ratio was found to degrade during galvanostatic switching after 1,400 cycles. EC-windows based on viologen-modified nanostructured TiO_2 films were first reported by using the redox chromophore $[N,N'\text{-(bis-2-phosphonoethyl)-4,4'-bipyridine}]^{2+}$ (Chrom 1), an electrolyte of $LiClO_4$ and ferrocene in γ-butyrolactone and a conducting glass as counter electrode [90, 91]. The device ($2.5 \times 2.5\ cm^2$) showed a transmittance change of 55% at 608 nm (65–10%), a colouration efficiency of

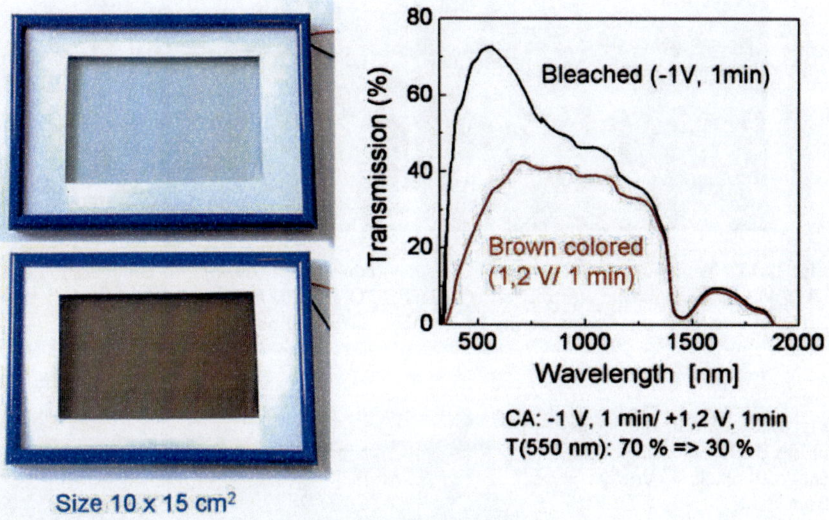

Fig. 12.12 $10 \times 15\ cm^2$ prototype brown coloured EC-window with the configuration glass/FTO/$NiO–TiO_2$/electrolyte (KOH)/$CeO_2–TiO_2$/FTO/glass in the *bleached* and *coloured state* (*left*). Transmission spectra of both states (*right*). The device was stable for about 7,000 cycles [57, 135]

170 cm^2C^{-1} (at 608 nm), a switching time of 1 s and a stability of more than 10,000 standard test cycles. The disadvantage of this system is that an electric current must be continuously supplied to maintain a given degree of colouration (Table 12.2).

Several devices have been prepared with Sol-Gel NiO-TiO_2 EC-layers with the configuration glass/FTO/$NiO–TiO_2$/electrolyte/$CeO_2–TiO_2$ or Nb_2O_5:Mo/FTO/glass [57, 135]; the electrolyte was 1 mm thick aqueous 1M KOH injected in the laminated devices. The device using the $CeO_2–TiO_2$ counter electrode coloured in brown has been tested up to 7,000 cycles; both the transmission at 550 nm in the bleached and coloured states slightly decreased with cycling from 72 to 60% and from 50 to 30%, respectively (Fig. 12.12). A device using a grey colouring cathodic Nb_2O_5:Mo counter electrode instead of $CeO_2–TiO_2$ coloured grey-brown and was tested during 5,000 cycles. The transmission in the bleached state decreased continuously from 60 down to 35% while that of the coloured state remained practically constant (20%).

12.4.3 EC-Displays

The feasibility of transparent EC-displays exhibiting high memory as well as paper-like displays working in reflective mode has been tested only by a few groups [93–95, 136]. At INM transparent displays with size up to $50 \times 80\ cm^2$

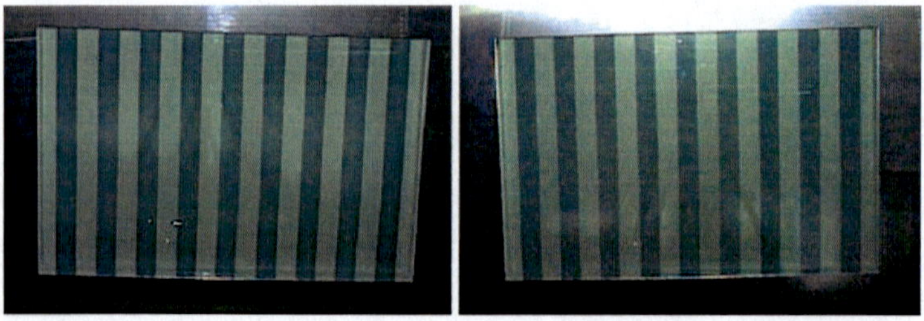

Fig. 12.13 Transparent 50 × 80 cm^2 EC-display (*patterned rows*) with the configuration glass/FTO/WO_3/nanocomposite electrolyte/CeO_2–TiO_2/FTO/glass. *Left*: after colouration of the *rows*, *Right*: after storage the display during 24 h without electrical power supply [109]

Fig. 12.14 Similar device as in Fig. 12.13 but with patterned pixels. Courtesy from INM

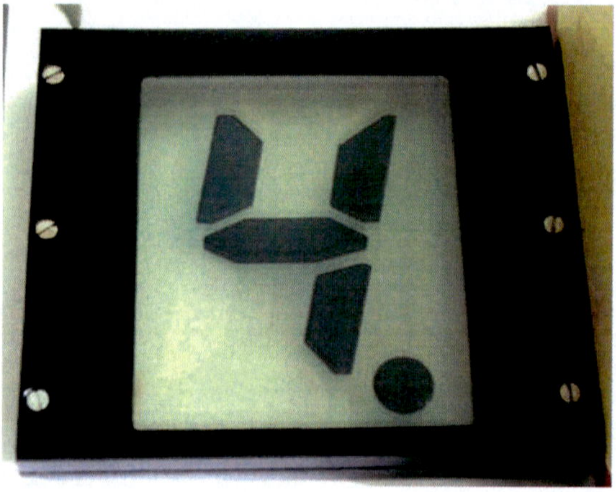

and 13 × 19 pixels showing a 40 s switching time and an excellent memory effect (colouration maintained >24 h without applying a voltage) using the configuration glass/FTO/WO_3/nanocomposite electrolyte/CeO_2–TiO_2/FTO/glass already described in Sect. 12.4.2 [109, 110]. The energy consumption for switching (40 s) and keeping of the information during 1 h was estimated to 0.03 Wh/m^2. The colouring electrode was patterned by laser ablation. Figure 12.13 and 12.14 show such a display obtained by patterning the EC- layer in the form of stripes or in form of pixels. When both the EC and the IS layers are patterned in form of stripes and mounting perpendicular to each other, pixels can be coloured independently and information can be programmed and changed at wish easily (e.g. prices for products, logos, etc.).

Ultrafast systems have been also reported: they consisted of a transparent nanostructured TiO_2 (4.0 μm thick) modified with a monolayer of Chrom 1 [90], a transparent nanostructured SnO_2 film (3.0 μm thick) modified by chemisorption of a monolayer of the redox chromophore [β-(10-phenothiazyl)propoxy]phosphonic

Fig. 12.15 Displays from NTERA, with permission from NTERA, [137]

acid (Chrom 2) and a 1 M $LiClOI_4$ in γ-butyrolactone electrolyte. The device had very fast colouration and bleaching (0.5 and 0.3 s, respectively over 10,000 switching cycles) and a typical transmittance change of 64/23%. By integrating a white reflector made of rutile titania or zinc oxide pigment particles between the electrodes of the above-described EC-window high quality paper-like displays have been realised [93, 94]. The oxide pigment particles provided a highly reflective background for the blue colouring electrode. The device diffuse reflectance at 600 nm was typically 45–50% in the bleached state and less than 5% in the coloured state giving a contrast ratio of 10–20 with a low applied voltage −1.3 V. Such system was trademarked under the name NanoChromics TM. The fabrication of paper-quality EC-displays based on the viologen-modified TiO_2 electrodes for the NanoChromics TM display was recently considerably improved using a microwave synthesis, a viable industrial process for the preparation of mesoporous nanocrystalline titania with high surface area up to 240 $m^2\ g^{-1}$ [136]. Such materials was used as the working electrode for an EC-display device with a SnO_2:Sb counter electrode deposited on an ITO-coated conducting glass. According to the authors, the working prototype was successfully fabricated. Apparently such materials can be printed on glass and plastic substrates [137] (Fig. 12.15).

By using different organic redox-chromophores (viologens) adsorbed on mesoporous anatase TiO_2 "paper-quality" EC-displays with different colours as blue, green, red and black have been obtained [94]. The device ($5 \times 5\ cm^2$) presented seven segmented cathodes, obtained by laser etching the FTO layer on which the mesoporous TiO_2 films (3 μm thick), and deposited by screen printing; the viologen compounds were attached by chemisorption from aqueous solution.

A related architecture device [138] described as an "electric-paint display" with carbon counter electrodes made with a nanoporous TiO_2 electrode modified with Chrom 1 [90] and a counter electrode consisting of two porous layers deposited on conductive glass: a 40 μm thick bottom layer of sintered carbon particles (graphite and carbon black) coated with a 20 μm thick white reflector made of large light-scattering TiO_2 pigment particles. Solutions of either lithium triflate $Li(F_3CSO_3)$ or tetrabutylammonium (TBA) triflate $((C_4H_9)_4N(F_3CSO_3))$ in MPN (3-methoxypropionitrile) were used as liquid electrolyte. The reflectance at 550 nm changes from 52% in the bleached state to 12% in the coloured state in less than 1 s. The bleached state is permanent in open circuit whereas the measured open-circuit memory of the coloured state ranges from about 15 min to >1 h.

12.4.4 Gasochromic Devices

It is also worthwhile to mention gasochromic devices. They are based on the optical modulation of WO_3 films obtained by the introduction of atomic hydrogen (H°) evolved either from a HCl/Zn mixture or using Pd spillover and hydrogen gas (H_2). The effect is obtained by exposing WO_3 films to hydrogen gas (H_2) where at the atomic hydrogen (H°) is provided by a thin layer of catalyst (Pt or Pd) in contact with the WO_3 film. The switching device is simple and inexpensive because it involves only a single WO_3 film covered with a catalyst, sufficient to obtain colouring/bleaching changes. The original transmittance state of the films can be recovered by flushing the device with air. Such devices have been made by the Sol-Gel process using WO_3 layers prepared by the ion exchange or peroxo-polytungstic acid method and a review is given by Georg et al. in [139].

12.4.5 Outlook

Many coatings and materials presenting functionalities required to build electrochromic devices, being either electrochromic, electronic or ionic conductors have been developed using the Sol-Gel process. However, when tested even in a small device requiring a rather long-term and stable use, most of them remain just for a scientific interest. Today there are only three promising electrochromic materials based on pure or doped WO_3 and Nb_2O_5 and mixed $NiO–TiO_2$. Unfortunately no Sol-Gel materials can be proposed today for the transparent electronic conducting layers as their sheet resistance is too high or the number of coatings that should be deposited and sintered at high temperature is too high. Concerning the ion storage layer, the mixed material $CeO_2–TiO_2$ appears very promising as a cathodic layer in anodic colouring EC-devices. Recent developments have also shown that such materials can be deposited as single layers, a great advantage for industrial application. Several hybrid materials containing H^+ or Li^+ ions that can be prepared in a fluid form but that can be polymerised by a mild temperature treatment or UV irradiation have been shown to have adequate ionic transparent conductors; however such a material is yet to be found for EC-devices involving promising $NiO–TiO_2$ electrochromic brown coloured layers in order to improve their long cycle behaviour.

An advantageous, non-Sol-Gel electrolyte system was developed by Gesimat in Berlin [140]: an ion-conducting polymer foil was developed as electrolyte using polyvinyl butyral (PVB). The ion-conducting PVB foil acts not only as electrolyte but shows also safety glass properties and is used as electrolyte foil in EC-windows fabricated with electrodeposited EC-layers [140]. Such a PVB electrolyte foil could in principle also be used as electrolyte for EC-layers fabricated by Sol-Gel technique.

Concerning the devices themselves, the Sol-Gel coating technology (basically the dip-coating process) was proved to be adequate to coat quite large substrates. Nevertheless, printing techniques, recently developed successfully for other Sol-Gel applications, can be an asset in the future especially to obtain patterned EC-layers (pixels) for the development of EC-displays. The drastic requirements necessary for industrial application have been only partly tested by very few developers. Nevertheless, several device configurations have shown that they work practically with quite small degradation up to more than 60,000 electrochemical cycles at room temperature but no device has been yet tested extensively under large temperature variation and extended sun exposure. Till now, only prototypes have been built and to our knowledge no industrial developments have been done for Sol-Gel made devices. In contrast to that, using coatings deposited by physical processes (PVD and electrodeposition) several prototypes have been developed in pilot plants (e.g. Gesimat, Berlin) and products with PVD fabricated EC-layers are currently in the market (e.g. EControl® glass by EControl-Glas GmbH & Co. KG, Germany [141], INFRASELECT® glass by Flachglas MarkenKreis GmbH, Germany [142], Velux-skylight windows by Sage Electrochromics, USA). Probably the most promising application using Sol-Gel made coatings lies in the development of displays (rigid or flexible) working in the reflection mode where the main coating can be made from Sol-Gel inorganic nanoparticles to which organic redox molecules can be attached. It is however doubtful that such material can be used for windows.

The future for EC-devices made using the Sol-Gel process is therefore still open but real improvements are still needed if one wishes to propose commercial applications. Although the Sol-Gel technology is rather simple and not expensive at a laboratory scale (as e.g. in university laboratories), industrial investments are important and will only be done if devices developed in pilot plans and fulfilling all severe tests on the long-term durability under several conditions can be proposed.

References

1. Granqvist CG (1995) Handbook of inorganic electrochromic materials. Elsevier, Amsterdam
2. Monk PMS, Mortimer RJ, Rosseinsky DR (1995) Electrochromism-fundamentals and applications. VCH Verlagsgesellschaft mbh, Weinheim
3. Lampert CM (1999) The world of large-area glazing and displays. In: Proceedings of SPIE, Switchable Materials and Flat Panel Displays, Denver, Colorado, July 21–22. SPIE, vol 3788. Bellingham, Washington, USA, pp 2–11.

4. Lampert CM (2003) Large-area smart glass and integrated photovoltaics. Sol Energy Mater Sol Cells 76:489–499
5. Granqvist CM (2007) Transparent conductors as solar energy materials: a panoramic review. Sol Energy Mater Sol Cells 91:1529–1598
6. Granqvist CG (2001) Electrochromic windows: toward an energy efficient architecture. Interface 3:18–19
7. Azens A, Granqvist CG (2003) Electrochromic smart windows: energy efficiency and device aspects. J Solid State Electrochem 7:64–68
8. Aegerter MA (1996) Sol-Gel chromogenic materials and devices. In: Reisfeld R, Jorgensen CK (eds) Structure and Bonding, vol 85. Springer, Berlin, pp 149–194
9. Heusing S, Aegerter MA (2005) Sol-Gel coatings for electrochromic devices. In: Sakka S (ed), Handbook of Sol-Gel Science and Technology, vol 3. Kluwer Academic Publishers, The Netherlands, pp 719–760
10. Heusing S, Aegerter MA (2006) Stand der Anwendung der Elektrochromie in der Architektur, Proc. des 6. Symposiums Zukunft Glas—Von der Tradition zum High-Tech-Produkt. Otti, Zwiesel, p 72
11. Rauh RD (1999) Electrochromic windows: an overview. Electrochim Acta 44:3165–3176
12. ÓBrian NA, Gordon J, Mathew H, Hichwa BP (1999) Electrochromic coatings-applications and manufacturing issues. Thin Solid Films 345:312–318
13. Yoshiaki I, Osamu N, Hideyuki K (1998) All-solid electrochromic anti-glare mirror. Murakami Kaimeido Co, US Patent 6, 06, 1168
14. Dornan CA, Habibi H, Lynam NR, McCabe IA (1994) Electrochromic mirrors and devices. Donnelly corporation, WO Patent 95 30 495
15. Bauer FT, Bechtel JH (1984) Automatic rearview mirror for automotive vehicles. Gentex Corporation, US Patent 4, 443, 057
16. Bechtel JH, Byker HJ (1990) Automatic rearview mirror system for automotive vehicles. Gentex Corporation, US Patent 4, 917, 477
17. Byker HJ (1992) Variable reflectance motor vehicle mirror. Gentex Corporation, US Patent 5, 128, 799
18. Byker HJ (1990) Single-compartment, self-erasing, solution-phase electrochromic devices, solutions for use therein, and uses thereof. Gentex Corporation, US Patent 4, 902, 108
19. Puetz J, Aegerter MA (2004) Transparent conducting oxide coatings in Sol-Gel technologies for glass producers and users. In: Aegerter MA, Mennig M (ed) Sol-Gel technologies for glass producers and users, Kluwer, The Netherlands
20. Bard AJ, Faulkner LR (2000) Electrochemical methods: fundamentals and applications. Wiley, New York
21. Agrawal A, Cronin JP, Zhang R (1993) Review of solid state electrochromic coatings using Sol-Gel techniques. Sol Energy Mater Sol Cells 31:9–21
22. Vroon ZAEP, Spee CIMA (1997) Sol-Gel coatings on large area glass sheets for electrochromic devices. J Non-Cryst Solids 218:189–195
23. Bessière A, Badot JC, Certiat MC, Livage J, Lucas V, Baffier N (2001) Sol-Gel deposition of electrochromic WO_3 thin film on flexible ITO/PET substrate. Electrochim Acta 46: 2251–2256
24. Kim C-Y, Lee M, Huh S-H, Kim E-K (2010) J Sol-Gel Sci Technol 53:176–183
25. Cronin JP, Tarico DJ, Agrawal A, Zhang RL (1993) Method for depositing electrochromic layers, US Patent 5, 252, 354
26. Cronin JP, Tarico DJ, Tonazzi JCC, Agrawal A, Kennedy SR (1993) Microstrucure and propewrties of Sol-Gel deposited WO_3 coatings for large area electrochromic windows. Sol Energy Mater Sol Cells 29:371–386
27. Cronin JP, Tarico DJ, Agrawal A, Zhang RL (1994) Method for depositing high performing electrochromic layers, United States Patent 5, 277, 986
28. Schmidt H, Krug H, Merl N, Moses A, Judeinstein P, Berni A (1994) Electrochromic thin-film systems and components thereof. Patent WO 95/28663

29. Munro B, Krämer S, Zapp P, Krug H (1998) Characterization of electrochromic WO_3-layers prepared by Sol-Gel nanotechnology. J Sol-Gel Sci Technol 13:673–678
30. Munro B, Conrad P, Krämer S, Schmidt H, Zapp P (1998) Development of electrochromic cells by the Sol-Gel process. Sol Energy Mater Sol Cells 54:131–137
31. Heusing S, Munro B, Koch T, Zapp P, Mennig M, Schmidt H (1999) Weiterentwicklung elektrochromer Dünnschichtsysteme auf Glas über naßchemische Verfahren. In: Proceedings of the 73th Glastechnische Tagung, Halle (Saale), Germany, pp 40–43
32. Bell JM, Matthews JP, Skryabin IL, Wang J, Monsma BG (1998) Sol-Gel deposited electrochromic devices. Renew Energy 15:312–317
33. Bell JM, Skryabin IL, Koplick AJ (2001) Large area electrochromic films—preparation and performance. Sol Energy Mater Sol Cells 68:239–247
34. Lefheriotis G, Papaefthimiou S, Yianoulis P (2004) Sol Energy Mater Sol Cells 83:115–124
35. Schmitt M, Heusing S, Aegerter MA, Pawlicka C, Avellaneda CO (1998) Electrochromic properties of Nb_2O_5 Sol-Gel coatings. Sol Energy Mater Sol Cells 54:9–17
36. Schmitt M, Aegerter MA (1999) Electrochromic properties of Nb_2O_5 and Nb_2O_5 :X (X=Sn, Zr, Li, Ti, Mo). In: Proceedings of the SPIE conference on switchable materials and flat panel displays, Denver, Colorado, July 1999. SPIE, vol 3788. pp 93–102
37. Schmitt M, Aegerter MA (2001) Electrochromic properties of pure and doped Nb_2O_5 coatings and devices. Electrochim Acta 46:2105–2111
38. Sun DL, Heusing S, Puetz J, Aegerter MA (2003) Influence of water on the electrochromic properties of Nb_2O_5:Mo, WO_3 and $(CeO_2)_x(TiO_2)_{1-x}$ Sol-Gel coatings and electrochromic devices. Solid State Ionics 165:181–189
39. Schmitt M, Aegerter MA (1999) Properties of electrochromic devices made with Nb_2O_5 and Nb_2O_5:X (X=Li, Ti or Mo) as coloring electrode. In: Procedings of the SPIE conference on switchable materials and flat panel displays. Denver, Colorado, July 1999. SPIE, vol 3788, pp 75–83
40. Dhanasankar M, Purishothaman KK, Muralidharan G (2010) Effect of tungsten on the electrochromic behavior of Sol-Gel dip coated molybdenum oxide thin films. Mater Res Bulletin 45:542–545
41. Li Y, Kudo T (1995) Electrochromic properties of spin-coated thin films from peroxo-polymolybdovanadate solutions. J Electrochem Soc 142:1194–1199
42. Dhanasankar M, Purishothaman KK, Muralidharan G (2010) Enhanced electrochromism in cerium doped molybdenum oxide thin films. Mater Res Bull 45:1969–1972
43. Azens A, Kullmann L, Vaivars G, Nordborg H, Granqvist CG (1998) Sputter-deposited nickel oxide for electrochromic applications. Solid State Ionics 113–115:449–456
44. Svensson JSEM, Granqvist CG (1986) Electrochromic hydrated nickel-oxide coatings for energy-efficient windows—optical-properties and coloration mechanism. Appl Phys Lett 49:1566–1568
45. Moser FH, Lynam NR (1990) US Patent 4, 959, 247
46. Miki T, Yoshimura K, Tai Y, Tazawa M, Jin P, Tanemura S (1995) Electrochromic properties of nickel oxide thin films prepared by the Sol-Gel method. Proc SPIE 2531:135–142
47. Šurca A, Orel B (1997) Sol-Gel derived hydrated nickel oxide electrochromic films: optical, spectroelectrochemical and structural properties. J Sol-Gel Sci Technol 8:743–749
48. Cerc Korošec R, Bukovec P, Pihlar B, Padežnik Gomilšek J (2003) The role of thermal analysis in optimization of the electrochromic effect of nickel oxide thin films, prepared by the Sol-Gel method: part I. Thermochimica Acta 402:57–67
49. Cerc Korošec R, Bukovec P (2004) The role of thermal analysis in optimization of the electrochromic effect of nickel oxide thin films, prepared by the Sol-Gel method: part II. Thermochimica Acta 410:65–71
50. Cerc Korošec R, Bukovec P (2006) Sol-Gel prepared NiO films for electrochromic application. Acta Chim Slov 53:137–147
51. Sharma PK, Fantini MCA, Gorenstein A (1998) Synthesis characterization and electrochromic properties of NiOxHy thin film prepared by a Sol-Gel method. Solid State Ionics 113–115: 457–463

52. Sharma PK, Mracia MCA, Fischer H, Craievich AF, Gorenstein A (1999) Factors influencing the electrochromic properties of nickel oxide thin films derived from Sol-Gel methode by dip-coating. Mat Res Soc Symp Proc 547:351–356
53. Moser FH, Lyman NR (1989) US Patent 4855166 and US Patent 4855161
54. Martini M, Brito GES, Fantini MCA, Craievich AF, Gorenstein A (2001) Electrochromic propertie of NiO-based thin films prepared by Sol-Gel and dip-coating. Electrochim Acta 48:2275–2279
55. Al-Kalhout A, Heusing S, Aegerter MA (2006) Electrochromism of NiO–TiO_2 Sol-Gel layers. J Sol-Gel Sci Technol 39:195–206
56. Al-Kalhout A, Aegerter MA (2007) Coloration mechanisms of Sol-Gel NiO–TiO_2 layers studied by EQCM. Sol Energy Mater Sol Cells 91:213–223
57. Al-Kalhout A, Pawlicka A, Aegerter MA (2006) Brown coloring electrochromic devices based on NiO-TiO_2 layers. Sol Energy Mater Sol Cells 90:3583–3601
58. Švegl F, Orel B, Kaučič V (2000) Electrochromic properties of lithiated Co-oxide (Li_xCoO_2) and Ni-oxide (Li_xNiO_2) thin films prepared by the Sol-Gel route. Sol Energy 68:523–540
59. Baudry P, Rodrigues ACM, Aegerter MA, Bulhoes LO (1990) Dip-coated TiO_2–CeO_2 films as transparent counter electrode for transmissive electrochromic devices. J Non-Cryst Solids 121:319–322
60. Štangar UL, Orel B, Grabec I, Ogorevc B, Kalcher K (1993) Optical and electrochemical properties of CeO_2 and CeO_2–TiO_2 coatings. Sol Energy Mater Sol Cells 31:171–185
61. Orel Z, Orel B (1994) Electrochemical and optical properties of Sol-Gel derived CeO_2 and mixed CeO_2/SnO_2 coatings. In: Proceedings of SPIE optical materials technology for energy efficiency and solar energy conversion XIII, Bellingham, Washington, USA. SPIE 2255:285–296
62. Pawlicka A, Avellaneda CO (2000) Thin film Sol-Gel of CeO_2–ZrO_2: the candidate for counter electrode in electrochromic devices. Mol Cryst Liq Cryst 354:1051–1061
63. Berton MAC, Avellaneda CO, Bulhoes LOS (2003) Thin film of CeO_2–SiO_2: a new ion-storage layer for smart windows. Sol Energy Mater Sol Cells 80:443–449
64. Opara Krašovec U, Orel B, Reisfeld R (1998) Electrochromism of $CeVO_4$ and Ce/V-oxide ion-storage films prepared by the Sol-Gel route. Electrochem Solid-State Lett 1:104–106
65. Avellaneda CO, Pawlicka A (1998) Preparation of transparent CeO_2–TiO_2 coatings for electrochromic devices. Thin Solid Films 335:245–248
66. Kim C-Y, Cho S-G, Lim T-Y, Choi D-K (2009) Anomalous lithium diffusion into CeO_2–TiO_2 thin film by film thickness. J Solid Sate Electrochem 13:1165–1170
67. Sun D-L, Puetz J, Heusing S, Aegerter MA (2002) Influence of water on the electrochemical properties of CeO_2–TiO_2 Sol-Gel coatings and electrochromic devices. Proc SPIE Sol-Gel Opt VI 4804:17–25
68. Sun D, Heusing S, Aegerter MA (2007) Li^+ion exchange in CeO_2–TiO_2 Sol-Gel layers studied by electrochemical quartz crystal microbalance. Sol Energy Mater Sol Cells 91:1037–1050
69. Verma A, Samanta SB, Bakhshi AK, Agnihotry SA (2004) Optimization of CeO_2–TiO_2 compositions for fast switching kinetics and improved Li ion storage capacity. Solid State Ionics 171:81–90
70. Verma A, Goyal A, Sharma RK (2008) Microstructural, photocatalysis and electrochelical investigations on $CeTi_2O_6$ thin films. Thin Solid Films 516:4925–4933
71. Verma A, Bakhshi AK, Agnihotry SA (2006) Effect of different precursor sols on the properties of CeO_2–TiO_2 films for electrochromic window application. Electrochim Acta 51:4639–4648
72. Berton MAC, Avellaneda CO (2001) Electrochemical properties of CeO_2–SnO_2 and CeO_2–SnO_2:X (X = Li, C, Si) films. Mater Res 4:241–244
73. Opara Krašovec U, Šurca Vuk A, Orel B (2002) Comparative studies of "all Sol-Gel" electrochromic windows employing various counter electrodes. Sol Energy Mater Sol Cells 73:21–37

74. Šurca A, Orel B, Opara Krašovec U, Lavrenčič Štangar U (2000) Electrochromic and structural studies of nanocrystalline Fe/V (1:2)-oxide and crystalline $Fe_2V_4O_{13}$ films. J Electrochem Soc 147:2358
75. Opara Krašovec U, Orel B, Hočevar S, Muševič I (1997) Electrochemical and spectroelectrochemical properties of SnO_2 and SnO_2/Mo transparent electrodes with high ion-storage capacity. J Electrochem Soc 144:3398–3409
76. Yang Y, Zhu Q, Jin A, Chen W (2008) High capacity and contrast of electrochromic tungsten-doped vanadium oxide films. Solid State Ionics 179:1250–1255
77. Jin A, Chen W, Zhu Q (2009) High Li^+-ion storage capacity and multi-electrochromism behaviour of electrodeposited molybdenum doped vanadium oxide films. Advan Mater Res 79–82:799–802
78. Kim S, Taya M, Xu C (2009) Contrast, switching speed and durability of V_2O_5–TiO_2 film-based electrochromic windows. J Electrochem Soc 156:E40–E45
79. Bell JM, Skryabin IL (1999) Failure modes of Sol-Gel deposited electrochromic devices. Sol Energy Mater Sol Cells 56:437–448
80. Coleman JP, Lynch AT, Madhukar P, Wagenknecht JH (1999) Antimony-doped tin oxide powders: electrochromic materials for printed displays. Sol Energy Mater Sol Cells 156:375–394
81. Doeuff S, Sanchez C (1989) Electrochromic properties of anatase TiO_2 films prepared by the Sol-Gel process. CR Acad Sci II 309:531–534
82. Özer N (1992) Reproducibility of the coloration processes in TiO_2 films. Thin Solid Films 214:17–24
83. Verma A, Basu A, Bakhshi AK, Agnihotry SA (2005) Structural, optical and electrochemical properties of Sol-Gel derived TiO_2 films: annealing effects. Solid State Ionics 176:2285–2295
84. Verma A, Kar M, Agnihotry SA (2007) Aging effect of diethanolamine stabilized sol on different properties of TiO_2 films. Electrochromic appl Sol Energy Mater Solar Cells 91:1305–1312
85. Zelazowska E, Rysiakiewics-Pasek E (2009) Thin TiO_2 films for an electrochromic system. Opt Mater 31:1802–1804
86. O'Regan B, Grätzel M (1991) A low-cost, high-efficiency solar cell based on dye-sensitized colloidal TiO_2 films. Nature 353:737–740
87. Hagfeld A, Vlachopoulos N, Gilbert S, Grätzel M (1994) Electrochromic switching with nanocrystalline TiO_2 semiconductor films. In: Proceedings of SPIE, optical materials technology for energy efficiency and solar energy conversion XIII, SPIE, vol 2255. Bellingham, Washington, USA pp 297–303
88. Marguerettaz X, ÓNeill R, Fitzmaurice DJ (1994) Heterodyads—electron-transfer at a semiconductor electrode liquid electrolyte interface modified by an adsorbed spacer acceptor complex. J Am Chem Soc 116:2629–2630
89. Hagfeld A, Vlachopoulos N, Grätzel M (1994) Fast electrochromic switching with nanocrystalline oxide semiconductor films. J Electrochem Soc 141:L82–L84
90. Cinnsealach R, Boschloo G, Rao SN, Fitzmaurice D (1998) Electrochromic windows based on viologen-modified nanostructured TiO_2 films. Sol Energy Mater Sol Cells 55: 215–233
91. Fitzmaurice D, Rao SN, Cinnsealach R, Enright B (1998) Eur Pat Applications 98/9032735
92. Cummins D, Boschloo G, Ryan M, Corr D, Rao SN, Fitzmaurice D (2000) Ultrafast electrochromic windows based on redox-chromophore modified nanstructured semiconducting and conducting films. J Phys Chem B 104:11449–11459
93. Bach U, Corr D, Lupo D, Pichot F, Ryan M (2002) Nanomaterials-based electrochromics for paper-quality displays. Adv Mater 14:845–848
94. Corr D, Bach U, Fay D, Kinsella M, McAtamney C, ÓReilly F, Rao SN, Stobie N (2003) Coloured electrochromic "paper-quality" displays based on modified mesoporous electrodes. Solid State Ionics 165:315–321

95. Xiong S, Phua SL, Dunn BS, Ma J, Lu X (2010) Covalently bonded polyaniline–TiO_2 hybrids: a facile approach to highly stable anodic electrochromic materials with low oxidation potentials. Chem Mater 22:255–260
96. Hwang T, Lee H, Kim H, Kim G, Mun G (2010) Enhancement of electrochemical durability of a film made of silica-polyaniline core-shell nanoparticles. Surf Review Lett 17:39–44
97. Shiping L, Lin X, Bingbing X (2009) Electrochromism of polyoxometalates. Prog Chem 21:1458–1464
98. Ghodsi FE, Tepehan FZ, Tepehan GG (2008) Electrochromic properties of heat-treatment thin fims of CeO_2–TiO_2–ZrO_2 prepared by Sol-Gel route. Sol Energy Mater Sol Cells 92:234–239
99. Akhavan D, Tohidi H, Moshlegh AZ (2009) Synthesis and electrochromic study of Sol-Gel cuprous oxide nanoparticles accumulated on silica thin films. Thin Solid Films 517: 6700–6706
100. Granqvist CG (1993) Electrochromics and smart windows. Solid State Ionics 60:213–214
101. Vaivars G, Furlani M, Mellander B-E, Granqvist CG (2003) Proton-conducting zirconium phosphate/ poly(vinyl acetate)/glycerine gel electrolytes. J Solid State Electrochem 7:724–728
102. Baetens R, Jelle BP, Gustavsen A (2010) Properties, requirements and possibilities of smart windows for dynamic daylight and solar energy control in buildings: a state- of- the- art review. Sol Energy Mater Sol Cells 94:87–105
103. Özer N, He Y, Lampert CM (1994) Ionic conductivity of tantalum oxide films prepared by the Sol-Gel process for electrochromic devices. In: Proceedings of SPIE optical materials technology for energy efficiency and solar energy conversion XIII 2255:456–466
104. Özer N, Lampert CM (1997) Structural and optical properties of Sol-Gel deposited proton conducting Ta_2O_5 films. J Sol-Gel Sci Technol 8:703–709
105. Hirano S, Yogo T, Sakamoto W, Takeichi Y, Ono S (2004) Processing of highly oriented $LiNbO_3$ thin films through a metal-organic precursor solution. J Eur Ceramic Soc 24:435–440
106. Granqvist CG, Avendaño E, Azens A (2003) Electrochromic coatings and devices: survey of some recent advances. Thin Solid Films 442:201–211
107. Dahmouche K, Atik M, Mello NC, Bonagamba TJ, Panepucci H, Aegerter MA, Judeinstein P (1997) Investigation of new ion-conducting ORMOLYTES: structure and properties. J Sol-Gel Sci Technol 8:711–715
108. Judeinstein P, Titman J, Stamm M, Schmidt H (1994) Investigation of ion-conducting ormolytes: structure-property relationships. Chem Mater 6:127–134
109. Heusing S, Niegisch N, Zapp P, Mennig M, Schmidt H, Krings LHN, Aartsen HJ (2000) Zur Entwicklung eines großflächigen elektrochromen Displayfensters aus Glas. In: Proceedings 74th Glastechnische Tagung Ulm, Germany, pp 278–281
110. Mennig M, Fink-Straube C, Heusing S, Kalleder A, Koch T, Munro B, Zapp P, Schmidt H (1999) Large area decorative and functional Sol-Gel coatings on glass. Thin Solid Films 1:343–344
111. Mennig M, Heusing S, Zapp P, Niegisch N, Schmidt H (2000) "Fabrication of large area, curved electrochromic modules for automotive application". In: Proceedings 3rd International Conference on Coatings on Glass (ICCG), Maastricht, The Netherlands, p 787
112. Orel B, Opara Krašovec U, Lavrenčič Štangar U, Judeinstein P (1998) All Sol-Gel electrochromic devices with Li^+ ionic conductor, WO_3 electrochromic films and SnO_2 counter-electrode films. J Sol-Gel Sci Technol 11:87–104
113. Grošelj N, Gaberščėk M, Opara Krašovec U, Orel B, Dražič G, Judeinstein P (1999) Electrical and IR spectroscopic studies of peroxopolytungstic acid/organic-inorganic hybrid gels. Solid State Ionics 125:125–133
114. Orel B, Šurca Vuk A, Jese R, Lianos P, Stathatos E, Judeinstein P, Colomban Ph (2003) Development of Sol-Gel redox I_3^-/I^- electrolytes and their application in hybrid electrochromic device. Solid State Ionics 165:235–246
115. Souza FL, Aegerter MA, Leite ER (2007) Solid hybrid polyelectrolyte with high performance in electrochromic devices: electrochemical stability and optical study. Sol Energy Mater Sol Cells 91:1825–1830 (also Electrochimica Acta 53:1635–1642)

116. Barbosa P, Rodrigues L, Silva M, Smith M, Gonçalves A, Fortunato E (2010) Application of di-ureasil ormolytes based on lithium tetrafluoroborate in solid-state electrochromic displays. J Mater Chem 20:723–730 (See also Electrochimica Acta (2009) 54:1002–1009
117. Costa RGF, Avellaneda CO, Pawlicka A, Heusing S, Aegerter MA (2008) Optoelectrochemical characterization of electrochromic devices with starch based solid electrolytes. Molec Cryst Liq Cryst 447:363–371
118. Avellaneda CO, Vieira DF, Al-Kalhout A, Heusing S, Leite ER, Pawlicka A, Aegerter MA (2008) All solid-state electrochromic devices with gelatine-based electrolyte. Sol Energy Mater Sol Cells 92:228–233 (also Electrochimica Acta (2007) 53:1648–1654)
119. Al-Kalhout A, Vieira DF, Avellaneda CO, Leite ER, Aegerter MA, Pawlicka A (2010) Gelatin-based protonic electrolyte for electrochromic windows. Ionics 16:13–19
120. Raphael E, Avellaneda CO, Aegerter MA, Silva MM, Pawlicka A (2012) Agar-based gel electrolyte for electrochromic device application. Mol Cryst Liq Cryst 554:1–9
121. de Vries GC (1999) Electrochromic variable transmission glass for picture tubes. Electrochim Acta 44:3185–3195
122. Nagai J, McMeeking GD, Saitoh Y (1999) Durability of electrochromic glazing. Sol Energy Mater Sol Cells 56:309–319
123. Czanderna AW, Benson DK, Jorgensen GJ, Zhang J-G, Tracy CE, Deb SK (1999) Durability issues and service lifetime prediction of electrochromic windows for buildings applications. Sol Energy Mater Sol Cells 56:419–436
124. Lynam NR, Agrawal A (1988) Automotive applications of chromogenic materials. In: Lampert CM, Granqvist CG, (eds) Proceedings of SPIE large-area chromogenics: materials and devices for transmittance control, Bellingham, Washington, USA, IS vol 4. pp 46–84
125. Lynam NR (1990) Smart windows for automobiles. In: International Congress and Exposition Detroit, Michigan 1990, SAE Technical Paper series (900419)
126. Judeinstein P, Livage J, Zarndiansky A, Rose R (1988) An "all gel" electrochromic device. Solid State Ionics 28–30 (part 2):1722–1725
127. Özer N, Tepehan F, Bozkurt N (1992) An "all-gel" electrochromic device. Thin Solid Films 219:193–198
128. Macêdo MA, Aegerter MA (1994) Sol-Gel electrochromic device. J Sol-Gel Sci Technol 2:667–671
129. Avellaneda CO, Dahmouche K, Bulhoes LOS, Pawlicka A (2000) Characterization of an all Sol-Gel electrochromic device WO_3/ormolyte/CeO_2–TiO_2. J Sol-Gel Sci Technol 19:447–451
130. Heusing S, Munro B, Zapp P, Mennig M, Schmidt H (1998) Effect of ITO and FTO conductive layers on switching properties of large area Sol-Gel electrochromic devices. In: Proceedings of International Meeting on Electrochromism, IME-3 (abstract)
131. Heusing S, Sun D-L, Otero-Anaya J, Aegerter MA (2006) Grey, brown and blue colouring Sol-Gel electrochromic devices. Thin Solid Films 502:240–245
132. Orel B, Šurca A, Opara Krašovec U (1998) Recent progress in Sol-Gel derived electrochromic devices. Acta Chim Slov 45:487–506
133. Orel B, Opara Krašovec U, Maček M, Švegl F, Lavrenčič Štangar U (1999) Comparative studies of "all Sol-Gel" electrochromic devices with optically passive counter-electrode films, ormolyte Li^+ ion-conductor and WO_3 or Nb_2O_5 electrochromic films. Sol Energy Mater Sol Cells 56:343–373
134. Özer N, Lampert CM (1998) Electrochemical characterization of Sol-Gel deposited coatings. Sol Energy Mater Sol Cells 54:147–156
135. Al-Kalhout A, Heusing S, Aegerter MA (2006) Brown colouring electrochromic devices based on Sol-Gel NiO-TiO_2 layers. In: Aegerter MA, Kirchoff V (ed) Proceedings 6th International Conference on Coatings on Glass and Plastics, Dresden, pp 161–164
136. Penyat P, Leyland N, McCormack DE, Colreavy J, Corr D, Pilai SC (2010) Rapid microwave synthesis of mesoporous TiO_2 for electrochromics displays. J Mater Chem 20:3650–3655
137. See www.ntera.com

138. Edwards MOM, Boschloo G, Gruszecki T, Petterson H, Sohlberg R, Hagfeldt A (2001) "Electric-paint displays" with carbon counter electrode. Electrochim Acta 46:2187–2193
139. Georg A, Graf W, Opara Krasovec U, Schulz J, Orel B, Wittwer V (2004) Gasochromic coatings, in Sol-Gel technologies for glass producers and users. In: Aegerter MA, Mennig M (ed) Sol-Gel technologies for glass producers and users, Kluwer, The Netherlands
140. Kraft A, Rottmann M, Heckner KH (2006) Large-area electrochromic glazing with ion-conducting PVB interlayer and two complementary electrodeposited electrochromic layers. Sol Energy Mater Sol Cells 90:469. http://www.gesimat.de
141. http://www.econtrol-glas.de/home/
142. http://www.flachglas-markenkreis.de/deu/data/content2seite.php?menu_id=542

Chapter 13
The Merits of Sol-Gel Processing for Electrochromic Windows: A Commercial Perspective

Anoop Agrawal and John P. Cronin

Abstract Electrochromic windows are reversible electrochemical devices. For architectural windows, typical electrochromic device structures comprise of several layers of coatings, where oxidation and reduction of electrode coatings result in actively controllable optical properties by application of electrical potential. These windows reduce the energy cost of a building envelope by dynamically changing the solar transmission through the windows which are used to optimize both the lighting costs and thermal gains with changing outdoor conditions. Commercial electrochromic windows are now becoming available; however, to obtain broader market acceptance the cost/benefit ratio of these windows must be improved. The windows being introduced into the marketplace do not use sol-gel technology in their manufacturing process. Judicious use of sol-gel technology can assist in obtaining a more favorable cost/benefit ratio. Sol-gel processes offer the opportunity of making the cost attractive, given the unique aspects of the coatings used in this application, which include multiple metal oxide compositions, significant thickness, microstructural requirements, incorporation of mobile ions and the ability to effectively use low-cost transparent conductors.

Keywords Bleached state · Dip coating · Electrochromic · Fluorine-doped tin oxide · Indium tin oxide · Insulated glass unit · Intercalation · Physical vapor deposition (PVD) · Plasma enhanced chemical vapor deposition (PECVD) · Reversible electrochemical devices · Sputtering · Switchable devices · Tungsten oxide · WO_3 film

A. Agrawal (✉) · J. P. Cronin
AJER LLC, 4541 East Fort Lowell Road, Tucson 85712, AZ, USA
e-mail: aagrawal@qwestoffice.net

J. P. Cronin
e-mail: jcronin@qwestoffice.net

M. Aparicio et al. (eds.), *Sol-Gel Processing for Conventional and Alternative Energy*, Advances in Sol-Gel Derived Materials and Technologies,
DOI: 10.1007/978-1-4614-1957-0_13,

List of abbreviations

EC	Electrochromic
IGU	Insulated glass unit
FTO	Fluorine doped tin oxide
ITO	Indium tin oxide
ASTM	American society for testing materials
UCPC	User controlled photochromic devices
PVD	Physical vapor deposition
PECVD	Plasma enhanced chemical vapor deposition

13.1 Background

With increased necessity for energy savings and continued emphasis on a reduction in fossil fuel consumption, much attention is being paid to energy efficient building structures. In this regard, electrochromic (EC) windows have been shown to be particularly energy efficient [1, 2]. Since the solar energy transmission of these windows can be dynamically changed by applying an electrical impulse, the optical properties of such windows can be optimized by the user or an intelligent control system. This control allows one to take into account changes in climate, weather, time of the day, etc. which may affect energy consumption in a building. The energy conservation benefits of EC windows are obtained by reductions in cooling, heating, and lighting costs in comparison to conventional windows that have fixed optical properties [3–5].

Electrochromic technology gained initial market traction in a number of applications [6], many of which are unrelated to energy conservation and is currently being used on a large scale for automotive mirrors (about 17 million in 2008) [7]. EC technology was also introduced in limited numbers for automotive sunroofs in 2005 [8, 9] and for aircraft windows in 2009 [10]. Today, improvements in EC performance and more efficient and cost-effective manufacturing processes such as sol-gel are opening the door to large-scale use of EC windows in architectural applications.

In this chapter we will discuss the advantages of incorporating sol-gel (or wet chemically deposited) thin film(s) into EC windows for architectural applications. We will not consider other switchable devices [11] for windows that include liquid crystal or suspended-particles smart windows where the application of sol-gel technology is marginal. A detailed description of the electrochromic technology, materials, mechanisms, and use of sol-gel processing is provided in Chap. 12. In addition, reviews on sol-gel technology as it applies to electrochromic technology have been published before [12–14]. In this chapter the merits of using sol-gel processing for architectural windows will be discussed, as well as why the merits

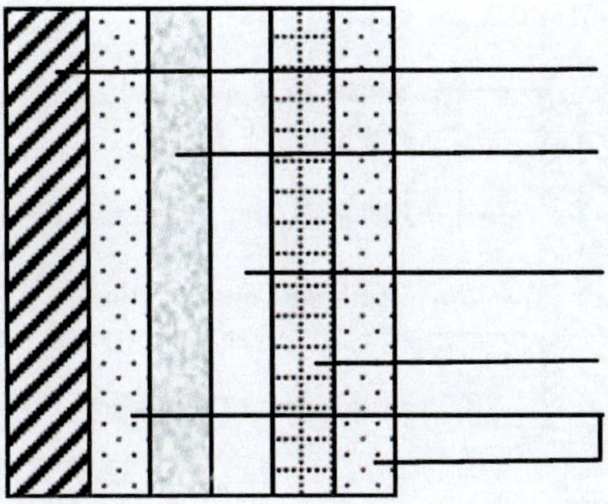

Fig. 13.1 Structure of EC thin film stack deposited on a single substrate

of such processing may lead to important commercial products. In order to provide a backdrop for this discussion, a brief primer on some of the popular EC device structures, particularly those found on recently commercialized EC windows, is presented. In addition, some of the materials and design considerations for EC windows are discussed prior to returning to the main topic of this chapter.

13.2 Electrochromic Device Structures and Electrochromic Materials

Electrochromic devices are comprised of multiple layers of coatings (current collectors, redox electrodes, and an electrolyte layer akin to the structure of a battery). Electrochromic devices designed for architectural applications must have superb durability with respect to resistance to solar radiation (including UV) in colored (dark) and bleached (clear) states, and additionally need to exhibit the ability to change color repeatedly over an operating lifetime of 20 or more years. Schematics of typical EC device structures that have been used for large-area window applications are shown in Figs. 13.1 and 13.2.

Figure 13.3 illustrates how these EC structures are incorporated in architectural windows as insulated glass units (IGUs) which are formed using two glass panes, one of which is used to accommodate the EC structure. There are several more comprehensive discussions of the various EC structures and materials employed in IGUs in the literature [11, 15].

The EC device in Fig. 13.1 uses a single substrate (e.g., glass) on which the EC layers are deposited. The device in Fig. 13.2 uses two substrates which are combined in a laminate type of arrangement with the electrolyte layer forming the connecting medium between the substrates. In either method, these devices for energy efficient windows are next assembled in an IGU type of arrangement. As shown in the two IGU configurations in Fig. 13.3, for energy efficiency, the EC panel assembly is in each case the outer of the two panel assemblies. The arrow

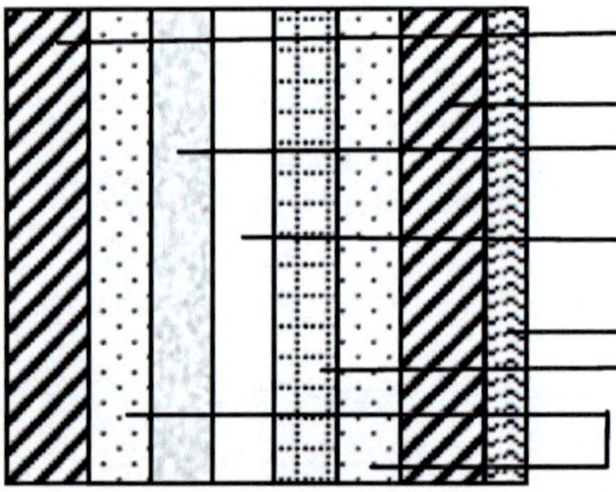

Fig. 13.2 Structure of laminated EC device

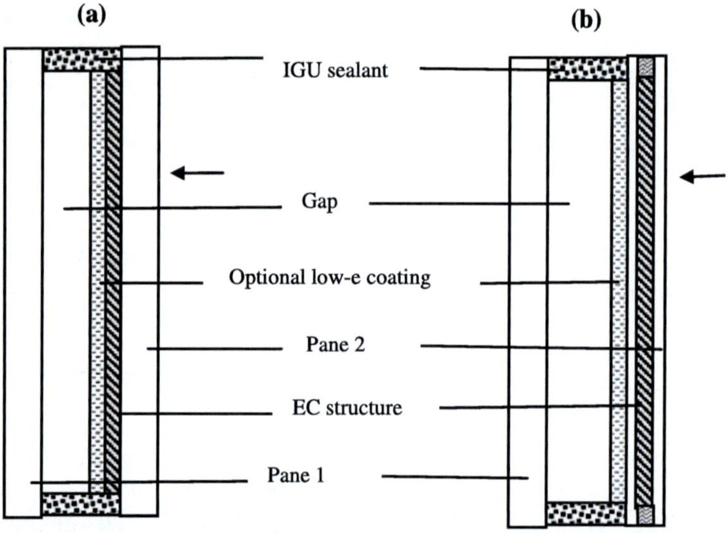

Fig. 13.3 Insulated glass unit (IGU) window formed with two elements, one of which is an Electrochromic pane. The *arrow* illustrates the path of sunlight through the IGU assemblies. **a** Illustrates an EC device that is formed on a single substrate on pane 2 (Per Fig. 13.1) and **b** at right depicts a laminated EC device which is also pane 2 (Per Fig. 13.2). The gap between the two panes of the IGU is evacuated or is filled with air, transparent aerogel, or a low conductivity gas such as sulfur hexafluoride and argon

illustrates the path of sunlight as it passes through the outer EC glass panel, the gap between the two panel assemblies, and finally through the transparent panel. If the devices are assembled as in Fig. 13.1, the EC coatings stack is on the outer panel of the IGU and faces the gap between the two panes of the IGU. If the EC devices are assembled as in Fig. 13.2, again with the coatings stack on the outer panel of the IGU, better thermal management occurs if the more absorptive coating is on that substrate of the EC laminate that faces the outside of the building. Since all

Table 13.1 Characteristics of TEC glass available from Pilkington

Product	Visible transmittance (%)	Sheet resistance (Ω/sq)	Haze (%)
Tec 7	80–82	6–8	5
TEC 8	80–81.5	6–9	12
TEC 15	82–84.5	12–14	≤0.74
TEC 35	82–84	32–48	≤0.65

TEC 10 (about 10 Ω/square) is not shown in the table as it has been discontinued recently.

commercial EC materials attenuate light by absorption, on a bright sunny day, the EC panes can get extremely hot and could mechanically fail. For example, it has been shown that an EC panel attenuating solar radiation from 60% in bleached state to 12% in colored state can reach 65°C when the ambient outdoor temperature is only 40°C [16]. These temperature increases will become more severe when the solar attenuation by the EC panel increases and more of the absorbed light converts to heat. This gives rise to a couple of issues that need to be addressed. First, to reduce the chances of mechanical failure due to the thermal expansion of glass, the substrate used for the EC side should be strengthened, e.g., tempered or laminated. Second, the heat produced by the EC window radiates into the building and thus reduces the thermal performance of the window. To address the issue of radiated heat, particularly in the warmer climates, it is highly desirable that the last surface of the EC panel facing the gap of the IGU should be fabricated from a low-e surface so that most of the heat generated by absorption is rejected to the outside of the building. It can be seen from the IGU structure shown in Fig. 13.3a that it uses an EC structure as shown in Fig. 13.1. If the transparent conductor on the outside surface (the surface facing the IGU gap) is judiciously chosen it could provide low-e functionality. For the IGU using the laminate type of structure shown in Fig. 13.3b, a separate optional low-e coating will be required that faces the gap. This coating is shown as part of the device in Fig. 13.2 and it could also be an added coating or a substrate with this coating which would be laminated to the EC device structure. As a comparison, the emissivity of an un-doped glass surface is 0.84, and for a commercially produced glass surface with a fluorine-doped tin oxide coating (TEC 15 with surface resistivity of 15Ω/square) this drops to 0.12 (see Table 13.1). Some of the best low-e glasses with multiple coatings have a surface emissivity of 0.04 or lower [17]. A set of durability test specifications for EC window devices and when integrated with IGU units are respectively provided in American Society for Testing Materials (ASTM) test methods E2141 and E2354, respectively. Commercialized EC windows are available from Sage Electrochromics, Inc. [18] with a structure as shown in Fig. 13.1 and from EControl-Glas GmbH and Co. [19] with an EC structure as shown in Fig. 13.2. Both devices use tungsten oxide (WO_3) as the electrochromic layer which absorbs most of the solar radiation when colored. Neither of these commercial devices incorporates sol-gel technology but represent systems where sol-gel technology could be used fruitfully. A comparison of electrical and solar performance of EC windows from these companies has been conducted [20].

EC devices work by intercalation and deintercalation of mobile ions; the mechanism is explained below using a model system with tungsten oxide (cathodic or colors upon reduction) and vanadium oxide (anodic) layer:

(Transparent) $WO_3 + xM^+ + xe^- \leftrightarrow M_xWO_3$ (Colored)
$V_2O_5M_x \leftrightarrow V_2O_5 + xM^+ + xe^-$ (both states of vanadium oxide are lightly colored)
$M^+ = H^+, Li^+, Na^+, K^+$ and e^- denotes electrons

If the anodic layer also colors when it is oxidized then it is called a complimentary coloring material, some examples of complimentary materials that can be used with tungsten oxide are iridium oxide and nickel oxide. However, in EC devices with counterelectrodes, the mobile ions "M^{+}" must be intercalated in one of the layers (EC layer or the counterelectrode) during the device assembly. The ions then shuttle between the EC electrode and the counterelectrode for coloration and bleaching. This process of intercalation can be expensive for layers that are deposited by physical vapor deposition (PVD), as layers in a partially constructed device are subject to reduction by electrochemical or other means before continuing to deposit or assemble other layers. For devices constructed as shown in Fig. 13.1, in-line PVD methods to intercalate ions in underlying EC layers have been introduced [21, 22] during the fabrication of the devices. In these methods, the electrolyte deposition step involves deposition of a lithium-rich material that intercalates the underlying electrochromic layer. Such processes are difficult to control. These methods also deplete lithium from the electrolyte layer, resulting in non-uniform distribution of the material and the consequent necessity for additional processes. Further, during manufacturing, since the electrical driving forces for ion intercalation are absent, it takes longer to complete the intercalation of a coating, which slows the manufacturing process and increases cost. In any case, when a PVD method is used to deposit all of the EC device layers, this step adds complexity and costs get added to the fabrication of the finished device.

Other related electrochemical devices that are in the development stage and result in superior thermal performance are those where the attenuation of the solar radiation is mostly by reflection, or in terms of improved functionality are those windows where the solar energy is used to power the windows so that the complexity of providing power to the windows can be reduced. Some of the reflective devices function by reversibly depositing a reflective metal from the electrolyte on one of the electrodes [23], while others function in a fashion similar to EC layer windows, where intercalation of ions in the EC layer results in properties changing from transparent to reflective [24–28]. Devices that are powered by energy from the solar spectrum can be configured in several ways; in any event, only a part of the solar energy is harvested as one needs to preserve high visibility in the bleached state. Preferably the solar energy harvested for powering the device should be mainly in UV and near infrared (NIR). Although UV solar radiation typically makes up less than 3% of the solar spectrum, its high energy level provides sufficient potential to power these devices. In one example of this design approach, multiple layers which together comprise a solar cell are deposited, and are then followed by multiple layers forming an EC structure [29]. These tandem

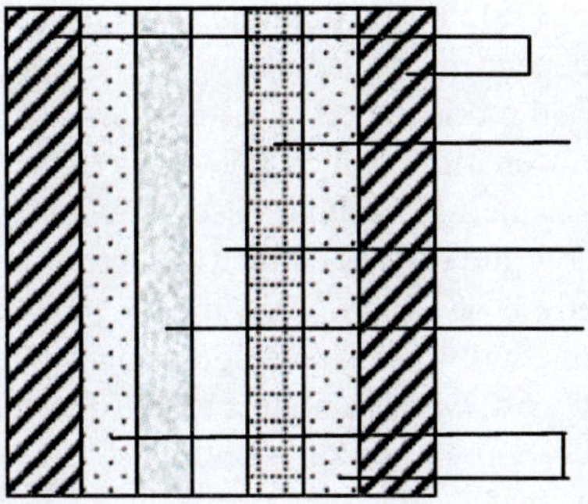

Fig. 13.4 User controllable photochromic device (UCPC) structure [36]

structures are difficult to form economically. An interesting concept as shown in Fig. 13.4 was introduced by Donnelly Corporation, where the counterelectrode in the EC cell (see Fig. 13.2) was replaced by titanium oxide and redox species were introduced in the electrolyte. Titanium dioxide harvested the electrical energy from the UV part of the solar spectrum making these windows exceptionally clear in the bleached state. In these devices it is only when the two electrodes of the cell are shorted by being exposed to solar radiation that the cell colored. By controlling the electrical resistance between the two electrodes, the device coloring kinetics could be controlled. These devices were initially called User Controlled Photochromic Devices (UCPC) [30, 31]. The light sensitive TiO_2 layer was deposited via a sol-gel process as was the tungsten oxide layer used as the chromogenic material. A new term was later coined for these devices—"Photo-Electrochromic Devices"—and additional development was conducted by varying both the composition of the redox materials in the electrolyte and placement and composition of the light sensitive electrodes and chromophores [32–34]. More recently, a complete substructure representing a solar cell has been used to replace the light sensitive layer [35].

13.3 Merits of Sol-Gel Technology for EC Windows

Sol-gel can play an important role in terms of both reducing the cost of finished EC products and enhancing the attributes and performance of electrochromic devices. The current cost of commercial EC windows is estimated to range from \$700 to \$1200/m^2 [37]. To achieve competitive retail pricing, it will be necessary to reduce this cost to about half, as calculated below. Part of the high cost is due to integration of the system including electronics and power connections in existing structures and new building designs, but a significant portion of the cost is represented by the EC devices themselves. As an example, for a large installation, multilayer, non-EC, low-e windows in an IGU configuration are available at a retail price of \$300 m^2 [38]. For a building, energy savings from a given EC window structure is dependent on several factors, for example, heat-or-cold dominated climate, window orientation, wall to glass area, building aspect ratio, etc. Under the best case scenario in which EC windows are used in climates which are dominated by hot weather, the annual energy savings is expected to be about 171 kW-hr/m^2 of window area [4, 5], which at current electricity prices in US (\$0.11 kW-hr) is monetized at \$19/yr. Assuming a payback of five years, a reasonable cost of EC windows in IGU configuration would be about \$95/m^2 in excess of the low-e IGU cost, which amounts to an EC window retail cost in large installation at \$395/m^2 or a premium of about 30%. The market potential for the EC windows can be estimated from the current annual global market for IGUs which is about 400 million square meters [39]. If durable EC comprising IGU could be produced and sold at an additional premium of 30%, the expected annual market size could be about 20% of the total IGU market or about 80 million square meters. As discussed below, sol-gel can play an important role in reducing the cost of the EC element in the product in order to meet the pricing target.

Sol-gel technology can favorably impact the production cost of EC devices through:

1. depositing sol-gel layers of mixed metal oxides without any processing cost penalty,
2. depositing sol-gel coatings in the required thickness and microstructure without adversely impacting the processing rate,
3. using sol-gel technology to facilitate the use of low-cost transparent conductors,
4. using sol-gel methods to potentially avoid a separate intercalation step in the formation of electrodes,

Sol-gel coatings may be used for both of the EC device configurations shown in Figs. 13.1 and 13.2. The sol-gel process may be used to deposit either a single layer or a sequence of coatings with different compositions. Three-layer anti-reflection dip coatings have been commercially produced for over two decades for architectural applications [40] in sizes up to 3.8 m × 1.8 m, and such coatings are also available on tempered glass. Since these are interference coatings, both the control of their thicknesses and their uniformity has to be extremely high. When multilayered devices as shown in Fig. 13.1 are produced by sol-gel, the chemistry of the solution becomes more challenging as one need's to have chemical compatibility and good adhesion with the underlying layers, and in addition the solution of the coating being deposited must not become contaminated from the material deposited in prior steps.

13.4 Deposition of Mixed Metal Oxides

In terms of attributes and performance, increasing use is being made of EC electrodes and counter electrodes which are comprised of metal oxides of more than one element. As an example, the most common electrochromic material used in existing EC devices is tungsten oxide. This material colors to a deep blue tint which dominates the color of the electrochromic windows. One of the important attributes of an EC window is reduction in lighting costs by modulating the visible part of the solar radiation through a window during the day to minimize the use of electrical lights. However, when these windows carry deep blue tints it shifts the visual perception of color in a work area which may not be desirable. In a study by Lawrence Berkeley National Laboratory, the blue systems scored 3.4 on a scale of 5 in terms of natural lighting in the work area [3]. To change the color of such coatings, the most viable method is to use metal oxide dopants. To achieve a more neutral color, molybdenum oxide [41] and also niobium oxide [42] have been suggested as dopants for tungsten oxide EC layers. When exposed to UV, tungsten oxide also possesses a photochromic tendency that interferes with its reversibility and durability, and it has been shown that doping with chromium oxide and lithium oxide reduces this tendency [43]. In another example, the addition of titanium oxide [44] increases the electrochemical reversibility of tungsten oxide. Addition of dopants also affects the coloring potential, thus one can use this method to change the coloring potential of the devices in which they are incorporated and match them with the potential requirements of the counterelectrode and/or the electrolyte used. For example, tungsten oxide coatings with niobium oxide and titanium oxide dopants showed that their coloration became more anodic by 0.2 V [45]. However, to incorporate a multitude of these attributes in a single coating it is likely that more than one dopant will be required, which will become even more technically challenging if PVD deposition methods are used. Most of the counterelectrodes that are used in EC devices that give rise to high reversibility, charge capacity, and durability have also incorporated mixed metal oxides such as nickel oxide doped with tungsten oxide [46] (as in the case of Sage devices); in EControl-Glas devices are likely to include a binary mixture of oxides chosen from titanium oxide, vanadium oxide, niobium oxide, and tantalum oxide [47]. In a publication, EControl-Glas disclosed the use of titanium–vanadium oxide as a counterelectrode [48, 49]. As another example, several mixed metal oxides were disclosed for counterelectrodes by Donnelly Corporation [43]. It is almost certain that mixed metal oxide counterelectrodes will be necessary in order to match the electrochemical capacity and to be compatible with the optical characteristics of the selected EC electrode.

Coating large areas with mixed metal oxides using physical vapor deposition is challenging and expensive. The sputtering cross-section of various materials or their vapor pressure differ, and maintaining the right stoichiometry through the film cross-section over a period of time is more difficult than is the case with a single metal oxide coating. However, in sol-gel, the process complexity associated with depositing a coating without a dopant or with one or any number of mixed oxides is the same although sol-gel processing requires considerable chemical work at the front end to form a solution with the desired dopant precursors.

13.5 Deposition of the Coatings of the Required Thickness and Microstructure

The cost of capital equipment for EC devices becomes significant because the thickness of coatings used in EC devices is rather large. From the patent literature, it is seen that the Sage devices [50] use tungsten oxide in a thickness of 400 nm and the thickness of the counter electrode with a tungsten oxide-doped nickel oxide is 200 nm. For the EC window manufactured by EControl-Glas, the thickness of the various layers can be estimated from their recent publication on a window that changes from about 55 to 15% and uses 15 mC/cm^2 of charge [51]. This would use tungsten oxide in a thickness of 200nm and titanium-vanadium oxide in a thickness of about 75 nm [50]. In EC devices where the counter electrode also colors, the thickness of the tungsten oxide is closer to the lower value. An economic analysis was conducted [52] on the cost of depositing tungsten oxide in a thickness of 400 nm using sputtering (a physical vapor deposition (PVD) process) and plasma-enhanced chemical vapor deposition (PECVD) on a roll-to-roll process. This analysis showed that the cost of deposition of the sputtered tungsten oxide coating (including the raw materials) was \$15/m^2. For PECVD the cost was about one-third of that of the PVD process due to higher throughput of such systems, even though the material cost for PECVD was about 50% higher. Most of the cost in sputtering was dominated by labor and depreciation of capital costs as the sputtering rate for the coating was 0.6 nm/s, versus. 10 nm/s for PECVD. The utilization rate of the target in this model was taken to be at 40% (i.e., 60% of the material coated the coater walls, etc.).

In the sol-gel method both of these shortcomings, i.e., coating speed and low material yield, can be addressed. For sol-gel processes to be economical it is imperative that two factors be considered. First, to reduce process costs, the coating must be deposited in the desired thickness in a single step. This also implies that the single-step coatings must not crack or lose adhesion to the substrate when they are further processed at elevated temperatures to yield the final desired composition and microstructure. If sol-gel processes are used where multiple coating processes are necessary to get the desired coating thickness, then sol-gel methods lose their attractiveness rapidly. Use of sol-gel processes for the deposition of a single coating using multiple passes will tremendously increase the cost of facilities and capital equipment (coatings have to be cooled after every firing step and then re-routed into the coatings chamber which decreases the throughput) while also decreasing the yields. Second, the stability of the solutions has to be very high. A low stability solution will require frequent replacement of the coating solution in the dip tank, and the materials cost will become excessive due to low yields. A desirable time period of solution stability should be several months so that material yields in excess of 80% can be maintained. In addition, one should be able to compensate for the loss of solvent with time by adding solvent without losing any of the solution's properties. Other methods of deposition, such as spraying, do not result in the same optical uniformity (within ±5%). In addition, because of waste the solution yield is low.

The thickness of sol-gel coating using a dip process increases with increasing withdrawal rate, which makes the process economically attractive with increased coating thickness. The coating thickness in dip process is controlled by the speed of withdrawal and the solution properties according to the following relationship [53]:

$$t = (2V_s\eta/dg)^{1/2}$$

where, t = wet coating thickness, V_s = withdrawal rate, η = coating solution viscosity, d = coating solution density, and g = gravitational constant.

In order for the electrochromic and the counter electrode layers to respond rapidly, it has been found that the porosity [54–57] and interface with the electrode and the electrolyte [58, 59] of these layers are important. Controlling porosity using sol-gel technology is much easier, as one can incorporate materials (also called molecular templates) in the coating solution; such materials include oxalic acid and polyethylene glycol [60–64]. The templates can be removed during subsequent coatings processing to leave a porous structure. Since the consolidation of the coating is carried out by increasing the temperature after the solution is coated on the substrate, controlling the temperature profile is also important to yield a desired morphology. It has been found that the oxygen content (or oxygen stoichiometry) is important to get the desirable electrochemical properties. For example, in tungsten oxide, oxygen content is important for reversibly intercalating lithium ions [65], and is also important for charge capacity [66]. Since in the sol-gel method the coatings are given a heat treatment after deposition, one can easily incorporate oxygenating additives and/or provide environmental and heat conditions in this step to attain the desired stoichiometry [67].

As an example, some of the coating solutions that yield porous WO_3 crack-free coatings in thickness of up to 1 μm in a single step, and have solution stability in excess of several months have been published [68]. Corresponding counter electrodes, matching them in charge capacity and that are capable of being deposited by sol-gel are also known [69].

13.6 Facilitating the Use of Low-Cost Transparent Conductors

EC devices can consume significant current when their optical state is changed by applying a voltage (about 15–25 mC/cm^2 is required for a typical device, which can be converted to the current requirements based on the device size and the kinetics desired). Therefore with increasing area, more charge (hence current) is required to pass from the electrical connections at the perimeter to the center of the device. This current causes the voltage to drop rapidly from the perimeter to the center of the device, which causes the device kinetics to be sluggish. Thus, the conductivity of the transparent conductors, such as indium tin oxide (ITO) have to be sufficiently high to have reasonable kinetics, i.e., for the change to be completed in a time period of less than 5 min in a window size of about one square meter. As an example, if one needs an ITO coating with surface resistance of 10 Ω/square, then using an ITO material with bulk resistance of 2×10^{-4} Ω cm would require a thickness of 200 nm. The range of resistance for transparent conductors suitable for practical EC window devices is typically between 6 and 15 Ω/square. This thickness is quite high as compared to that required to field powered devices, e.g., liquid crystal devices which may only use about a one-tenth of the conductor thickness. In order to increase the conductivity of transparent conductors, one may also use highly conductive metal meshes [70], but they result in substantial increase in cost.

ITO coatings have become expensive due to the worldwide shortage of indium as a consequence of its use in flat panel displays. The price of the metal has increased tenfold to $1000/kg in the past six years. At current indium prices, ITO coatings on glass with a sheet resistance of 12 Ω/square are expected to cost $30 and $40/m^2 in volume. If the EC windows realize their market potential of 80 million m^2/year, very significant price pressures will be placed on indium. The only low-cost transparent conductor commercially available at present is fluorine-doped tin oxide (FTO). There are several manufacturers of such glass; in the US FTO-coated glass is manufactured by Pilkington (located in Toledo, Ohio) and sold as TEC™; in Europe it is available as "K" glass. The FTO is applied on the float line process by Atmospheric Pressure Chemical Vapor Deposition (APCVD) [71]. The coating is deposited while the glass is still hot (about 600°C) and is in the process of cooling down. The price of FTO-coated glass in large volumes is less than $7.5 m^2.

One problem with the chemistry of the FTO deposition process is that it results in hazy coatings due to scattering of light from sodium chloride crystals formed at the glass surface from a reaction of sodium in the glass with chloride from the precursor. The process also results in a rough surface which additionally contributes to the haze. Both these factors increase with increasing conductivity or increasing thickness or decreasing sheet resistance. For large-area EC glazing, transparent conductor resistivities of $\leq$15 Ω/sq are required to minimize the resistive drop across the surface and to ensure uniform coloring and bleaching of the cell. Table 13.1 lists some of the characteristics of TEC™ glass available from Pilkington [72]. TEC 10 (about 10 Ω/square) is not shown in the table as it has been discontinued recently.

Generally, the level at which haze can be detected by the human eye for a window application is about 2%. For skylight applications one may use a glass with higher haze. As explained above, part of this haze is due to surface roughness; when PVD methods are used to deposit coatings on such glass, then part of the roughness converts to entrapped voids which do not help in reducing the haze. However, when coatings are applied using sol-gel, a marked reduction in haze is noticed as shown in Table 13.2 [73] which makes this high conductivity—high haze glass usable for EC windows. This is demonstrated using TEC 10 substrate in

Table 13.2 Reduction of surface roughness and haze in FTO coated glass using a sol-gel derived WO_3 layer

Sample	Sol-gel WO_3 Thickness (nm)	Surface Roughness RA (Å)	% Haze
Uncoated 10Ω/sq FTO	N/A	70	5.0
10Ω/sq FTO + WO_3	75	25	2.9
Microscope Slide	N/A	15	0.09
ITO 12 Ω/sq	N/A	25	0.15

Table 13.2. The data is compared to a glass slide and commercial ITO with a surface resistance of about 12 Ω/sq. Further, an advantage of FTO glass is that it can be bent and tempered without destroying the conductivity of the FTO coating.

Such transparent conductors may be used for both substrates of the EC device shown in Fig. 13.2. However, it can only be used as the first conductor on the substrate. For the second conductor, it is challenging to use a sol-gel coating, as most transparent conductors require high temperatures (typically ≥350°C) to process, which may compromise the desirable properties of some of the previously deposited layers. A transparent metal layer (such as thin layer of gold of about 8 to 10 nm [74]) with a surface resistance in a range of about 6 to 10 Ω/square may be used instead. Such thin layers can be deposited at reasonable process cost using PVD methods. A recent alternative to solution-deposited transparent conductors have been use of conductive nanowire solutions to deposit a fine random mesh which is both transparent and conductive [75, 76].

13.7 Avoidance of a Separate Intercalation Step

As discussed earlier, EC devices work by reversibly shuttling the mobile M^+ ions (e.g., H^+, Li^+) from the counterelectrode to the electrochromic electrode via the electrolyte (i.e., the ion conductor). Thus, these mobile ions need to be added to at least one of these electrodes during processing. This is an expensive step in the electrode deposition process using conventional methods. Either the process sequence is broken so that the ions can be added via a chemical or an electrochemical process before continuing with the device assembly, or one must incorporate evaporation or a sputtering process to intercalate a removable ion (usually lithium) into the layer. The latter processes when combined with the physical vapor deposition of the electrode are amenable to a continuous process, which is preferable, but it is still an expensive process to implement and conduct as discussed earlier.

When sol-gel or wet-chemical methods are used, it is possible to include the addition of mobile ions in the coatings deposition process without any additional step. This implies that such incorporation will not add extra costs when the devices are produced. One of the first uses of this technique was in the deposition of polyaniline coatings in electrochromic devices which, when formed were in a reduced state due to the incorporation of protons [77]. This was accomplished via the addition of reducing agents to the polyaniline coating solutions. In another instance, a counterelectrode coatings with a composition of lithium–nickel–manganese oxide was formed using sol-gel methods [78]. This counterelectrode was combined in a device with a tungsten oxide-based EC electrode. It was reported that none of the layers required any additional reduction, as lithium from the counterelectrode could be reversibly extracted in the finished device.

13.8 Summary

Sol-gel methods can be used very effectively to reduce the cost of EC devices for energy efficient windows. However, the sol-gel methodology will have to be carefully designed so that it can influence the cost structure in a significant way. Some of the important considerations in sol-gel processes are the following which need to be exploited fully in order to gain the most impact:

1. Coatings with multiple oxides can be deposited without any added process complexity.
2. Coatings must be deposited in the required thickness in one coating step, and the coatings solution must have a high temporal stability.
3. Addition of precursors to the coating solutions and processing conditions can be optimized in order to get the desired microstructure.
4. Inexpensive transparent conductors must be used, as sol-gel coatings are able to reduce the haze of the finished EC device.
5. The coating solution must incorporate suitable chemistry to allow the reduction step (incorporation of reversibly removable ions) within the coating process.
6. Solution chemistry and processing must be compatible with the underlying layers, and the material from the underlying layers must not contaminate the solution of the coating being deposited.

References

1. Warner JL, Reilly MS, Selkowitz SE, Arasteh DK, Ander GD (1992) Utility and economic benefits of electrochromic smart windows. Lawrence Berkeley Laboratory report LBL-32368
2. Selkowitz SE, Rubin M, Lee ES, Sullivan R, Finlayson E, Hopkins D (1994) A review of electrochromic window performance factors. Lawrence Berkeley Laboratory, report LBL-35486
3. Lee E (2006) A design guide for early-market electrochromic windows. Report prepared for California Energy Commission by Lawrence Berkeley Laboratory, CA, report# CEC-500-2006-052-AT16, http://windows.lbl.gov/comm_perf/Electrochromic/refs/LBNL-59950.pdf. Accessed 16 Feb 2011
4. Sullivan R, Rubin M, Selkowitz S (1996) Energy performance analysis of prototype electrochromic windows. Lawrence Berkeley National Laboratory report # LBNL 39905

5. Papaefthimiou S, Syrrakou E, Yianoulis P (2006) Energy performance assessment of an electrochromic window. Thin Solid Films 502:257–264
6. Pawlicka A (2009) Development in electrochromic devices. Recent Patents Nanotechnol 3:177–181
7. Gentex Corporation, Zeeland, Michigan, USA (2008) Annual report
8. Ferrari cars, Superamerica. http://www.ferraricars.org/ferrari-superamerica/. Accessed 23 Feb 2011
9. Giron J-C, Schütt, J, Pender D, BéteilleF, Fanton X (2003) Proceedings of Glass Processing Days. pp 460–461
10. Gentex Corporation, Zeeland, Michigan, USA (2009) Annual report
11. Lampert CL (2004) Chromogenic materials. Mater Today 7(3):28
12. Agrawal A, Cronin JP, Zhang R (1993) Review of solid state electrochromic coatings produced using sol-gel techniques. Sol Energy Mater Sol Cells 31:9–21
13. Livage J, Ganguli D (2001) Sol-gel electrochromic coatings and devices: a review. Sol Energy Mater Sol Cells 68:365–381
14. Heusing S, Aegerter M (2005) Sol-gel coatings for EC devices. In: Sakka S (ed) handbook of sol-gel science and technology, Chap. 35. Kluwer Academic Publishers, New York
15. Baetens R, Jelle BP, Gustavsen A (2010) Properties, requirements and possibilities of smart windows for dynamic daylight and solar energy control in buildings: a state-of-the-art review. Sol Energy Mater Sol Cells 94:87–105
16. Lampert CL, Agrawal A, Nagai J (1999) Durability Evaluation of Electrochromic Devices—an Industry Perspective. Sol Energy Mater Sol Cells 56:449
17. Siegel JD (2002) The MSVD Low E 'Premium Performance' Myth-Actual energy conservation performance of different types of Low E glazings in residential windows. Int Glass Rev 1:55
18. Sage Electrochromics, Fairbault, MN, USA. www.sage-ec.com. Accessed 10 Mar 2011
19. EControl-Glas, Germany. www.econtrol-glas.de. Accessed 10 Mar 2011
20. Mack I, Steiner R, Oelhafen P (2009) Electrically controlled windows: perforamance of new products. In: Proceedings CISBAT Swiss federal institute of technology in Lausanne (EPFL), p 26
21. Widjaja E, Delporte G (2010) Method of making an ion-switching device without a separate lithiation step. U.S. Patent No. 7830585
22. Pitts JR, Lee S-H, Tracy C E, Gillaspie D (2009) Synthesizing thin films of lithiated transition metal oxide for use in electrochemical and battery devices. Published U.S. Patent Application No. 20090057137
23. Tench DM, Warren LF, Cunningham MA (2001) Diffusely reflecting reversible electrochromic mirror. U.S. Patent No. 6256135
24. Slack JL, Locke JCW, Song S-W, Ona J, Richardson TJ (2006) Metal hydride switchable mirrors: factors influencing dynamic range and stability. Sol Energy Mater Sol Cells 90: 485–490
25. Richardson TJ (2003) Electrochromic materials, devices and process of making. U.S. Patent No. 6647166
26. Huiberts JN, Griessen R, Rector JH, Wijngaarden RJ, Dekker JP, de Groot DG, Koeman NJ (1996) Nature-London 380:231
27. Kazuki Y, Yasusel Y, Kazuki T (2010) All-solid-state reflective dimming electrochromic device having buffer layer and dimmer member using the same. Published U.S. Patent Application No. 20100188726
28. Tajima K, YamadaY, Okada M, Yoshimura K (2010) Accelerated degradation studies on electrochromic switchable mirror glass based on magnesium–nickel thin film in simulated environment. Sol Energy Mater Sol Cells :1716–1722
29. Bullock JN, Xu Y, Benson DK, Branz HM (1995) Tandem self-powered photovoltaic-electrochromic window coatings. In: Lampert CM, Deb SK, Granqvist CG (eds) Optical materials technology for energy efficiency and solar energy conversion XIV. Proc SPIE 2531:35–41

30. Teowee G, Allemand P-M, Cronin JP, Tonazzi TCL, Agrawal A (1997) Novel photochromic devices. U.S. Patent No. 5604626
31. Teowee G, McCarthy K, Agrawal A, Allemand P-M, Cronin JP (1999) User controllable photochromic (UCPC) devices. Electrochim Acta 44:3017–3026
32. Gregg BA (1997) Photoelectrochromic cells and their applications. Endeavor 21(2):52–55
33. Hauch A, Georg A, Baumgärtner S, Opara-Krasovec U, Orel B (2001) New photoelectrochromic device. Electrochim Acta 46:2131–2136
34. Lam D, Branda NR (2010) Variable transmittance optical filter and uses thereof. Published PCT Patent Application No. WO/2010/142019
35. Huang L-M, Chen C-H (2010) Photosensitive electrochromic device. US patent 7855822
36. Cronin JP, Kennedy SR, Agrawal A, Gudgel TJ, Uhlmann DR (1999) Electrochromic glazing. Mater Res 2(1):1–9
37. Market data obtained from current electrochromic window suppliers and distributors
38. Pilkington Corporation. http://www.pilkington.com/resources/faqenergikarelegacy2.pdf. Accessed 13 Mar 2011
39. Skinner N, Insulating glazing in a warm climate-even more important than in a cold climate, http://www.glassfiles.com/library/article.php?id=783&search=skinner&page=1. Accessed 9 Mar 2011 (in year 2000, 280 million sq meters of IGU were produced, assuming a 4% growth rate/year for flat glass (Pilkington and the flat glass industry 2010, http://www.nsg.com/resources/pfgi2010.pdf), would result in 400 million square meters in 2010
40. Amiran Glass Product Guide, Schott Glass, Elmsford, NY. http://www.us.schott.com/2009_architecture/english/download/amiran_brochure_-_2008.pdf. Accessed 14 Feb 2011
41. Yamada S, Kitao M (1990) Modulation of absorption spectra by the use of mixed films of MocW1-cO3. In: Lampert CM, Granqvist CG (eds) Large area chromogenics: materials and devices for transmittance control. SPIE optical engineering Press, Bellingham, Washington USA, p 246
42. Heusing S, Sun D-L, Otero-Anaya J, Aegerter MA (2006) Grey, brown and blue coloring sol-gel electrochromic devices. Thin solid films 502(1–2):240–245
43. Allemand PM, Ingle A, Cronin JP, Kennedy SR, Yao Y, Tonazzi JCL, Boulton J, Agrawal A (2001) Electrochromic devices. U.S. Patent No. 6266177
44. Hashimoto S, Matsuoka S (1991) J Electrochem Soc 138:2403
45. Gillet PA, Fourquet JL, Bohnke O (1992) Proc. SPIE 1728:82
46. Burdis MS, Weir DGJ (2006) Electrochromic devices and methods. Published U.S. Patent Application No. 20060209383
47. Jodicke D (2009) Electrochromic elements using antioxidadnts to suppress self-discharging. Published U.S. Patent Application No. 20090225393
48. Jödicke D, Wittkopf H (2007) The 2nd generation of an electrochromic solar control glazing—ready for projects. In: Proceedings of glass performance days, pp 394–395
49. Burdis MS, Siddle JR, Batchelor RA, Gallego JM (1998) V0.5Ti0.5Ox thin films as counterelectrodes for electrochromic devices. Sol Energy Mater Solar Cells 54(1–4):93–98
50. Burdis MS, John DG (2008) Electrochromic devices and methods. Published U.S. Patent Application No. 20080169185
51. Wittkopf H (2010) Elektrochrome Beschichtungen. Sonnenchutzglaser der neuen Generation Vakum in Forchung and Praxis 22(3):26–30
52. Garg D, Henderson PB, Hollingsworth RE, Jensen DG (2005) An economic analysis of the deposition of electrochromic WO3 via sputtering or plasma enhanced chemical vapor deposition. Mater Sci Eng B119:224–231
53. Brinker CJ, Scherer GW (1990) Sol-gel science:the physics and chemistry of sol-gel processing, Chap. 13. Academic Press, New York
54. Crandall RS, Faughan BW (1976) Dynamics of coloration of amorphous electrochromic films of WO3 at low voltages. Appl Phys Lett 28(2):95–97
55. Mohapatra SK (1978) Electrochromism in LixWO3. J Electrochem Soc Solid State Sci Technol 125(2):284–288

56. Randin J-P, Viennet R (1983) Proton diffusion in tungsten trioxide films. J Electrochem Soc: Solid State Sci Technol 129(10):2349–2354
57. Kamimori T, Nagai J, Mizuhashi M (1983) Transport of Li+ ions in amorphous tungsten oxide films. SPIE 428:51–56
58. Zhang J-G, Tracy CE, Benson DK, Deb SK (1993) The influence of microstructure on the electrochromic properties of LixWO3 thin films: part I. Ion diffusion and electrochromic properties. J Mater Res 8(10):2646–2656
59. Zhang J-G, Tracy CE, Benson DK, Deb SK (1993) The influence of microstructure on the electrochromic properties of LixWO3 thin films: part II. Limiting mechanisms in coloring and bleaching processes. J Mater Res 8(10):2657–2667
60. Cronin J P, Tarico D J, Agrawal A, Zhang R (1994) Method for depositing high performing electrochromic layers. Published U.S. Patent No. 5277986
61. Cronin JP, Kennedy SR, Agrawal A, Gudgel TJ, Yao YJ, Tonazzi JCL (1999) Properties of WO3 coatings for large area electrochromic devices. In: Sundaram SK, Bickford DF, Hornyak EJ Jr (eds) Electrochemistry of glass and ceramics (Ceramic Transitions), vol 92. American Ceramic Society, USA, pp 175–193
62. Cheng W, Baudrin E, Dunn B, Zink JI (2001) Synthesis and electrochromic properties of mesoporous tungsten oxide. J Mater Chem 11:92–97
63. Ozkan E, Lee S-H, Liu PC, Tracy E, Tepehan FZ, Pitts JR, Deb SK (2002) Electrochromic and optical properties of mesoporous tungsten oxide films. Solid State Ionics 149:139
64. Deepa M, Singh DP, Shivaprasad SM, Agnihotri SA (2007) A comparison of electrochromic properties of sol-gel derived amorphous and nanocrystalline tungsten oxide films. Curr Appl Phy 7:220–229
65. Vink TJ, Boonekamp EP, Verbeek RGFA, Tamminga Y (1999) Lithium trapping at excess oxygen in sputter-deposited a-WO3 films. J Appl Phy 85(3):1540–1544
66. Burdis M (1997) Properties of sputtered thin films of vanadium-titanium oxide for use in Electrrochromic windows. Thin Solid Films 311:286–298
67. Cronin JP, Tarico DJ, Tonazzi JC, Agrawal A, Kennedy SR (1993) Microstructure and properties of sol-gel deposited WO_3 coatings for large area electrochromic windows. Sol Energy Mater Sol Cells 29:371–386
68. Cronin JP, Tarico DJ, Agrawal A, Zhang RL, Tonazzi JCL (1996) Precursor solutions for forming coatings. U.S. Patent No. 5,525,264
69. Allemand PM, Ingle A, Cronin JP, Kennedy SR, YaoY, Tonazzi JCL, Boulton J, Agrawal A (2001) Electrochromic devices. US Patent No. 6266177B1
70. Agrawal A, Tonazzi JCL LeCompte R, Cronin JP, Kennedy SR. McCarthy K, Denesuk M, Teowee G (2001) Busbars for electrically powered cells. US Patent No. 6317248
71. Gerhardinger P, Stickler D (2008) Fluorine doped tin oxide coatings-over 50 years and going strong. Key Eng Mater 380:169–178
72. Pilkington TEC Glass Product informationhttp://Products.construction.com/swts_content_files/1179/313761.pdf. Accessed 8 Mar 2011
73. Cronin JP, Agrawal A, Trosky M (1999) Method for reducing haze in tin oxide transparent conductive coatings. US Patent No. 5900275
74. Holland L, Siddall G (1953) The properties of reactively sputtered metal oxide films. Vacuum 3(4):375–391
75. Alden JS, Dai H, Knapp MR, Na S, Pakbaz H, Pschenitzka F, Quan X, Spaid M A, Winoto A, Wolk J (2007) Nanowires based transparent conductors. Published U.S. Patent Application No. 2007/0074316
76. Agrawal A, Cronin JP, Metal coatings, conductive nanoparticles and applications of the same. Published US Patent Application No. 20100002282
77. Allemand P-M Grimes FR, Bigelow BA, Agrawal A (2001) Electrochromic devices with improved processability and methods of preparing the same. US Patent No. 6,327,069
78. Agrawal A, Zhang R, Boulton J (2002) Chromogenic glazing for automobiles and display filters. US Patent No. 6373618

Chapter 14
Ferroelectric Thin Films for Energy Conversion Applications

Barbara Malič, Alja Kupec, Hana Uršič and Marija Kosec

Abstract The chapter attempts to describe the potential of sol-gel derived ferroelectric thin films in energy conversion applications in two fields, namely in cooling by exploiting the electrocaloric effect and in energy harvesting. First, the basics of the sol-gel processing of ferroelectric thin films with the $Pb(Zr,Ti)O_3$ solid solution as the representative material are described. The electrocaloric effect is described as a reversible temperature change in a material under an applied electric field at adiabatic conditions. Sol-gel derived relaxor ferroelectric thin films demonstrate a giant electrocaloric effect and they could be applied for solid-state cooling devices in microelectronics. In energy harvesting, the energy is derived from external sources, captured, stored and used, for example in small autonomous devices. In ferroelectric thin films, the piezoelectric effect, i.e. the generation of an electric charge upon applied stress, is exploited. Due to the miniaturization of electronic components, thin films with lateral dimensions in nanometer to micrometer range are appropriate candidates for future miniature applications related to energy.

B. Malič (✉) · A. Kupec · H. Uršič · M. Kosec
Jožef Stefan Institute, Jamova 39, 1000, Ljubljana, Slovenia
e-mail: barbara.malic@ijs.si

B. Malič
Centre of Excellence on Nanoscience and Nanotechnology, Ljubljana, Slovenia

B. Malič
Centre of Excellence SPACE-SI, Ljubljana, Slovenia

M. Kosec
Centre of Excellence NAMASTE, Ljubljana, Slovenia
e-mail: marija.kosec@ijs.si

M. Aparicio et al. (eds.), *Sol-Gel Processing for Conventional and Alternative Energy*, Advances in Sol-Gel Derived Materials and Technologies,
DOI: 10.1007/978-1-4614-1957-0_14, © Springer Science+Business Media New York 2012

Keywords Annealing • Crystallization • Electrocaloric effect • Energy harvesting • Ferroelectric–paraelectric (F–P) phase transition • Lanthanum-modified lead zirconate titanate (PLZT) • Lead zirconate titanate (PZT) • Microelectromechanical systems (MEMS) • Perovskite • Piezoelectric effect • Polarization • Pyrolysis • Relaxor ferroelectric • Spin coating • Thin film deposition

14.1 Introduction

Sol-gel derived ferroelectric (FE) thin films have been extensively studied since the first reports in mid-1980s [1, 2]. The drive for research has been the miniaturization of components in micro-electronics and micro-electro-mechanical systems (MEMS).

Ferroelectric materials are characterized by the nonlinear, hysteretic, dependence of polarization P versus applied electric field E, which stems from their polar structure [3]. The switchable remanent polarization has been exploited in non-volatile random access memories [4–6]. Furthermore, they possess high values of dielectric permittivity and they have been explored for thin film capacitor applications [7, 8]. Electric field-induced ordering of dipoles resulting in a temperature change, denoted as the electrocaloric effect could be applied in heating and cooling devices. [3, 9, 10].

Ferroelectrics are a subgroup of piezoelectrics. The applied stress results in the generation of an electric charge on the surface of the material, which renders them useful for energy harvesting applications [11, 12]. In the inverse piezoelectric effect, an external electric field, applied to a piezoelectric material, produces a deformation of the material [13].

The lead zirconate titanate solid solution [$Pb(Zr,Ti)O_3$, PZT] has been among the most widely studied materials for FE thin films. The material crystallizes in perovskite ABO_3 structure and its polarization stems from the off-center position of the B-site cations, octahedrally coordinated by oxygen ions. Platinized silicon, usually with the TiO_2 adhesion layer, has been the commonly used substrate [14]. It serves as the bottom electrode in the MIM geometry (metal-insulator-metal, the top electrode may be Pt or Au) and furthermore, it allows integration of FE thin films in microelectronic and MEMS components [15–17].

Figure 14.1a shows the cross-section microstructure of a $Pb(Zr_{0.53}Ti_{0.47})O_3$ thin film on platinized silicon substrate with a sol-gel derived nucleation layer of $PbTiO_3$, consisting of columnar grains. The film is single-phase perovskite with a (100) orientation (Fig. 14.1b). In Fig. 14.1c the polarization–electric field dependence reveals a hysteresis loop.

The description of the PZT thin film sol-gel processing is given in the next section.

In further text, two areas of implementation of FE thin films in energy conversion as in electrocaloric effect and in energy scavenging or harvesting are introduced.

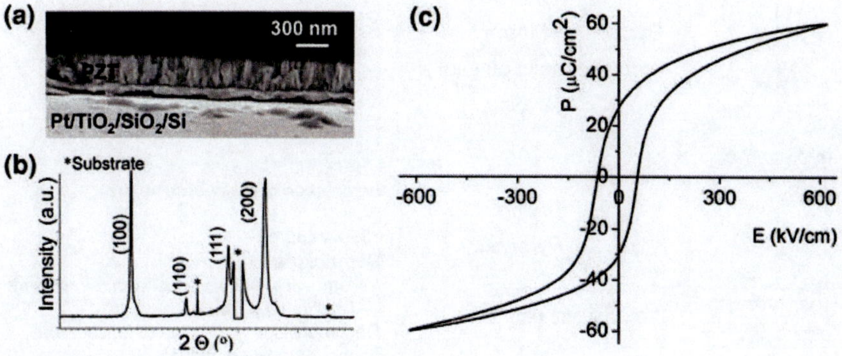

Fig. 14.1 **a** Columnar microstructure of solution derived $Pb(Zr_{0.53}Ti_{0.47})O_3$ thin film (cross section field-emission scanning electron microscope (FE-SEM) image), **b** phase composition and **c** ferroelectric loop of the film, obtained at 100 Hz. (Courtesy of Electronic Ceramics Department, Jožef Stefan Institute)

14.2 Sol-Gel Processing of FE Thin Films

The sol-gel processing of FE thin films, referred to also as the Chemical Solution Deposition (CSD), in analogy with Physical Vapor Deposition (PVD) and Chemical Vapor Deposition (CVD) processes, has been discussed in a number of reviews, e.g. [8, 18]. Figure 14.2 contains a general scheme of the film processing steps, together with their temperature ranges and chemical reactions taking place in the films, which are described in further text. They include:

- synthesis of the coating solution or sol,
- deposition of the solution on the substrate, the as-deposited film has been referred to as the 'gel' film,
- drying to remove residual solvent, and pyrolysis to remove residual functional groups, resulting in an amorphous film,
- annealing step (crystallization), yielding a crystalline film.

The deposition, and further drying, pyrolysis and/or heating steps may be repeated until the required thickness is achieved. Typically, the thickness of less than 100 nm is obtained in one deposition step. For piezoelectric applications, the thicknesses of up to a few μm are produced to achieve the required performance, e.g. [16].

14.2.1 Synthesis of the Sols

The commonly used sol-gel processing routes for PZT-based thin films are based on the procedures introduced by Gurkovich and Blum in 1984 for powders [1] and by Budd, Dey and Payne for thin films [2]. In Fig. 14.3 the flow chart for the processing

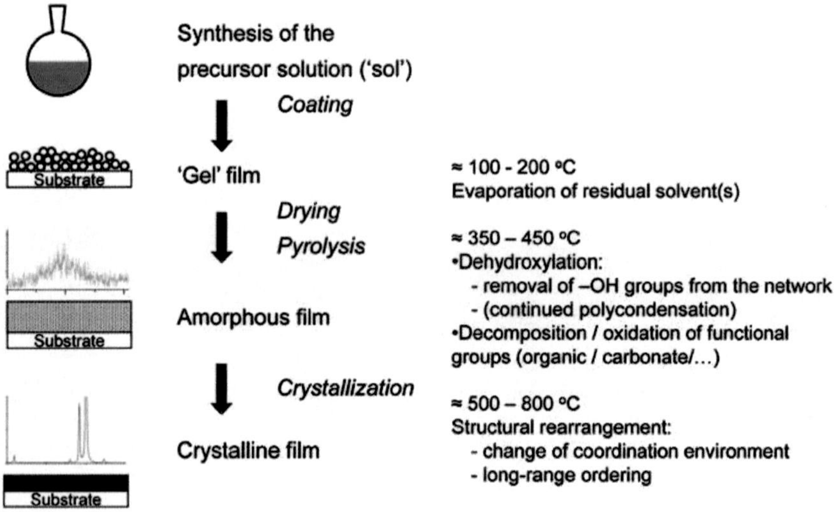

Fig. 14.2 Schematic representation of the sol–gel processing of thin films

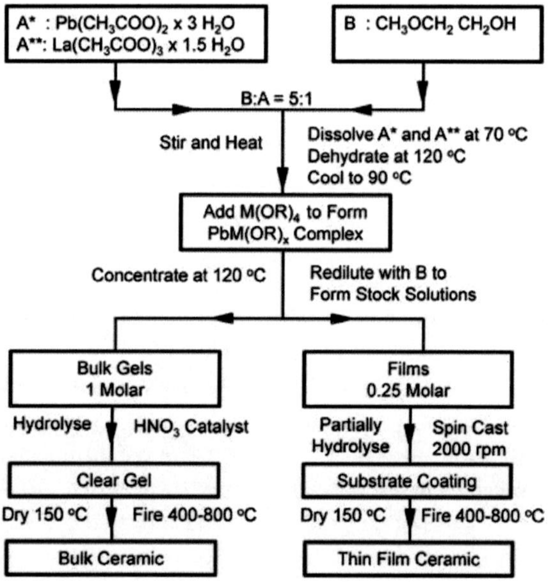

Fig. 14.3 Flow chart of the processing of PLZT gels and films, M = Zr, Ti (from [2])

of La-modified PZT $\left[\text{PLZT}, (\text{Pb}_{1-x}\text{La}_x)(\text{Zr}_{0.65}\text{Ti}_{0.35})_{1-x/4}\text{O}_3,\ x = 0.09\right]$, as described by Budd et al. [2], is included. The reagents are lead and lanthanum acetates and zirconium and titanium alkoxides, propoxides or butoxides. The solvent is an ether-alcohol, 2-methoxyethanol (MOE), which reacts with the transition metal

(TM) alkoxides by the transalcoholysis reaction, i.e., by the exchange of alkoxide groups resulting in a less reactive mixed alkoxide [19, 20].

During heating and distillation the reaction between the lead acetate (and in the case of PLZT, also lanthanum acetate) and TM alkoxides results in the –Pb–O–M– bond. The reaction is called the ester-elimination, as the by-product is an ester. The Eq. 14.1 shows a simplified reaction between one acetate and one alkoxide group. In reality the process is more complex, as for example reported in [21–23].

$$-[\mathrm{Pb}-\mathrm{OAc}] + [\mathrm{RO}-\mathrm{M}]- \;\rightarrow\; -[\mathrm{Pb}-\mathrm{O}-\mathrm{M}]- +\mathrm{ROAc} \qquad (14.1)$$

In addition to the MOE-based synthesis, the hybrid or chelate routes have been introduced, where the Ti and Zr alkoxides are first reacted with acetic acid resulting in a more homogeneous distribution of B-site cations as obtained by the conventional MOE route [24, 25]. A higher level of chemical homogeneity in the PZT sol in the MOE route could be obtained by modification of Zr-propoxide by acetic acid [26]. In addition to MOE, other solvents, as for example *1,3*-propanediol, have been introduced [27].

A common problem related to the processing of lead-based perovskites is the expected PbO loss during heating. The vapor pressure of PbO above the PZT depends on the Zr/Ti ratio and is higher for the Zr-rich than for the Ti-rich solid solutions [28]. An excess of lead reagent in the amount of 10–20 mol % has been usually added already in the solution [29] or alternatively, introduced as an additional layer during the film processing [30, 31].

14.2.2 Deposition, Drying and Pyrolysis

Spin-coating has been the common method for deposition of FE thin films [19]. The as-deposited film contains functional groups, which may undergo further reactions of hydrolysis and polycondensation. In the drying and pyrolysis steps, in the approximate temperature range between 200 and 400°C, the following processes occur: evaporation of residual solvent(s), dehydration and dehydroxylation and organics decomposition. Due to the low mass of the film, the majority of the studies have been performed on the as-dried amorphous precursors, and the processes have been often followed by combined mass spectrometry (MS) or Fourier transform infrared spectrometry (FT-IR) and thermal analysis (thermogravimetry-TG) [32–39]. Nevertheless, some solutions for analyzing thermal decomposition of the films are proposed in [40–43].

As an example, the thermal decomposition of PZT as-dried precursor by TG/FT-IR is shown in Fig. 14.4. The thermal decomposition of the precursor occurred upon heating to about 500°C. The in situ IR analysis of evolved gases provided information on the temperature ranges of solvent evaporation, dehydroxylation and organic group thermal oxidation [35].

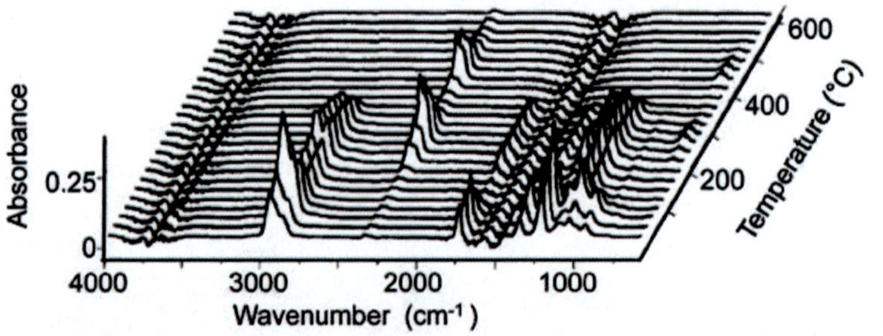

Fig. 14.4 FT-IR reflection spectra of the volatile species evolved upon thermal decomposition of the alkoxide-derived $Pb(Zr_{0.3}Ti_{0.7})O_3$ precursor as a function of temperature. The spectra were obtained by coupling the FT-IR spectrometer via a heated capillary. Characteristic bands of hydroxyl groups, CO_2, ester, ether and alcohol, are 3600–3700 cm^{-1}, 2300–2400 cm^{-1}, 1750 cm^{-1} and 1260 cm^{-1}, 1130 cm^{-1}, 1060 cm^{-1}, respectively. The analysis was performed with a heating rate of 10 K/min in air (50 ml/min). From [35] reproduced by permission of Elsevier

Fig. 14.5 Temperature ranges of perovskite (P) and pyrochlore (Py) phases in the PZT system depending on the composition. From [44] reproduced by permission of Cambridge University Press

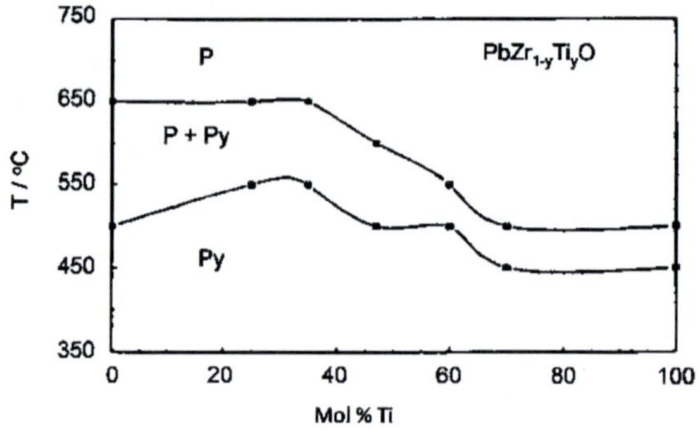

14.2.3 Crystallization

The crystallization of the thermodynamically stable perovskite PZT phase occurs between 500 and 700°C, usually via a transient non-ferroelectric pyrochlore-type phase, which is lead-deficient. The latter is stable between approximately 450 and 550°C [44–48]. By rapid annealing over the temperature region of stability of the pyrochlore phase, the fraction of the perovskite phase could be increased [49]. Lead oxide excess contributes to lowering of the perovskite crystallization temperature: PZT 53/47, prepared from the stoichiometric and 20 mol % excess solutions crystallized at 600 and 525°C, respectively [50].

The crystallization temperature of the perovskite phase depends on the Zr/Ti ratio and it is higher for Zr-rich PZT compositions, as shown in Fig. 14.5. It was found that the activation energy for the nucleation was larger than for the growth, in other words, nucleation is the rate-limiting step for the perovskite crystallization. The activation energy for nucleation is decreasing with increasing Ti-content in PZT [44, 51–53]. Consequently, a 45 nm thick $PbTiO_3$ layer was successfully

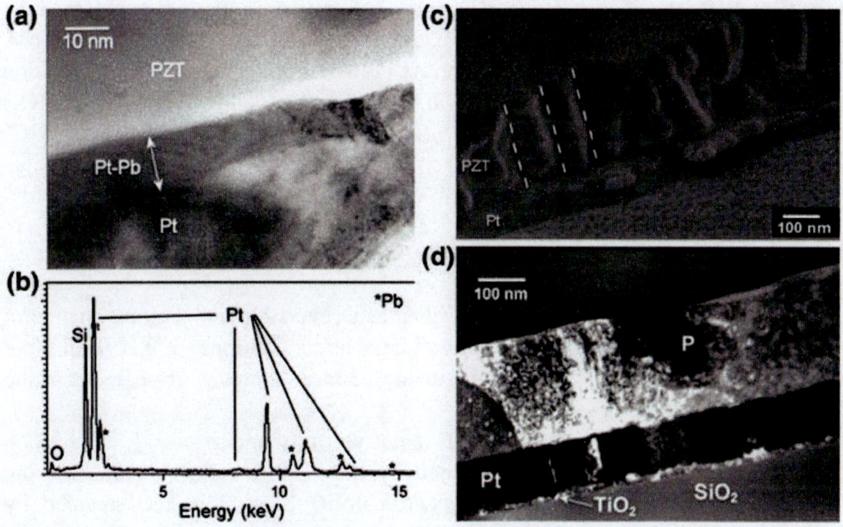

Fig. 14.6 **a** Cross-section TEM micrograph of $Pb(Zr_{0.3}Ti_{0.7})O_3$ film heated to 400°C in air and quenched. **b** The energy dispersive X-ray spectroscopy (EDXS) spectrum of the Pt–Pb phase. Cross section (**c**) FE-SEM and (**d**) TEM images of the (111) oriented films, heated at 400°C for 30 min and 60 min, respectively. The films shown in **a**–**c** were prepared by the modified MOE-based route from Pb-oxide and TM butoxides, while the film shown in **d** was synthesized from Pb-acetate. **a** and **b** from [61], reproduced with permission from Cambridge University Press, **c** and **d** from [62], reproduced with permission of the author

introduced as the nucleation layer for the PZT 53/47 films on sapphire substrates, resulting in a decrease of perovskite crystallization temperature for 100°C to 500°C [44].

Due to application requirement the PZT films have been mainly deposited on platinized silicon substrates. The main obstacles for successful integration of the films into silicon technology are high processing temperatures. At temperatures above 500°C, interface reactions may contribute to a decreased functional response of PZT films [54–56]. Perovskite crystallization temperature could be lowered by a design of solution chemistry, which contributed to a more homogeneous distribution of constituent metal atoms already in solution [26, 57]. Another possibility is to introduce UV absorbing species, such as β-diketonates in the coating solution and promote thermal decomposition of the as-deposited film by UV-irradiation in the photochemical solution deposition (PCSD) approach [58, 59]. Further possibilities of low-temperature thin film processing are discussed in [60].

Different nucleation layers have been successfully implemented to obtain PZT films with either (111) or (100) orientation of the perovskite phase.

The (111) perovskite orientation could be obtained through an 'in-situ' generated nucleation layer of Pt–Pb intermetallic phase, as a consequence of a partial reduction of Pb^{2+} species in the as-deposited film [57, 63, 64]. In Fig. 14.6 the transmission electron microscope (TEM) micrograph of the Pt–Pb phase in the PZT film quenched after heating to 400°C and the columnar microstructure obtained via heterogeneous nucleation on the Pt–Pb phase after heating at 400°C are shown [61]. A thin, 1–2 nm, layer of TiO_2 rutile phase (110) was also an efficient nucleation layer for (111) perovskite orientation, while on the other hand the PbO (or $PbTiO_3$) acted as nucleation layer for the (100) orientation [16, 65, 66].

14.3 Electrocaloric Effect

The electrocaloric effect (ECE) is described as a reversible temperature change of the material under applied electric field at adiabatic conditions [3, 9, 10]. In other words, the material heats up or cools down due to an increase or decrease of the applied electric field, respectively.

The mechanism underlying the ECE is not yet fully understood. It is based on the entropy change, which is stimulated by the electric field-induced ordering of dipoles. The polarization of the electrocaloric material is accompanied by an increase of temperature under adiabatic conditions, and opposite, the depolarization causes the temperature to drop. This phenomenon allows performing state-changing cycles in a similar manner as in a Carnot cycle [67–70]. This thermodynamic cycle describes a reversible change of an ideal gas, which enables to convert a given amount of thermal energy into work or, conversely, to use a given amount of work for cooling. The cycle consists of four phases of operation; two adiabatic and two isothermal, where the adiabatic change refers to a process where no heat is transferred from the environment or vice versa.

The pressure–volume (*P*–*V*) diagram from Fig. 14.7a illustrates the changes in pressure, volume and temperature of a gas, confined in a cylinder. The point 1 describes the low temperature, low pressure and high volume state. On the path from 1 to 2, the gas is compressed by the work from outside—by compression (+*P*), causing the temperature to rise from "T_{low}" to "T_{high}", (adiabatic compression). On the path from 2 to 3, the heat is ejected to the heat sink and the gas is isothermally compressed. On the path from 3 to 4 the system is insulated from the environment and the pressure decreases ($-P$). Since no heat is transferred from the system, the pressure reduction is compensated by a decrease in temperature from "T_{high}" to "T_{low}" (adiabatic expansion). On the path from 4 to 1, the heat is transferred from the environment, and the gas expands isothermally. The system returns to the initial state and the cycle can start again.

In Fig. 14.7b the electrocaloric cooling cycle is presented within the electric field–entropy (*E*–*S*) diagram. The cycle starts with the application of the electric field (+*E*) to the material, insulated from the environment (path from 1 to 2). This causes a partial alignment of dipoles, and consequent decrease of entropy. Since the material is insulated from the environment, the reduction of entropy is compensated by an increase of temperature from "T_{low}" to "T_{high}" (adiabatic polarization). On the path from 2 to 3 the heat is transferred to the surroundings ($-Q$) and the electric field increases further to prevent the dipoles from reabsorbing the

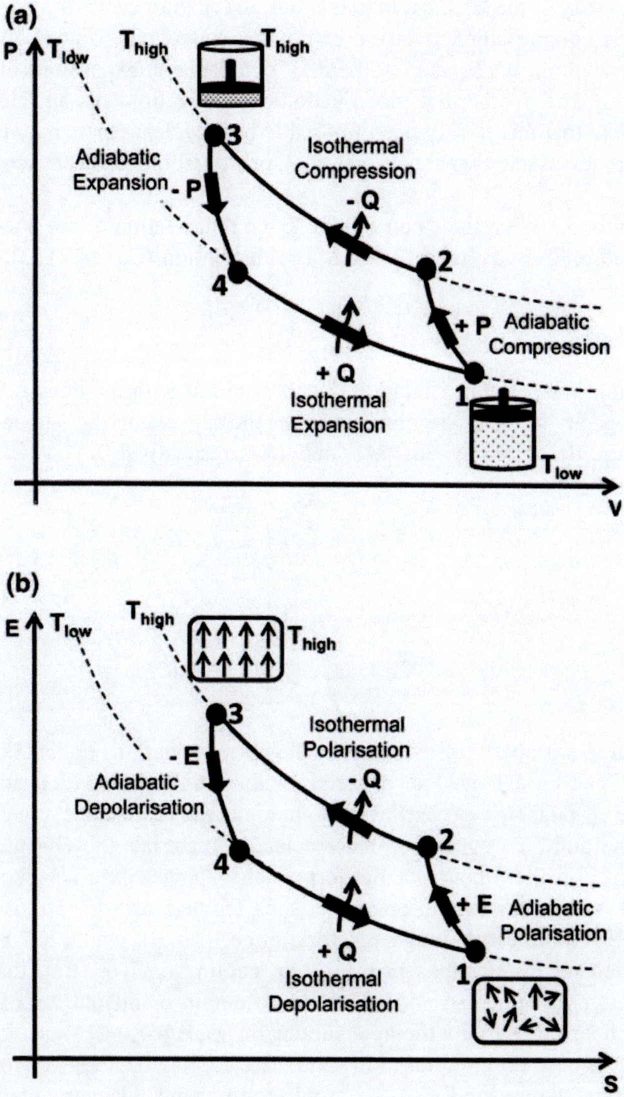

Fig. 14.7 **a** (*P–V*) diagram of the Carnot cycle in a refrigerator and **b** electric field-entropy (*E–S*) diagram of the cooling cycle in the electrocaloric material, after [68–71]

heat (isothermal polarization). On the path from 3 to 4, the material is insulated from the environment and the applied electric field is reduced ($-E$). The electric dipoles in the material return to disordered state and thus, the sample temperature drops from "T_{high}" to "T_{low}". This adiabatic depolarization is followed by isothermal depolarization on the path from 4 to 1 [67–70].

The similarity of the ECE cycle to the thermodynamic cycle of a gas is evident: the adiabatic compression/expansion of a gas is analogous to adiabatic polarization/depolarization, while the isothermal compression/expansion of a gas is analogous to the isothermal polarization/depolarization of an electrocaloric material. Note that this is only one of possible heat cycles; others may involve two adiabatic processes and two at constant field, or two isothermal and two at constant field [69].

The relation between the electric field E, the polarization of the material P and the temperature T is governed by the Maxwell relation (Eq. 14.2) [3].

$$\left(\frac{\partial P}{\partial T}\right)_E = \left(\frac{\partial S}{\partial E}\right)_T \tag{14.2}$$

The reversible temperature change ΔT for a material with the density ρ and with the heat capacity C, and the electrocaloric entropy change ΔS under adiabatic conditions are described by Eqs. 14.3 and 14.4, respectively.

$$\Delta T = -\frac{1}{C\rho}\int_{E_1}^{E_2} T\left(\frac{\partial P}{\partial T}\right)_E dE \tag{14.3}$$

$$\Delta S = -\int_{E_1}^{E_2}\left(\frac{\partial P}{\partial T}\right)_E dE \tag{14.4}$$

The ECE is a property of dielectrics. The above equations indicate that a large ΔS and ΔT can be achieved in materials with a high polarization and a strong dependence of dielectric properties on temperature. Ferroelectric materials apply to both conditions. Furthermore, in ferroelectric materials the change of polarization is large in the vicinity of the ferroelectric–paraelectric (*F–P*) phase transition, so it is advantageous to operate at such temperatures [3, 10, 67].

The ECE resembles the magnetocaloric effect (MCE), which has been extensively studied for applications in solid-state cooling devices. But, the magnetic fields are more complicated and expensive to obtain compared with the electric fields, which severely limits the applications of MCE [67, 72].

The ECE is of great importance in solid-state cooling devices for a broad range of applications, such as cooling of electronic components, like computer chips, and temperature regulation of sensors [10, 68, 73]. A refrigerator based on ECE could provide environmentally friendlier and more efficient alternative for conventional vapor-compression refrigeration technology [67, 68, 70]. Moreover, the ECE could enable recovery of electrical power from waste-heat [10, 70, 73].

The earliest experimental measurement of the ECE was performed in 1930 [9] and investigations of suitable electrocaloric materials intensified 1960s and 1970s. The effect was small and insufficient for practical applications. For example, the largest adiabatic temperature change $\Delta T \sim 2.6$ K was found in $Pb_{0.99}Nb_{0.02}(Zr_{0.75}Sn_{0.20}Ti_{0.05})_{0.98}O_3$ ceramics upon application of $E = 300$ kV/cm

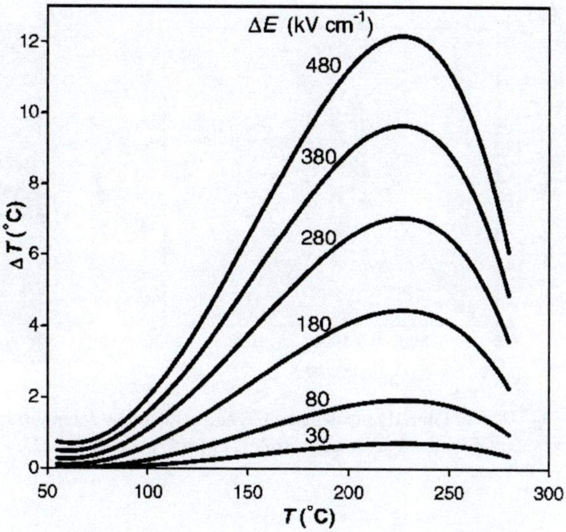

Fig. 14.8 Temperature changes ΔT at different applied electric fields ΔE as a function of temperature T. From [10] reproduced by permission of American Association for the Advancement of Science

(voltage of 7.5 kV) at $T = 160°C$, which is close to the *F–P* phase transition of this material; $T_c = 161°C$ [74].

Higher electric fields should induce larger temperature changes, however, the limitation is the breakdown field of the material. Thin films can withstand much higher electric fields. As their thickness is in sub-μm range, the applied voltages are low, for example, the voltage of 5 V across a 100 nm thick film corresponds to the electric field of 500 kV/cm. Therefore thin films represent an interesting option also in terms of device miniaturization.

In 2006, Mischenko et al. [10] reported a large ECE in the solution derived $Pb(Zr_{0.95}Ti_{0.05})O_3$ (PZT 95/5) thin films, prepared by the chelate based route (for details refer to Sect. 14.2.1). The about 350 nm thick films on $Pt(111)/TiO_x/SiO_x/Si(100)$ substrates were annealed at 650°C for 10 min on a hot plate in air. Figure 14.8 shows the temperature changes ΔT under applied electric field at different temperatures. The results were obtained indirectly, namely, by measuring the temperature dependence of the polarization *P(T)* and the ΔT was deduced by Eq. 14.3.

The observed adiabatic temperature change was $\Delta T = 12$ K at 25 V (776 kV/cm) at 226°C, which is close to the temperature of *F–P* phase transition. The ECE at such, rather high, working temperatures could allow cooling application in the automotive, aerospace or food industries [75], but lower working temperatures would be advantageous for the room-temperature cooling solutions.

In the search of materials with a giant ECE, the entropy difference between the dipole-disordered and ordered state should be considered [76]. Relaxors are characterized by broad, frequency-dependent dielectric permittivity maxima and slim polarization-electric field hysteresis loops [76, 77].

Relaxor ferroelectric 0.9 $PbMg_{1/3}Nb_{2/3}O$–0.1 $PbTiO_3$ (0.9PMN–0.1PT) has got a frequency-dependent permittivity maximum at ~60°C [78]. Mischenko et al.

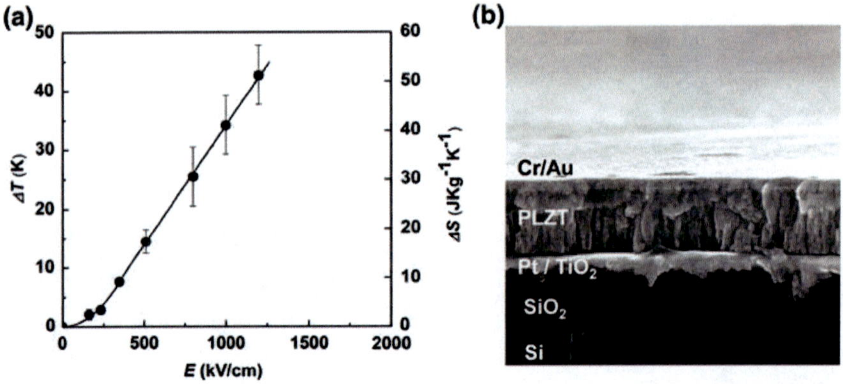

Fig. 14.9 **a** Directly measured ECE and **b** the cross-section microstructure of 8/65/35 PLZT thin films on platinised silicon substrates. From [76] reproduced by permission of American Institute of Physics

investigated ECE in 0.9PMN–0.1PT thin films prepared from an alkoxide-acetate based solution [75]. The films on Pt(111)/Ti/SiO_2/Si(100) substrates were annealed at 750°C and consisted of a randomly oriented perovskite phase with traces of pyrochlore. The 260 nm thick films achieved the temperature change of $\Delta T = 5$ K at 25 V (960 kV/cm) at 75°C. Similarly as in PZT 95/5 films [10], the temperature change was indirectly determined from the *P(T)* measurements.

Correia et al. reported electrocaloric properties of 0.93PMN–0.07PT thin films exhibiting the maximum temperature change at as low as 25°C [79]. The authors prepared perovskite 0.93PMN–0.07PT thin films on a $Pb(Zr_{0.8}Ti_{0.2})O_3$ nucleation layer [80]. The indirectly determined $\Delta T = 9$ K was obtained at $E = 723$ kV/cm (15 V) for the 210 nm thick films [81].

Most of the ECE studies have relied on indirect measurements to deduce the ECE of a material, assuming relatively small hysteresis losses and sufficiently small coercive fields [82]. Lu et al. reported the directly measured ECE in relaxor ferroelectric lantanum-modified lead zirconate titanate $(Pb_{0.88}La_{0.08})(Zr_{0.65}Ti_{0.35})O_3$ (PLZT 8/65/35) thin films [76]. A high resolution calorimeter [71, 82] was utilized for direct measurements of the ECE. The PLZT thin films were prepared by a modified MOE-based route [26]. About 450 nm thick films deposited on a nucleation layer of $PbTiO_3$ and rapid thermally annealed at 650°C crystallized in perovskite phase with preferred (100) orientation. The dielectric permittivity maximum of bulk ceramics with the same composition is at 110°C [83]. Figure 14.9a shows the ECE of the PLZT films measured at 45°C, and comparable results were obtained also at higher temperatures. The direct electrocaloric measurement confirmed the giant electrocaloric effect, with the temperature change exceeding 40 K at the field amplitude of around 1100 kV/cm.

Table 14.1 summarizes the electrocaloric properties of sol-gel derived thin films. Since both, ΔT and ΔS are important for electrocaloric materials, the value

Table 14.1 Comparison of ECE in selected thin films (from Ref. [76])

Thin film	T (°C)	E (kV/cm)	ΔS (J/kgK)	ΔT (K)	ΔT*ΔS (J/kg)
PZT 95/5[a]	226	48	7.9	12	95
0.93PMN–0.07PT[b]	25	72.3	9.7	9	87.3
PLZT[c]	45	120	50	40	2000

[a] [10], [b] [79], [c] [76]

of ΔT times ΔS as a figure of merit was also calculated. The optimum electrocaloric properties were found in the PLZT thin films, where $\Delta T{*}\Delta S$ is 2000 J/kg; a value suitable for niche applications [76].

14.4 Energy Harvesting

Energy harvesting (EH) or scavenging offers another possibility of applications of ferroelectric thin film structures. It is the process by which the energy is derived from external sources, captured, stored and used in small autonomous devices. In EH the energy source is available in the environment and is therefore free, in contrast to conventional fuels, such as oil and coal. The sources for EH include the energy of ambient radiation sources, mechanical vibrations or even human energy such as energy produced by walking, etc. [84–87]. The EH devices provide small amounts of power and they can be used for low-energy electronics such as mobile phones, MP3 players, pagers and voice recorders [84].

The EH is based on different physical principles, for example, piezoelectric effect, electromagnetic induction or electrostriction in dielectric elastomers. In this part of the chapter a short overview on current status and further research issues in ferroelectric thin films, where the piezoelectricity of the films is exploited for energy harvesting applications, is given.

In piezoelectric materials the mechanical stress T and the strain S are related to the dielectric displacement D and the electric field E, as indicated in the constitutive equations:

$$\{S\} = \left[s^E\right] \{T\} + [d] \{E\} \tag{14.5}$$

$$\{D\} = [d]^T \{T\} + \left[\varepsilon^T\right] \{E\} \tag{14.6}$$

where $[s^E]$ is the compliance matrix evaluated at a constant electric field, $[\varepsilon^T]$ is the permittivity matrix evaluated at a constant stress and $[d]$ is the matrix of the piezoelectric coefficients.

The piezoelectric coefficients of the films differ from the coefficients of the bulk ceramics with the same composition. One of the main reasons is that the films are clamped by the substrates. For a clamped film, the ratio D_3/T_3 does not represent the piezoelectric coefficient d_{33} of the free sample, but an effective piezoelectric coefficient d_{33}^{eff} [88].

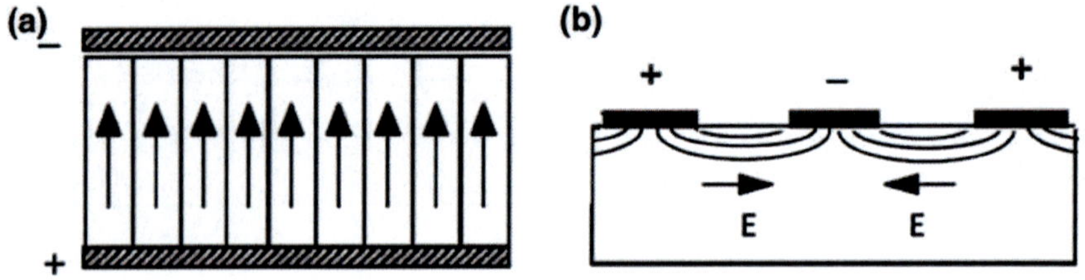

Fig. 14.10 Basic electrode systems for driving piezoelectric films; **a** planar capacitor structure with parallel top and bottom electrodes, **b** structure with interdigitated electrodes on the surface of the film. From [16] reproduced by permission of John Wiley and Sons

$$d_{33}^{eff} = d_{33} - 2d_{31}\frac{\frac{v_s}{Y_s} + s_{13}^E}{(s_{11}^E + s_{12}^E)}, \tag{14.7}$$

where d_{33} and d_{31} are the direct and the transverse piezoelectric coefficients, respectively, (C/N), s_{13}^E, s_{11}^E, s_{12}^E are the elastic compliance coefficients at a constant electric field (m^2/N), v_s is the Poisson's ratio of the substrate, and Y_s is the Young's modulus of the substrate (N/m^2).

Since in ferroelectric materials $d_{31} < 0, s_{13} < 0$ and d_{31} is relatively large, the effective coefficient measured for the films is lower than that of the unclamped material $\left(d_{33}^{eff} < d_{33}\right)$. The non-zero piezoelectric tensor elements for polycrystalline ferroelectric thin film are d_{33}, d_{31} and d_{15} [13, 16, 88].

The basic electrode configuration for driving a piezoelectric film is a planar structure with top and bottom electrodes as shown in Fig. 14.10a. The arrows symbolize the average polarization per grain in a poled ferroelectric material. Polar films also offer the possibility to work with the shear d_{15} piezoelectric contribution. Interdigitated electrodes are applied for this purpose. In this configuration the field between the electrodes is more or less parallel to the film plane (Fig. 14.10b). In EH, the use of interdigitated electrodes is considered suitable, since the harvested voltage should be as large as possible [16].

A typical EH device based on thin film technology is a film-substrate bimorph structure [16, 86, 87, 89, 90]. The structure is designed so that the bimorph is at one end mounted in the rigid base and at the other one a free-standing structure is loaded by a proof mass. The latter is added to improve the electrical output power generation as studied in [87]. The schematic diagram of the film-substrate bimorph structure is shown in Fig. 14.11. The structure is designed to transform the mechanical vibrations from ambient vibration sources to electrical energy through the piezoelectric effect [87, 89–91]. It can be used in the environment, where parasitic mechanical vibrations are present, for example in workshops where mechanical vibrations are produced by engines and machines.

The first report on EH micro-generators processed from ferroelectric films was fabricated by screen-printing technology [92]. A tapered cantilever beam was capable of generating the power of 3 μW at a frequency of 80 Hz [92, 93]. Most of

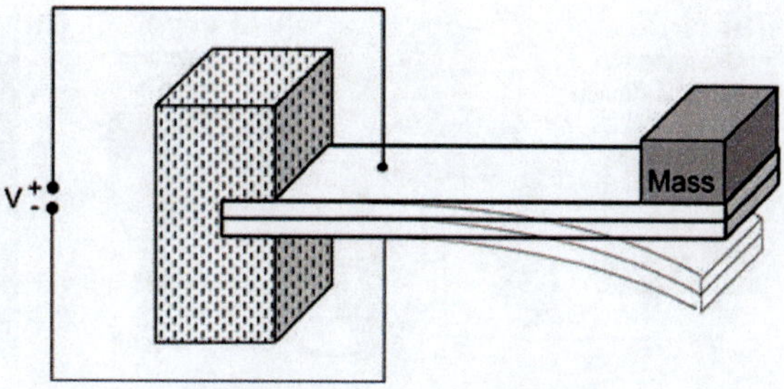

Fig. 14.11 A schematic diagram of a film-substrate bimorph structure mounted on a rigid base. The free-standing structure is loaded with a proof mass (after Refs [86, 87, 89])

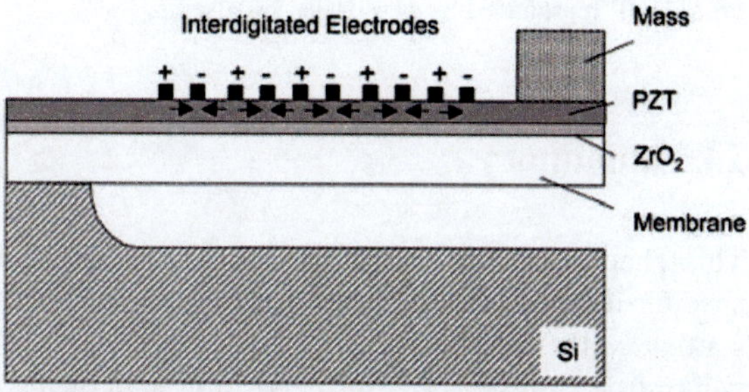

Fig. 14.12 PZT micro-generator based on the thin film technology. From [95] reproduced by permission of Elsevier

the available environmental vibrations have a high acceleration magnitude at around 100 Hz, for example, resonance frequencies of washing machines are around 110 Hz [94]. Later, thin films and micromachined technology have been used for fabricating EH devices [91, 95].The sol-gel derived PZT thin films were prepared on silicon-based substrates to form the bimorph cantilever structures [91, 95, 96]. The fabricated device exploiting the d_{31} piezoelectric mode produced the power of 2.15 μW at the frequency of 461 Hz [96]. The thin film devices based on d_{33} piezoelectric mode were also designed [91, 95]. The schematic diagram of the structure is shown in Fig. 14.12. The Pt/Ti electrodes were patterned into an interdigitated shape on top of the thin film.

The proof mass was added to the free-end of the cantilever. The maximum continuous power output of this device was 1 μW with the output voltage of 2.4 V [91, 95]. The power, produced by the EH devices based on sol-gel derived PZT thin films, is in order of microwatts [91, 97, 98], which is enough for hand-held devices.

Another application of piezoelectric-film devices is in conversion of human motions, such as movements of legs and arms, into electrical power. Such devices can be integrated into clothes, backpacks and shoes [84, 99, 100]. In Fig. 14.13a the concept of the EH-intelligent clothing with flexible piezoelectric layers is presented. The state-of-the-art development is in new and more efficient thin film devices for MEMS applications and for integration into clothes and shoes. One of the first systems for measuring the electrical output of a piezoelectric film, integrated into a shoe under realistic excitation, is shown in Fig. 14.13b [100].

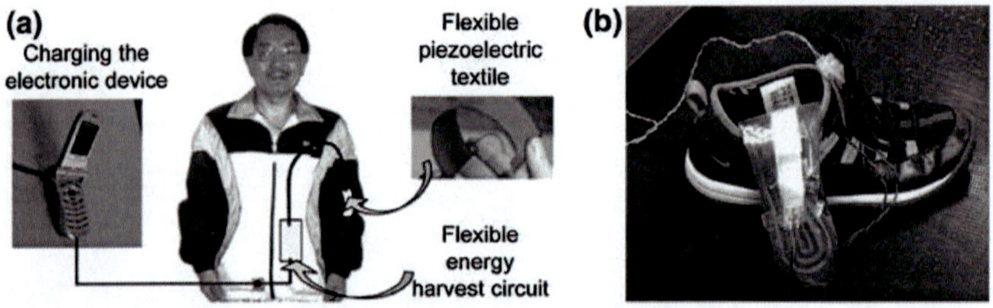

Fig. 14.13 **a** Concept of EH-intelligent clothing with flexible piezoelectric textile. From [99] reproduced by permission of Springer. **b** Prototype of the piezoelectric film inserted in a shoe. From [100] reproduced by permission of Elsevier

14.5 Summary

This chapter provides a short overview of research of ferroelectric thin films, specifically those based on $Pb(Zr,Ti)O_3$ (PZT). The steps of thin film processing, starting with the commonly used sol-gel synthesis route, based on the early works from mid-1980s, film deposition and stages of the thermal treatment are discussed.

The basics of two areas of applications involving energy conversion are presented, namely the electrocaloric effect and energy harvesting. While the former one is based on the electric field-induced dipole ordering, which results in temperature changes, the latter explores the piezoelectric effect, namely the generation of electric charge upon application of stress.

Acknowledgments This work was supported by the Slovenian Research Agency projects P2-0105, J2-1227 and 1000-09-310144 and by the European Union, European Regional Development Fund and Ministry of Higher Education, Science and Technology of Slovenia. The authors would like to thank Zdravko Kutnjak, Brigita Rožič and Sebastjan Glinšek for valuable advice in the process of writing.

References

1. Gurkovich SR, Blum JB (1984) Preparation of monolithic $PbTiO_3$ by a sol-gel process. In: Hench LL, Ulrich DR (eds) Ultrastructure processing of ceramics, glasses and composites. Wiley, New York, pp 152–160
2. Budd KD, Dey SK, Payne DA (1985) Sol-gel processing of $PbTiO_3$, $PbZrO_3$, PZT and PLZT thin films. Brit Cer Pr 36:107–121
3. Lines ME, Glass AM (1977) Principles and applications of ferroelectrics and related materials. Clarendon Press, Oxford
4. Scott JF, Paz De Araujo CA (1989) Ferroelectric memories. Science 246(4936):1400–1405
5. Auciello O, Scott JF, Ramesh R (1998) The physics of ferroelectric memories. Phys Today 51(7):22–27

6. Boettger U, Sommerfelt SR (2003) Ferroelectric random access memories. In: Waser R (ed) Nanoelectronics and information technologies advanced electronic materials and novel devices. Wiley VCH, Weinheim, pp 527–606
7. Dimos D, Mueller CH (1998) Perovskite thin films for high-frequency capacitor applications. Annu Rev Mater Sci 28(1):397–419
8. Brennecka GL, Ihlefeld JF, Maria JP, Tuttle BA, Clem PG (2010) Processing technologies for high-permittivity thin films in capacitor applications. J Am Ceram Soc 93(12): 3935–3954. doi:10.1111/j.1551-2916.2004.01459.x
9. Kobeko PP, Kurchatov IV (1930) Dielectric properties of ferroelectrics. Z Phys 66:192
10. Mischenko AS, Zhang Q, Scott JF, Whatmore RW, Mathur ND (2006) Giant electrocaloric effect in thin-film $PbZr_{0.95}Ti_{0.05}O_3$. Science 311(5765):1270–1271. doi:10.1126/science.1123811
11. Anton SR, Sodano HA (2007) A review of power harvesting using piezoelectric materials (2003–2006). Smart Mater Struct 16(3):R1–R21
12. Cook-Chennault KA, Thambi N, Sastry AM (2008) Powering MEMS portable devices—a review of non-regenerative and regenerative power supply systems with special emphasis on piezoelectric energy harvesting systems. Smart Mater Struct 17(4):043001–043033. doi:10.1088/0964-1726/17/4/043001
13. Damjanovic D (1998) Ferroelectric, dielectric and piezoelectric properties of ferroelectric thin films and ceramics. Rep Prog Phys 61(11):1267–1324
14. Sreenivas K, Reaney I, Maeder T, Setter N, Jagadish C, Elliman RG (1994) Investigation of Pt/Ti bilayer metallization on silicon for ferroelectric thin film integration. J Appl Phys 75(1):232–239
15. Trolier-McKinstry S, Muralt P (2004) Thin film piezoelectrics for MEMS. J Electroceram 12(1–2):7–17
16. Muralt P (2008) Recent progress in materials issues for piezoelectric MEMS. J Am Ceram Soc 91(5):1385–1396. doi:10.1111/j.1551-2916.2008.02421.x
17. Muralt P, Polcawich RG, Trolier-McKinstry S (2009) Piezoelectric thin films for sensors, actuators, and energy harvesting. MRS Bull 34(9):658–664
18. Schwartz RW, Schneller T, Waser R (2004) Chemical solution deposition of electronic oxide films. C R Chimie 7(5):433–461. doi:10.1016/j.crci.2004.01.007
19. Brinker CJ, Scherer GW (1990) Sol-gel science: The physics and chemistry of sol-gel processing. Academic Press, San Diego
20. Turova NY, Turevskaya EP, Kessler VG, Yanovskaya MI (2002) The chemistry of metal alkoxides. Kluwer Academic Publishers, Norwell
21. Dekleva TW, Hayes JM, Cross LE, Geoffroy GL (1988) Sol-gel processing of lead titanate in 2-methoxyethanol: Investigations into the nature of the prehydrolyzed solutions. J Am Ceram Soc 71(5):C280–C282
22. Beltram T, Kosec M, Stavber S (1993) Reactions taking place during the sol-gel processing of PLZT. Mater Res Bull 28(4):313–320
23. Hubert-Pfalzgraf LG (1992) Heterometallic aggregates as intermediates on the molecular routes to multicomponent oxides. In: Hampden-Smith MJ, Klemperer WG (eds) Better ceramics through chemistry V, vol 271. Materials Research Society, Pittsburg, pp 15–25
24. Yi G, Wu Z, Sayer M (1988) Preparation of $Pb(Zr, Ti)O_3$ thin films by sol gel processing: Electrical, optical, and electro-optic properties. J Appl Phys 64(5):2717–2724. doi:10.1063/1.341613
25. Assink RA, Schwartz RW (1993) Proton and carbon-13 NMR investigations of lead zirconate titanate ($Pb(Zr, Ti)O_3$) thin-film precursor solutions. Chem Mater 5(4):511–517. doi:10.1021/cm00028a019
26. Malič B, Arčon I, Kodre A, Kosec M (2006) Homogeneity of $Pb(Zr, Ti)O_3$ thin films by chemical solution deposition: extended X-ray absorption fine structure spectroscopy study of zirconium local environment. J Appl Phys 100(5):051612-1–051612-8
27. Tu YL, Calzada ML, Phillips NJ, Milne SJ (1996) Synthesis and electrical characterization of thin films of PT and PZT made from a diol-based sol-gel route. J Am Ceram Soc 79(2):441–448. doi:10.1111/j.1151-2916.1996.tb08142.x

28. Haerdtl KH, Rau H (1969) PbO vapor pressure in the $Pb(Zr_xTi_{1-x})O_3$ system. Solid State Commun 7:41–45
29. Sato E, Huang Y, Kosec M, Bell A, Setter N (1994) Lead loss, preferred orientation, and the dielectric properties of sol-gel prepared lead titanate thin films. Appl Phys Lett 65(21): 2678–2680. doi:10.1063/1.112600
30. Tani T, Payne DA (1994) Lead oxide coatings on sol-gel-derived lead lanthanum zirconium titanate thin layers for enhanced crystallization into the perovskite structure. J Am Ceram Soc 77(5):1242–1248
31. Brennecka GL, Huebner W, Tuttle BA, Clem PG (2004) Use of stress to produce highly oriented tetragonal lead zirconate titanate (PZT 40/60) thin films and resulting electrical properties. J Am Ceram Soc 87(8):1459–1465
32. Coffman PR, Barlingay CK, Gupta A, Dey SK (1996) Structure evolution in the $PbO–ZrO_2–TiO_2$ sol-gel system: Part II—pyrolysis of acid and base-catalyzed bulk and thin film gels. J Sol-gel Sci Techn 6(1):83–106
33. Merkle R, Bertagnolli H (1998) Investigation of the pyrolysis of lead zirconate titanate gels with coupled differential thermal analysis, thermogravimetry and infrared spectroscopy. J Mater Sci 33(17):4341–4348
34. Merkle R, Bertagnolli H (1998) Investigation of the crystallization of lead titanate and lead zirconate titanate with X-ray diffraction and differential thermal analysis. Bericht Bunsen Gesell 102(8):1023–1031. doi:10.1002/bbpc.19981020807
35. Nouwen R, Mullens J, Franco D, Yperman J, Van Poucke LC (1996) Use of thermogravimetric analysis—Fourier transform infrared spectroscopy in the study of the reaction mechanism of the preparation of $Pb(Zr, Ti)O_3$ by the sol-gel method. Vib Spectrosc 10(2):291–299
36. Fe L, Norga G, Wouters DJ, Nouwen R, Van Poucke LC (2000) Absorption-reflection infrared spectroscopy studies of sol-gel prepared ferroelectric $Pb(Zr, Ti)O_3$ thin films on Pt electrodes. J Sol-Gel Sci Techn 19(1–3):149–152
37. Malič B, Cilenšek J, Mandeljc M, Kosec M (2005) Crystallization study of the alkoxide-based $Pb(Zr_{0.30}Ti_{0.70})O3$ thin-film precursor. Acta Chim Slov 52(3):259–263
38. Malič B, Calzada ML, Cilenšek J, Pardo L, Kosec M (2010) Thermal analysis study of 0.7 $Pb(Mg_{1/3}Nb_{2/3})O_3$–0.3 $PbTiO_3$ diol-based precursors for chemical solution deposition of thin films. Adv Appl Ceram 109(3):147–151. doi:10.1179/174367509X12502621261415
39. Malič B, Kupec A, Kosec M (2012) Thermal Analysis and its role in chemical solution deposition of functional oxide thin film. In: Waser R, Kosec M, Payne DA (eds) Chemical solution deposition of functional oxide thin films. Springer (To be published)
40. Gallagher PK (1992) Applications of thermoanalytical methods to the study of thin films. J Therm Anal Calorim 38(1):17–26. doi:10.1007/bf02109105
41. Leskelä M, Leskelä T, Niinistö L (1993) Thermoanalytical methods in the study of inorganic thin films. J Therm Anal Calorim 40(3):1077–1088. doi:10.1007/bf02546867
42. Gallagher PK, Gyorgy EM, Schrey F (1987) The use of thermomagnetometry to follow reactions of thin-films. Thermochim Acta 121:231–239
43. Hirashima H, Imai H, Miah MY, Bountseva IM, Beckman IN, Balek V (2004) Preparation of mesoporous titania gel films and their characterization. J Non-Cryst Solids 350:266–270. doi:10.1016/j.jnoncrysol.2004.09.009
44. Kwok CK, Desu SB (1993) Low temperature perovskite formation of lead zirconate titanate thin films by a seeding process. J Mater Res 8(2):339–344. doi:10.1557/JMR.1993.0339
45. Chapin LN, Myers SA (1990) Microstructure characterization of ferroelectric thin films used in non-volatile memories—optical and scanning electron microscopy. In: Myers ER, Kingon AI (eds) Ferroelectric thin films, vol 200. Materials Research Society, Pittsburgh, pp 153–158
46. Faure SP, Barboux P, Gaucher P, Ganne JP (1992) Synthesis of ferroelectric thin films and ceramics from solution processes. Ferroelectr 128(1):19–24. doi:10.1080/00150199208015060
47. Carim AH, Tuttle BA, Doughty DH, Martinez SL (1991) Microstructure of solution-processed lead zirconate titanate (PZT) thin films. J Am Ceram Soc 74(6):1455–1458. doi:10.1111/j.1151-2916.1991.tb04130.x

48. Griswold EM, Weaver L, Sayer M, Calder JD (1995) Phase transformations in rapid thermal processed lead zirconate titanate. J Mater Res 10:3149–3159. doi:10.1557/JMR.1995.3149
49. Tuttle BA, Schwartz RW, Doughty DH AVJ (1990) Characterization of chemically prepared PZT thin films. In: Myers ER, Kingon AI (eds) Ferroelectric thin films, vol 200. Materials Research Society, Pittsburgh, pp 159–165
50. Suzuki H, Kondo Y, Kaneko S, Hayashi T (1999) Orientation control of low-temperature processed Pb(Zr, Ti)O_3 thin films. T MRS Jap 24:39–42
51. Chen KC, Mackenzie JD (1990) Crystallization kinetics of metallo-organics derived PZT thin film. In: Zelinski BJ, Brinker CJ, Clark DE, Ulrich DR (eds) Better ceramics through chemistry IV, vol 180. Materials Research Society, Pittsburg, pp 663–668
52. Lee JS, Kim CJ, Yoon DS, Choi CG, Kim JM, No K (1994) Effects of seeding layer on perovskite transformation, microstructure and transmittance of sol-gel-processed lanthanum-modified lead zirconate titanate films. Jpn J Appl Phys 33:260–265. doi:10.1143/JJAP.33.260
53. Peng CH, Desu SB (1992) Investigation of structure development in MOD Pb(Zr_xTi_{1-x})O_3 films by an optical method. In: Kingon AI, Myers ER, Tuttle BA (eds) Ferroelectric thin films II, vol 243. Materials Research Society, Pittsburg, pp 335–340
54. Brooks KG, Reaney IM, Klissurska R, Huang Y, Bursill L, Setter N (1994) Orientation of rapid thermally annealed lead zirconate titanate thin films on (111) Pt substrates. J Mater Res 9(10):2540–2553
55. Park KH, Kim CY, Jeong YW, Kwon HJ, Kim KY, Lee JS, Kim ST (1995) Microstructures and interdiffusions of Pt/Ti electrodes with respect to annealing in the oxygen ambient. J Mater Res 10(7):1790–1794. doi:10.1557/JMR.1995.1790
56. Atsuki T, Soyama N, Sasaki G, Yonezawa T, Ogi K, Sameshima K, Hoshiba K, Nakao Y, Kamisawa A (1994) Surface morphology of lead-based thin films and their properties. Jpn J Appl Phys 33:5196–5200. doi:10.1143/JJAP.33.5196
57. Malič B, Mandeljc M, Dražič G, Škarabot M, Muševic I, Kosec M (2008) Strategy for low-temperature crystallization of titanium-rich PZT thin films by chemical solution deposition. Integr Ferroelectr 100(1):285–296
58. Calzada ML, Bretos I, Jimenez R, Guillon H, Pardo L (2004) Low-temperature processing of ferroelectric thin films compatible with silicon integrated circuit technology. Adv Mater 16(18):1620–1624. doi:10.1002/adma.200306401
59. Bretos I, Jimenez R, Garcia-Lopez J, Pardos L, Calzada ML (2008) Photochemical solution deposition of lead-based ferroelectric films: avoiding the PbO-excess addition at last. Chem Mater 20(18):5731–5733. doi:10.1021/cm801269g
60. Malič B, Glinšek S, Kosec M (2012) Low temperature processing of functional oxide thin films. In: Waser R, Kosec M, Payne D (eds) Chemical solution deposition of functional oxide thin films. Springer, (To be published)
61. Mandeljc M, Malič B, Kosec M, Dražič G (2005) Influence of the lead-compound on PZT 30/70 thin film orientation. In: Ferroelectric thin films XIII, vol 902E. Materials Research Society, Pittsburg, p 0902-T03-37. doi:10.1557/PROC-0902-T03-37
62. Mandeljc M (2006) Študij kristalizacije tankih plasti na osnovi Pb(Zr, Ti)O_3 (Study of crystallization of Pb(Zr,Ti)O_3 thin films). PhD Thesis, Uni Ljubljana, Ljubljana
63. Chen SY, Chen IW (1994) Temperature-time texture transition of Pb($Zr_{1-x}Ti_x$)O_3 thin films: I, role of Pb-rich intermediate phases. J Am Ceram Soc 77(9):2332–2336
64. Huang Z, Zhang Q, Whatmore RW (1999) Structural development in the early stages of annealing of sol-gel prepared lead zirconate titanate thin films. J Appl Phys 86(3):1662–1669
65. Muralt P, Maeder T, Sagalowicz L, Hiboux S, Scalese S, Naumovic D, Agostino RG, Xanthopoulos N, Mathieu HJ, Patthey L, Bullock EL (1998) Texture control of $PbTiO_3$ and Pb(Zr, Ti)O_3 thin films with TiO_2 seeding. J Appl Phys 83(7):3835–3841
66. Muralt P (2006) Texture control and seeded nucleation of nanosize structures of ferroelectric thin films. J Appl Phys 100:051605–051616. doi:10.1063/1.2337362
67. Lu SG, Rožič B, Kutnjak Z, Zhang Q (2010) Electrocaloric effect (ECE) in ferroelectric polymer films. In: Coondoo I (ed) Ferroelectrics. InTech, Rijeka, pp 99–118

68. Lu SG, Zhang Q (2009) Electrocaloric materials for solid-state refrigeration. Adv Mater 21(19):1983–1987. doi:10.1002/adma.200802902
69. Scott JF (2011) Electrocaloric materials. Annu Rev Mater Res 41(1):229–240. doi:10.1146/annurev-matsci-062910-100341
70. Epstein RI, Malloy KJ (2009) Electrocaloric devices based on thin-film heat switches. J Appl Phys 106(6):064501–064507. doi:10.1063/1.3190559
71. Kutnjak Z, Petzelt J, Blinc R (2006) The giant electromechanical response in ferroelectric relaxors as a critical phenomenon. Nature 441(7096):956–959
72. Tishin M, Spichkin YI (2003) The magnetocaloric effect and its applications. Institute of Physics Publishing, London
73. Neese B, Chu B, Lu SG, Wang Y, Furman E, Zhang QM (2008) Large electrocaloric effect in ferroelectric polymers near room temperature. Science 321(5890):821–823. doi:10.1126/science.1159655
74. Tuttle BA, Payne DA (1981) The effects of microstructure on the electrocaloric properties of Pb(Zr, Sn, Ti)O_3 ceramics. Ferroelectr 37(1–4):603–606
75. Mischenko AS, Zhang Q, Whatmore RW, Scott JF, Mathur ND (2006) Giant electrocaloric effect in the thin film relaxor ferroelectric 0.9 $PbMg_{1/3}Nb_{2/3}O_3$–0.1 $PbTiO_3$ near room temperature. Appl Phys Lett 89(24):242912-1–242912-3. doi:10.1063/1.2405889
76. Lu SG, Rožič B, Zhang QM, Kutnjak Z, Li X, Furman E, Gorny LJ, Lin M, Malič B, Kosec M, Blinc R, Pirc R (2010) Organic and inorganic relaxor ferroelectrics with giant electrocaloric effect. Appl Phys Lett 97(16):162904-1–162904-3. doi:10.1063/1.3501975
77. Cross LE (1987) Relaxor ferroelectrics. Ferroelectr 76:241–267
78. Ye ZG, Dong M (2000) Morphotropic domain structures and phase transitions in relaxor-based piezo-ferroelectric (1-x)Pb($Mg_{1/3}Nb_{2/3}$)O_3-x$PbTiO_3$ single crystals. J Appl Phys 87(5):2312–2319
79. Correia TM, Young JS, Whatmore RW, Scott JF, Mathur ND, Zhang Q (2009) Investigation of the electrocaloric effect in a $PbMg_{1/3}Nb_{2/3}O_3$—$PbTiO_3$ relaxor thin film. Appl Phys Lett 95(18):182904-1–182904-3. doi:10.1063/1.3257695
80. Zhang Q, Whatmore RW (2001) Sol-gel PZT and Mn-doped PZT thin films for pyroelectric applications. J Phys D Appl Phys 34(15):2296–2301
81. Correia TM, Zhang Q (2010) The impact of PZT seed layer on microstructure, dielectric and ferroelectric properties of 0.93PMN-0.07PT thin films. Ferroelectr 405(1):275–280
82. Rožič B, Malič B, Uršič H, Holc J, Kosec M, Neese B, Zhang QM, Kutnjak Z (2010) Direct measurements of the giant electrocaloric effect in soft and solid ferroelectric materials. Ferroelectr 405(1):26–31
83. Haertling GH, Land CE (1971) Hot-Pressed (Pb, La)(Zr, Ti)O_3 ferroelectric ceramics for electrooptic applications. J Am Ceram Soc 54(1):1–10
84. Jia D, Liu J (2009) Human power-based energy harvesting strategies for mobile electronic devices. Front Energy Power Eng China 3(1):27–46. doi:10.1007/s11708-009-0002-4
85. Shenck NS, Paradiso JA (2001) Energy scavenging with shoe-mounted piezoelectrics. Micro IEEE 21(3):30–42
86. Roundy S, Wright PK (2004) A piezoelectric vibration based generator for wireless electronics. Smart Mater Struct 13(5):1131–1142. doi:10.1088/0964-1726/13/5/018
87. Kok SL, White NM, Harris NR (2008) A free-standing, thick-film piezoelectric energy harvester. In: Sensors, 2008 IEEE 589–592. doi:10.1016/j.sna.2006.06.061
88. Lefki K, Dormans GJM (1994) Measurement of piezoelectric coefficients of ferroelectric thin films. J Appl Phys 76(3):1764–1767
89. Beeby S, Tudor M, White N (2006) Energy harvesting vibration sources for microsystems applications. Meas Sci Technol 17(12):R175–R195. doi:10.1088/0957-0233/17/12/R01
90. Zhu D, Tudor J, Beeby S (2010) Strategies for increasing the operating frequency range of vibration energy harvesters: a review. Meas Sci Technol 21(2):022001. doi:10.1088/0957-0233/21/2/022001
91. Choi W, Jeon Y, Jeong JH, Sood R, Kim S (2006) Energy harvesting MEMS device based on thin film piezoelectric cantilevers. J Electroceram 17(2):543–548. doi:10.1007/s10832-006-6287-3

92. White NM, Glynne-Jones P, Beeby SP (2001) A novel thick-film piezoelectric microgenerator. Smart Mater Struct 10(4):850–852
93. Glynne-Jones P, Beeby S, White N (2001) Towards a piezoelectric vibration-powered microgenerator. IEE P-Sci Meas Tech 148(2):68–72
94. Roundy S, Wright PK, Rabaey J (2003) A study of low level vibrations as a power source for wireless sensor nodes. Comput Commun 26(11):1131–1144
95. Jeon YB, Sood R, Jeong Jh, Kim SG (2005) MEMS power generator with transverse mode thin film PZT. Sensor Actuat A-Phys 122(1):16–22. doi:10.1016/j.sna.2004.12.032
96. Shen D, Park JH, Ajitsaria J, Choe SY, Wikle HC, Kim DJ (2008) The design, fabrication and evaluation of a MEMS PZT cantilever with an integrated Si proof mass for vibration energy harvesting. J Micromech Microeng 18(5):055017. doi:10.1088/0960-1317/18/5/055017
97. Fang HB, Liu JQ, Xu ZY, Dong L, Wang L, Chen D, Cai BC, Liu Y (2006) Fabrication and performance of MEMS-based piezoelectric power generator for vibration energy harvesting. Microelectr J 37(11):1280–1284. doi:10.1016/j.mejo.2006.07.023
98. Fang HB, Liu JQ, Xu ZY, Dong L, Chen D, Cai BC, Liu Y (2006) A MEMS-based piezoelectric power generator for low frequency vibration energy harvesting. Chin Phys Lett 23(3):732–734
99. Uchino K (2008) Piezoelectric actuators 2006. J Electroceram 20(3):301–311. doi:10.1007/s10832-007-9196-1
100. Mateu L, Moll F (2006) Appropriate charge control of the storage capacitor in a piezoelectric energy harvesting device for discontinuous load operation. Sensor Actuat A-Phys 132(1):302–310

Chapter 15
Nanoscale Oxide Thermoelectrics

Antonio Feteira and Klaus Reichmann

Abstract The renaissance of thermoelectrics has been bolstered by the contemporary call for energy harvesting technologies. The potential use of thermoelectrics for direct conversion of waste heat into electricity is gaining momentum and oxides are envisaged as the most promising materials for high temperature applications. Nevertheless, prior to the commercial deploying of this technology, the efficiency of thermoelectric oxides needs to be perfected. Inevitably, the large thermal conductivity of oxides needs to be reduced. Several strategies are currently being explored, including sol-gel processing of oxide thermoelectrics. The higher density of interfaces in nanoceramics fabricated from sol-gel processed powders is regarded as an effective approach to enhance phonon scattering, and thereby reduce thermal conductivity. In this chapter, the fundamentals of thermoelectrics are presented alongside the most promising oxides for the fabrication of thermoelectric modules for energy harvesting. Potential benefits of using sol-gel processed powders are highlighted and the current state-of-art all-oxide thermoelectric modules are presented. Finally, we proposed the exploration of hexagonal perovskites.

Keywords Aluminum-doped zinc oxide · Bismuth telluride · Calcium cobalt oxide · Clathrates · Energy harvesting · Lead telluride · Phonon propagation ·

A. Feteira (✉)
Christian Doppler Laboratory for Advanced Ferroic Oxides,
University of Birmingham, Birmingham, UK
e-mail: a.feteira@bham.ac.uk

A. Feteira
Department of Physics, University of Warwick, Coventry, UK
e-mail: a.feteira@warwick.ac.uk

K. Reichmann
Christian Doppler Laboratory for Advanced Ferroic Oxides,
Graz University of Technology, Graz, Austria
e-mail: k.reichmann@tugraz.at

M. Aparicio et al. (eds.), *Sol-Gel Processing for Conventional and Alternative Energy*, Advances in Sol-Gel Derived Materials and Technologies,
DOI: 10.1007/978-1-4614-1957-0_15,

Piezoelectric · Seebeck coefficient · Skutterudites · Sodium cobalt oxide · Strontium titanate · Waste heat · Zinc oxide · ZT value

15.1 Introduction

Traditionally, electrical power is generated in large, centralised plants powered by fossil fuels, nuclear fission or flowing water. For every kilowatt of power produced, there is an associated percentage of waste heat that simply escapes into the atmosphere. Indeed, it is estimated that two-thirds of the energy produced at power plants is wasted as heat. Staggering amounts of energy are also lost in heavy industrial processes. In the United States, this wasted energy it is equivalent to four times the energy output of all the country's coal-fired power plants. It was estimated that any technology able to recover waste heat at 10% efficiency could generate electricity valued at $50 billion in the United States alone and $200 billion worldwide (Alphabeth Energy, News Release [1]). Unquestionably, one of the major driving forces for energy harvesting concerns the economical benefits. Nevertheless, the recent research efforts and investments in this area have been further promoted by its "green" tag. CO_2 emissions are deemed responsible for global weather changes. In order, to contravene the potential catastrophic consequences of global weather changes, the G8 countries agreed to at least halve CO_2 emissions by 2050. Of course, this target can only be achieved if the consumption of fossil fuels is effectively reduced.

Despite the manifold appealing aspects of energy harvesting, some of the emergent technologies based on piezoelectric and thermoelectric materials have not reached sufficient levels of maturity and efficiency to be implemented on a large scale. In fact, the idea of using thermoelectric materials to harvest energy has been around for several decades, but until very recently, the use of this technology to turn waste heat into electricity was prohibitively too expensive to be deployed on a commercial basis. Indeed, when oil was only $20 a barrel, the use of thermoelectric modules was deemed economically unfeasible, however estimates that oil may reach $200 a barrel in the near future, has bolstered the renaissance of thermoelectrics. Moreover, following the 2011 near-nuclear disaster in Fukushima (Japan), some governments began reassessing their nuclear future, with Germany, for example, contemplating the potential of renewable sources to satisfy the country's energy requirements. Under current circumstances it is evident that any alternative source of energy associated with low levels of CO_2 emissions is definitively welcome.

Considering that thermal energy is being incessantly wasted in everyday ordinary activities such as driving, cooking, heating, washing, just to mention a few. It is no surprise that some efforts are already underway to harvest energy from those activities. For example, some automobile manufacturers are investigating the potential replacement of alternators by thermoelectric generators mounted on the exhaust stream system. In a conventional automobile powered by a typical internal combustion engine, most of the fuel energy is lost as waste heat. Indeed, only 25% of the fuel energy is used for vehicle mobility and to power accessories. Calculations have shown that the widespread use of thermoelectric converters for vehicular waste heat recovery may lead to a 10% improvement in fuel efficiency, translating to fuel savings of $150 per year for every automobile, and of course a concomitant reduction in CO_2 emissions. In September 2009, Amerigon Incorporated, a leader in advanced thermoelectric technologies, announced projects to install and test thermoelectric waste heat recovery generators in vehicles from BMW and Ford, in order to evaluate how much waste heat from engine exhaust system can be effectively converted into electrical power. Projects aimed at improving automobile fuel efficiency have received support from many governments, because they may help to meet objectives on reducing dependency on foreign energy imports (mainly oil) and reducing greenhouse gas emissions. BMW announced in their customer information a 5% potential reduction in fuel consumption, which is a conservative estimate. Especially, because recovering heat energy from braking has not been taken into account. Basically, this system would provide electrical energy during slowing down whereas the thermoelectric generator would produce electrical energy during acceleration, when the temperature at the exhaust increases rapidly [2].

In Japan, the Showa Denko K.K. (SDK) Company has been developing thermoelectric modules to recover waste heat from waste incinerators. Their initial laboratory tests, where the temperature on the hot side was at 600°C and the temperature on cold side was 50°C, a conversion efficiency of 6.2% and power density of 2.4 W/cm^2 was achieved. A maximum output of 21.6 W was also reported, which at the date of printing this book represented the highest level in the world for thermoelectric modules. Following these tests, SDK have started full-scale demonstration experiments in power generation based on waste heat in a waste incinerator (SDK News release [3]).

In 2010, Murata Corporation unveiled a prototype thermoelectric device, which generates up to 38 mW/cm^2 at 360°C. This device generates electricity using two semiconductive ceramics linked by a metal plate, which when heated to 90°C on one side and cooled to 20°C on the other produces 10 mW of electricity, enough to power a small plastic fan, such as those found in laptops. It is estimated that in 10 years similar devices will be found in cell phones, video cameras, laptops and other portable electronic devices, as the heat produced by these devices can then be used to partially recharge their internal batteries [4].

Beside these rather prototype examples of thermoelectric generators, it should be kept in mind, that there are well established applications of thermoelectric phenomena, which reached mass production several decades ago. The earliest and till nowadays most successful application makes use of the small but useful linear voltage of a thermocouple to measure temperature accurately between 3 and 2,000 K. The other quite prominent field of application is temperature control by thermoelectric elements. These so-called Peltier elements are part of medical equipment, analytical systems or electronic com-pounds. But also low cost devices like camping coolers find a huge market. The thermoelectric materials used in this mass market are mainly bismuth- or lead-tellurides or -selenides (Bi_2Te_3, PbTe), which are produced on a large scale.

Thermoelectrics were also employed in more sophisticated applications. For example, the Pioneer 10 launched in 1972 was electrically powered by a thermoelectric generator heated by a plutonium source (radioisotope thermoelectric generator, RTG). Voyager 1 and Voyager 2 (launched in 1977) used SiGe thermoelectric elements in their RTG powered by plutonium. This setup is still used in spacecraft e.g., the New Horizon launched in 2006. Due to the plutonium source such power systems are widely obnoxious to people, especially during swing-by manoeuvres (e.g., earth fly-by of the Cassini-Huygens spacecraft in 1999). Such RTGs are also employed terrestrially for powering remote lighthouses or beacons. If other heat sources like burning of oil or gas are available, thermoelectric generators are found in telecommunications, security, monitoring and surveillance. Global Thermoelectric, for example, provides thermoelectric generators to power remote monitoring stations for oil and gas pipelines together with cathodic corrosion protection. Other fields of applications are self-powered sensor systems. Such sensors connected wireless by RFID need a power supply in the microwatt to milliwatt range. This power can be provided by a thermoelectric generator making use of rather small temperature differences (e.g., ΔT from inside to the outside of an airplane). To the authors knowledge most of these types of generators (as well as those employed in automotive heat recovery systems) use the above mentioned bismuth- or lead-tellurides or —selenides, which have a maximum operating temperature of 200–400°C.

In future, thermoelectrics may also find applications in primary power generation. For example, it was estimated that in one single day, solar cells operating at 20% efficiency and covering 0.1% of the planet's surface would be sufficient to supply the worldwide yearly required energy. The Sun as energy source can also be used by thermoelectric modules which directly convert solar heat into electricity. The advantage of thermoelectric modules compared to photovoltaic solar cells is that they use the whole solar spectrum (IR, UV and visible radiation), while photovoltaic cells only use the UV–Vis part of the spectrum. For such applications in the power range of kilowatt there is a need for thermoelectric materials exceeding the operating temperature range of conventional materials. This is a strength of oxide thermoelectrics, which are given special attention in Sect. 15.3.

In summary, thermoelectric materials are capable of directly converting heat into electricity. Hence, they can play a significant role in both energy conservation (i.e., waste heat harvesting) and primary power generation using renewable sources such as sunlight. Thermoelectric devices offer unique features such as no moving parts, quiet operation, low environmental impact and high reliability. Customers for thermoelectric materials and devices include the energy sector, the automotive and consumer products industries, the military and space exploration agencies such as NASA. A current topic of discussion concerns the exact role of thermoelectrics in the future of power generation. Definitely, it should not be expected that thermoelectric conversion of energy will solve the global energy demands, as claimed by some, however if efficient thermoelectric materials are made available, this technology may play an important part in future energy production and management. This shows that materials research into thermoelectrics is expected to play a key role in the development of new energy harvesting technologies. The purpose of this chapter is to review some of the most promising oxide thermoelectrics and to highlight how sol-gel processing can be used to improve the performance of these materials. Prior to that, we will present important fundamental concepts on thermoelectric phenomena, in order to provide sufficient background for the subsequent sections, especially for the reason to pursue sol-gel processed thermoelectrics.

15.2 Thermoelectric Phenomena

In 1821, Seebeck discovered the principle of thermoelectric energy conversion when he found that introducing a temperature difference across a metal bar generated a voltage between its two ends. Thus, when electric loads are connected at both ends of a metal bar, electric current can be obtained. The inverse phenomenon, when an applied voltage leads to a temperature gradient is known as the Peltier effect, and it has been widely explored in refrigeration.

The performance of a thermoelectric device depends directly on the temperature gradient between the two ends of a solid and on the intrinsic properties of that solid [5]. The maximum energy conversion efficiency, $\eta_{\max}$, is then given by:

$$\eta_{\max} = \frac{T_{\text{hot}} - T_{\text{cold}}}{T_{\text{hot}}} \cdot \frac{(M-1)}{\left(M + \frac{T_{\text{cold}}}{T_{\text{hot}}}\right)} \tag{15.1}$$

where, $M = (1 + \mathrm{ZT}_{\text{ave}})^{1/2}$, with T_{ave} being the average temperature, whereas T_{hot} and T_{cold} are the temperatures of the hot and cold ends, respectively. This relation clearly shows that efficiency is higher for both larger temperature differentials and larger ZT values. Basically, ZT determines the fraction of the Carnot efficiency that can be theoretical achieved by a thermoelectric material. In terms of the intrinsic properties of a thermoelectric material, ZT is expressed as:

$$\mathrm{ZT} = \frac{S^2 \sigma T}{k} \tag{15.2}$$

where T is the absolute temperature, S the Seebeck coefficient, σ the electrical conductivity and k the thermal conductivity. This dimensionless figure of merit is useful to compare the performance of different thermoelectric materials. The first generation of bulk thermoelectric materials developed in the 1950 was based on homogeneous Bi_2Te alloys. These materials were characterised by ZT $\sim$ 0.8–1 near room temperature, as illustrated in Fig. 15.1, and once incorporated into devices could operate $\sim$5–6% conversion efficiency. A second generation of bulk materials based on nanostructuring concepts has been developed in the past 15 years. These materials reach ZT values ranging from 1.3 to 1.7 and are expected to be incorporated into power generation devices with conversion

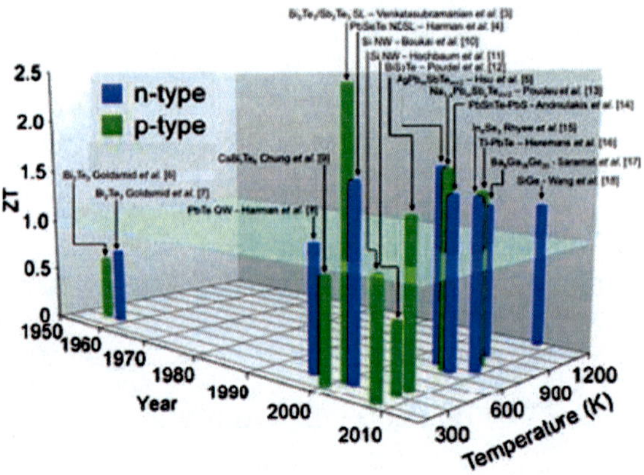

Fig. 15.1 Chronological evolution of traditional thermoelectric materials [6]

efficiencies of 11–15%. The thermoelectrics community has been actively pursuing materials ZT $\geq$ 3, which may reach 30% conversion efficiency, and therefore making these solid state systems competitive with traditional mechanical conversion systems [6].

In theory, high-ZT thermoelectric materials can be achieved by suitable manipulation of the Seebeck coefficient, the electrical conductivity or the thermal conductivity. In practice, the enhancement of ZT is faced by insurmountable difficulties, because these parameters are interrelated. For instance, the $S^2\sigma$ product (commonly referred to as the power factor) cannot be optimised independently, because thermopower and electrical conductivity are determined by details on electronic structure and scattering of charge carriers (electrons or holes). In fact, increasing thermopower of a material without depressing the electrical conductivity is a major challenge.

Since the overall efficiency of a thermoelectric module is proportional to the temperature gradient, as shown by Eq. 15.1, it becomes evident that thermal conductivity, *k*, must be low; otherwise the circuit will become short-circuited. In fact, to date most significant improvements on ZT values came primarily from reductions of the thermal conductivity rather than from power factor enhancements. Hereafter, we briefly explain the fundamentals of thermal conductivity.

Thermal conductivity coefficient, *k*, provides a quantitative measure of the rate at which thermal energy is transported along the thermal gradient. The two principal mechanisms of thermal conductivity in a solid are from conduction electrons and lattice vibration phonons.

$$k = k_{\text{ele}} + k_{\text{lat}} \tag{15.3}$$

From Table 15.1, it becomes evident that thermal conductivity, *k*, coefficients for solids range over approximately four orders of magnitude. The highest thermal conductivity is observed in metals like copper and covalent crystals with short chemical bonds such as diamond. In cooper, thermal conductivity involves free electrons, whereas in diamond, phonons propagate rather easily due to the tightly

Table 15.1 Thermal conductivities of different materials

Material	k_{11}	k_{22}	k_{33}
Rochelle salt	0.5	0.61	0.60
Sodium chloride	6.5	6.5	6.5
Rutile	9.3	9.3	12.9
Germanium	65	65	65
Silicon	175	175	175
Copper	400	400	400
Diamond	550	550	550

bonded carbon atoms. In contrast, ionic solids such as sodium chloride show a rather lower thermal conductivity [7].

In pure metals, the electronic contribution for the thermal conductivity prevails over the lattice contribution. Moreover, the Wiedemann–Franz law (Eq. 15.4) predicts that the electrical conductivity, σ, and the thermal conductivity are proportional to each other.

$$\frac{k}{\sigma T} = L = 2.45 \times 10^{8}\ \mathrm{W\Omega K^{-2}} \qquad (15.4)$$

where, L is called the Lorenz number. For copper at 0°C, $L = 2.23 \times 10^{8}$ WΩ. The Wiedemann–Franz law works reasonable well for most metals at temperatures of this order. The Wiedeman-Franz immediately highlights that an increase of the power factor, through the enhancement of the electrical conductivity results in an increase on the electronic thermal conductivity. Hence, the most effective approach to maximise ZT is through the manipulation of the lattice thermal conductivity, because this is the only parameter which is not determined by the electronic structure.

In an electrically insulating solid, the most important contribution to thermal conductivity arises from the propagation of phonons. In this case, the thermal conductivity is given by:

$$k = \frac{1}{3}\bar{v} l C_v \qquad (15.5)$$

where $\bar{v}$ is the average speed for phonons, l is the mean free path for phonon propagation and C_v is the heat capacity. Obviously, the temperature dependence these three parameters may impact directly on the thermal coefficient k at a given temperature. Hereafter, we summarised the temperature dependence of these parameters. For a more detailed explanation the readers can refer to classical textbooks on solid state physics such as "Introduction to Solid State Physics" by Charles Kittel [8]. The average speed for phonons, $\bar{v}$, propagating in a solid is approximately constant, however both the number density and the energy density are greater at the hot end. Basically, heat flow is primarily due to phonon flow with phonons being created at the hot end and destroyed at the cold end. Real materials contain imperfections such as dislocations, grain boundaries and impurities, which act as scatters for phonons, as schematically illustrated in Fig. 15.2.

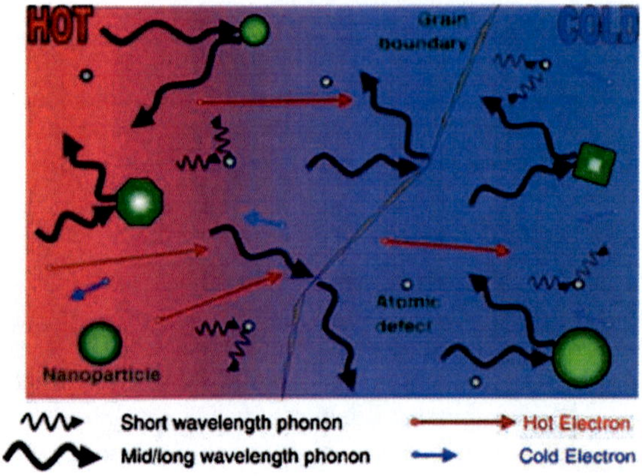

Fig. 15.2 Schematic representation of phonon propagation and interaction with defects [6]

At very low temperatures (under 40 K), the dominant phonon becomes so long, that these imperfections are no longer effective scatters, so thermal conductivity has always a T^3 dependence, as predicted by the Debye T^3 law:

$$C_v = \frac{12Nk_B\pi^4}{5}\left(\frac{T}{\theta_D}\right)^3 \qquad (15.6)$$

Hence, from Eq. 15.6, it follows that the temperature dependence of the thermal conductivity depends on the temperature dependence of the heat capacity. In summary, at very low temperatures, phonon scattering is insignificant because the low number of excited phonons and their very long wavelength, and thermal conductivity is rather low; however it increases proportionally to T^3 until the Debye temperature, θ_D. At temperatures greater than the Debye Temperature, θ_D, the heat capacity is given by the classical solution $Cv = 3\ Nk_B$ (where N = number of atoms and k_B = Boltzmann constant). In this case, the temperature dependence of thermal conductivity is dependent on the temperature dependence of the free path for phonon propagation, l, which is known to obey a T^{-1} dependence.

The most useful insights into thermal conductivity can be summarised as follows: (i) in the high temperature range k_{lat} follows a T^{-1} dependence (ii) a low melting point is consistent with low k_{lat} (iii) large interatomic distances are consistent with low k_{lat} and (iv) k_{lat} decreases with increasing atomic masses.

Up to now a certain decoupling of electric conductivity from thermal conductivity is only achieved in materials following the phonon-glass electron-crystal principle (PGEC), a concept proposed by Slack in 1995. Such compounds need a rigid structure, which provides the good electrical conductivity. Within this rigid structure one can find large cavities that can be occupied by heavy atoms. The size of the cavity and the low binding forces enable "rattling" of the heavy atom thereby increasing the scattering of phonons, which de-creases the lattice part of the thermal conductivity. This concept can be found in skutterudites (e.g. $CoSb_3$) Half-Heusler compounds (e.g. TiNiSn), Clathrates (e.g. $Ba_8Ga_{16}Ge_{30}$) or Zn_4Sb_3.

Disordered zinc in Zn_4Sb_3 with phonon-glass and electron-crystal thermoelectric properties [6].

The lowest thermal conductivity, *k*, occurs for an amorphous material, because the mean free path for phonon propagation, *l*, is on the order of the size of the randomly distributed coordination polyhedra (~0.2–0.5 nm). Hence, lowest achievable lattice thermal conductivity can be estimated as 0.25^{-1} $Wm^{-1}K^{-1}$.

15.3 Thermoelectric Oxides

Commercial thermoelectric modules for room temperature applications are traditionally made from $(Bi,Sb)_2(Te,Se)_3$ alloys, whereas for mid- (~650 K) and high- (>1000 K) temperature applications, the highest ZT values are observed for (Bi,Sn)(Te,Se) and SiGe alloys, respectively. Nevertheless, the wide application of these alloys is being limited by technical, environmental and economical factors. From the technical viewpoint, surface oxidation and vaporisation of these harmful elements is a major issue precluding their use in high temperature energy harvesting technologies. In addition, the scarcity of some of these elements makes their production costs rather prohibitive. Oxides, owing to their natural merits of low thermal conductivity and high temperature stability, are emerging as promising thermoelectrics. Moreover, some complex oxides provide a more effective way to independently control both the electron and phonon transport properties, as briefly described below from a chronologic point of view.

The first strategy used to enhance ZT in oxides, simply followed the philosophies previously employed to the canonical metal thermoelectric Bi_2Te_3. Basically, doping was employed with the purpose of introducing lattice point defects in order to enhance phonon scattering and thereby lower the thermal conductivity. Nevertheless, thermoelectric power and electrical conductivity often changed in opposite directions with doping, therefore doping offered no primary benefits over the enhancement of the power factor. Moreover, in a disorder lattice it is rather difficult to independently control both electron and phonon transport and often doping deteriorates the electronic performance (e.g., mobility). Hence, often an overall gain in ZT cannot be realised through doping.

Subsequently, researchers turned into natural superlattices. This approach is inspired on the periodic arrangement of the regions characterised by different electron and phonon transport characteristics. Basically, oxides occurring as natural superlattices afford an independent control of the transport properties. This ability to act independently over the electron and phonon systems, due to complex nature of some oxides provides a promising direction to improve ZT values of materials for high temperature energy harvesting.

More recently, some investigators have been exploring a so-called nanostructuring approach, which actually offers the potential to independently control transport properties, thereby contributing to an enhancement of ZT. Again this approach was originally applied with success to metallic alloys, by either

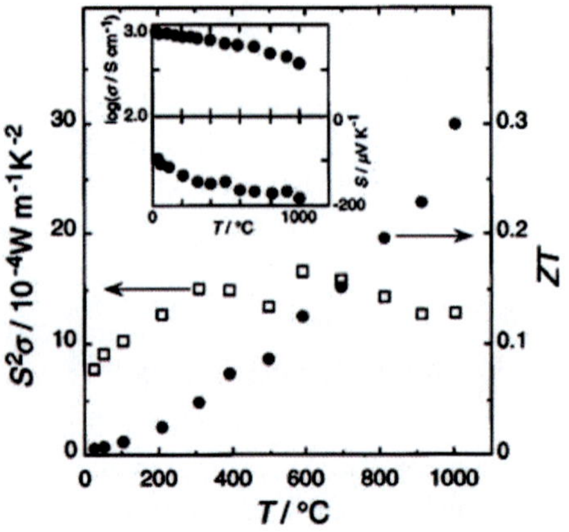

Fig. 15.3 Temperature dependence of Power factor and ZT for $Zn_{0.8}Al_{0.2}O_2$ ceramics [9]

exploring materials with self-formed inhomogeneities on the nanoscale driven by phase segregation phenomena such as spinodal decomposition or via comminuation to nanocrystalline sizes followed by sintering. Mercouri Kanatzidis group has recently reviewed and discussed the latest developments on nanostructured thermoelectrics, with special emphasis on the metallic systems. The reader is therefore referred to that excellent review [6].

The fabrication of all-ceramic thermoelectric modules requires both p- and n-type semiconductor oxides, as described in more detail in Sect. 15.5. Unfortunately, most n-type oxides exhibit rather low ZT values (<1), which is the major obstruction towards their wide application. The search for n-type thermoelectrics has revolved essentially around ZnO-based materials and Perovskite and Perovskite-related oxides. The purpose of the remaining of this chapter is to present some of the most promising families of thermoelectric oxides and then unveiled the benefits of using sol-gel processing to reduce thermal conductivity in bulk nanoceramics. Indeed, exploiting boundary scattering in nanoceramics is an additional and highly successful strategy to further reduce the thermal conductivity.

15.3.1 ZnO-Based Oxides

ZnO is ubiquitous in electronic applications, where it can be used either as a voltage surge protection element (varistor) or a gas sensor. Basically, ZnO is a wide band gap semiconductor with an optical band gap of 3.2 eV that exhibits n-type electrical conductivity. The high melting temperature (1,800°C) and chemical stability of ZnO combined with its high Seebeck coefficient have been put forward as the main selection criteria for its potential application as an n-type thermoelectric. Nevertheless, ZnO exhibits a relatively low electrical conductivity and a too high thermal conductivity. Several doping strategies have been employed to overcome these two major drawbacks and thereby improved ZT of ZnO ceramics.

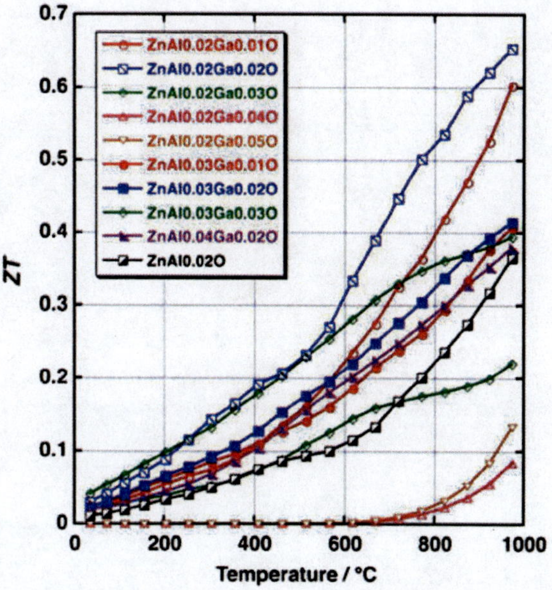

Fig. 15.4 ZT values for Al/Ga co-doped ZnO ceramics [10]

Ohtaki et al. [9] reported $Zn_{0.98}Al_{0.02}O$ to have promising thermoelectric performance, i.e., a ZT value of ~0.2 at 800°C, as illustrated in Fig. 15.3.

This result corroborated the potential of ZnO-based ceramics as promising *n*-type oxides for applications above 700°C. A remarkable improvement in the thermoelectric performance of dense ZnO ceramics was also latter reported Ohtaki et al. [10]. These researchers found Al and Ga codoping of ZnO to drastically decrease the thermal conductivity of the oxide, which is only accompanied by a small decrease in the electrical conductivity. Moreover, this double doping approach leads also to a significant enhancement in the thermopower, and consequently the overall ZT values are significantly improved, as illustrated in Fig. 15.4. For example, ZT reaches 0.65 at 975°C for the composition $Zn_{0.96}Al_{0.02}Ga_{0.02}O$. This result appears to be the highest ZT value reported so far for a bulk *n*-type oxide.

Layered In–Zn–O (IZO) compounds with the formula $In_2O_3(ZnO)k$, where k is an integer, are also promising thermoelectric oxides. These phases consist of alternating layers of In–O octahedra and In, Zn, and O in a wurtzite-like structure, as illustrated in Fig. 15.5. Basically, they form a natural superlattice. Similarly to high-ZT *p*-type Na-based and Ca-based cobalt oxide systems, the low thermal conductivity in this structure is believed to arise from the layered structure. This is supported by the fact that thermal conductivities for homologous $In_2O_3(ZnO)k$ have been reported to be as low as 2–2.5 $Wm^{-1}K^{-1}$ for k = 3, 4 and 5 at 1,000°C, while similarly doped specimens of the component oxides In_2O_3 and ZnO have greater thermal conductivity at the same temperature [11].

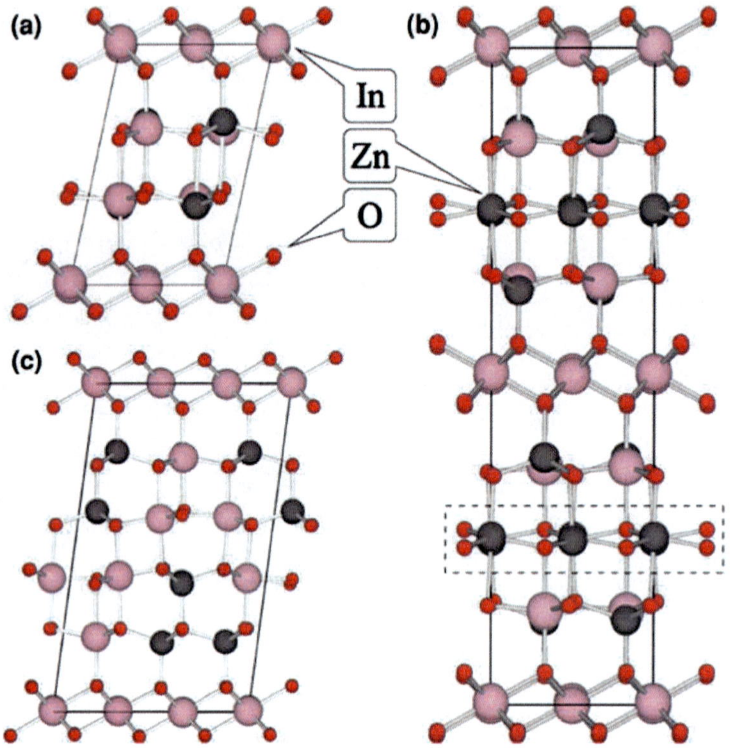

Fig. 15.5 Schematic representation of layered In-Zn-O, **a** $k = 1$, **b** $k = 2$ and **c** $k = 3$ [11]

Nanostructuring effects in bulk ZnO-based nanoceramics was investigated by Kinemuchi et al. [12], who demonstrated an effective reduction in thermal conductivity via nanostructuring in ZnO, as illustrated in Fig. 15.6.

This behaviour was successfully explained by the Callaway model, indicating that enhanced boundary scattering dominates the thermal conductivity in nanograined ZnO. On the other hand, the Seebeck coefficient appeared independent of the grain size and electrical conductivity decreased with decreasing grain size, as illustrated in Fig. 15.7a, overcompensating the reduction in thermal conductivity. Hence, ZT in nanograined bulk ZnO remained lower than in larger grains, as illustrated in Fig. 15.7b. Despite the unfortunate reduction of the power factor, this result shows that perhaps nanostructuring may be effective in other type of thermoelectric oxides.

15.3.2 Perovskite and Perovskite-Related Oxides

In general, the term perovskite refers to compounds represented by the formula ABO_3, which are isostructural with $CaTiO_3$. Nevertheless, the ideal cubic

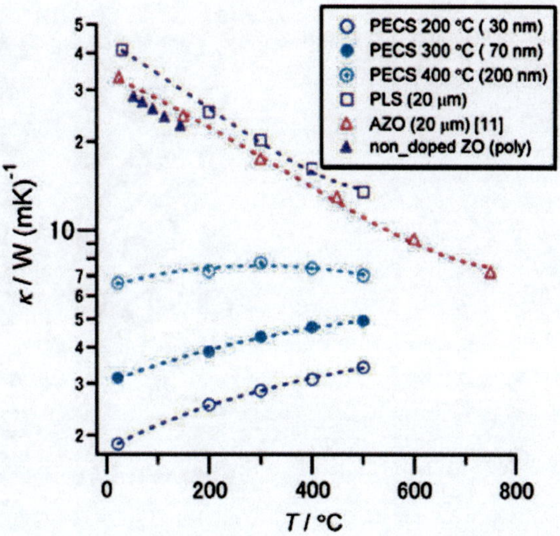

Fig. 15.6 Grain size depence of the thermal conductivity of ZnO ceramics [12]

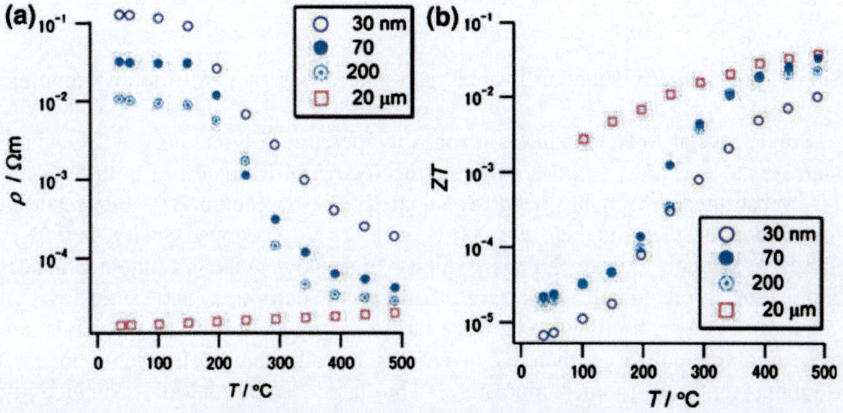

Fig. 15.7 Grain size dependence of **a** electrical conductivity and **b** ZT values for ZnO ceramics [12]

perovskite structure is better typified by $SrTiO_3$, which possesses a primitive cubic unit cell with a lattice constant a = 3.905 Å. This structure consists of closed-packed SrO_3 layers, with 1/4 of the octahedral interstices occupied by the Ti cations. This structure is quite flexible concerning chemical substitution, i.e., both Sr^{2+} and Ti^{4+} can be replaced by ions of similar ionic radii. Donor doping occurs when the replacing ion possesses a higher charge than the ion that it is replacing, for example when Sr^{2+} is replaced by La^{3+} or Ti^{4+} by Nb^{5+}. The thermoelectric properties of donor-doped $SrTiO_3$ single-crystals, ceramics and epitaxial thin films have received considerable attention and some investigators postulated that $SrTiO_3$-based oxides form a promising group of *n*-type thermoelectric, due to the stable electronic properties at high temperatures. Moreover, all the components of

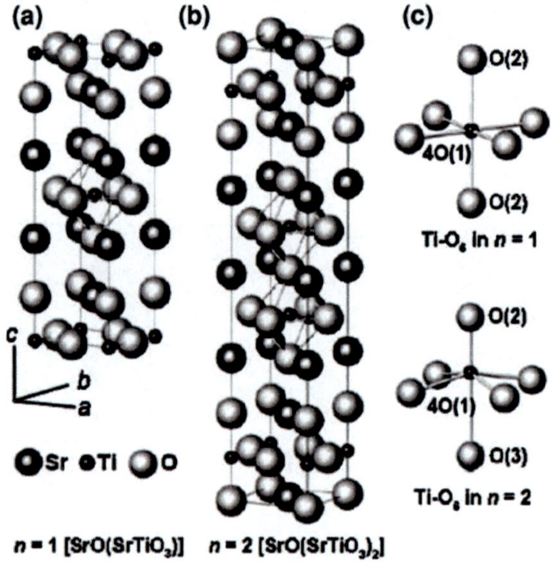

Fig. 15.8 Schematic representation of the crystal structure of Rudlesden-Popper phases [14]

$SrTiO_3$ are naturally abundant, which also makes it attractive from an economic viewpoint.

Single-crystal $SrTiO_3$ exhibits a room temperature ZT value of ~0.1, which increases to ~0.27 at 1,000 K. The thermoelectric performance of both Sr^{2+} and Ti^{4+} substituted $SrTiO_3$ has been investigated. The maximum ZT value reported for Nb-doped $SrTiO_3$ is 0.37 at 1,000 K, whereas for La-doped $SrTiO_3$ is 0.21 at 973 K. Although doped $SrTiO_3$ exhibits high power factors, their practical application is precluded by the large thermal conductivities. The rather low ZT value of $SrTiO_3$ is ascribed to its large lattice thermal conductivity, which is one order of magnitude larger than in conventional metallic thermoelectrics, whose ZT is about unity at room temperature. The large lattice thermal conductivity ($k = 12\ Wm^{-1}K^{-1}$ at 300 K and $3.1\ Wm^{-1}K^{-1}$ at 1,000 K) is due to high phonon frequencies of the major constituent oxygen. For this reason doping is not very effective at reducing k, however other approaches based on phonon scattering at interfaces appear more promising. For example, several researchers have been pursuing nanostructuring approaches to improve ZT of $SrTiO_3$ and other perovskites. A very comprehensive review is given by Koumoto et al. [5]. In addition, Lee et al. [13] suggested that *n*-type thermopower in perovskite oxides can be enhanced by reduction reactions, which also control the carrier concentrations for electrical conductivity. Nevertheless, due to the oxidation of Ti^{3+} to Ti^{4+} above 400°C, $SrTiO_3$ becomes insulating and ZT decreases dramatically. Hence, for high temperature energy harvesting, $CaMnO_3$ became the base *n*-type material of choice, because Mn^{3+} is stable to higher temperatures.

Alternatively, reduction of thermal conductivity can in principle be accomplished in natural superlattices, such as Rudlesdden-Popper (RP) phases, which

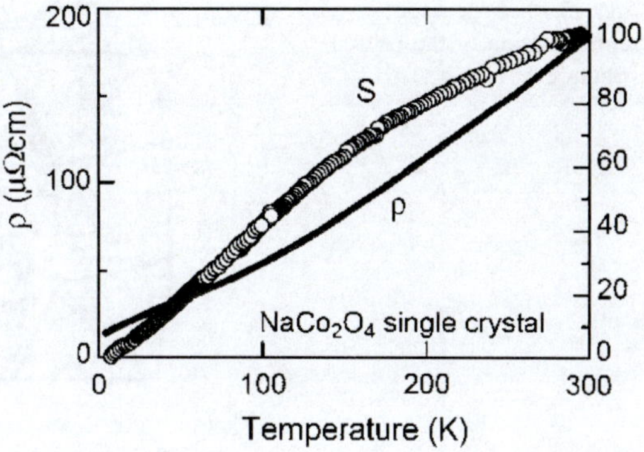

Fig. 15.9 Temperature dependence of the electrical resistivity and Seebeck factor for $NaCo_2O_4$ single crystal [15]

structure comprises perovskite blocks separated by rock-salt type layers, as illustrated in Fig. 15.8.

Lee et al. [14] compared the thermoelectric performance of $SrO(SrTiO_3)_n$ ($n = 1, 2$) RP phases against the cubic perovskite-type Nb-$SrTiO_3$ and found the lattice thermal conductivity to decrease by more than 50% at room temperature and 30% at 1,000 K, due to phonon scattering at the $SrO/(SrTiO_3)_n$ interfaces. Nevertheless, ZT at 1,000 K only reach 0.14, in comparison with 0.37 for the Nb-$SrTiO_3$. The main reason for lower ZT values exhibited by RP phases is due to the lower electrical conductivity imparted by the insertion of the insulating SrO layers between the conductive perovskite blocks.

15.3.3 Cobalt Oxide-Based Layered Structures

In contrast with *n*-type thermoelectric oxides, several *p*-type materials with ZT values greater than unity are available. Among those, Cobalt oxide layered structures such as $NaCo_2O_4$ and $Ca_3Co_4O_9$ are the most studied. In 1997, Terasaki et al. [15] reinvestigated the transport properties of $NaCo_2O_4$ single-crystals and confirmed its large thermopower and low electrical conductivity. The Seebeck coefficient, *S*, in $NaCo_2O_4$ increases with increasing temperature, as shown in Fig. 15.9, and reaches 100 μV/K at 300 K, which is nearly ten times larger than that of typical metals. The resistivity, ρ, is highly anisotropic between in- and out-of-plane directions. The in-plane resistivity is as low as 200 μΩ cm at 300 K.

Although, *S* is smaller in $NaCo_2O_4$ (100 μV/K) than in Bi_2Te_3 (200 μV/K), ρ is also lower (0.2 vs. 1 mΩ cm) and therefore their Power Factors ($S^2\sigma$) are comparable (50 vs. 40 μW/K^2 cm). Remarkably, the mobility, μ, in $NaCo_2O_4$ is one order of magnitude lower than in Bi_2Te_3, indicating that a “dirty” conductor can be as efficient as conventional thermoelectric materials. The thermal conductivity of $NaCo_2O_4$ at 300 K ranges from 4–5 W.m^{-1}K^{-1}, depending on sample quality. Recalling the Wiedemann–Franz law in the previous section, the lattice thermal conductivity can be estimated to range from 1 to 2 W.m^{-1}K^{-1}, which is

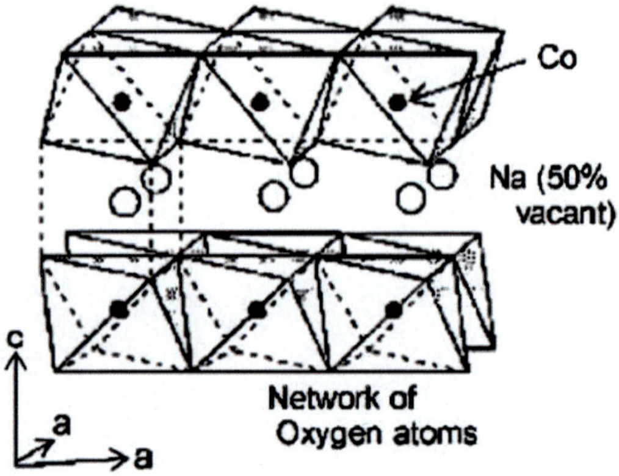

Fig. 15.10 Schematic representation of the crystal structure of $NaCo_2O_4$

comparable to that of most classic thermoelectric materials. ZT values greater than the unity have been reported for $NaCo_2O_4$ single-crystals at 800 K, whereas in polycrystalline ceramics the best ever reported is ZT $\sim$0.8 at 1,000 K. The lower ZT value was ascribed to the higher resistivity of the ceramic sample.

The crystal structure of $NaCo_2O_4$, Fig. 15.10, is composed of CdI-type CoO_2 layers formed by edge-sharing distorted octahedra with Na^+ randomly occupying 50% of the interlayer sites. In this natural structure the $[CoO_2]^{0.5+}$ layers act as the electronic-transport block, whereas the highly disorder Na^{+1} ions act as the phonon scattering block. The origin of the large thermopower is explained in terms of spin and orbital degrees of freedom in the strongly correlated electron $[CoO_2]^{0.5+}$ blocks.

In summary, Terasaki et al. [15] showed $NaCo_2O_4$ to be a potential thermoelectric material, whose power factor is even higher than that of Bi_2Te_3. Moreover, for the first time it was demonstrated that S and ρ could be changed independently, which represents a major breakthrough the theoretical limitations derived from the conventional theories. Following the exciting results obtained for $NaCo_2O_4$, other layers cobaltites such as $Ca_3Co_4O_9$ and $Bi_2M_3Co_2O_y$ (where M = Sr, Ca or Ba).

15.4 Sol-Gel Processing of Thermoelectric Oxides

For economical reasons most technological important oxides are synthesised by the conventional solid state route, which involves the mixing of raw materials in large tumbling mills followed by their reaction at high temperatures. Although, this process is economically very attractive, it brings serious limitations such as introduction of undesirable impurities during the milling process, and especially a limited reduction of particle size.

Optimisation of the thermoelectric efficiency can be carried out via microstructural design. Several authors proposed several approaches based on nanoscale features such as phase separation, orientated grain growth and fine-grained materials. In particular, finer grain sizes exhibit tremendous advantages over larger grained materials, because not only thermal conductivity may be reduced due to phonon scattering at the grain boundaries but also power factor may be increased due to electron filtering at the grain boundaries.

Sol-gel processing is regarded as an effective route for the fabrication of fine and size-selected powders. In addition, sol-gel processing is compatible with large volume outputs, a necessary condition for an effective transfer to mass production and commercialisation. Despite of the apparent potential benefits of synthesising thermoelectric oxides via sol-gel processing only a limited number of case studies can be found in the open literature. Most of those studies are devoted to the processing of typical thermoelectric oxides such those presented in Sect. 15.3. Hereafter, some of those investigations are critically reviewed. It will become evident that sol-gel processing does not always impart an improvement in the efficiency of thermoelectrics; however some results obtained for particular oxides are still encouraging to pursue sol-gel processing in order to improve thermoelectric oxides.

Sol-gel processing has been extensively employed for the synthesis of very fine ZnO powders used in the fabrication of varistors, because these powders lead to improved properties in comparison with those prepared by the conventional solid state reaction. Cai et al. [16] investigated the thermoelectric properties of Al-doped ZnO ceramics prepared by sol-gel using inexpensive zinc acetate as starting material. The experimental procedure can be summarised as follows. First, a 0.5 M $Zn(Ac)_2$ solution was prepared by dissolving $Zn(Ac)_2.2H_2O$ in distilled water. Al_2O_3 with an average particle size of ~ 0.5 μm was added to this solution while continuously stirring it. Drops of 0.5 M NH_3H_2O were added to this solution and when pH reached 6.0, fine $Zn(OH)_2$ particles began to precipitate until the pH value was about 8.1. Subsequently, the precipitate was filtered and dried at 100°C for 1 h in air. In order to remove all organics the powder was calcined at 300°C in air for 2 h. The particle size of the powder ranged from 0.1 to 2 μm. Ceramics were then fabricated by hot pressing at 950°C under 100 MPa for 30 min in vacuum, using a heating rate of 10°C/min. These ceramics reached a relative density of 95%. The pure ZnO sample possessed a high room temperature Seebeck coefficient of about 220 μV/K, which was three times larger than that of Al_2O_3-doped samples. On the other hand, the thermal conductivity decreased with increasing Al_2O_3 content. Basically, the disorder in the ZnO crystal lattice combined with a large amount of interfaces increases phonon scattering as expected. The dependence of the figure of merit, Z (not ZT in the original publication) for the samples with different Al_2O_3 contents is show in Fig. 15.11. In particular, figure of merit at 773 K for the 3 mol% is about 2.6×10^{-5} K^{-1}, which is lower than the value 1.15×10^{-4} K^{-1} reported by Ohtaki et al. [9], but higher than the value 1.5×10^{-6} K^{-1} reported by Tanaka et al.

La-$SrTiO_3$ is another *n*-type thermoelectric oxide that has been prepared by sol-gel methods. For example, Shang et al. [17] prepared $La_{0.08}Sr_{0.92}TiO_3$ using commercially available $Ti(OC_4H_9)_4$, $Sr(NO_3)_2$ and $La(NO_3)_3.6H_2O$ as the raw materials, deionised water and CH_3CH_2OH as the solvent and CH_3COOH as the

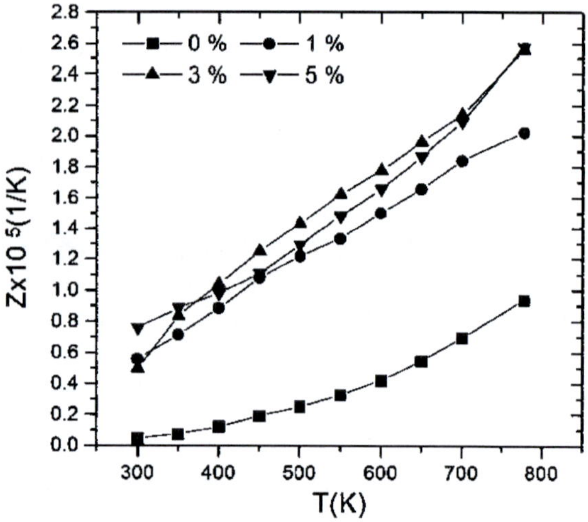

Fig. 15.11 Compositional and temperature dependence of Z for Al-doped ZnO ceramics [16]

catalyser and chelator, respectively. Initially, $Sr(NO_3)_2$ and $La(NO_3)_3.6H_2O$ were dissolved in the deionised water and $Ti(OC_4H_9)_4$ in CH_3CH_2OH, respectively. The two solutions were mixed and CH_3COOH was added until a sol was formed. After drying the sol at 50°C for 1 h, a gel was formed, which was further dried at 80°C for 24 h. The resulting power was finally calcined at 530°C in air for 1 h. Spark plasma sintering at 1,200°C for 5 min was sufficient to produce ceramics with a relative density of 91.6%. This sample shows a ZT value of 0.08 at ~680 K, which increases with temperature.

Although, some sol-gel routes may require costly metal alkoxides and involve complex laboratorial skills, Liu et al. [18] demonstrated that $NaCo_2O_{4-\delta}$ can be successfully synthesised using a simple aqueous sol-gel route. Basically, these workers dissolved $NaNO_3$ and $Co(NO_3)_3$ in deionised water and then mixed with urea. In order to compensate for Na loss during calcinations they used a Na:Co ratio of 1.15:2. The mixed solid solution was then heated at ~100–135°C until a gel was formed, followed by heating at 250–300°C to remove residual organics from the gel. This gel was subjected to different combinations of calcination temperatures and oxygen partial pressures. The electrical resistivity and thermopower was found to vary with sintering conditions. The largest resistivity and thermopower was observed for the samples with highest oxygen deficiency. The largest Power factor was observed for the sample prepared at 700°C under flowing oxygen as illustrated in Fig. 15.12.

$Ca_3Co_4O_9$ is one of the p-type materials of choice to fabricate of all-oxide thermoelectric modules as it will be shown in the next section. $Ca_3Co_4O_9$ is commonly prepared by heating a mixture of $CaCO_3$ and Co_3O_4 at 920°C for several hours. Despite of the simplicity of this process, it exhibits several drawbacks such as large particle size, limited degree of chemical homogeneity and low sinterability. In order to overcome some of these limitations, several investigators synthesised $Ca_3Co_4O_9$-based ceramics via sol-gel processing routes. For example,

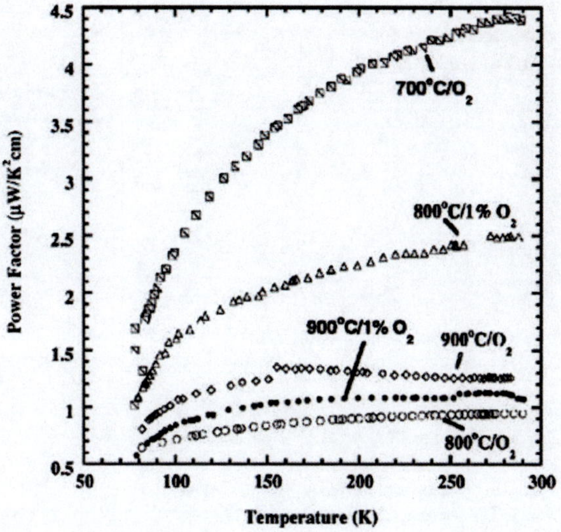

Fig. 15.12 Effects of processing conditions on the Power factor of $NaCo_2O_4$ ceramics [18]

Zhang et al. [19] developed a sol-gel citrate method to synthesise $Ca_3Co_4O_9$ using stoichiometric mixtures of nitrates of calcium and cobalt were dissolved in an aqueous solution of citric acid and 2% (volume ratio) Polyethylene glycol (PEG) 400. The resulting solution was heated at 80°C in order to obtain the gel. The gel was dried initially at 120°C for 12 h and further heat treated in air at temperatures between 350 and 900°C for 2 h. $Ca_3Co_4O_9$ nanopowders with a particle size ranging from 30 to 50 nm were synthesised through this simple sol-gel method. The substitution of Ca in $Ca_3Co_4O_9$ by heavier ions was found to decrease thermal conductivity by phonon scattering.

Similarly, Nan et al. [20] to synthesised the $(Na_xCa_{1-x})_3Co_4O_9$ ($x = 0.05$–0.2) series by dissolving $Ca(NO_3)_2$, $NaNO_3$ and $Co(NO_3)_2$ in citric acid. Their XRD results showed that single-phase powders isostructural with $Ca_3Co_4O_9$ and with an average particle size of 65 nm could be obtained by calcination of the dried gel at 600°C. Considering that $Ca_3Co_4O_9$ decomposes above 920°C, these researchers preferred to sinter their ceramics at 890°C for 5 h. Nevertheless, under these sintering conditions the relative densities remained lower than 80%. In order, to circumvent the sintering inertia of $Ca_3Co_4O_9$-based ceramics some authors employ hot-pressing sintering techniques, as it will be shown later. Figure 15.13 shows the temperature dependence of Z for the $(Na_xCa_{1-x})_3Co_4O_9$ x = 0.05 and 0.15 samples. Z increases with temperature reaching $1.2 \times 10^{-4}\ K^{-1}$ at 700°C (i.e., ZT $\sim$ 0.12).

In order to further enhance ZT values in $Ca_3Co_4O_9$-based systems, some investigators employed doping strategies based on replacement of Ca by heavier ions such as rare earths. $Ca_{3-x}RE_xCo_4O_9$, where RE = Yb or Er are two examples of novel systems prepared by sol-gel processing methods. Xu et al. [21] prepared a series of $Ca_{3-x}Yb_xCo_4O_9$ ($x = 0$, 0.2, 0.4 and 0.5) ceramics using a citrate

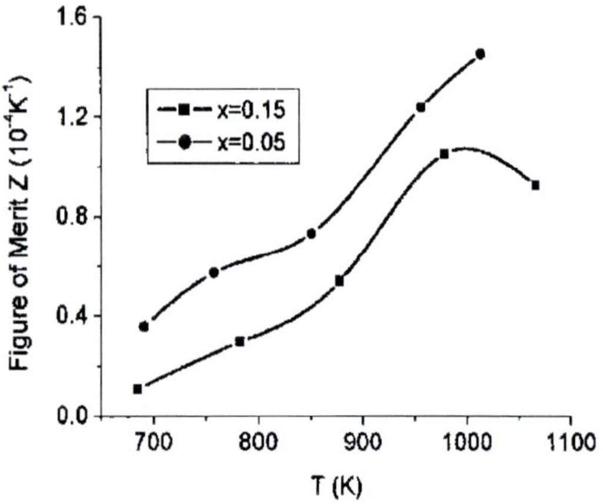

Fig. 15.13 Temperature dependence of Z for $(Na_xCa_{1-x})_3Co_4O_9$ with $x = 0.05$ and 0.15 [20]

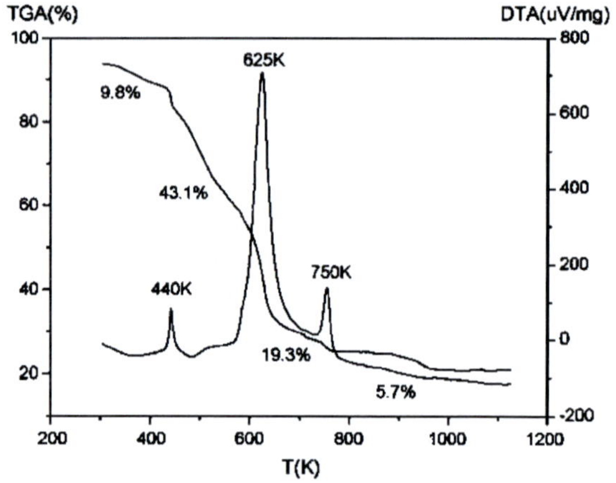

Fig. 15.14 Combined TGA and DTA for $Ca_{3-x}Yb_xCo_4O_9$ powders [21]

complexation sol-gel method. Their process, started by dissolving high purity Yb_2O_3 in nitric acid to form $Yb(NO_3)_3$. This product was subsequently mixed in distilled water with $Ca(NO_3)_2.4H_2O$, $Co(NO_3)_2.6H_2O$ and $C_6H_8O_7.H_2O$ according to the appropriated stoichiometry. The solution was then heated to 65°C for 1 h resulting in a purple transparent sol, which was subsequently dried at 130°C to form a gel. The dried gel was then calcined at 800°C in air for 3 h to obtain $Ca_{3-x}Yb_xCo_4O_9$ powders.

According to the DTA analysis, in Fig. 15.14, the formation process of $Ca_{3-x}Yb_xCo_4O_9$ can be summarised as follows. The endothermic peak at 440 K accompanied by weight loss is due to evaporation of adsorbing water and decomposition of citric acid. The strong exothermic peak between 600 and 770 K accompanied by a dramatic weight loss is ascribed to the vigorous combustion of nitrates and citrates. The exothermic peak at 750 K may be due to decomposition of residual citric acid. Basically, the process of formation of $Ca_{3-x}Yb_xCo_4O_9$ starts

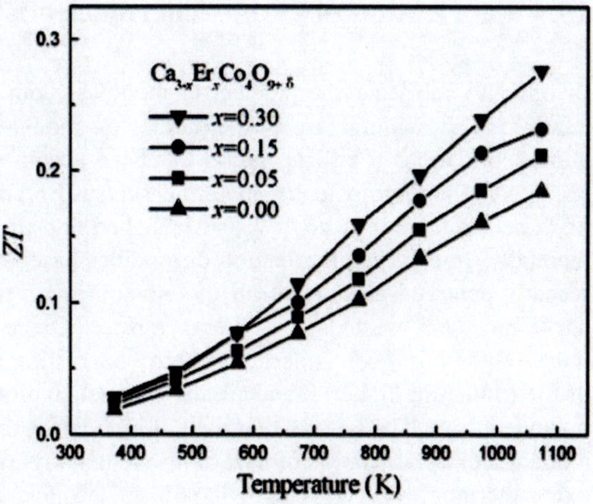

Fig. 15.15 Temperature and compositional dependence of ZT values for $Ca_{3-x}Er_xCo_4O_9$ ceramics [22]

with the removal of organics to form CaO, Yb_2O_3 and CoO. With increasing temperature CoO oxidises to Co_3O_4 and the reaction between these precursors is completed below 600°C. Sintering of $Ca_3Co_4O_9$-based ceramics is traditionally carried out in the 750–850°C temperature range, because above 920°C $Ca_{3-x}Yb_xCo_4O_9$ decomposes to $Ca_{3-x}Yb_xCo_2O_6$, which exhibits much higher resistivity. In order to enhance densification, hot-pressing sintering at 950°C under and applied pressure of 30 MPa was employed by these investigators. The best thermoelectric performance for the $Ca_{3-x}Yb_xCo_4O_9$ series was observed for $x = 0.02$, with ZT reaching 0.16 at 600°C. The homologous $Ca_{3-x}Er_xCo_4O_9$ (0, 0.05, 0.15 and 0.5) series was prepared by Pei et al. [22] using a very similar process. Their X-ray powder diffraction analysis showed single-phase materials to be obtained up to $x = 0.3$. In order to obtain dense ceramics hot-pressing sintering at 900°C for 10 h under an applied pressure of 35 MPa was employed. This sintering process provided ceramics with a relative density of 96%. The evaluation of the thermoelectric properties of these ceramics revealed the substitution of Ca^{2+} by Er^{3+} resulted in a reduction of both the electrical and thermal conductivity and an increase in the Seebeck coefficient for all compositions with the exception of $x = 0.05$. From Fig. 15.15, which shows the compositional dependence of ZT, it becomes evident that Er-doping enhances ZT, and this system exhibits marginally better ZT values than $Ca_{3-x}Yb_xCo_4O_9$.

Recent results have shown that a wide size distribution of nanograins can effectively scatter different phonon modes and thereby reduce thermal conductivity. In addition, enhanced power factors can be obtained from nanocomposite thermoelectrics consisting of granular regions. This is result suggests that ZT values of existing materials can be further enhanced by increasing power factors and consequently it appears to be a new promising philosophy to design thermoelectrics. Certainly, sol-gel processing should be considered for such approach.

15.5 Fabrication of Oxide Thermoelectric Modules

In order to fabricate thermoelectric modules, both *p*- and *n*-type materials are required. The manufacture of thermoelectric modules used especially in refrigeration applications (Peltier coolers) has been accomplished by the use of conventional metallic thermoelectric materials such as Bi_2Te_3 and PbTe. The manufacture of these modules it is now well established and they are widely available commercially. In contrast, fabrication of modules based on oxide materials it has only recently emerged, despite of all the advantages over metallic-based modules for harvesting heat waste at high temperatures. Oxide thermoelectric modules are indeed the natural contenders for energy harvesting at high temperatures because of their inherent higher chemical and thermal stability in relation to conventional metallic alloys. The latter either melt at high temperatures or are likely to undergo oxidation or volatilisation of hazardous elements. Nevertheless, the manufacture of oxide thermoelectric devices it is still at its infancy stage and very few reports exist, particularly describing their performance. Moreover, the development of all-oxide thermoelectric devices has been hampered by processing difficulties such as cracking or exfoliation because of large differences between the thermal expansion of oxides and metallic electrodes. The large high contact resistance at the junctions between the metallic electrode and the oxide legs also limits dramatically the output power, and often the performance of the modules is inferior from that expected from the properties of the starting materials.

Hence, the design of a thermoelectric device cannot be uniquely based on the selection of the most promising thermoelectric materials, but also needs to take into consideration potential integration issues. For example, it was found that the mechanical properties of the junctions between oxide legs and metallic electrodes can be perfected by mixing the metallic paste with the oxide powder. This approach appears also to beneficially reduce the contact resistance.

Matsubara et al. [23] reported the fabrication of an all-oxide thermoelectric power generator using spark plasma sintered $Ca_{2.75}Gd_{0.25}Co_4O_9$ and conventionally sintered $Ca_{0.92}La_{0.08}MnO_3$ as the *p* and *n* legs, respectively. A schematic representation of the device is illustrated in Fig. 15.16.

At 700°C in air, the power factors for *p* and *n* legs were 4.8×10^{-4} $Wm^{-1}K^{-2}$ and 2.23×10^{-4} $Wm^{-1}K^{-2}$, respectively. The device with 8 *p-n* pairs generated an output power of 63.5 mW at 773°C under a temperature differential, ΔT, of 390°C. Moreover, this device proved to be operating for more than two weeks without any performance decay, as show in Fig. 15.17.

Reddy et al. [24] described the fabrication of a four-leg prototype oxide thermoelectric modules consisting of two pairs of *p*-type $Ca_3Co_4O_9$ and *n*-type $Ca_{0.95}Sm_{0.05}MnO_3$ of either 5 or 10 mm height and a cross-section 16 mm^2.

The four thermoelectric elements were electrically connected in series by silver strips of 1 mm thickness using silver paste as the bonding agent. The assembly was placed between two alumina plates of 25×25 mm^2 and 1.5 mm thickness that stood in parallel thermally and acted as hot and cold ends for thermoelectric legs.

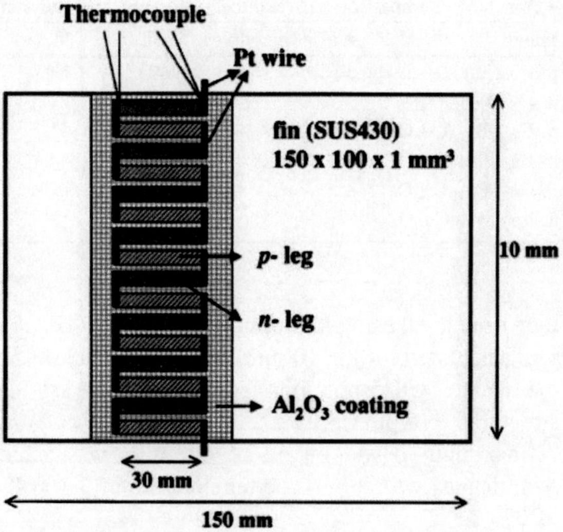

Fig. 15.16 Schematic of a $Ca_{2.75}Gd_{0.25}Co_4O_9$/$Ca_{0.92}La_{0.08}MnO_3$ thermoelectric module [23]

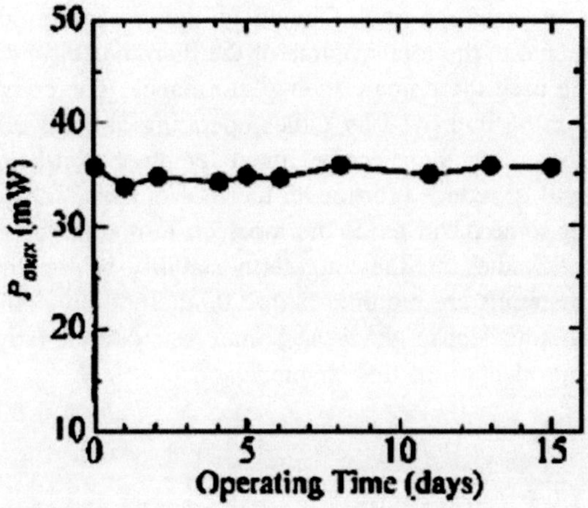

Fig. 15.17 Time dependent performance of a $Ca_{2.75}Gd_{0.25}Co_4O_9$/$Ca_{0.92}La_{0.08}MnO_3$thermoelectric module (Matsubara et al. [23])

The assembled module was then subjected to a heat treatment at 915°C for 5 h in order to hardening the silver paste. This process ensured both low Ohmic and good thermal contacts and also improved the mechanical strength of the module. Temperature differences of up to 925 K along the length of the thermoelectric legs were created by heating the top alumina plate using mica heaters embedded in copper plates, while the bottom alumina plate is maintained at room temperature by placing on a copper block cooled by water circulation. The performance was evaluated up to a maximum temperature of 1,000 K with a temperature differential, ΔT, up to 925 K between the hot and cold plates. An open circuit voltage, V_o, of 400 mV was generated at an operating temperature of 1,000 K and an internal resistance, R_i, of 1.27 Ω was measured. The maximum output power was

Table 15.2 Comparison between the performance of different all-oxide thermoelectric modules

Materials	# of *p–n* couples	ΔT (K)	V_o (v)	P_{max} (mW)	P_{max}/ΔT (mW/K)
p: $Ca_3Co_4O_9$ n: $(ZnO)_7In_2O_3$	44	673	1.8	423	0.629
p: $Ca_{2.7}Bi_{0.3}Co_4O_9$ n: $La_{0.9}Bi_{0.3}Co_4O_9$	140	551	4.5	150	0.272
p: $Ca_{2.7}Bi_{0.3}Co_4O_9$ n: $La_{0.9}Bi_{0.3}Co_4O_9$	8	565	1	170	0.301

then calculated using the formula $P_{max} = V_o^2/4R_i$. A maximum output of 31.5 mW was calculated for the 10 mm long legs module, which decreased to 24.5 mW for the module with 5 mm long legs. In 2011, the fabrication of a module using layer-structured *p*-type $Ca_3Co_4O_9$ and *n*-type $(ZnO)_7In_2O_3$ legs was reported [25]. A maximum power output of 423 mW was achieved in the module with 44 *p–n* couples with a hot-side temperature of 1,100 K and a temperature difference of 673 K.

More recent prototypes are becoming more sophisticated larger output powers have been reported. Obviously, the maximum power generated is almost proportional to the total volume of the thermoelectric devices and this characteristic can be used to compare their performance. The conversion efficiency of the modules can be improved by either operating at a larger temperature differentials and/or increasing the thermoelement length. Nevertheless, the maximum power output will be reduced for longer thermoelements. Table 15.2, compares the performance of some of all-oxide thermoelectric modules so far developed.

Studies on the long term stability of the thermoelectric legs and electrode materials are required before to attempt any commercial deployment of all-oxide thermoelectric devices. Some tests are already underway as mention in the introduction of this chapter.

15.6 Conclusions

Although thermoelectric modules were originally developed for cooling applications, recent studies revealed the potential of these devices for electrical power generation using waste heat in the temperature range 300–400 K. Nevertheless, thermoelectric modules operating at higher temperatures are much more appealing to harvest waste energy from high temperature processes. Without a doubt, the amount of electricity produced by thermoelectrics alone will only satisfy a portion of total global energy needs. Nevertheless, thermoelectrics alongside a variety of renewable sources such as wind, wave, solar, biomass will certainly play an important role to increase the present energy output without increasing CO_2 emissions. For high temperature energy harvesting, oxides are the materials of choice because of their thermal and chemical stability. Nevertheless, further improvement of the efficiency of thermoelectric modules is required. If efficiency levels of 30% are reached, thermoelectrics will competitive with current energy sources. Hence, the current challenges in the field of thermoelectrics fall into two categories. One is focused on materials property, whereas the other concerns the successful transfer of those properties into an efficient device. In terms of materials property, it is necessary to increase the thermopower without depressing the electrical conductivity; ideally the power factor needs to be enhanced by two to fourfold. Reduction of the intrinsic lattice thermal conductivity is already reaching the lowest theoretical value, which is expected for amorphous materials. Nevertheless, the thermal conductivity can also be reduced by boundary scattering in bulk nanoceramics, whereas power factor may also increase due to electron filtering at the grain boundaries. Indeed, a significant improvement of the thermoelectric performance in the most-studied existing thermoelectric materials has been achieved by creating nanograins and nanostructures in the grains using the combination of high-energy ball milling, direct-current-induced hot-press processes and sol-gel processing. sol-gel processing has tremendous advantages conventional ceramic processing to produce fine-grained ceramics.

In summary, in this chapter we cover key concepts in the field of thermoelectric oxides research, with special emphasis on the role of sol-gel processing to further enhance the efficiency of thermoelectric bulk nanoceramics. Finally, we would like to suggest the investigation of hexagonal perovskites as potential thermoelectric should be explored. One of the authors has previously highlighted the potential of use hexagonal structures with different number of face-shared octahedra to control of the electrical conductivity [26], now he is also suggesting that this feature may be useful to control the thermal conductivity.

Acknowledgments AF would like to acknowledge the Higher Education Funding Council for England (HEFCE) the award of a Senior Research Fellowship at Universities of Birmingham and Warwick.

References

1. Alphabeth Energy, News Release (2010) Alphabet energy awarded government grants totaling $320,000. www.alphabethenergy.com
2. Amerigon Incorporated, News Release (2009) Amerigon subsidiary BSST to test thermoelectric waste heat recovery system on BMW group and Ford vehicles. www.amerigon.com
3. SDK, News release (2010) SDK and PLANTEC test thermoelectric power generation modules in waste incinerator. http://www.showa-denko.com
4. Lemon S (2008) Murata working on turning laptop heat into power, IDG News Service
5. Koumoto K, Wang Y, Zhang R, Kosuga A, Funahashi R (2011) Oxide thermoelectric materials: a nanostructuring approach. Ann Rev Mater Res 40:363–394. doi:10.1146/annurev-matsci-070909-104521
6. Vineis CJ, Ali Shakouri A, Majumdar A, Kanatzidis MG (2010) Nanostructured thermoeletrics: big efficiency gains from small features. Adv Mater 22:3970–3980. doi:10.1002/adma.201000839
7. Newnham R (2004) Properties of materials: anisotropy, symmetry and structure. Oxford University Press, New York

8. Kittel C (1995) Introduction to solid state physics, 7th edn. Wiley, New York
9. Ohtaki M, Tsubota T, Eguchi K, Arai H (1996) High-temperature thermoelectric proper-ties of $(Zn_{1-x}Al_x)O$. J Appl Phys 79:1816–1818
10. Ohtaki M, Araki K, Yamamoto K (2009) High thermoelectric performance of dually doped ZnO ceramics. J Electr Mater 38:1234–1238. doi:10.1007/s11664-009-0816-1
11. Hopper EM, Zhu Q, Song JH, Peng H, Freeman AJ, Mason TO (2011) Electronic and thermoelectric analysis of phases in the $In_2O_3(ZnO)k$ system. J Appl Phys 109:013713–013714. doi:10.1063/1.3530733
12. Kinemuchi Y, Mikami M, Kobayashi K, Watari K, Hotta Y (2010) Thermoelectric properties of nanograined ZnO. J Elec Mat 39:2059–2063. doi:10.1007/s11664-009-1009-7
13. Lee S, Yang G, Wilke RHT, Trolier-McKinstry S, Randall CA (2009) Thermopower in highly reduced *n*-type ferroelectric and related perovskite oxides and the role of heterogeneous non-stoichiometry. Phys Rev B 79:134110
14. Lee KH, Kim SW, Ohta H, Koumoto K (2006) Ruddlesden-Popper phases as thermoelectric oxides: Nb-doped $SrO(SrTiO_3)n$ ($n = 1, 2$). J Appl Phys 100:063717
15. Terasaki I, Sasago Y, Uchinokura K (1997) Large thermoelectric power in $NaCo_2O_4$ single crystals. Phys Rev B 56:12685–12687. doi:0163-1829/97/56$\sim$20/12685
16. Cai KF, Müller E, Drašar C, Mrotzek A (2003) Preparation and thermoelectric properties of Al-doped ZnO ceramics. Mat Sci Eng 104:45–48. doi:10.1016/S0921-5107(03)00280-0
17. Shang PP, Zhang BP, Li JF, Ma N (2010) Effect of sintering temperature on thermoelectric properties of La-doped $SrTiO_3$ ceramics prepared by sol-gel process and spark plasma sintering. Solid State Sci 12:1341–1346. doi:10.1016/j.solidstatesciences.2010.05.005
18. Liu CJ, Liao JY, Wu TW, Jen BY (2004) Preparation and transport properties of aqueous sol-gel synthesized $NaCo_2O_{4-\delta}$. J Mater Sci 39:4569–4573
19. Zhang YF, Lu QM, Zhang QY (2006) Synthesis and characterization of $Ca_3Co_4O_9$ nanoparticles by citrate sol-gel method. Mater Lett 60:2443–2446. doi:10.1016/j.matlet.2006.01.013
20. Nan J, Wu J, Deng Y, Nan CW (2003) Synthesis and thermoelectric properties of $(Na_x\text{-}Ca_{1-x})_3Co_4O_9$ ceramics. J Eur Ceram Soc 23:859–863
21. Xu J, Wei C, Jia K (2010) Thermoelectric performance of textured $Ca_{3-x}Yb_xCo_4O_{9-\delta}$ ceramics. J Alloys Compd 500:227–230. doi:10.1016/j.jallcom.2010.04.014
22. Pei J, Chen G, Zhou N, Lu DQ, Xiao F (2011) High temperature transport and thermoelectric properties of $Ca_{3-x}Er_xCo_4O_{9+\delta}$. Phys B 406:571–574. doi:10.1016/j.physb.2010.11.043
23. Matsubara I, Funahashi T, Takeuchi T, Sodeoka S, Shimizu T and Ueno K Fabrication of an all-oxide thermoelectric power generator. Appl Phys Lett 78:3627–3629. doi: 0003-6951/2001/78(23)/3627/3
24. Reddy ES, Noudem JG, Hebert S, Goupil C (2005) Fabrication and properties of four-leg ox-ide thermoelectric modules. J Phys D Appl Phys 38:3751–3755. doi:10.1088/0022-3727/38/19/026
25. Choi SM, Lee KH, Lim CH, Seo WS (2011) Oxide-based thermoelectric power generation module using p-type $Ca_3Co_4O_9$ and n-type $(ZnO)_7In_2O_3$ legs. Energy Convers Manag 52:335–339. doi:10.1016/j.enconman.2010.07.005
26. Feteira A (2009) Negative temperature coefficient resistance (NTCR) ceramic thermistors: an industrial perspective. J Am Ceram Soc 92:967–983. doi:10.1111/j.1551-2916.2009.02990.x

Chapter 16
Sol-Gel Processes for Nuclear Fuel Fabrication

K. Nagarajan and V. N. Vaidya

Abstract Sol-gel based processes have been developed for the fabrication of nuclear fuel materials, namely, the oxides, carbides and nitrides of uranium, plutonium and thorium as well as their solid solutions. These processes have the advantage of being free from using radioactive powder and being amenable for adaptation to remote handling. The fuel microspheres prepared through sol-gel processes can be vibrocompacted to form sphere-pac fuel pins or compacted into pellets for stacking in a fuel pin. Among the sol-gel processes, the external gelation and internal gelation processes are being used in plant scale for the preparation of kernels of coated particle fuels for high temperature gas cooled reactors. They also have the potential to be used for the fabrication of fast reactor fuels as well as fuels containing minor actinides for the transmutation reactors.

Keywords Coated particle fuels · Enriched uranium dioxide · Fuel cladding · Internal gelation · Microsphere · Nuclear fission · Nuclear fuel · Peptization · Plutonium oxide · Silicides · Thorium oxide · Uranium carbide · Uranium nitride · Uranium oxide · Water extraction gelation

K. Nagarajan (✉)
Indira Gandhi Centre for Atomic Research, Kalpakkam 603 102, India
e-mail: knag@igcar.gov.in

V. N. Vaidya
Formerly with Bhabha Atomic Research Centre, Mumbai 400 085, India
e-mail: vaidyavin@yahoo.co.in

M. Aparicio et al. (eds.), *Sol-Gel Processing for Conventional and Alternative Energy*, Advances in Sol-Gel Derived Materials and Technologies,
DOI: 10.1007/978-1-4614-1957-0_16,

16.1 Introduction

In the current scenario of fast depleting fossil energy resources in the world, nuclear energy is an important source that needs to be developed to its full potential. Nuclear energy is one of the cleanest forms of energy, as it generates no greenhouse gases and leaves no residual effect on the environment. The nuclear reactors used currently for producing power are all based on nuclear fission. These nuclear fission-based reactors are large-sized machines and reactors of up to 1600 MWe generation capacity are being operated. Today, 441 nuclear reactors are operating in the world producing 375 GWe that corresponds to about 14% of electricity production. Some countries such as France, Belgium etc. have more than >70% of their energy needs met by using nuclear sources. All the energy generated by fission reaction is deposited within the nuclear fuel in the reactor and needs to be transported via a suitable coolant to form steam, which in turn, is used to generate electricity. It is known that the efficiency of conversion of heat to electricity depends on the outlet temperature of the coolant. Nuclear fuels used in these reactors thus have to withstand high temperatures and a very highly hostile radiation environment inside the reactor. Therefore nuclear fuels used for the power reactors are generally ceramic materials which are stable in this environment, though metals and molten salts have also been used as fuels in some reactors.

Ceramic materials have been known from the very early ages. Baked earthen pots were the first ceramic products and the sophisticated china pottery followed later. Generally, ceramic materials are associated with high chemical stability at elevated temperatures without any large change in the physical properties. These properties of ceramic materials make them highly attractive for their use as nuclear fuel materials as stated above. Oxides, carbides, nitrides and silicides of uranium, thorium and plutonium and their solid solutions have been used as nuclear fuels in various reactors.

Most of these reactors use the fuel in the form of pellets that are stacked inside a metal cladding. Natural or slightly enriched uranium dioxide pellets have been used as fuels for commercial thermal neutron-based reactors all over the world and powder metallurgical routes for pellet fabrication process are also commercially well established. Mixed oxides of uranium and plutonium are used as fuels in fast reactors and in some thermal reactors. Mixed oxides of uranium and thorium are being considered for use in thermal reactors. Use of thorium and plutonium along with uranium helps to sustain the nuclear fuel resources for longer periods. Some of the isotopes of Pu give high gamma and neutron doses. The ^{233}U isotope from thorium reactors invariably contains the ^{232}U isotope which also results in high gamma dose. Hence the facilities for the fabrication of fuels containing these isotopes necessitate the entire operation to be carried out remotely inside concrete hot cells that are alpha tight and gamma shielded. Powder-based routes are very difficult to be carried out remotely as they involve a large number of mechanical steps. The fabricated fuel pellets, especially the smaller pellets of fast reactors, have to meet stringent quality control requirements with respect to dimensional tolerances and surface defects. As more than a million pellets need to be fabricated for a fast reactor with stringent specification requirements, remote fabrication is very difficult and results in lowering of throughputs and generation of rejects. Solution-based fuel manufacturing processes are aimed at eliminating powder handling and the problems associated with it.

'Sphere-pac' fuel is an alternate form of fuel design that has been developed to make the fuel fabrication process simpler and more amenable for remotisation. In the 'sphere-pac' fuel pin, fuel in the form of spherical particles or microspheres is packed into a metal cladding by the vibrocompaction process wherein mechanical vibrations aid packing. Although vibrocompaction has not been industrially employed for commercial fuel fabrication, in the recent years, there is renewed interest in the process, as 'sphere-pac' fuel is considered ideally suited for the transmutation reactors. Transmutation reactors aim either at the reduction of the weapon grade plutonium stock piles or at the transmutation of the minor actinides (MA), neptunium, americium and curium that are partitioned or recovered from the high-level nuclear waste into lighter nuclides. Considering the fact that a large amount of helium gas is produced due to the alpha decay of americium, the sphere-pac fuel design that can accommodate the helium gas is preferred over the pellet form for the MA containing fuels. Further, the large amount of heat produced due to the decay of curium and high levels of gamma and neutron dose necessitate the fabrication of the MA containing fuels to be carried out remotely in hot cells. Solution-based fabrication methods with subsequent vibrocompaction for packing make sphere-pac fuel design to be well suited for these fuels.

16.1.1 Advantages of Sol-Gel Methods

Development of solution-based fabrication methods dates back to the early sixties. They were first investigated for the production of (Th-^{233}U) oxide-based kernels of coated particle fuels for the high temperature gas cooled reactors (HTGR). Carbonaceous coatings are subsequently deposited on the spherical kernels of fuel materials. Later, these processes were developed for the fabrication of fast reactor fuels. The name sol-gel process [1] for ceramic nuclear fuels (to be referred to as sol-gel process hereafter) is a generalised heading for solution/sol-based routes. Sol-gel processes involve gelation of the droplets of a sol (colloidal solution), or solution by hydrolysis resulting into condensation and polymerisation of the desired fuel material into gel microspheres. The gel microspheres, after further processing, are converted into high density microspheres by sintering and are used in sphere-pac fuel pins. Alternatively, they can be converted into soft microspheres, suitable for pressing and sintering to form high density pellets by sol-gel microsphere pelletisation process (SGMP) [2].

Sol-gel processes provide the following advantages over the conventional powder-based pellet making processes. The sol-gel processes

(i) use only sol/solutions and avoid handling of powders containing U, Pu.
(ii) involve handling of fluids and fluid-like microspheres. The transfer of fluids is much simpler than that of powders of fuel materials especially in a remotised facility. Thus sol-gel processes are more easily amenable to remote handling.
(iii) can use the actinide nitrate solutions from the reprocessing plants as the feed which reduces the number of process steps such as actinide oxalate precipitation, conversion to actinide oxides etc. involving powder handling.
(iv) SGMP process provides the advantage of elimination of powder handling and the associated problems and at the same time, the fuel is fabricated in the form of pellets, the well-established fuel form.

The disadvantages of the sol-gel processes are:

(i) Large amount of liquid waste is produced in these processes.
(ii) Large amount of irradiation experience is available around the world for the pellet fuels and their irradiation behaviour is well established, but the irradiation experience for sphere-pac fuel is very limited.

16.2 Sol-Gel Processes for Nuclear Fuel Fabrication

16.2.1 General Flow Sheet of Sol-Gel Processes

There are many sol-gel process routes that have been developed for nuclear fuel fabrication. A general flow sheet common to these processes is given in Fig. 16.1. Droplets of the solution or sol of the fuel material are generated and these droplets are then converted into hard gel particles by the gelation process (physical or chemical). The gel particles are washed, dried, calcined and sintered into high density fuel microspheres. In all the process steps, the fuel materials undergo shrinkage resulting in large stresses that could damage the particles. But, the process parameters are optimised to produce crack-free particles. In the sphere-pac fuel pin, the sintered microspheres are packed into metal cladding using mechanical vibrations. In the SGMP process, the calcined gel particles are pressed into fuel pellets and are sintered. In the coated particle fuel, the sintered microspheres are used as kernels on which layers of carbon and silicon carbide/zirconium carbide are coated.

16.2.2 Water Extraction Gelation Process (ORNL Process)

This process was developed at Oak Ridge National Laboratory (ORNL), USA in the sixties. Haas et al. [3] and McBride [4] describe the process in detail. It is based on the extraction of water from the droplets of a sol, when the droplets are

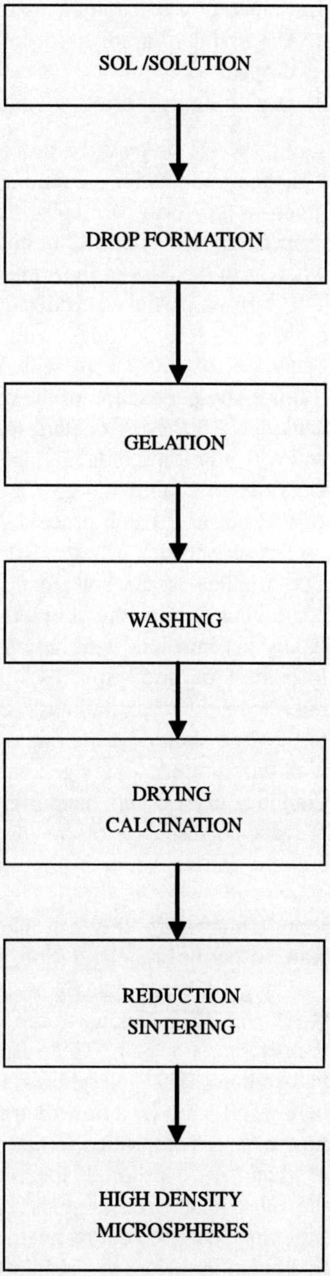

Fig. 16.1 General flow sheet of the sol-gel processes

suspended in an organic liquid, '2-ethyl-1-hexanol (2EH) leading to the formation of solid spheres in the form of gel. This gelation comes under the category of physical gelation. The method involves the following three principal steps:

1. Preparation of an aqueous oxide sol.
2. Dispersing the sol as droplets into 2EH, while water is extracted from the droplets.
3. Drying and sintering at controlled conditions.

All the sol preparation procedures start with nitrate solutions of fuel materials. The requirements for the preparation of sol depend on the system and are generally given in the form of recipes that are reliable for getting the products of desired properties. For example, uranium and plutonium require adjustment of valency whereas in the case of thorium, which is only tetravalent, the step is not necessary. The various partial neutralisation methods used for sol preparation include precipitation of hydroxides with ammonium hydroxide followed by peptisation, extraction of nitric acid with liquid amines, hydrothermal denitration to oxide followed by peptisation of the oxide mixed with water with residual or added nitric acid, etc. All these procedures, unless they are carried out under inert atmosphere and with a holding reducing agent like hydrazine, are not suitable for U systems as U(IV) is not stable and gets oxidised to U(VI). In general, the sol preparation is carried out as a batch process.

The second step of water extraction from the sol droplets is a very slow process. The droplets of the sol are fluidized in a tapered column containing 2EH in a continuous mode; the droplets shrink with water removal, become dense and finally gel into hard gel particles which drop out of the gelation column. For the formation of good spheres, the mass transfer of water across the gel–liquid interface rather than the mass transfer of water from the inner side to the outer side of the drop should be the rate controlling step. A gel surface that is in equilibrium with unsaturated 2EH is generally smooth and non-sticking and thus amenable for draining, drying and handling. Surfactants like Span 80 are added to 2EH to prevent coalescence of the drops, clustering or sticking of gel spheres to the column walls. Another important aspect of the process is that the maximum amount of water in 2EH is controlled at 1% throughout the process (although at room temperature 2EH can take up to 2.6% of water), by partially removing the water using online distillation.

The gel spheres from the water extraction step are first washed with halocarbons to recover 2EH sticking to the surface of the spheres and are dried in air at room temperature for 24 h. This results in the removal of residual water and also the halocarbons. The gel spheres are then slowly heated from room temperature to below 500 K under a flow of argon gas and steam mixture to remove volatiles. The flow rates are controlled so that it does not cause cracking of spheres. The purpose of steam is to promote the removal of organic substances and nitrates before the subsequent process steps. Argon gas is used to prevent ignition of organics. Sintering is carried out by heating to 1400 K under a flowing gas stream containing hydrogen, in the case of uranium containing systems.

The process has been used for the preparation of microspheres of UO_2, ThO_2, $(U,Th)O_2$ and $(U,Pu)O_2$. It was successfully scaled up to engineering scale and has been used for the manufacture of coated particle fuel for the HTGR and sphere-pac fuel for fast reactors. However, the main drawback of the process for uranium systems is the need to have the reduced form of uranium, [U(IV)] in solution, which is not very stable, and the long contact time required for dehydration of droplets. For making fuels containing uranium, plutonium and MA, controlling the colloidal gel formation is a real challenge, since their chemical behaviours are very different. Some of them having oxidation characteristics are suitable for hydrolysis at low pH, while others exhibit the opposite behaviour.

In the water extraction process of gelation, too many physical and chemical parameters are needed to be controlled including tedious heat treatment steps for getting good quality microspheres. Hence the process has been discontinued. However, using this process, more than 100 kg of thoria microspheres were produced on a pilot plant scale at ORNL.

16.2.3 External Chemical Gelation Process

The external gelation process (EGP) was developed in Italy and is known as the SNAM process [5–7]. It is also known as the gel-supported precipitation method. The process was developed with the objective of making microspheres of up to 500 μm to be used as kernels in coated particle fuels and even today many countries interested in coated particle fuel development are pursuing this process. The flow sheet for EGP is shown in Fig. 16.2. In this process, a water soluble polymer is added to the heavy metal nitrate solution or sol the droplets of which are exposed to ammonia gas. The objective of the polymer addition is to support the spherical shape of the particle, when the ammonia gas penetrates the droplets to precipitate the heavy metals and thus harden the spheres. Since ammonia has to diffuse through the droplet for neutralisation, the heavy metal concentration is kept low (0.7 M). The spherical shape of the droplets may undergo deformation at the gas–liquid interface or liquid–liquid interface. Exposure to ammonia gas strengthened the surface and avoided distortion. One of the attractive features of the process for urania microsphere preparation is that no pretreatment or pre-neutralisation is needed. The surface hardened spheres are then allowed to age in an ammonia solution to improve their integrity. The principal steps are:

1. Sol or broth preparation.
2. Sphere formation and gelation.
3. Ageing, washing and dehydration.
4. Calcination and sintering.

Addition of a modifier along with the polymer during the preparation of broth yielded good results. The prerequisites for the polymer are: [8]

1. It should be soluble in water and be compatible with the high ionic content of the sol or broth.

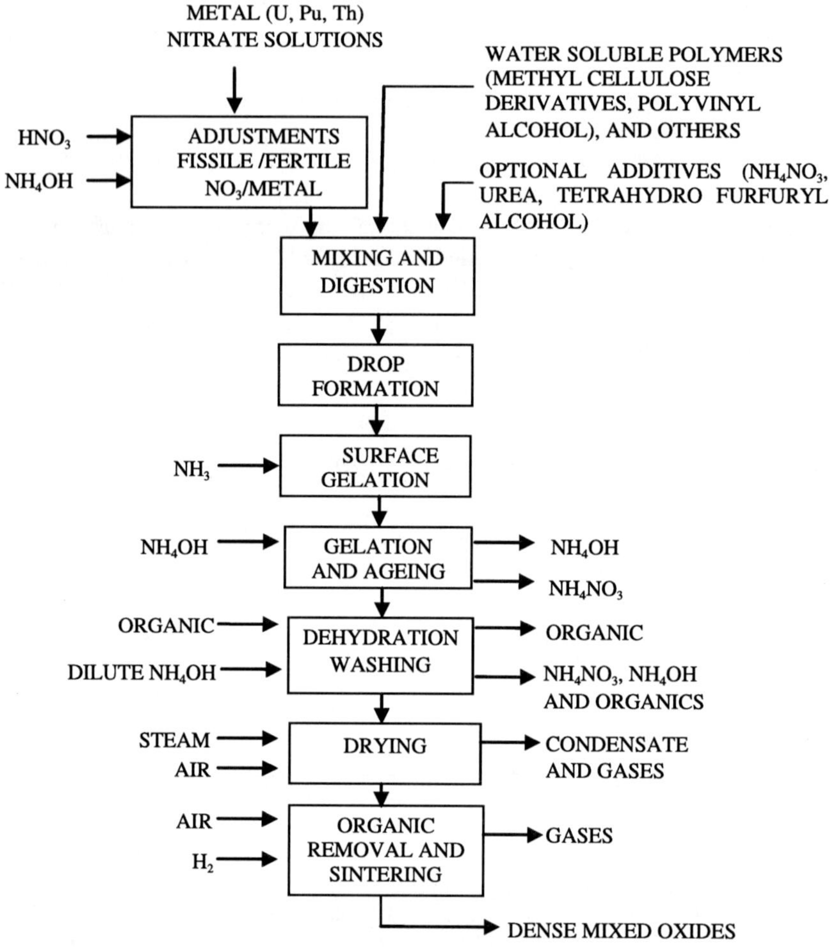

Fig. 16.2 Flow sheet for EGP for nuclear fuel fabrication

2. It should thicken the sol or broth along with the modifier to enhance drop formation, but at the same time the sol or broth should have enough fluidity to make it flow through the nozzle for forming droplets.
3. It should ensure quick formation of a gel structure to withstand impact at the aqueous interface; it should form a gel structure that is strong enough to support the heavy metal in the spherical shape until it solidifies, then be capable of debonding without damaging the inorganic structure.

Although urania sol preparation does not involve any pretreatment in this method, denitration of thorium nitrate and plutonium nitrate solutions have to be carried out to prepare thorium nitrate sol or plutonium (IV) containing polymer. Hydroxy propyl methyl cellulose (Methocel) was the polymer and Tetra-HydroFurfuryl Alcohol (THFA) was the modifier used in the original SNAM process. A variety of polymers and modifiers have been used in the development of the process. However, methocel and THFA are the most popular ones.

In the second step of drop formation, a two-fluid nozzle was used in the SNAM process with air and sol or broth as the two fluids. The air flow rate and the nozzle diameter controlled the drop size and the air flow also helped to prevent ammonia from causing gelation of the solution in the nozzle that would lead to clogging of the tip of the nozzle. As described above, the droplets were allowed to gel in an ammonia gas stream.

The ageing solution was the same as the gelation medium (1–30% ammonia) and the ageing time varied from several minutes to overnight. Washing was done with dilute ammonia solution to remove the salts (ammonium nitrate) and at the same time to avoid repeptization of the heavy metals. For drying, various methods like azeotropic distillation, wash with carbon tetrachloride (CCl_4), air drying, vacuum drying etc. have been used. Calcination was done in air at 723–823 K to remove the residual organic compounds and convert the metals into their oxides. Sintering was done under reducing atmosphere at temperatures ranging from 1,500 to 2,500 K.

Although the process was developed in Italy, it has been adapted by many countries, including Germany [9–12] and United Kingdom [13, 14] for the production of coated particle fuels and fuels for fast reactors. The process was modified by Kern Forschung Anlage (KFA), Julich, for the preparation of thoria containing spheres. The objective of the modification was to eliminate the addition of organic compounds [15]. In the modified procedure, a higher heavy metal concentration of 2 mol/L has been adopted compared to 0.5–0.7 mol/L used in the SNAM process. Further, the degree of preneutralisation was increased to 87.5 ± 0.5% from 70 to 80%. Since the resulting sols had high viscosity, no organic additives were needed. In the next step of dispersion of the sol into droplets, a horizontal nozzle vibrating with a constant frequency was used through which a laminar flow of the sol was maintained. Then for gelation there were two variations in the KFA process flow sheet. In one version, droplets were dispersed into hexone or methylisobutyl ketone containing ammonia where precipitation occurred. In the other version, the droplets were exposed to ammonia gas stream. For making $(Th,U)O_2$ microspheres, in the original SNAM process, the thorium and uranium feeds were prepared separately and mixed before the droplet formation. But in the KFA process, thorium and uranium were processed together for making Th-U sol for preparing ceramic ThO_2-UO_2 spheres. The KFA process was designed to produce sintered particles of <600 μm in diameter (T.D.). However, when the process was used as such for larger particles (825 μm) ellipsoids rather than spherical particles were obtained due to higher impact damage. An organic polymer [polyvinyl alcohol (PVA)] was added to the sol to overcome this problem. But it resulted in softening of the gel. In the KFA process, isopropanol was used for washing as it removed the salts and PVA. The rest of the process steps are similar except that UO_3 in kernels must be reduced to UO_2 between 873 and 1073 K with H_2 or Ar/H_2 before sintering.

In the flow sheet developed by the British, uranium hexa fluoride converted into uranyl fluoride was used as the starting solution. Another modification used by them was the use of a foam technique in which the ammonia gas is bubbled through ammonium hydroxide solution to form the foam to cause gelation. The foam helped to cushion the impact on the droplet and at the same time harden the spheres. In the recent years, the process has been used for the preparation of large-sized (800–900 μm) and small-sized UO_2 (50–70 μm) microspheres in Japan [16]. The process is under consideration for preparing $(U,Pu)O_2$ microspheres for use in fast breeder reactors.

The Institute of Nuclear Energy Technology, China, has used a modified version of EGP incorporating some features of the internal gelation process (IGP) which will be discussed in the following sections, for the preparation of UO_2 kernels for the coated particle fuels of HTR-10. The process is called total gelation process (TGP). The procedure followed is mainly that of external gelation. However, hexamethylene tetramine (HMTA) which is generally used in the IGP is added to the broth along with urea, PVA and THFA. Addition of HMTA enabled increasing the concentration of U in the broth to 1.26 M, resulting in less shrinkage compared to EGP and thus better sphericity and less cracking of the spheres was achieved. Using this process, kernels having an average diameter of 498 μm needed for making coated particles of 600 μm for HTR-10 up to 3 kg of U per batch have been prepared [17].

All the versions of the EGP end up with large quantities of aqueous waste solution containing ammonium nitrate and organics and hence special processes are needed to be developed to treat the waste solution to make it suitable for waste management/disposal.

16.2.4 Internal Gelation Process

The IGP was developed by Kanij et al. [18] and Brugghens et al. [19] at KEMA laboratories, Netherlands in 1967 for the production of UO_2 microspheres. The flow sheet for the internal gelation process is shown in Fig. 16.3. A schematic diagram of the internal gelation setup is shown in Fig. 16.4. The method uses a water soluble chemical, HMTA (also known hexa or urotropin) that releases ammonia on heating to cause precipitation of heavy metal to form a gel. The droplets of the solution are allowed to fall in a hot immiscible liquid to provide the heating. Since the ammonia donor and the heavy metal ions are dissolved in the same solution, gelation occurs rapidly and nearly uniformly throughout the drop typical of a homogeneous precipitation. The rate of gelation depends on the concentration of the chemicals and the solution as well as the heat transfer from the hot liquid into the droplet. The process comprises the following major steps:

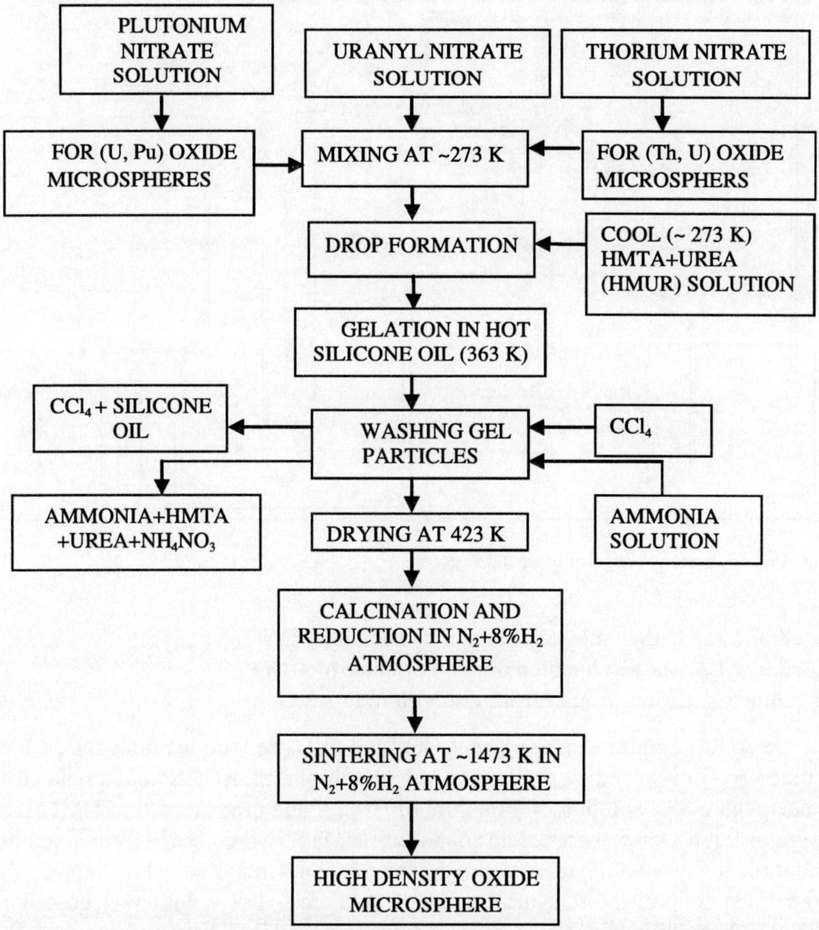

Fig. 16.3 Flow sheet for IGP for nuclear fuel fabrication

1. Broth preparation.
2. Sphere formation and gelation.
3. Washing and drying.
4. Calcination and sintering.

In the case of urania microspheres, preparation of Acid Deficient Uranyl Nitrate (ADUN) solution with a NO_3^-/UO_2^{2+} mole ratio of 1.5–1.7 is the first step of the flow sheet, as it has higher solubility in water. It is done in one of the following ways:

1. adding UO_3 to a sub-stoichiometric amount of nitric acid (HNO_3) or to a stoichiometric solution of uranyl nitrate.

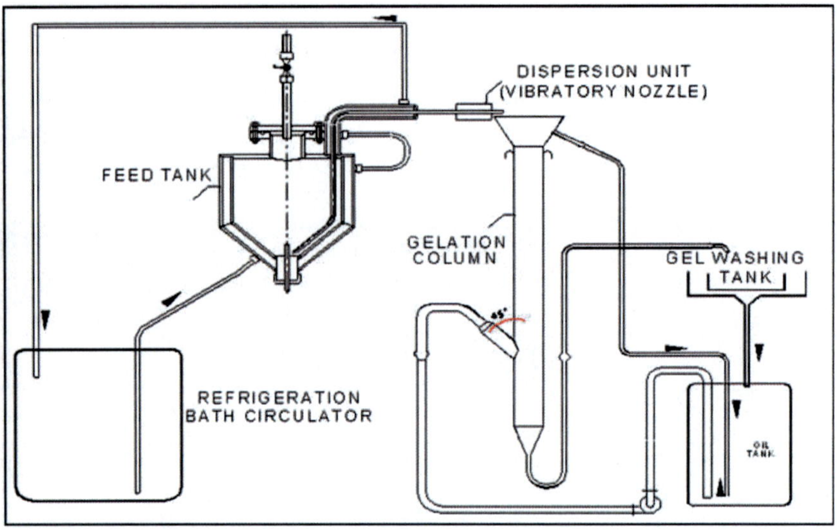

Fig. 16.4 Schematic diagram of an IGP setup

2. adding U_3O_8 to a sub-stoichiometric amount of HNO_3.
3. adding UO_2 to a sub-stoichiometric amount of HNO_3.
4. amine extraction of acid from uranyl nitrate solution.

The ADUN solutions made using UO_2 and U_3O_8 require aeration for 24 h to remove NO_2 dissolved in the broth. A 3 M solution of ADUN thus prepared is mixed with a 3 M solution of a mixture of HMTA and urea (known as HMUR) in the ratio 1:1.4. Under these acidic conditions, HMTA hydrolyses to form ammonia and formaldehyde which leads to premature gelation. Urea is added to complex the uranyl ion to prevent this reaction with ammonia. The solution is cooled to 273–272 K to reduce the rate of decomposition of HMTA and to increase the stability of the uranyl-urea complex. Hence broth preparation is carried out by mixing cold uranyl nitrate solution with cold HMUR solution. Although gelation has been reported to occur over a wide range of solution compositions (uranium molarity from 0.7 to 1.6) and a range of molar ratios of HMUR to Uranium 0.75–2.3, the window of operating parameters become narrower for making hard, opaque gels which behave well in the subsequent steps to result in good quality, crack-free hard microspheres. Detailed studies carried out at Bhabha Atomic Research Centre (BARC) Mumbai, India have resulted in the gelation field diagram shown in Fig. 16.5 [20] for the preparation of urania microspheres. Feed compositions from the gelation field diagram resulting in hard, opaque gel are suitable for making very good quality UO_2 microspheres which can withstand washing, drying, reduction and sintering procedures with a very small (1%) process loss. Regions of uranium molarity between 1.0 and 1.4 M have been found to yield good quality UO_2 microspheres of 500–700 μm size and also require lower

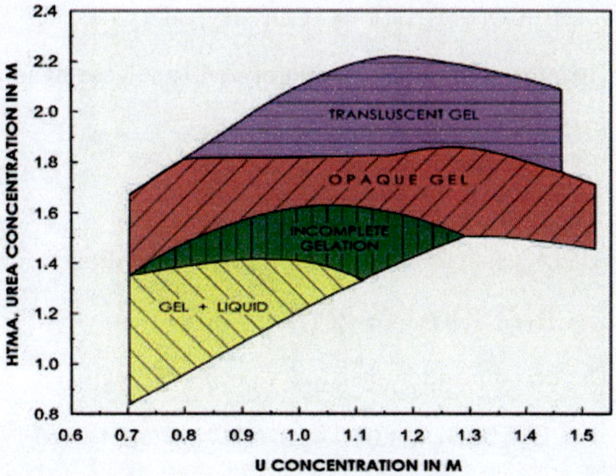

Fig. 16.5 Gelation field diagram for the preparation of urania microspheres

amounts of HMTA and urea per kg of urania microspheres produced. Regions of lower molarity of Uranium (0.7–0.9 M) have been used for the preparation of fine microspheres of 70–100 μm size which enables the use of nozzles of reasonable size for making microspheres in this size range.

The organic liquid to be used as the medium for gelation should have high density close to that of the spheres formed and high volatility for enabling the removal from spheres. Although trichloroethylene (TCE) has a boiling point (bp) of 360 K, it was used as the organic medium at temperatures ranging from 338 to 353 K. Perc (bp = 384 K) and silicone oil have also been used as the media. The high density of perc is a disadvantage requiring the uranium concentration of the reaction mixture to be greater than 1.55 M in order to avoid the floating of microspheres. Further, the solubility of HMTA and urea is very high in the chlorinated hydrocarbons, especially perc and hence perc is not useful. Silicone oil with a specific gravity of 0.9 and a high viscosity of 36 cP (DC200 of M/s. Dow Corning, USA having a viscosity of 100 cp at room temperature) has been used as the medium in many countries like Switzerland and India. The viscous nature of silicone oil helps in maintaining the spherical shape of the droplets and providing adequate contact time even in small columns. The chemistry of IGP has been investigated in detail by Collins et al. [21] and Haas et al. [22].

16.2.4.1 Chemistry of the Gelation Process

In the broth made by mixing the ADUN with HMUR, the pH varies between 3.5 and 6 depending on the NO_3^-/U ratio and uranium precipitation occurs above a pH of 3.25. The presence of urea helps to complex the uranium to form $UO_2[CO(NH_2)_2]_2^{2+}$. When the droplets of the broth are introduced in the hot silicone oil, the following reactions occur which control the extent of gelation. At first, temperature-induced decomplexation of ADUN-urea complex occurs.

$$UO_2[CO(NH_2)_2]_2^{2+} \Leftrightarrow 2CO(NH_2)_2 + UO_2^{2+} \quad (16.1)$$

Then the metal ion gets hydrolyzed to release hydrogen ion

$$UO_2^{2+} + H_2O \Leftrightarrow UO_2(OH)^+ + H^+ \quad (16.2)$$

$$n\ UO_2(OH)^+ \Leftrightarrow UO_2(OH)_n^{n+} \quad (16.3)$$

HMTA gets protonated consuming this hydrogen ion:

$$(CH_2)_6N_4 + H^+ \Leftrightarrow [(CH_2)_6N_4H]^+ \quad (16.4)$$

The decomposition of HMTA occurs through

$$[(CH_2)_6N_4H]^+ + 3H^+ + 4NO_3^- + 6HOH \Leftrightarrow 4NH_4^+ + 4NO_3^- + 6HCHO \quad (16.5)$$

Reactions (16.1) to (16.3) are the initiating reactions for the gelation. Since all these reactions are highly temperature dependent and the degree of acid deficiency affects the total concentration of hydrogen ion produced during hydrolysis, temperature and acid deficiency greatly influence the gelation process. Further, the neutralising effect of reaction (16.4) or (16.5) to form the metal ion polymer, ($UO_2(OH)_n^{n+}$) is governed by the HMUR/U ratio. At large HMUR/U ratios, the protonation reaction (16.4), which is ionic in nature and hence very fast, is involved in the hydrolysis. The fast kinetics does not allow the crystallites to grow and the resultant gel is transparent or translucent. The gels formed with lower molar ratio of HMUR/U are predominantly formed by the relatively slower reaction (16.5) resulting in the large crystallites and opaque gel.

Washing is an important step in the internal gelation process. Washing is done at first with an organic solvent like CCl_4 or trichloro ethylene (TCE) to remove the organics and then with ammonia to remove the urea and ammonium nitrate ensuring that the pH is at or above 8.5. Incompletely washed spheres fail during the subsequent steps. Opaque gels are easier to wash than transparent or translucent gels.

Drying is done in air between room temperature and 425 K in an oven or in a belt drier in a continuous mode. Different calcination conditions have been used by different laboratories. Generally, after drying calcination is done at 773–823 K in air. For urania microspheres, reduction is carried out at 823–873 K to form UO_2 followed by sintering under reducing atmosphere at 1473–1523 K. Reduction step is not required for preparing thoria microspheres.

IGP has also been used for the preparation of $(U,Pu)O_2$ microspheres at KFA, and Gesellschaft zur Wideraufarbeitung von Kernbrennstoffen GmbH (GWK), Germany, EIR, Switzerland and the then European Transuranium Institute at Karlsruhe and BARC, India.

Studies have also been carried out in many countries on the preparation of thorium oxide and $(U,Th)O_2$ microspheres. A large number of studies have been carried out in India on these systems and Kumar et al. [23] have reported the gelation field diagram for thorium system shown in Fig. 16.6.

Fig. 16.6 Gelation field diagram for the preparation of thoria microspheres

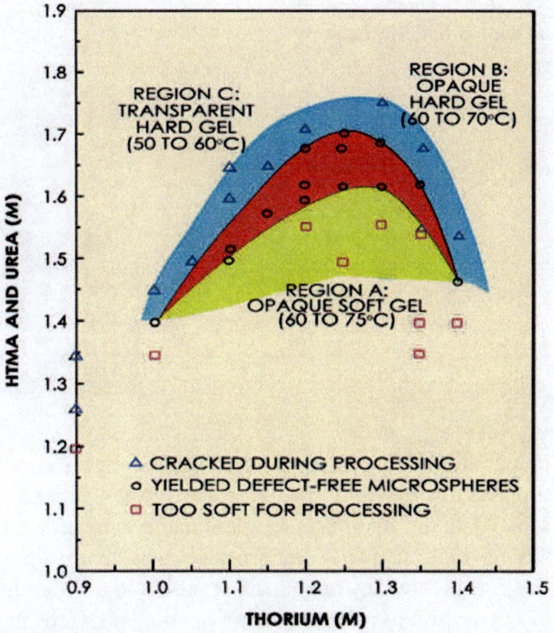

16.2.4.2 Internal Gelation Method for Making Fine Microspheres

Fuel pins of thermal reactors are fabricated with smear densities in the range of 91–95% of theoretical density (TD), whereas fast reactor oxide fuel pins have relatively lower smear densities (80–85% TD) and the smear densities of carbide or metal fuel pins are still lower to accommodate swelling due to fission products. Sphere-pac fuels should also have similar densities. A maximum smear density of 66% TD can be obtained by using vibrocompaction of a single size fraction microspheres and hence packing of multiple size fractions is needed. Extensive studies on packing of spherical particles have been carried out by McGeary [24]. It is known that with two size fractions smear densities of up to 82% can be achieved in practice and for higher densities three size fractions are needed. For ensuring optimum binary packing, the diameters of the two size fractions must differ by a factor of 7 and further the ratio of the diameter of the coarse microspheres to that of inner diameter of the clad tube must be at least 7. For a typical fast reactor fuel pin of ~6 mm internal diameter, the coarse size fraction will be in the range of 700–800 μm and the fine fraction microspheres in the size range of 100–125 μm.

The internal gelation equipment shown in Fig. 16.4 is the conventional gelation set up used for making microspheres of >500 μm and so the coarse fraction is made by this method. In the oil column, the flow of gelation medium (silicone oil) is maintained counter current to the flow of droplets so as to increase the residence time of the droplets in the gelation medium, in view of the gelation time required which is in the order of 20 s. However, the same method cannot be adopted for the

Fig. 16.7 Jet entrainment set up used in IGCAR, India

preparation of fine fraction microspheres for the reasons described here. The small droplets formed from fine nozzles arrive at a very fast pace (number of droplets per sec). Also, the small droplets are unable to penetrate the surface of the oil which is having high viscosity and surface tension resulting in the coalescence of the droplets before they are gelled. It results in a wider distribution of sizes. Therefore, in the method for the production of smaller microspheres, it should be ensured that the spheres are well separated before the completion of gelation. Two different approaches are adopted for the preparation of smaller microspheres. One is a variant of the equipment used for internal gelation, known as jet entrainment technique [25, 26]. The other is an entirely different method known as the resin loading method [27, 28]. In the jet entrainment technique, the fine broth droplets are dispersed on to a flowing jet of hot silicone oil, rather than an oil column, which prevents the droplets from coalescing. The nitrate to U ratio and the HMUR/U ratio are also different for the smaller microspheres. The jet entrainment set up for preparation of smaller microspheres of UO_2 at the Indira Gandhi Centre for Atomic Research (IGCAR), Kalpakkam, India is shown in Fig. 16.7.

Washing is done in a column mode under suction and the time required for washing is also drastically less than that for the coarse fraction microspheres. The wash liquid is further removed by a solvent like isopropyl alcohol. Finally after draining of all wash liquid, drying is carried out by keeping the washed filter funnel in the oven. The other process parameters for calcination, reduction and sintering are the same as those for the coarse fraction. However, sintering temperature and duration are chosen so as to avoid possible sticking of the microspheres with each other.

The resin loading method, successfully used for many years at ORNL for the preparation of actinide targets for heavy element production [29] involves loading of the resin with metal ions and heat treatment of the metal loaded resin. As this is not a gelation process, it will not be dealt with further here.

16.2.4.3 Microwave-Assisted Internal Gelation

In the conventional internal gelation process, as discussed above, the broth containing metal nitrate solution (Metal = U, Pu, MA, etc.) and HMUR is cooled to 271 K to prevent premature gelation occurring in the broth tank itself. Then the broth droplets are heated externally using a hot organic medium such as silicone oil. Alternatively, it is possible to heat the droplets internally using a microwave heating device. Many studies on microwave-assisted gelation have been reported. [30–32]. One of the major advantages of internal heating emanates from the fact that the gelation reaction occurs in a shorter time (<0.1 s) compared to the external heating using oil (20 s). Therefore, the length of the gelation column can be considerably reduced to be accommodated in a conventional glove box of $\sim$1 m height. Another added advantage is that the gelled spheres are made in any medium and hence there is no organic solvent to be removed. This considerably reduces the waste volume generated during the process. However, rapid gelation poses other problems. It was found that the broth with lower concentrations of U ($\approx$0.7 M) successfully gelled whereas in the case of higher concentrations of U (>1 M) the gelled particles cracked during drying and other heat treatment steps [33]. This is due to the fact that microwave gelation leads to lower crystallites due to the fast kinetics of the process and the problems arise in washing and cracking during drying and other heat treatment steps. However, when these gelled particles are allowed to age for longer time under dilute ammonia solution, the crystallites in the gel particles start growing and the above-mentioned problems are overcome. These improvements enabled preparation of defect free microspheres using internal heating method even with high concentration of U in the broth. Yamagishi [33] developed an apparatus by which internal gelation using internal heating method could be adopted even without pre-cooling the broth solution. This was done by precise mixing of very small quantities of broth solutions (metal nitrate and HMUR) just before drop formation. It was also reported [33] that hot and flowing ammonium nitrate (0.1 mol/L and weakly acidic) served as a better ageing solution and yielded better results.

16.2.4.4 Internal Gelation for Advanced Ceramic Fuels

Internal gelation process is very well suited for the preparation of uranium carbide or nitride microspheres [34–36]. Large number of studies have been carried out at EIR, Switzerland in IGP for the preparation of these microspheres. The process flow sheet for the oxide microspheres is slightly modified to add carbon powder to the feed solution of nitrate and HMUR. Gelation is carried out in the same way as for the oxide fuel preparation. After the washing step, the gel spheres are first dried in air up to 373 K and further drying is done in argon atmosphere to prevent any residual carbon getting ignited. The resultant dried gel particles, (UO_3 + C), are heat treated in vacuum at 1073 K wherein UO_3 is converted to UO_2 as per reaction (16.6).

$$UO_3 + C \rightarrow UO_2 + CO_2 \quad (16.6)$$

When it is further heated at 1473 K, carbothermic reduction occurs as per (16.7) to form UC and sintering is carried out by heating at 1973 K under inert atmosphere to get high density UC microspheres. Reaction (16.7) has two parts; reactions (16.7a) and (16.7b).

$$UO_2 + 3C \rightarrow UC + 2CO \quad (16.7)$$

$$UO_2 + 4C \rightarrow UC_2 + 2CO \quad (16.7a)$$

$$UO_2 + 3UC_2 \rightarrow 4UC + 2CO \quad (16.7b)$$

At first UC_2 is formed and in the absence of enough carbon, it reacts with remaining UO_2 to form UC. Thus initial C/U ratio is controlled to prepare UC or UC_2. Similarly, UN microspheres are prepared by carrying out the carbothermic reduction under flowing nitrogen atmosphere at 1773 K wherein UN gets formed as per (16.8).

$$UO_2 + 2C + 0.5N_2 \rightarrow UN + 2CO \quad (16.8)$$

Similar procedure has been used for the preparation of ThC_2 microspheres as well. Mixed uranium–plutonium monocarbide (UPuC) and mononitride (UPuN) are considered as advanced fast reactor fuels owing to their higher breeding potential as well as higher thermal conductivity. They also have the added advantage of having excellent compatibility with the sodium coolant used in the fast reactors. Extensive information and review articles are available in the literature on the properties and performance of these fuels [37, 38]. The advantage of using sol-gel process for these fuels derives from the fact that these fuel materials are highly pyrophoric in powder form and the sol-gel methods obviate the need for powder handling. Further, in the sol-gel processes the heat treatment is done in a single step and any exposure to oxygen/moisture is avoided during the process. Hence contamination of oxygen in nitride and nitrogen contamination in carbide are avoided. (U, Pu) C microspheres have been prepared in EIR, Switzerland and used in many experimental pins in several campaigns.

16.2.4.5 Recovery of Waste Streams

One of the problems of the IGP is the generation of large quantities of low level liquid waste streams. The success of IGP will depend on the treatment of these waste streams. The best way is to recycle them back to the process.

CCl_4 waste, which contains only silicone oil, is distilled and the CCl_4 and silicone oil are recycled. The aqueous waste contains ammonia solution, large amount of ammonium nitrate, HMTA, urea and trace quantities of actinides. A scheme has been worked out at BARC, Mumbai, India to recycle the chemicals like HMTA, urea, major quantity of ammonia solution and actinides [39]. The flow sheet that has been

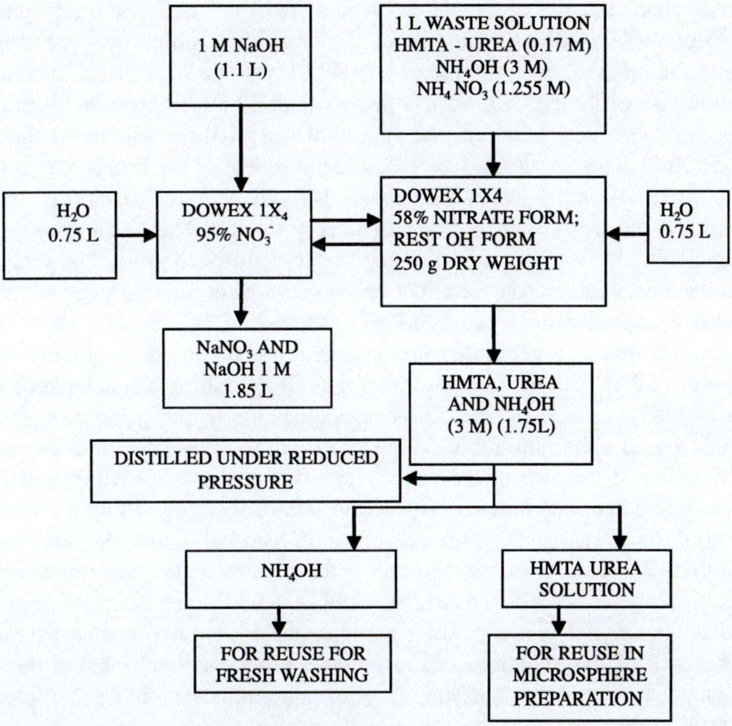

Fig. 16.8 Flow sheet for the treatment of waste from IGP

evolved for the treatment of waste is shown in Fig. 16.8. In this process, the nitrate from the waste solution is recovered using an anion exchange resin in OH form and the resultant effluent containing HMTA, urea and ammonium hydroxide is vacuum distilled to separate solution containing HMTA and urea solution as well as ammonium hydroxide which can be recycled in the IGP process. The anion exchange resin is washed with an alkali solution to remove the nitrate. The process rejects entire quantity of nitrate as sodium nitrate waste. The process also eliminates the actinides as they are loaded on the anion exchange resin in the very first stage.

Looking at the details of IGP, it becomes obvious that it is the most suitable process for preparing microspheres of a wide range of sizes, namely, 100–850 μm for use in fast reactors, coated particle fuel and SGMP for thermal reactor fuel. It also provides for the waste treatment from the process and thus offers a comprehensive package.

16.2.4.6 General Comments on the Sol-Gel Process Parameters

All the process steps of the flow sheets of the various sol-gel processes play an important role in getting crack free microspheres of desired density. One of the important characteristics of the gel that would yield such qualified microspheres is the size of the crystallites. The size of the crystallites primarily depend on the nature of the process. For example, in the ORNL process of physical gelation, the crystallite size of the gel will always be less than 20 nm whereas in all the other processes of chemical gelation, the size of the crystallites will be in the range of 20–500 nm. The crystallite size is also determined by the kinetics of gelation. As was discussed earlier, when the gelation is rapid, as in the case of microwave-assisted IGP, the crystallites formed will be very small and hence an ageing step in ammonia solution is a must for allowing the crystallites to grow. The size of the crystallites has a bearing on the effect of the subsequent process steps of washing and other heat treatment steps on the gel particles.

The crystallites in a gel matrix are arranged to generate the capillaries (pores) which are smaller than the crystallites in size. The washing characteristics of the gels are clearly defined by the diffusion coefficients of the gelation agents and the product ammonium nitrate into the wash solution. These diffusion coefficients are a function of the size of the capillaries; smaller the size smaller will be the diffusion coefficient and smaller will be the rate of washing. Thus to ensure that washing of the components mentioned above in reasonable time periods, the pore size of the gel should be at least 50 nm. If the pore diameters are small, some of the gelation agents, especially nitrates, remain behind even after washing for a long time. Incompletely washed gel particles do not survive the heat treatment steps because of the evolution of the gaseous products. Additionally, the small diffusion coefficients cause building up of osmatic imbalance in the particle. This may result in the damage of the particle during the washing step itself.

The morphology of the gel material is also important in the heat treatment of these particles, as mentioned above. Transparent gel materials normally have crystallite size $<$20 nm. These gel particles after drying will have a very large surface area in the range of 200 m^2/g. Thus a large amount of surface energy is stored in these gel particles. During the heat treatment, when this energy is released, the particles are likely to get damaged (cracking, shattering etc.). It is also observed that such microspheres are also highly chemically active and in the case of uranium dioxide, they would oxidise even at room temperature and get damaged. Opaque gel microspheres normally have surface area below 50 m^2/g and an average crystallite size greater than 100 nm. Such particles can be washed very easily (normally within an hour) and they are also stable during the heat treatment. Thus in sol-gel process development there is an endeavour to find feed compositions and other parameters such as temperature, etc. which yield opaque gel particles. In the absence of the same, the process flow sheet is modified to include an ageing step to increase the crystallite size so as to facilitate washing and heat treatment steps. Although particles with high surface area sinter at lower temperatures, damage to the particles is increased during the heat treatment steps, especially if the particle size is large.

The diameter of the gel particle also plays an important role in all these steps, as smaller the particle size, easier will be the washing and better will be the behaviour during the heat treatment. However while washing very small particles, $<$200 μm, the draining of wash liquid is rendered very difficult and special washing procedures like washing with isopropyl alcohol or methyl alcohol is resorted for removing the last traces of ammonia solution. Similarly, the transfer of these soft gel particles is avoided and the filter container itself is transferred to an oven for drying.

Among the heat treatment steps, the reduction step plays an important role in the preparation of urania microspheres. If the reduction is done below 723 K, the carbon and nitrogen removal will be incomplete. Calcination at temperatures >873 K results in the decomposition of UO_2 into U_3O_8. This in turn causes the density of the sintered product to be low. Thus it is important to limit the calcination temperature to 773 K and the reduction of the UO_3 gel product must be started at 773 K. The completion of the reduction process also should be below 873 K.

All the sol-gel processes are designed to bring down the time required for the gelation of the droplet. This is achieved either by pre-treatment of the solution like addition of urea or addition of ammonia to bring the pH of the broth to 3.1 (in the case of uranium solution) or making the sol with 87% neutralisation (for ThO_2/ThO_2-UO_2 sol preparation in KFA process) before the droplets are exposed to ammonia. In case of IGP, cooling of the solution is done to keep the solution ready for the fast gel formation by heating in the oil. The broth would gel immediately, if exposed to higher temperatures.

Also the acidity of the metal nitrate solution before mixing with HMUR is brought to minimum. In case of oxide microspheres, ADUN solutions are used with NO_3^-/U ratio of 1.6. Similarly, partial neutralisation of thorium nitrate solution is carried out by using ammonia to obtain a virtual nitrate to Th ratio equal to 3.

The kinetics of gelation is not very fast in the processes like ORNL process which are sol-based. However, an attempt is made to keep the water content in the sol to minimum and a sol with >3 M is used for making the droplet. In view of the slow kinetics, special equipment is used to keep the gel droplets in contact with water extraction medium 2EH, for a longer time using a counter current fluidized column. However, for the gel particles to be formed properly, the water extraction rate has to be controlled so as to leave minimum strain in the gel formed. This is achieved by keeping the water content of 2EH to about 1%, although the saturation solubility of water is 2.6%. This achieved by on line distillation of water from part of the 2EH stream, in a continuous mode.

Special sintering procedures have been adopted for the sintering of UO_2/(U,Pu)O_2 microspheres called low temperature oxidative sintering. This procedure is based on the increase of the rate of diffusion of U^{5+} in U^{4+} matrix and hence the O/M is maintained in the region of 2.1 instead of the stoichiometric value of 2.0. This is achieved by having a cover of CO_2 gas in the sintering environment. The CO_2 gas at that temperature has an oxygen potential which is equivalent to that of $UO_{2.1}$. Similarly, a drastic reduction in the sintering temperature of (U,Th)O_2 is achieved (from 1973 to 1623 K) by sintering in air. Again it is because of the formation of U^{5+} in the solid matrix of ThO_2. This is very useful for sintering of SGMP $(U,Th)O_2$ pellets.

16.3 Sol-Gel Microsphere Pelletisation Process

As mentioned earlier, though large amount of experience exists in the fabrication of thermal and fast reactor fuels in the form of pellets, difficulties are encountered in the fabrication of pellets containing ^{233}U or Pu. It would be of great advantage, if one could eliminate handling the powders of these fuel materials in a shielded remotely operated fuel fabrication facility. Thus some early attempts were made in ORNL, USA and KFA, Julich to make soft microspheres using the sol-gel process that can be compacted into pellets. The process developed by Zimmer at et al. at KFA, Julich [40] using EGP and aimed at making defect free pellets of fuel materials using sol-gel microspheres as feed material was named as the SGMP process.

Generally, sol-gel microspheres of fuel materials, after the drying step are hard and do not get crushed completely during pressing for making pellets through SGMP. The microstructure of such sintered SGMP pellets showed a blackberry structure due to the residual spherical structure of microspheres used as feed. The sintered pellets were also of low density [41]. Microspheres that were too soft resulted in lamination of the pellet flat ends [42]. Mathews and Hart [43] found that the UO_2, ThO_2 and $(U_{0.25}Th_{0.75})O_2$ microspheres prepared by the SNAM EGP cracked and crumpled during handling. Many attempts have been made to overcome this problem. The original work at ORNL suggested oxidation reduction cycle treatment for dry UO_3 microspheres prepared using IGP with standard process conditions ($U = 1.25$ M, $HMUR = 1.75$ M) to yield microspheres that could be compacted into pellets. The flow sheet for the SGMP process based on IGP is shown in Fig. 16.9. However, they reported that good SGMP pellets could be made only if the dry microspheres were of <400 μm diameter.

Another method developed at KFA, Julich was to add carbon powder as pore former in the oxide gel fuel material during external gelation (KFU) process. These carbon containing gel microspheres were carefully oxidised to remove carbon to yield soft, pressable microspheres. The pellets made using this process were defect free with no blackberry structure. The process was extended for making pellets of various oxides [44–47]. This process had two main drawbacks. Firstly, the addition of carbon in the feed broth made droplet gelation process very difficult to control because of frequent choking up of the nozzle used for droplet formation. Secondly, the heat treatment step required a very careful control of the gas composition as well as the total time of the oxidation cycle was more than 48 h.

Suryanarayana et al. [48] were successful in developing SGMP for making defect free UO_2 fuel pellets using carbon free internal gelation process. The SGMP using internal gelation process (IGP) has further been successfully extended for

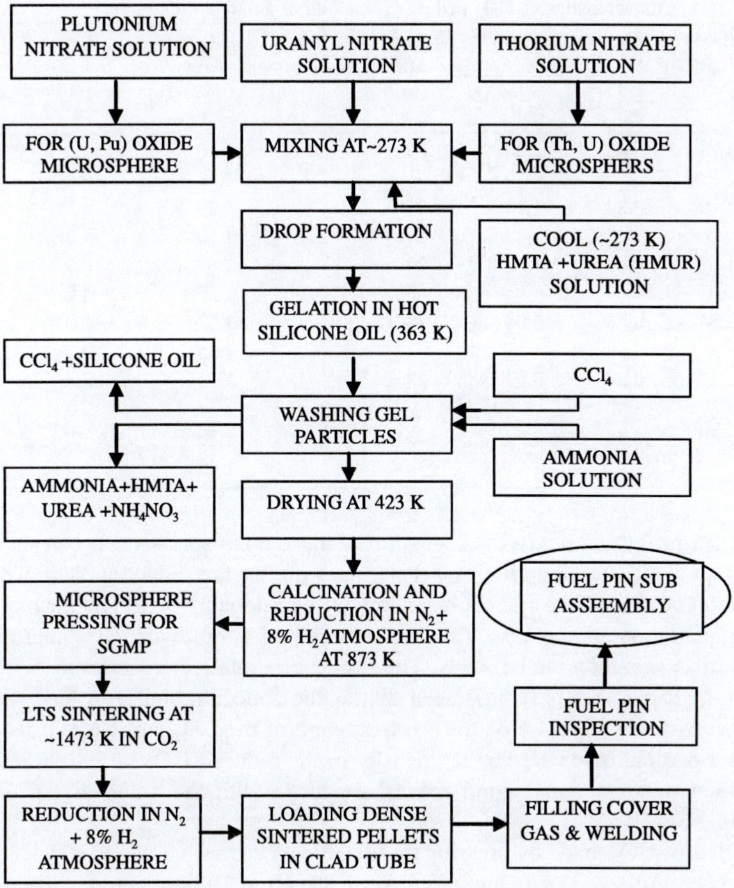

Fig. 16.9 Flow sheet for SGMP process

making defect free ThO_2, $(Th,U)O_2$ and $(U,Pu)O_2$ pellets suitable for thermal reactors.

The process development was arrived at by studying the gelation field diagram for UO_2 as reported by Vaidya et al. [20] shown in Fig. 16.5. It was found by Vaidya et al. [20] that the gels made using high concentration of U (>1.4 M) in feed solution had large crystallites of >500 nm as compared to those made using low concentration of uranium (<1.1 M) with crystallite size <50 nm. The UO_2 microspheres made using gels of large crystallite size (U concentration >1.4) were heat treated in air at 1073 K, reduced in Ar + 8% H_2 at 823 K and stabilized in CO_2 at same temperature before cooling were found to be soft and crushable. The heat treatment at 1073 K of dry UO_3 microspheres in air was designed to make a phase change of the gel particles from UO_3 to U_3O_8 resulting in the softening of the microspheres. The surface area of such calcined microspheres is relatively

Table 16.1 Characteristics of UO_2 pellets as function of gelation conditions

[U] (M)	[HMTA] [UREA] (M)	Tap density UO_3 gel (g/cm^3)	Tap density UO_3 dried (g/cm^3)	Surface area UO_3 dried (m^2/g)	Tap density UO_2 (g/cm^3)	Crush strength UO_2 (N/Particle)
1.25	1.75	0.83	1.20	39	2.70	5.0
					2.50	4.8
1.30	1.69	0.84	1.10	31	1.80	4.5
					1.70	4.3
1.35	1.62	0.85	1.02	29	1.67	4.0
					1.60	4.0
1.40	1.54	0.86	0.94	28	1.52	3.3
					1.55	3.5
1.45	1.52	0.87	0.93	28	1.50	2.5
					1.51	3.0
1.50	1.50	0.88	0.90	23	1.50	2.0
					1.45	1.8

large (>5 m^2/g) and hence the reduction of these microspheres was carried out at 823 K to avoid any sintering of gel particles during the reduction step. The stabilization of these UO_2 microspheres in CO_2 was done to avoid any air oxidation during pellet-making process. The crush strength of 1.5 mm dia. UO_2 microsphere per particle was found to be <3 N. The results of a relative crush strength study is shown in Table 16.1 [48]. It is seen that as the concentration U in feed solution increased from 1.25 to 1.5 M the crush strength of the UO_2 particle reduced from 5 N per particle to <3 N. The tap density of the reduced UO_2 microspheres is an important parameter and good pellets are made with the value of tap density around 1.6–1.7.

Pellets of UO_2 made by pressing these particles were sintered at 1423 K in CO_2 atmosphere followed by reduction in Ar + 8% H_2 at 1073 K before cooling (low temperature sintering process or LTS). These pellets were found to have defect free microstructure with 97% TD. Using the SGMP LTS route, more than 100 kg of UO_2 pellets were fabricated at BARC, India for their use in pressurised heavy water reactors (PHWR). Kumar et al. [49] extended IGP-based SGMP process for making the $(U_{0.95}\,Pu_{0.05})O_2$ pellets with excellent microstructure and density. The process was further extended by Pai et al. [50] for making PHWR grade pellets of ThO_2 and $(Th_{0.97}\,U_{0.03})O_2$. A low temperature air atmosphere sintering procedure was used at 1623 K for getting $(Th_{0.97}U_{0.03})O_2$ pellets of >97% TD. This process development has great significance in the exploitation of thorium in nuclear reactors. The SGMP process for manufacture of $(U_{0.95}Pu_{0.05})O_2$ fuel pellets is important for the recycle of plutonium in thermal reactors.

Fast reactor pellet fuels should be of lower density (80–85% TD) as compared to those of thermal reactor fuels (>90% TD). SGMP process using the hard microspheres resulting in pellets with blackberry structure and lower density has been exploited for making (U,Pu)O_2, (U,Pu)C and (U,Pu)N [51] pellets containing ~15% Pu. The process flow sheet for making the carbide and nitride pellets included a carbothermic reduction step similar to the procedure discussed earlier for making carbide and nitride microspheres.

Another variant of the SGMP process is the microsphere impregnation process. It is also known as gel impregnation process. It was developed for simplifying the fabrication of $(U,Th)O_2$ pellets containing ^{233}U [52]. In this process, dried thoria gel microspheres prepared by internal gelation process were impregnated with uranyl nitrate solution. Impregnated thoria microspheres were dried and calcined to obtain free flowing porous microspheres. Irregular shaped pores obtained within microspheres due to dissociation of uranyl nitrate made the microspheres easily crushable. These microspheres were cold pelletised at 300 MPa to a green density of 60% T.D. The pellets were sintered at 1923 K in Ar + 8% H_2. Pellets of high density (97% TD) have been prepared by this process. The advantage of this process for making $(U,Th)O_2$ pellets containing ^{233}U is that preparation of thoria microspheres by the gelation process can be carried out in an unshielded facility and further process steps that are simpler but involve handling ^{233}U in a shielded facility. Another advantage of this method is that the uranium content of the mixed oxides can be varied by changing the molarity of the uranyl nitrate solution. The gel impregnation process has been developed in laboratory scale only and needs further development before implementation in larger scale.

16.4 Fabrication of Coated Particle Fuels by Sol-Gel Processes

Coated particles with the different kernel materials, namely, UO_2, UC_2, UO_2 + UC_2, $(Th,U)O_2$, (Th,U)C have been tested as nuclear fuel for HTGR [21]. The research and development of these fuels has been in progress in several countries right from the early sixties [10]. The reference design of coated particles consists of a kernel having low enriched uranium with tri-isotropic (TRISO) coating over the kernel. The word TRISO describes the coating structure over the fuel particles consisting of an inner porous pyrocarbon layer and two dense pyrocarbon coatings on the outside interspaced by a silicon carbide layer. A typical coated fuel particle is shown in Fig. 16.10. The coatings act as the pressure can to withstand the fission gases and also to prevent the fission products from coming into contact with the coolant. The carbon in these layers act as the moderator. Many countries use a UO_2 kernel of about 500 μm diameter. UO_2 kernels, compared to UC_2 kernels, have better retaining ability for some fission products, e.g. the rare earths, and higher stability at temperatures of 2273–2773 K that might be encountered in a core heat up accident. However under high burn up conditions production of excessive CO in UO_2 kernels can lead to mechanical failure of the coatings and accelerate what is known as “the amoeba effect” [53]. Hence in USA, a kernel termed ‘UCO’, a mixture of UC_2 and UO_2 is used with the objective of controlling the formation of CO and the associated problems.

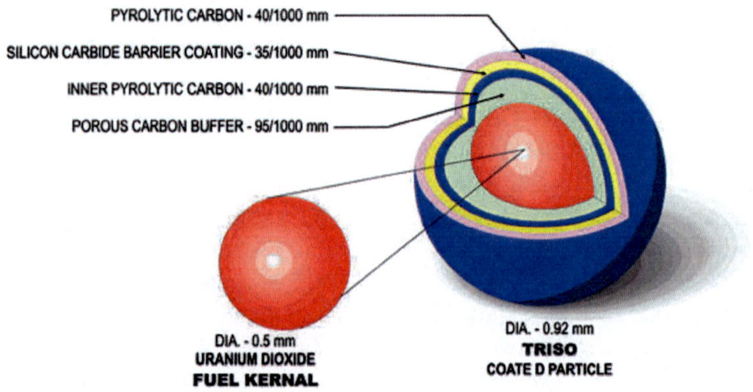

Fig. 16.10 A typical coated fuel particle

Different countries use different sizes of kernel and the layers of coating. With respect to fuel element design, two variations are known. In the design adopted by KFA, Julich, Russia and China, the particles are embedded in a graphite matrix to form a spherical fuel element of 60 mm diameter with a uranium free outer layer of 5 mm. Japan and USA use a graphite block with holes as fuel element in which the fuel rods or compacts, made of the coated particles, are inserted [54].

The coated particle fuels would experience temperatures above 1873 K and burn up higher than 9 at.%, and hence it should be ensured that the particles are spherical, smooth and crack free. The starting point of the fuel fabrication is the kernel production through the sol-gel process. External gelation has been adopted for production of UO_2 kernels by many countries. As was mentioned above, China uses the TGP for producing the fuel for the HTR-10 reactor. In USA, internal gelation is used for the preparation of UCO kernels. The TRISO coating over the kernels is done using chemical vapour deposition within a fluidized bed. For higher temperature utilisation of the HTGR fuels, zirconium carbide coating is considered to be better than SiC. Development programmes for ZrC coating are in progress in USA, Japan and Russia. The particles with ZrC coating are termed ZrC-TRISO coated fuel particles. External gelation is being used for the production of TRISO coated fuels with UO_2 kernels for the pebble bed modular reactor (PBMR) in South Africa [55].

Coated particle fuels are also being proposed for the gas cooled fast reactors also, one of the Generation IV reactor concepts. Okano et al. [56] described the concept of a large scale (2400 MWTh) helium cooled fast reactor with coated particles having mixed (U,Pu) nitride kernels and TiN coatings.

16.5 Irradiation Behaviour of the Sphere-Pac Fuels

Nuclear fuel is expected to stay inside the reactor to generate good quality heat (high temperature operation for high thermal efficiency) and should not fail prior to its discharge from the reactor. Irradiation testing is an important step in nuclear fuel development programme. Test irradiations are carried out to establish the parameters such as the amount of heat that can be generated by the fuel, the temperature profile inside the fuel, corrosion of the cladding by the fuel and the fission products, the total heat (burn up) that can be generated by the fuel, etc. These tests are essential for determining the quality of the fuel as well as the economics of the fuel. Three types of irradiations have been used: irradiation of thermal reactor fuels in thermal reactors, fast reactor fuels in fast reactors and fast reactor fuels in thermal reactors. These differences must be kept in mind while interpreting the results. Sphere-pac test fuel irradiations have been carried out at ORNL and Argonne National Laboratory (ANL), USA, Paul Scherrer Institute (formerly EIR) in Switzerland, Harwell in UK and Comitato Nazionale per l'Energia Nucleare (CNEN) in Italy and a water irradiation programme was carried out in Netherlands. The irradiation studies on sphere-pac fuels have been reviewed by Lacky and Selle [1] as well as Icenhour [57]. The results of the in-pile and post-irradiation examination of the sphere-pac oxide and carbide fuels have been very encouraging.

The irradiation results compare the performance of the sphere-pac fuels with that of the pellet fuels which is well known. Most of the results show that the sphere-pac carbide and oxide fuels perform as good as or better than pellet fuels. The oxide microspheres of the sphere-pac fuels get sintered and fused together at very high temperatures and high linear power resulting in significant changes in the structure. They have been reported to get restructured and the features of the restructured fuel, except at the periphery, are identical to that of restructured pellet fuel. Though the region adjacent to the clad retains its original structure, the microspheres in the region are connected together. The restructuring occurs in the very beginning of life, within first 100 h of irradiation, as shown by the results of Morihara et al. [58] and Bart et al. [59]. Sphere-pac fuels have been reported to have less Fuel Clad Mechanical Interaction (FCMI) and less Fuel Clad Chemical Interaction (FCCI) than the pellet fuels [60]. The fission gas release behaviour has been reported to be the same in both types of fuels. Under the US-Swiss collaboration, irradiation of sphere-pac (U, Pu)C fuel fabricated in EIR, Wurlingen, Switzerland was carried out in FFTF [61] to a peak burn up of 8% FIMA. The results published recently show that the performance of sphere-pac fuel is as good as the pellet fuel [62].

Though sphere-pac fuels have been reported to be as good as the pellet fuels in the irradiation tests, sphere-pac fuels are yet to find application in industrial scale. In general, there are two important concerns regarding the sphere-pac fuel. One is the possible loss of fuel in case of clad breach and contamination of the coolant. The other is the relocation or segregation of the fuel particles during handling and irradiation. In the irradiation experiment carried out by EIR, Switzerland with hyperstochiometric carbide fuel, though there was clad breach due to carburisation and the size of the breach was many times larger than the fuel particles, no fuel material was found outside the fuel pin. It is on the expected lines, since the fuel restructures very early in life and the fuel column becomes integrated. However, as indicated by Peddicord et al. in their review [60], fuel loss could occur, only if the clad breach occurs, before the initial start up, but not after the start up. The second problem of segregation could be solved by choosing the proper size fraction of the microspheres especially the finer fraction to be at least 1/7 of the diameter of the coarser ones. However, more number of irradiation tests are needed before sphere-pac fuels are accepted for industrial production.

16.6 Sol-Gel Processes for Transmutation Fuels

The need for transmutation of MA, neptunium, americium and curium, into lighter nuclides stems from the major contribution of these elements to the radiotoxicity of the nuclear waste and to their long half lives. One of the options to transmute these elements is to incorporate them as minor components of the regular uranium oxide or MOX fuels for fast reactors. However, during irradiation, fresh MA will be produced by neutron absorption by uranium and plutonium. Hence the other option is to use the inert matrix fuels (IMF) by incorporating the MA in inert matrices such as yttria, zirconia, etc. IMF gives other advantages [63] as well. They simplify the fabrication process. Amenability of the sol-gel processes to remote handling, as discussed earlier, makes them suitable candidates for the fabrication of the fuel targets containing MA for the transmutation reactors.

Preparation of (U, Pu, Np) oxide microspheres containing 75% U, 20% Pu and 5% Np has been discussed by Pouchen et al. [64] for FUJI collaborative programme among Japan Nucear Cycle Development Institute (JNC, Japan), PSI, Switzerland and Nuclear Research and consultancy Group (NRG), Netherlands). These microspheres were produced by IGP using a flow sheet based on the nitrate solutions of uranium, plutonium and neptunium, similar to the one shown in Fig. 16.4 for (U,Pu) oxide microspheres. The large microspheres of 800 μm diameter were produced using a silicone oil column at 376–379 K and the fine microspheres of 70 μm diameter using an oil jet, as has been discussed earlier. Preparation of fuels with Zr nitride matrix has also been investigated by PSI, Switzerland [65] using IGP-based SGMP.

Institute of TransUranium elements (ITU), Karsruhe has prepared the CERMET fuels of $PuAmO_{2-x}$-Mo and $PuAmZrO_{2-x}$-Mo for irradiation in Phenix reactor [66]. For the European Facility for Industrial Transmutation (EFIT), which is an Accelerator Driven System (ADS), CERMET fuels based on Mo is considered the reference fuel and CERCER fuels based on MgO matrix as the back up. CERMET fuels have the advantages of higher thermal conductivity. The fabrication flow sheet used at ITU for the preparation of CERMET fuels is based on the preparation of PuO_2 or $PuZrO_2$ porous beads through EGP followed by infiltration with solution containing minor actinide. The beads are then compacted into pellets along with Mo. Other fuels prepared by ITU are: $(U_{0.74}Pu_{0.24}Am_{0.02})O_2$, $(Zr_{0.8}Y_{0.134}Am_{0.066})O_2$, $(Zr_{0.767}Y_{0.127}Pu_{0.038}Am_{0.068})O_2$. Their irradiation behaviour has also been found to be similar to that of the conventional fuels without the MA.

Pillon et al. [67] of CEA have discussed the implications of the curium management on industrial scale fabrication of targets containing americium and curium. The proposed flow sheet envisages the use of IGP rather than the EGP, used in the SUPERFACT project, for making microspheres of (Am, Cm) oxides due to the amenability of IGP for using microwave heating instead of hot silicone oil and the associated reduction in waste volume. The flow sheet envisages the use of nitrate solutions of americium and curium from reprocessing plants having a concentration of 0.45 g/L and subjecting them to a two stage concentration step of distillation to achieve a concentration of 360 g/L before feeding the same to sol-gel process for making microspheres of 100 μm sizes. The (Am, Cm) oxide microspheres are compacted into pellets along with MgO microspheres to form the IMF fuel pellets. They also discuss the problems that could be encountered while using sol-gel processes for MA. Hydrolysis of trivalent americium and curium could be more difficult than those of U, Pu and Th in their tetra or hexavalent states. Organic reagents such as HMTA or urea could undergo degradation due to the radiolysis or decay heat of curium. Recycling of liquid waste could also pose a problem. Attempts are being made to overcome this problem.

16.7 Conclusions

Sol-gel-based processes provide powderless routes for the fabrication of nuclear fuels. The fuel materials are prepared in the form of microspheres which can be used for fabrication of sphere-pac fuel pins and pellet fuel pins. They also offer the possibility of integrating fuel fabrication plant with fuel reprocessing plant thus eliminating the need for transport of fissile material across the domain boundaries of fuel movement. The safety and proliferation resistance aspects of fuel cycle are thus enhanced. The sol-gel processes are also advantageous for the fabrication of fuel containing minor actinide fuels because of their amenability for remotisation. Of the two sol-gel processes that are widely used, EGP is useful for the fabrication of microspheres of <500 μm whereas IGP is useful for microspheres of a wider range of sizes. Sol-gel processes are being used for preparing the kernels of coated particle fuels for HTGR, PBMR, etc. and are being considered for fabrication of transmutation targets. They are one of the most promising routes for the fabrication of $(U,Th)O_2$ fuel pellets containing ^{233}U. However, they are yet to find industrial applications for fabrication of fast reactor fuels. More irradiation experience is needed for the sol-gel processes and sphere-pac fuels to be accepted for industrial scale fabrication of fast reactor fuels.

References

1. Lackey WJ, Selle JE (1978) Assessment of gel-sphere-pac fuel for fast breeder reactors. Oak Ridge National Laboratory Report, ORNL-5468. Oak Ridge, Tennesse
2. Ganguli C, Basak U, Vaidya VN, Sood DD, Roy PR (1989) Sol-gel microsphere pelletisation of UO_2 and UO_2-PuO_2 pellets of PHWR fuel specifications using internal gelation process. In: Hastings IJ (ed) Proceedings of second international conference on CANDU fuel, pp 108–124
3. Haas PA, Haws CC Jr, Kitts FG, Ryon AD (1968) Engineering development of sol-gel process at Oak Ridge National Laboratory. Oak Ridge National Laboratory Report, ORNL/TM-1978. Oak Ridge, Tennesse
4. McBride JP (1966) Preparation of UO_2 microspheres by a sol-gel technique. Oak Ridge National Laboratory ORNL-3874. Oak Ridge, Tennesse
5. Brambilla G, Gerontopoulos P, Neri D (1970) The SNAM process for the preparation of ceramic nuclear fuel microspheres: laboratory studies. Energ Nucl 17:17–24
6. Facchini AG, Gerontopoulos P (1974) Present state of the art of the SNAM process for preparation of ceramic nuclear fuel microspheres. In: Sol-gel process for fuel fabrication, International Atomic Energy Agency Report, IAEA-161, Vienna, pp 227–245
7. Cogliati G, Facchini A (1970) A pilot plant for continuous preparation of U-Pu oxide spheres. In: Proceedings of symposium on sol-gel processes and reactor fuel cycles, CONF-700502, U.S. Atomic Energy Commission, Washington, pp 210–226
8. Beatty RL, Norman RE, Notz KJ (1979) Gel-sphere-pac fuel for thermal reactors—assessment of fabrication technology and irradiation performance. Oak Ridge National Laboratory Report, ORNL/TM-5469. Oak Ridge, Tennesse
9. Huschka H, Vygen P (1977) Coated fuel particles: requirement and status of fabrication technology. Nucl Technol 35:239–245
10. Kadner M, Baier J (1976) Production of fuel kernels for high temperature reactor fuel elements. Kerntech 18:413–420
11. Zimmer E, Naefe P, Ringel H (1978) Aqueous chemical processes for the preparation of high temperature reactor fuel kernels. Radiochim Acta 25:161–169
12. Ringel H, Zimmer E (1979) The external gelation process for preparation of ThO_2 and (Th, U)O_2 fuel kernels. Nucl Technol 45:287–298
13. Hardy CJ, Lane ES (1970) Gel process development in the United Kingdom. In: Proceedings of symposium on sol-gel processes and reactor fuel cycles, CONF-700502, U.S. Atomic Energy Commission, Washington, pp 137–149
14. Taylor HA (1974) Gel precipitation process development in the UK. In: Sol-gel process for fuel fabrication, International Atomic Energy Agency Report, IAEA-161, Vienna, pp 1–7
15. Ganguly C, Langen H, Zimmer E, Mertz ER (1986) Sol-gel microsphere pelletisation process for fabrication of high density ThO_2-2%UO_2 fuel for advanced pressurized heavy water reactors. Nucl Technol 73:84–95
16. Tomita Y, Morihira M, Kihara Y, Tamaki Y (2005) Fuel microsphere fabrication tests for sphere-pac fuel by the external gelation process. In: Proceedings of conference GLOBAL 2005, Paper No. 198
17. Fu X, Liang T, Tang Y, Xu Z, Tang C (2004) Preparation of UO_2 kernel for HTR-10 fuel element. J Nucl Sci Tech 41:943–948
18. Kanij JBW, Noothout AJ, Votocek O (1974) The KEMA U(VI) process for the production of UO_2 micropsperes, In: Sol-gel process for fuel fabrication, International Atomic Energy Agency Report, IAEA-161, Vienna, pp 185–195
19. Brugghens FW, Noothout AJ, Hermans MEA, Kanij JBW, Votocek O (1970) A U(VI)-process for microsphere production. In: Proceedings of symposium on sol-gel processes and reactor fuel cycles, CONF-700502, U.S. Atomic Energy Commission, Washington, pp 252–263

20. Vaidya VN, Mukerjee SK, Joshi JK, Kamat RV, Sood DD (1987) A study of chemical parameters of the internal gelation based sol-gel process for uranium dioxide. J Nucl Mater 148:324–331
21. Collins JL, Hunt RD, Del Cul GD, Williams DF (2004) Production of depleted UO_2 kernels for the advanced gas-cooled reactor program for use in TRISO coating development. Oak Ridge National Laboratoty Report, ORNL/TM-2004/123. Oak Ridge, Tennesse
22. Haas PA, Begovich JM, Ryon AD, Vavruska JS (1980) Chemical flow sheet conditions for preparing urania spheres by internal gelation. Ind Eng Chem Prod Res Dev 19:459–467
23. Kumar N, Sharma RK, Ganatra VR, Mukerjee SK, Vaidya VN, Sood DD (1991) Studies of the preparation of thoria and thoria-urania microspheres using internal gelation process. Nucl Technol 96:169–177
24. McGeary RK (1961) Mechanical packing of spherical particles. J Am Ceram Soc 44:513–522
25. Ganatra VR, Kumar N, Suryanarayana S, Bamankar YR, Reghu N, Vaidya VN, Mukerjee SK (2008) J Radioanal Nucl Chem 275:515–522
26. Krishnan RV, Nagarajan K, Ravichandar SC, Prabhu TV, Ravisankar G, Kasiviswanathan KV (2010) Sol-gel development activities at IGCAR, Kalpakkam. J Sol-Gel Sci Technol. doi :10.1007/s10971-010-2315-5
27. Del Cul GD, Mattus CH, Icenhour AS, Felker LK, Williams DF (2005) Fuel fabrication development for the surrogate sphere-pac rodlet. Oak Ridge National Laboratory Report, ORNL/TM-2005/108. Oak Ridge, Tennesse
28. Johnson DR, Lackey WJ, Sease JD (1975) The effects of processing variables on HTGR fuel kernels fabricated from U-loaded cation exchange resins. Oak Ridge National Laboratory Report, ORNL-TM-4989. Oak Ridge, Tennesse
29. Chattin FR, Benker DE, Lloyd MH, Orr PB, Ross RG, Wiggens JT (1981) Transplutonium elements—production and recovery. In: Navratil JD, Schulz W (eds) ACS Symposium Series 161. American Chemical Society, Washington, pp 173–185
30. Knotik K, Leichter P, Bonek E (1981) Herstellung von HTR-brennstoffkugeln durch mikrowellenhärtung von wässrigen metallsalz,-kunstharztröpfchen. Atomenerg Kerntech 39:191–194
31. Ledegerber G (1982) Improvements of the internal gelation process. Trans Am Nucl Soc 40:55–56
32. Jungo Ch, Ledergerber G (1984) Process and apparatus for producing microspheres. US Patent 4,431,164
33. Yamagishi S (1998) A new internal gelation process for fuel microsphere preparation without cooling initial solutions. J Nucl Mater 254:14–21
34. Mukerjee SK, Dehadraya JV, Vaidya VN, Sood DD (1990) Kinetics study of carbothermic synthesis of uranium monocarbide microspheres. J Nucl Mater 172:37–46
35. Ledergerber G, Kopajtic Z, Ingold F, Stratton RW (1992) Preparation of uranium nitride in the form of microspheres. J Nucl Mater 188:28–35
36. Sood DD, Mukerjee SK, Vaidya VN, Venugopal V (1993) Uranium nitride: its preparation, oxidation and vaporisation. J Met Mater Process 5:13–22
37. Alder HP, Ledergerber G, Stratton RW (1988) Advanced fuel for fast breeder reactors produced by gelation methods. In: Advanced fuel for fast breeder reactors: fabrication and properties and their optimization, International Atomic Energy Agency Report, IAEA-TECDOC-466, Vienna, pp 81–95
38. Federer JI, Tennery VJ (1978) Synthesis of (U, Pu)C by carbothermic reduction of mixed oxides and evaluation of sintering behaviour. Oak Ridge National Laboratory Report, ORNL/TM-6089. Oak Ridge, Tennesse
39. Kumar A, Vittal Rao TV, Mukerjee SK, Vaidya VN (2006) Recycling of chemicals from alkaline waste generated during preparation of UO_3 microspheres by sol-gel process. J Nucl Mater 350:254–263
40. Zimmer E, Ganguly C, Borchardt J, Langen H (1988) SGMP-an advanced method for fabrication of UO_2 and MOX pellets. J Nucl Mater 152:169–177

41. Tiegs SM, Haas PA, Spencer RD (1979) The sphere-cal process for fabrication of fuel pellets from gel microspheres, Oak Ridge National Laboratory Report, ORNL/TM-6906. Oak Ridge. Tennesse
42. Kumar A, Kutty PS, Basak U (2009) Development of sol-gel based processes for fabrication of nuclear fuel pellets. In: Vaidya VN et al (eds) Proceedings of the international conference on sol-gel processes for advanced ceramics (SGPAC 2009), pp 90–95
43. Mathews RB, Hart PE (1980) Nuclear fuel pellet fabrication from gel-derived microspheres. J Nucl Mater 92:207–216
44. Ganguly C, Linke U, Kaiser E (1987) Characterization of (U, Ce)O_2 pellets prepared by the sol-gel microsphere pelletization process. Metallography 20:1–14
45. Ganguly C, Langen H, Zimmer E, Merz E (1986) Sol-gel microsphere pelletisation process for fabrication of high density ThO_2-2%UO_2 fuel for advanced pressurised heavy water reactors. Nucl Tech 73:84–95
46. Ganguly C (1990) Sol-gel microsphere pelletisation process for fabrication of conventional and advanced ceramic nuclear fuels. Met Mater Process 1:253–274
47. Basak U, Mishra S, Nair MR, Ramachandran R,Majumdar S, Kamath HS (2004) Process development for the fabrication of Thoria and Thoria based nuclear fuel pellets. In: Ganguly C, Jayaraj RN (eds) Proceedings of symposium on characterization and quality control of nuclear fuels (CQCNF 2004), pp 196–205
48. Suryanarayana S, Kumar N, Bamankar YR, Vaidya VN, Sood DD (1996) Fabrication of UO_2 pellets by gel pelletisation technique without addition of carbon as pore former. J Nucl Mater 230:140–147
49. Kumar N, Pai RV, Joshi JK, Mukerjee SK, Vaidya VN, Venugopal V (2006) Preparation of (U, Pu)O_2 pellets through sol-gel microsphere pelletization technique. J Nucl Mater 359:69–79
50. Pai RV, Mukerjee SK, Vaidya VN (2004) Fabrication of (Th, U)O_2 pellets containing 3 mol% of uranium by gel pelletisation technique. J Nucl Mater 325:159–168
51. Ganguly C, Hegde PV (1997) Sol-gel microsphere pelletisation process for fabrication of (U,Pu)O_2, (U, Pu)C and (U,Pu)N fuel pellets for the prototype fast breeder reactor in India. J Sol-Gel Sci Technol 9:285–294
52. Pai RV, Dehadraya JV, Bhattacharya S, Guprta SK, Mukerjee SK (2008) Fabrication of dense (Th, U)O_2 pellets through microsphere impregnation technique. J Nucl Mater 381: 249–258
53. Verfondern K, Nabielek H, Kendall JM (2007) Coated particle fuel for high temperature gas cooled reactors. Nucl Eng Technol 39:603–616
54. Verfondern K (1997) Fuel performance and fission product behaviour in gas cooled reactors. International Atomic Energy Agency Report, IAEA-TECDOC-978, Vienna
55. Muller A (2006) Establishment of the technology to manufacture uranium dioxide kernels for PBMR fuel. In: Proceedings of the third international topical meeting on high temperature reactor technology (HTR 2006), pp 1–7
56. Okano Y, Naganuma M, Ikeda H, Mizuno T, Konomura M (2005) Conceptual design study of helium cooled fast reactor in the "Feasibility Study" in Japan. In: Proceedings of GLOBAL 2005 conference, Paper No. 412
57. Icenhour AS, Williams DF (2005) Sphere-pac evaluation for transmutation, Oak Ridge National Laboratory Report, ORNL/TM-2005/41. Oak Ridge, Tennesse
58. Morihara M, Nakamura M, Hellwig C, Bakker K, Ozawa T, Bart G, Kihara Y (2005) PIE results of comparative irradiation tests in HFR for sphere-pac fuel, pellet fuel and vipac fuel. In: Proceedings of GLOBAL 2005 conference, Paper No. 109
59. Bart G, Bakker K, Hellwig C, Kihara Y, Ozawa T, Wallin H, Shigetome Y (2007) FUJI, initial sintering comparison test for pelletized, sphere-pac and vipac-fast breeder reactor mixed oxide fuel. J Nucl Sci Technol 44:329–336
60. Peddicord KL, Stratton RW, Thomas JK (1986) Analytical and experimental performance of sphere-pac nuclear fuels. Prog Nucl Energ 18:265–299
61. Stratton RW, Ledergerber G, Ingold F, Latimer TW, Chidester KM (1990) Fuel fabrication processes, design and experimental conditions for the joint US–Swiss mixed carbide test in FFTF. J Nucl Mater 204:39–49

62. Bart G, Botta FB, Hoth CW, Ledergerber G, Mason RE, Stratton RW (2008) AC-3-Irradiation test of sphere-pac and pellet (U, Pu)C fuel in the US fast flux test facility. J Nucl Mater 376:47–59
63. Fernandez A, Haas D, Somers J (2006) Fabrication of plutonium based coated particle fuel at the institute for transuranium elements. In: Proceedings of conference on high temperature reactors (HTR2006), pp 1–8
64. Pouchan MA, Ingold F, Kopajtic Z, Tomita Y, Kono S (2003) Fabrication and characterization of MOX microspheres for FUJI project. In: Proceedings of global 2003 conference, pp 653–657
65. Burghartz M, Ledergerber G, Hein H, van der Laan RR, Konings RJM (2001) Some aspects of the use of ZrN as an inert matrix for actinide fuel. J Nucl Mater 288:233–236
66. Haas D, Fernandez A, Nästren C, Staicu D, Somers J (2005) Properties of CERMET fuels for minor actinides transmutation in ADS. In. Proceedings of 12th international conference on emerging nuclear energy systems (ICENES'2005), Brussels, Belgium, SCK.CEN, Mol, Belgium, pp 1–10
67. Pillon S, Grandjean S, Drain F, Schmidt N, Escleine J-M, Ravenet A, Valentin B, Jean CS (2002) Impact of the curium management on the fabrication of MA—bearing targets at an industrial scale in the frame of a mixed PWR and FR P&T scenario. In: Proceedings of seventh information exchange meeting on actinide and fission product partitioning and transmutation, Jeju, pp 537–547

Chapter 17
Sol-Gel Packaging for Electrochemical Devices

Andrei Jitianu, Louis Gambino and Lisa C. Klein

Abstract Packaging is an essential part of manufacturing electrochemical devices, such as micro-batteries and super-capacitors. Hermetic packaging is required to prevent humidity and gases from degrading components and interconnects. Typically, packaging is the last step in assembling such devices. Consequently, packaging materials have to be applied at low processing temperatures. Sol-gel processed hybrid organic–inorganic materials are convenient low temperature materials for packaging. This chapter, which surveys the use of sol-gel hybrids in packaging of electronic devices, is based, in part (Jitianu and Klein In Hybrid nanocomposites for nanotechnology: electronic optical magnetic and bio/medical applications. Springer, Berlin, pp. 429–453, 2009), which appeared in 2009. The survey has been updated and focused on electrochemical systems.

Keywords Alkoxysilanes · Coatings · Critical thickness · Diffusivity · Dimethyl diethoxysilane (DMDES) · Diphenyl-diethoxysilane (DPhDES) · Electrochemical devices · Hermetic barrier · Hermeticity · Humidity · Hybrid gels · Hybrid materials · Hydrolysis · Hydrophobic · Melting gels · Methyl triethoxysilane (MTES) · Methyltrimethoxysilane (MTMS) · Moisture · Molecular chains ·

A. Jitianu (✉) · L. Gambino · L. C. Klein
Department of Materials Science and Engineering, Rutgers University, 607 Taylor Road, Piscataway, NJ 08854, USA
e-mail: andrei.jitianu@lehman.cuny.edu

A. Jitianu
Department of Chemistry, Lehman College, City University of New York, Davis Hall, 250 Bedford Park Boulevard West, Bronx, NY 10468, USA

L. Gambino
e-mail: lougami@gmail.com

L. C. Klein
e-mail: licklein@rci.rutgers.edu

M. Aparicio et al. (eds.), *Sol-Gel Processing for Conventional and Alternative Energy*, Advances in Sol-Gel Derived Materials and Technologies,
DOI: 10.1007/978-1-4614-1957-0_17,

Organic-inorganic gels · Oxygen transmission rate (OTR) · Packaging · Permeation · Phenyl-triethoxysilane (PhTES) · Polycondensation · Sol-gel · Tetraethyl orthosilicate (TEOS) · Thin films · Water vapor transmission rate (WVTR)

17.1 Introduction

New devices for energy storage and energy conversion present a challenge to packaging materials. Materials used for fabrication of these devices have strict requirements, with respect to chemical and physical stability, and hermeticity to water, oxygen and other gases [1, 2]. Meeting these requirements leads to better performance and longer lifetime of the electronic devices. In particular, chemical and physical stability means that there are no interactions between components during fabrication or assembly of the devices. Hermeticity describes the property of a package that prevents the components from interacting with the external environment. The packaging is designed to protect the components from the invasion of water and gases, and from contamination from the external environment. Also, the packing prevents the release of products of the electrochemical processes. The success of creating a hermetic barrier is a function of the bulk permeability and the quality of sealing [3].

Barrier materials in packaging are classified by their function as barriers against permeation of oxygen and/or humidity. Their efficacy is measured by oxygen transmission rate (OTR) $cm^3/m^2{\cdot}day$ and water vapor transmission rate (WVTR) $g/m^2{\cdot}day$. Chantham [4] showed that there are two mechanisms responsible for gas transport through materials under the influence of a pressure gradient applied across opposing faces of a thick barrier layer: bulk diffusive flow via the solubility-diffusion mechanism (material property), and flow through defects in the materials (function of microstructure, meaning inhomogeneities, pinholes, micro-channels, micro-cracks, or grain boundaries). The mechanism of solubility-diffusion consists of several steps such as absorption of permeate onto barrier surface, solution of permeate into the barrier material, diffusion of it through the material and desorption of the permeate onto the opposite surface. In addition, permeation is a function of permeate characteristics, such as size of molecules, molecular mass, chemical and physical interaction with barrier materials.

Besides being hermetic, many packaging systems serve as a dielectric insulator. Packaging also should have a coefficient of thermal expansion that matches other components in the device. If the coating initially shows viscous flow, it should be rigid after curing. The coating should be shape conforming. Finally, the packaging should be abrasion resistant and durable.

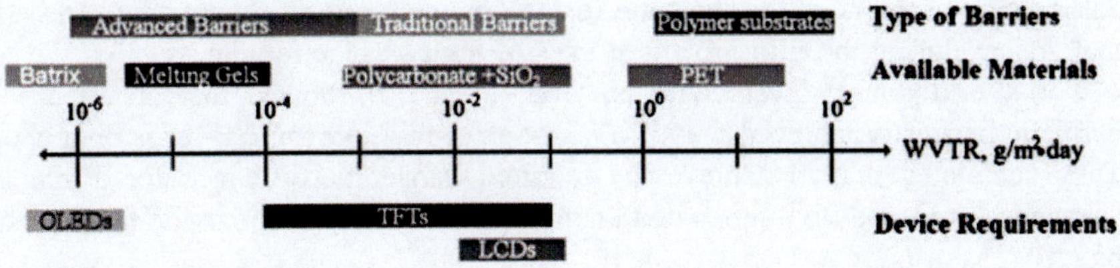

Fig. 17.1 Schematic scale of water vapor transmission rate (WVTR) requirements for common flexible electronic devices and the barrier performance provided by available materials (adapted from reference [2])

17.2 Humidity Barriers

One of the strictest requirements for packaging is that it serves as a hermetic barrier to humidity. Generally, thin films have been investigated to achieve hermetic barriers. For example, a thin film hermetic barrier for organic light-emitting diode (OLED) applications has to have a water vapor transmission rate (WVTR) below $\sim 10^{-6}$ g/m^2·day at 25°C and 40% relative humidity [1, 5]. Figure 17.1 schematically shows the order of magnitude required for various electronic devices and the actual protection given by the various materials [1]. It is known that traditional polymeric materials are not able to provide the necessary hermeticity for an extended operating lifetime [2].

One approach to increase hermetic performance adopted by the food industry is to use metal- or metal oxide-coated plastic films. The barrier performance of these hybrid materials is limited by intrinsic defects [3, 5] and roughness [2]. In fact, it has been demonstrated that aluminum films reduce the permeability of atmospheric gasses 1000-fold compared to the polymeric substrate alone. However, these metallized plastic materials do not meet the measure of hermeticity, and have electrical properties that are undesirable for electronic packaging, meaning they are not electronically insulating or transparent. DaSilva et al. [6] showed that SiO_2 and Si_xN_y transparent thin films deposited on polyester or polyethylene terephthalate (PET) by plasma-enhanced chemical vapor deposition (PECVD) provide an adequate hermetic barrier with insulator properties. The main problem for these materials is that they are brittle.

Another approach is to sputter alternating aluminum oxide and polyacrylate layers, using plasma-enhancement under vacuum [6, 7]. These hybrid transparent films obtained by PECVD [5, 6] are extremely effective as barriers even though they are very thin. It was shown [5, 8] that the inorganic layer has a critical coating thickness where the oxygen transmission rate (OTR) and water vapor transmission rate (WVTR) decrease as much as 1000-fold. This critical thickness is a function of the type of inorganic layer. For example, the critical thickness is 15 nm for SiO_2 and ~ 8 nm for Si_xN_y coatings on polyester. Generally, for these films the permeation is attributed to the microscopic defects that may have their origin in the dust particles present on the substrate surface, in geometric shadowing and stresses that appear during the film growth at sites of high surface roughness.

For hybrid double-layer coatings, Graff et al. [5] showed that an effective hermetic barrier is achieved only if a near-defect-free inorganic layer is obtained. They calculated that to achieve the required barrier performance for a single inorganic layer on a PET substrate the effective diffusivity of the oxide layer must be $<10^{-18}$ cm^2/s.

To address these issues, a multilayer barrier of polyacrylate–aluminum oxide pairs using the same PECVD technique was developed [2, 5]. The polymer-AlO_x pairs were repeated five times and a WVTR $\sim 8 \times 10^{-5}$ g/m^2·day was achieved. For the multilayer hybrid barrier, it was calculated that the defects in the AlO_x layers of the stacks are far enough apart that the vapor diffusion vector is mostly parallel to the polymer P_1 layer, resulting in a very long effective diffusion pathway between AlO_x and PET layers. Assuming a parabolic rate law behavior for transport through a film, an explanation for the effectiveness of the multilayer structure is referred to as the lag time effect. For a simple polymer layer PET-P_1 the lag time was demonstrated to be 1.72 h, while for four layers PET-P_1-AlO_x-P_2 the lag time was 22.1 h. For a barrier with five pairs and with a defect spacing of 100 μm the lag time was calculated to be 1,752 h ($\sim$72 days).

OLEDs present the most stringent requirements for packaging. While they are not energy storage or conversion devices, some facts about their packaging are instructive. Several authors [9–11] point out that OLEDs have a limited lifetime, reflected in the decreasing of electro-luminescence (EL) efficiency, an increase in drive voltage and the appearance of "dark spots" on the surface. It was shown that OLEDs prepared using $[Al(q)_3]$, where q is 8-hydroxyquinoline, lose 50% of EL intensity in 100 h of operation in dry argon at an intensity of I $\sim$5 mA/cm^2. Blue light OLEDs, prepared using distyryl arylene derivatives, lose 90% EL intensity, when operated in dry N_2 at I = 6 mA/cm^2. In comparison, OLEDs that operate in air lose 99% of EL initial intensity in 150 min.

Burrows et al. [2, 10–13] studied OLEDs on glass and on silicon substrates for rigid flat displays [10, 13] and on polymeric substrates for flexible OLEDs (F-OLED) [2, 11, 12]. They showed that to achieve an optimum operating lifetime, a hermetic isolation of the OLEDs from atmospheric oxygen and moisture is necessary, mainly because Ca and Mg used as the cathodes oxidize in the presence of O_2 or H_2O, and the organic semiconductor itself is damaged. Oxidation of 10% of the cathode resulted in the degradation of the EL properties. Losing even 5Å of metal at the metal–organic compound interface is very significant in the decay of the initial properties. It was calculated that the maximum leakage tolerated for an OLED that contains an Mg cathode is 1.5×10^{-4} g/m^2·day for a one-year lifetime.

To achieve better performance for their OLEDs, Burrows et al. [2, 11, 12, 14] engineered organic–inorganic multilayer flexible films. For this hybrid material the permeation rate of water is $\sim 10^{-5}$ g/m^2·day. These organic–inorganic films were vacuum-deposited by roll-coating technology, commercially called Batrix^{TM}. They have a sandwich structure with multiple layers of aluminum oxide between polyacrylate layers. The vacuum roll-coating technology leads to encapsulation of the OLED active devices in a single step together with Batrix™ preparation. This approach produced very good results, in terms of lifetime stability of these devices.

As stated above, defects such as pinholes, micro-pores, micro-channels, and micro-cracks in the layers have an important role in the level of hermeticity. Recently, Groner et al. [15] and Carcia et al. [16] introduced atomic layer deposition (ALD) as a new method for hermetic barrier preparation. The ALD process is known for growing thin, conformal pinhole-free films with featureless microstructure and low stress. They showed [15, 16] that by using this ALD method, it is possible to obtain 25 nm layers of Al_2O_3 on polyethylene naphthalate and Kapton™ substrates with a WVTR $\sim 1.5\ 10^{-3}$ g/m^2·day. Briefly, this new process consists of deposition of an Al_2O_3 layer onto a polymer substrate in a hot-wall ALD flow reactor at 120°C using sequential self-limiting exposures to trimethylaluminum and water with intermediate nitrogen gas purging steps. In this way defect-free films with ultra low permeability can be prepared. Table 17.1 summarizes the main types of barriers presented in the literature, the method used for their fabrication and their effectiveness.

17.3 Performance Tests for Hermetic Barriers

In order to quantify the properties of materials for hermetic barriers it is necessary to have reliable methods to measure the WVTR and OTR accurately. These measurements are routine for the food packaging industry and the pharmaceutical packaging industry. The goals for new hybrid materials are to have an OTR below 1×10^{-5} cm^3/m^2·day and WVTR approximately 1×10^{-6} g/m^2·day at 39°C and 95% relative humidity (RH). In order to measure such small rates (WVTR and OTR) traditional, commercial equipment used for the food industry (available from MOCON [17, 18]) cannot be used. Commercial equipment is limited to 5×10^{-3} cm^3/m^2·day for O_2 and 5×10^{-3} g/m^2·day.

In order to check the permeation rate of oxidizing species (O_2 and H_2O) through the hybrid barriers, new techniques have been developed. Permeation is defined as mass transfer of a substance (permeate) through a solid [19]. Permeation rate is a function of diffusivity of permeate through solid and the solubility of the permeate in the solid. Nilsson and Forssell [20] suggested calcium as an indicator, and Nisato et al. [21, 22] designed a calcium test to evaluate WVTR and OTR and high performance barriers such as Batrix™ and the data are summarized in Table 17.2.

The cathodes for OLEDs consist of reactive metals that degrade very rapidly, making the calcium test a suitable one since calcium also has a high reactivity in contact with water and oxygen. The reactions that mainly take place are (Reactions (17.1)–(17.3))[23]:

$$2Ca + O_2 \rightarrow 2CaO \tag{17.1}$$

Table 17.1 Overview of barriers, thickness and their performances

Substrate	Coatings materials	Method of coating	Thickness (nm)	WVTR (g/m²·day)	OTR (cm³/m²·day)	Ref.
PET	6 pairs polyacrylate-Al_2O_3	Evaporation in a roll-coating configuration	5×10^{-3}	$<10^{-5}$	$>5 \times 10^{-3}$	[2, 5, 11, 22]
PET	SiO_2	PECVD	8–200	–	1×10^{-1}	[6]
PET	SiN	PECVD	8–15	–	$< 5 \times 10^{-3}$	[6]
Silicon wafers	Up to 500 pairs of polyacrylate-Al_2O_3	Vacuum deposition and Sputter with energetic plasma	500	1×10^{-5}	–	[7]
PET	5 pairs polyacrylate-Al_2O_3	Flash evaporation in vacuum; UV curing	10–30	2×10^{-6}	–	[14]
PEN	Al_2O_3	ALD	25	1×10^{-5}	5×10^{-3}	[15, 16]
Kapton™	Al_2O_3	ALD	25		5×10^{-3}	[15, 16]
Polycarbonate	SiO_2 + ITO Al_2O_3 + ITO	PECVD	–	$2–8 \times 10^{-3}$	–	[56]

Table 17.2 The WVTR calculates using the Ca test for different materials

Sample	Storage temperature (°C)	Effective WVTR (g/m^2·day)	Reference
Glass	20	5×10^{-5}	[22]
Flexible glass 500	20	8×10^{-5}	[22]
Batrix™	20	2×10^{-6}	[23]
	50	2×10^{-5}	[23]
	80	9×10^{-5}	[23]

$$Ca + H_2O \rightarrow CaO + H_2 \tag{17.2}$$

$$CaO + H_2O \rightarrow Ca(OH)_2 \tag{17.3}$$

Water vapor and oxygen have distinct transmission rates due to the fact that these two permeates have different permeation mechanisms. The various methods for measuring water vapor and oxygen transmission rates are summarized and illustrated in Ref. [24] (Table 17.2).

17.4 The Sol-Gel Process for Organic–Inorganic Hybrids

While PECVD and ALD methods lead to hermetic coatings, the equipment necessary for these processes is complicated and expensive, involving high temperatures and/or very low pressure. An alternative way to develop new protection materials is the sol-gel process, involving alkoxides and alkoxysilanes.

Ideally, materials for electronic devices are flexible. Avoiding rigidity in a sol-gel process is addressed by using hybrid organic–inorganic gels. Principally, hybrid gels offer lightweight, mechanical flexibility, ease of processing and, in many cases, transparency.

In a sol-gel process carried out with a precursor containing all identical alkoxy groups, for example, tetraethyl orthosilicate (TEOS) $Si(OC_2H_5)_4$, the hydrolysis and condensation polymerization reactions generate an inorganic polymer with Si–O–Si links. In a sol-gel process with a precursor such as methyltrimethoxysilane (MTMS) $(CH_3)Si(OCH_3)_3$, there is a direct link between Si and C that does not undergo hydrolysis. The presence of the Si–C bond decreases the functionality of the precursor and has as a consequence the formation of linear molecular chains. For systems where TEOS is used the molecular chains formed are tri-dimensional. By combining these precursors, the product of the sol-gel process is a hybrid.

A wide variety of silica-based precursors with one, two or three non-hydrolytically active functional groups, are commercially available [25, 26]. Some precursors have reactive sites capable of undergoing inorganic reactions (hydrolysis/polycondensation) and other reactive sites capable of undergoing UV polymerization or thermally induced cross-linking.

The main reactions of hydrolysis–polycondensation that take place during the sol-gel processes with the organically modified alkoxides are the following (Chemical schemes (17.4)–(17.6)):

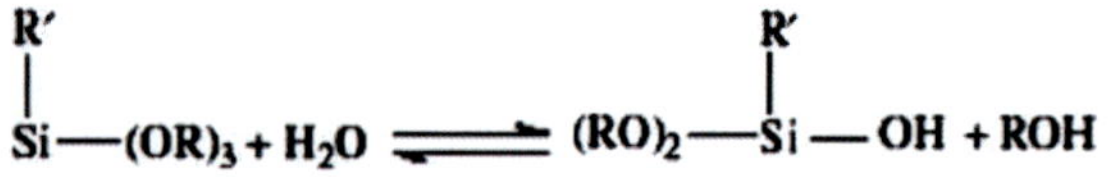

(17.4)

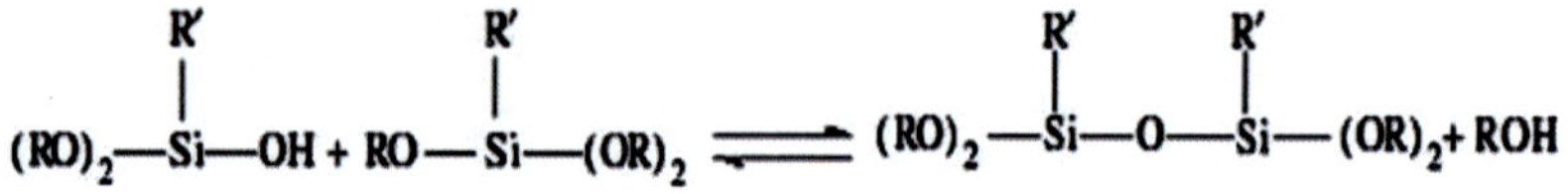

(17.5)

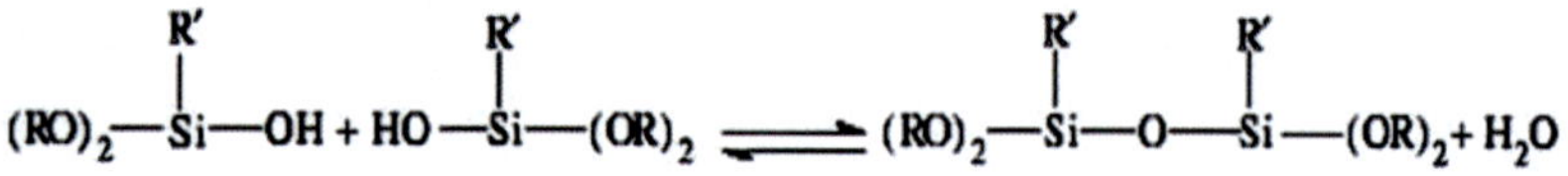

(17.6)

where R is -O–C_2H_5 and R′ is –CH_3 in the case of methyltriethoxysilane (MTES).

The first reaction (17.4) is a hydrolysis reaction, while reactions (17.5) and (17.6) are condensation reactions. Many studies [25, 26] report that during the hydrolysis–polycondensation processes the silicon-carbon bonds are not involved, and that silicon–carbon bonds remain in the final hybrid materials. The presence of the organic modifier groups R' decreases the degree of cross-linking. The sol-gel processes are very complex, since these involve competitions between different reactions including re-esterification, depolymerization and transesterification [27].

While hybrid materials are often classified by the fact that there is a direct covalent bonding between the organic and inorganic parts, it is also possible to have hybrids without direct linking by covalent bonds, but where hydrogen bonding between organic and inorganic parts occurs, with anywhere from weak to relatively strong interactions. Another way of classifying hybrids is either physical hybrids or chemical hybrids. In physical hybrids, the organic polymer and the inorganic polymer can form simultaneously or sequentially. An organic monomer may be the starting point for a thermal or UV-initiated polymerization. In physical hybrids, the interaction between the organic polymer and the inorganic polymer is through hydrogen bonding. Categorizing physical hybrids is approximate at best, because of the processing factors that influence their structures, for example, phase separation, solubility as a function of molecular weight, and relative speeds of hydrolysis, polymerization and gelling. In chemical hybrids, the assumption is that the organic and inorganic components are covalently bonded. Many low temperature hybrids have been synthesized using the sol-gel process and methacrylate compounds [28, 29]. Other hybrids have been discussed in detail [30, 31].

Poly(methyl methacrylate) (PMMA)/SiO_2 hybrids have been synthesized by a number of sol-gel processes. Typically, they have been prepared by mixing methyl methacrylate monomer with tetraethyl orthosilicate (TEOS). Their polymerization is carried out simultaneously. Alternatively, TEOS has been polymerized first and then infiltrated with MMA monomer. In such hybrids Van der Waals forces, hydrogen bonding or electrostatic attraction between organic and inorganic moieties are expected to influence the behavior of the final gel. In comparison to the sol-gel approach of using alkoxysilyl-containing organic precursors or coupling agents, PMMA/SiO_2 hybrids that do not contain covalent bonding are simpler to prepare. One recent [32] process used vinyl triethoxysilane (VTEOS) to generate an oxide network with polymerizable vinyl groups. The difference in this case of VTEOS over TEOS is the presence of an inorganic and organic network.

17.5 Organic/Inorganic Hybrid Flexible Films

Amberg-Schwab et al. [33, 34] first used hybrid films to obtain an oxygen barrier film for use in the food industry to prevent the degradation of different flavors. The two hybrid systems were (a) 3-mercaptopropyl-trimethoxysilane and 1,1,1-tris-(hydroxymethyl)-propan-triacrylate (UV curable) and (b) tetramethoxysilane, zirconium propoxide or aluminum-tri-sec-butoxide, and 3 amino-propyl-triethoxysilane (thermally curable). As supports for the above hybrid systems, they investigated biaxially oriented polypropylene (BOPP), PET and regenerated cellulose. For the first system, UV-curable deposited on BOPP, the effectiveness of the barrier against penetration of the flavors was very high. Only two flavors (diphenyl oxide and cis-3-hexanol) out of seven were detected through this barrier, and these only in traces. To reduce the permeability of the water, oxygen and flavors further, they used a vacuum deposition of SiO_x between the polymer layer and thermally curable hybrid coating. By using this method, the OTR was lower than the detection limit 0.05 $cm^3/g{\cdot}day$ [34]. These PET-SiO_x-hybrid polymer-PET and BOPP-SiO_x-hybrid polymer structures (OTR $< 0.05\ cm^3/m^2{\cdot}day$) present excellent adhesion and barrier properties even under high mechanical and thermal stress.

In another thermally curable system, with tetramethoxysilane (TMOS), zirconium propoxide or aluminum-tri-sec-butoxide, and 3 aminopropyl-triethoxysilane, 3-glycidoxypropyltrimethoxysilane (GLYMO)(Chemical Scheme (17.7)):

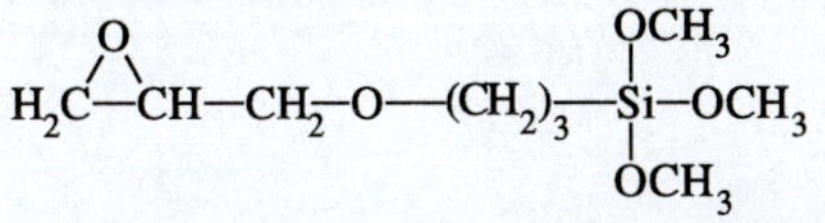

(17.7)

was introduced [35, 36]. The epoxy groups from GLYMO contributed to the formation of an organic network when thermally cured (100°C/2 h). In order to form the organic network, the desired ring-opening reaction of GLYMO is with an epoxy from another oxirane. The result is formation of oligo- or poly-(ethylene-oxide) derivates. It was demonstrated that the presence of aluminum-tri-sec-butoxide catalyzed the ring-opening reactions of GLYMO [37]. The amount of inorganic component TMOS was varied from 0 to 40 mol%. The OTR decreased with the increasing concentration of inorganic component on a polypropylene substrate. This result was attributed to increasing the density of the inorganic network, which hindered the penetration of the oxygen molecules. For industrial applications, thermal curing at 100°C for 2 h is not an economical approach, so an IR curing treatment for 3 min was developed. There are many supports, e.g. polyolefins (polypropylene, PP) and plasticized PVC, which cannot be thermally cured [37]. For these supports a photochemically curable barrier was developed.

Recently "Passive-Active barrier" coatings prepared by a sol-gel method have been developed [38]. Previously, it was shown that the combinations between SiO_x or AlO_x inorganic layers with polymer layers gave WTR < 10^{-3} g/m^2·day or OTR << 10^{-3} cm^3/m^2·day. By combining these passive layers with active oxygen barrier layers, these barriers could be improved further. Basically, the concept is design an "oxygen scavenger system", which is based on photo-initiation of a metal oxidation cyclo-olefin directly bonded to a silicate molecular chain. This process prevents the formation of low molecular weight oxidation products. Instead, the active layers consume any oxygen trapped during the packaging process, using UV activation. The inorganic network is formed by a simple sol-gel process, and adheres to many substrates. The main reaction that takes place to form these scavenger films is (Chemical schemes (17.8)–(17.10)):

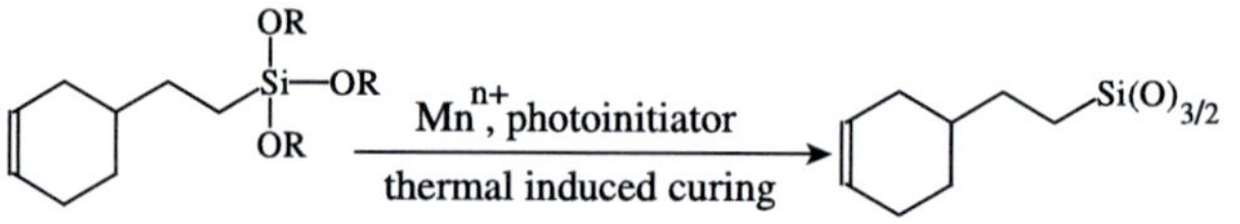

(17.8)

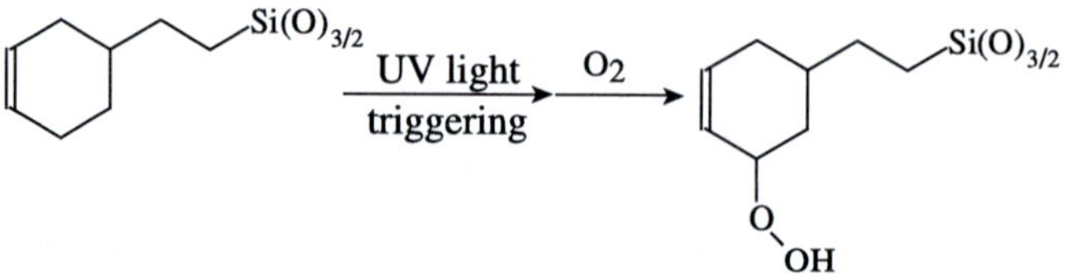

(17.9)

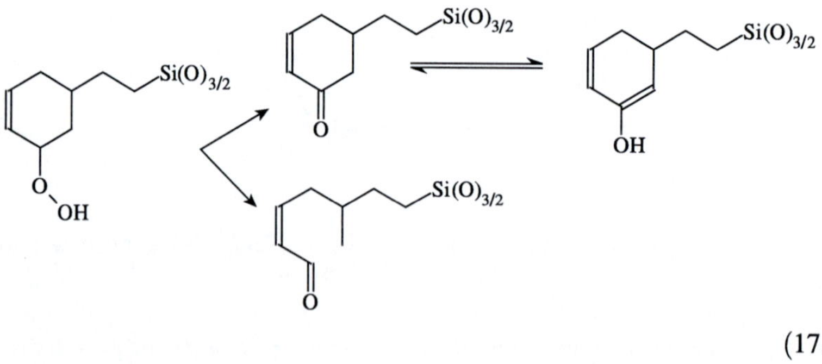

(17.10)

Amberg-Schwab et al. [38] using this procedure developed flexible thin films for dye-sensitized solar cells. The flexible thin films were useful on non-planar surfaces, as well as flat surfaces. One feature of these coatings is that they are compatible with solar cells, which might be used as construction components, such as roof tiles or facades.

As mentioned before, permeation in a nonporous medium is controlled mainly by the diffusivity of the permeant and by the solubility of the permeant in the material. Organic polymers and especially silicones [39] are permeable to some gases, where thin inorganic glasses have very low permeability. Hybrid organic–inorganic materials should fall somewhere in between the organic polymers and inorganic glasses, but synergetic effects are observed. In some systems with high hydrophobicity and very little porosity, the permeability is low. In other cases, hybrids result in highly permeable materials with selective permeation, for example in ion or proton conductive polymers [40].

An example of a hydrophobic coating is one prepared from TEOS and MTES by a sol-gel process. When this combination was deposited on a commercial ionomer Surlyn (Chemical Scheme (17.11)), the presence of the film increased the contact angle of Surlyn® with respect to a drop of water. Surlyn™ is an ionomer, which is chemically stable toward a variety of chemical agents, and is simple to apply to different substrates by lamination [41]. The main applications of Surlyn™ are for layered packaging [42]. Atomic force microscopy images showed that sol-gel films were crack-free and continuous. Chemical interaction between the sol-gel coating and Surlyn™ was demonstrated by spectroscopic methods. By Raman spectroscopy, it was shown that the methyl groups were concentrated at the surface of the films, leading to the observed increase of the contact angle [43]. Alternating Surlyn™ and hybrid layers may be one approach to hermetic barriers using a material that is already common in packaging.

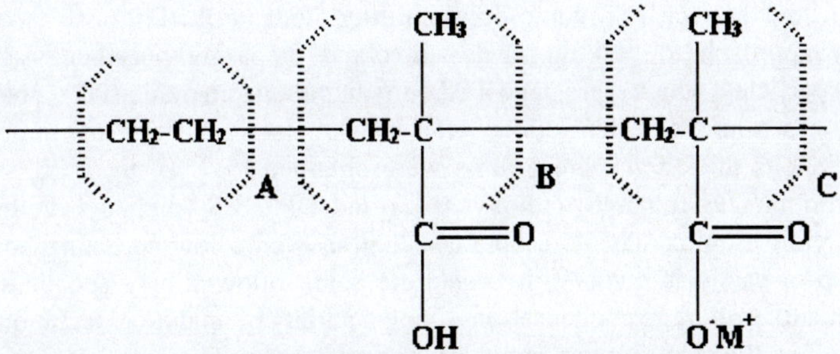

(17.11)

Other applications for hybrid systems are anti-adhesive and antistatic surfaces by controlling the hydrophobicity and oleophobicity of the surfaces.

Fluoroalkyl modified silicon-based alkoxides [44] have been incorporated in hybrids to change the surface properties. While the primary focus has been hermeticity, it is also necessary to consider the mechanical stability of coatings. Hybrid coatings have been developed to address abrasion resistance, particularly when it comes to coatings for optical fibers [57]. The requirements for this application are extremely demanding because these coatings must be applied to fibers during the fiber drawing operation, allowing the fiber to be spooled and cabled in high volume manufacturing operation.

The features of this coating system are that the coating (i) be a one-component system, (ii) have sufficient viscosity, and (iii) show good wetting to the fiber substrate. Once the coating is applied to the fiber, the coating must UV cure quickly. Other properties of the coating are hardness to resist scratching, flexibility, toughness to resist cracking and optical transparency [45]. In addition, these coatings need to protect the fibers from fatigue, which is a time-dependent process resulting from environmental exposure. Typically, UV-curable acrylates have been used. A variety of acrylates have been modified with a sol-gel process to introduce chemical links between organic and inorganic components. Functional group substituted alkoxysiloxanes in the coatings promote adhesion to the fiber substrate [46].

17.6 Melting Gel Barriers

A special category of hybrid gels constitutes the so-called "melting gels" [47–50]. Melting gels are a class of organically modified silica gels that are rigid at room temperature, flow at temperature T_1 and consolidate at temperature T_2 ($T_2 > T_1$), when cross-linking is complete. The process of (a) softening, (b) becoming rigid and (c) re-softening can be repeated many times. So-called melting gels were developed originally to replace low-melting inorganic glasses. These hybrids have been developed to replace low-melting temperature sealing glasses that melt around 600°C. This temperature is too high for most electronic packaging, especially newer devices such as organic light-emitting diodes (OLED).

An early approach to "melting gels" involved preparing poly(benzyl-silsesquioxane) particles, which were deposited onto an indium-tin oxide (ITO) coated substrate by electrophoretic deposition. After heat treatment at very low temperatures, continuous thick transparent films were obtained [51]. Using a different approach with PhTES (phenyl-triethoxysilane) and DPhDES (diphenyl-diethoxysilane), poly-silsesquioxanes were obtained that showed a low softening point. The first step of catalysis involved hydrochloric acid, followed by a second step with ammonia to encourage condensation. Using a variety of analytical techniques, it was shown that the resulting molecular structure was three dimensional, with the organic groups having weak bonds between molecular chains [51–53]. PhTES-DPhDES hybrids could be formed with or without ethanol, and the glass transition temperature of these hybrids varied with ratio of PhTES and DPhDES [53, 54].

In general, melting gel behavior is observed in polysiloxane polymers that contain cross-links between di-substituted and mono-substituted siloxanes. The reversible behavior of becoming rigid and then softening by cycling from room temperature to about 110°C can be repeated many times. However, once the gel is heated to its consolidation temperature, it loses the ability to soften. The consolidation temperature increases with an increase in the number of non-hydrolytic groups. Gels with larger numbers of residual methyl groups have overall higher weight loss. Based on differential scanning calorimetry, it appears that the consolidation temperature facilitates further cross-linking of the polysiloxane network [55]. The increase in cross-linking results in an increase in the measured glass transition temperature, which is a reflection of a stiffer polymer.

To test the hermeticity of melting gel, a syringe needle was filled with melting gel of composition 75%MTES–25%DMDES (dimethyl diethoxysilane). Using capillary action, the gel filled the inside of a stainless steel syringe needle with OD = 0.5 mm and ID = 0.4 mm (49). Limiting the time the tube was immersed in the melting gel controlled the height of gel in the needle. Several filled tubes were treated at 140°C for 24 h. The gel inside the tube appeared to have no cracks, under optical microscope inspection. The samples were sent to a commercial test lab for helium leak tests. The resulting helium permeability was reported to be 10^{-9} atm cm^3/s of helium, or Helium Transmission Rate $\sim 8.6 \times 10^{-5}$ cm^3/day.

Another test of hermeticity was to evaluate how long encapsulated lithium samples appeared shiny under different conditions (T < 10°C, room temperature, 70°C in oven). Similar to calcium, lithium reacts with water and oxygen according to:

$$4Li + O_2 \rightarrow 2Li_2O \tag{17.12}$$

$$Li_2O + H_2O \rightarrow 2LiOH \tag{17.13}$$

Since Li_2O is white and powdery, it is easy to follow the reaction visually.

Lithium was sandwiched between two slides of mica. A plastic ring was used as a mold and melting gel was poured into the ring to cover the mica-lithium metal-mica sandwich, on a polycarbonate substrate. This is pictured in Fig. 17.2. The 75%MTES–25%DMDES composition was used. The mica side of the sandwich is uppermost in the picture, with a thick melting gel covering, ~2 mm. Samples kept in ambient conditions for 4 weeks are shown in Fig. 17.2 (a: 0 h, b: 24 h, c: 336 h and d: 672 h).

Pictures of the samples were taken at various times. The area of shiny lithium was determined and recorded as a function of time. By following the area dulled by oxidation, the percentage degradation was measured at room temperature and 70°C, on polycarbonate and glass substrates. These data are plotted over the 4-week period in Fig. 17.3. Each of the data sets shows parabolic rate law behavior. A parabolic rate law coefficient was calculated and the results are listed in Table 17.3. Using the calculated coefficients, a time to 100% degradation is estimated and included in the Table 17.3. Between the two samples on glass, the better results are measured at room temperature. Between the two samples on polycarbonate, the better results are at 70°C, possibly because of better bonding between the polycarbonate and the melting gel. In fact, the sample on

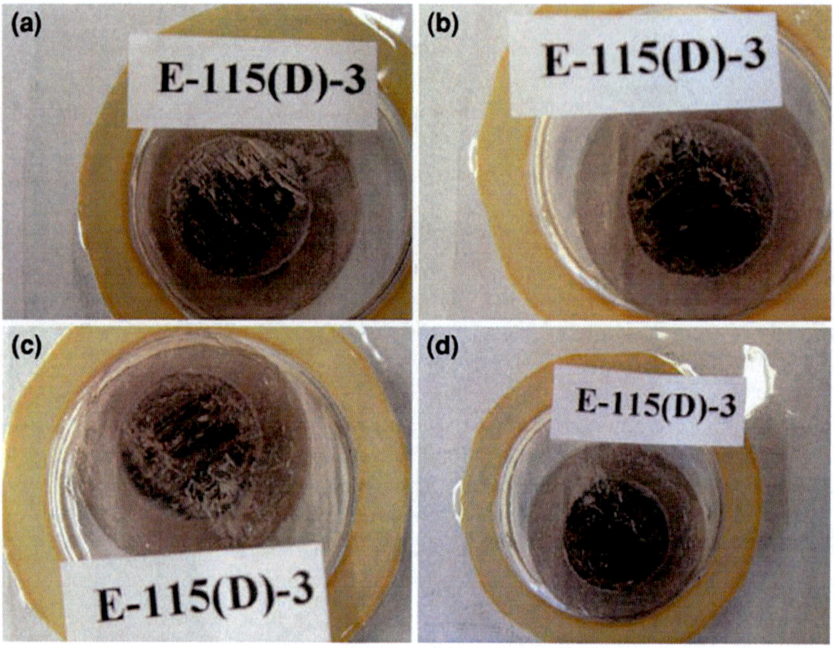

Fig. 17.2 Series of lithium samples sealed with melting gel 75%MTES/25%DMDES after various times of storage at room temperature in ambient air (**a** 0 h, **b** 24 h, **c** 336 h and **d** 672 h)

Fig. 17.3 Measured areal degradation (in %) for four samples of 75%MTES/25%DMDES versus time. The curves were used to estimate parabolic rate constants for degradation

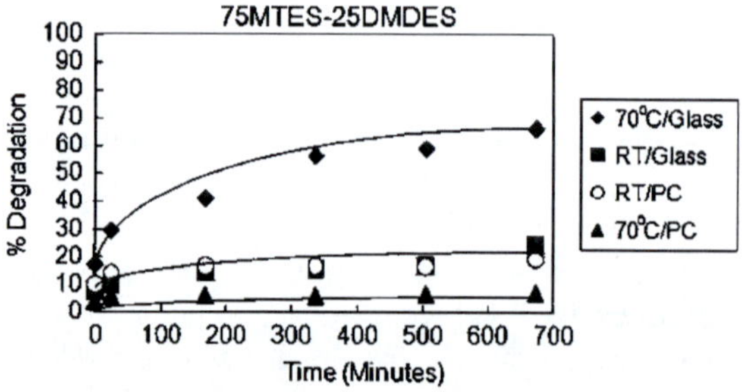

polycarbonate is estimated to take longer than 60 years before all of the lithium has reacted.

While the results with melting gels are preliminary, they do look promising. Their compatibility with lithium is a good sign for battery components. Also, their fluidity at around 100°C makes it possible to use them as a substitute in applications where epoxies are commonly used. Further studies of their dielectric properties and optical properties are in progress.

Table 17.3 Calculated Parabolic Rate Constants for Lithium Stored under 75%MTES–25%DMDES Melting Gels, from data in Fig. 17.3

Substrate	Temperature (°C)	Calculated rate constant (%/sec$^{1/2}$)	Estimate of time to 100% degradation (days)
Glass	Ambient	0.0108	1,005
Glass	70	0.03155	116
Polycarbonate	Ambient	0.00584	3,389
Polycarbonate	70	0.00212	>25,000

17.7 Conclusions

In this chapter, the main methods of preparation of hybrid hermetic packaging are summarized. The main focus is on hybrid materials obtained by a sol-gel method, since this does not use expensive equipment, as PECVD or ALD requires. An advantage of the sol-gel method is the ready availability of the precursors. In addition, hybrid coatings are somewhat flexible and have low processing temperatures. The organic groups bonded directly to the silica network block increase the hydrophobicity of the surface and fill up porosity in the film. Through these mechanisms, gas transport across the films is reduced.

References

1. Lews J (2006) Material challenge for flexible organic devices. Mater Today 9:38–45
2. Burrows PE, Graff GL, Gross ME, Martin PM, Shi MK, Hall M, Mast E, Bonham C, Bennett W, Sullivan MB (2001) Ultra barrier flexible substrates for flat panel display. Display 22:65–69
3. Ely K (2000) Issues in hermetic sealing of medical products. MDDI 186
4. Chantham H (1996) Oxygen diffusion barrier properties of transparent oxide coatings on polymeric substrates. Surf Coat Tech 78:1–9
5. Graff GL, Williford RE, Burrows PE (2004) Mechanisms of vapor permeation through multilayer barrier films: Log time versus equilibrium permeation. J Appl Phys 96:1840–1849
6. da Silva Sobrinho AS, Czeremuszkin G, Latreche M, Wertheimer MR (2000) Defect-permeation correlation for ultrathin transparent barrier coatings on polymers. J Vac Sci Technol A 18:149–157
7. Vogt BD, Lee H-J, Prabhu VM, DeLongchamp DM, Lin EK, Wu W-I (2005) X-ray and neutron reflectivity measurements of moisture transport through model multilayered barrier films for flexible displays. J Appl Phys 97:114509-1–114509-7
8. da Silva Sobrinho AS, Latreche M, Czeremuszkin G, Klemberg-Sapieha JE, Wertheimer MR (1998) Transparent barrier coatings on polyethylene terephthalate by single and dual-frequency plasma-enhanced chemical vapor deposition. J Vac Sci Technol A 16:3190
9. Tang CW, van Slyke SA (1997) Organic electroluminescent diodes. Appl Phys Lett 51: 913–915
10. Burrows PE, Bulovic V, Forrest SR, Sapochak LS, McCarty DM, Thomson ME (1994) Reliability and degradation of organic light emitting devices. Appl Phys Lett 65: 2922–2924

11. Burrows PE, Graff GL, Gross ME, Martin PM, Hall M, Mast E, Bonham C, Bennett W, Michalski L, Weaver M, Brown JJ, Fogarty D, Sapochak LS (2001) Gas permeation and lifetime test on polymer-based barrier coatings. Proc SPIE 4105:75–83
12. Gu G, Burrows PE, Venkatesh S, Forrest SR, Thompson ME (1997) Vacuum-deposited non-polymeric flexible organic light-emitting devices. Opt Lett 22:172–174
13. Baldo MA, Lamannsky S, Burrows PE, Thompson ME, Forrest SR (1999) Very high-efficiency green organic light-emitting devices based on electro-phosphorescence. Appl Phys Lett 75:4–6
14. Waver MS, Michalski LA, Rajan K, Rothman MA, Silvernail JA, Brown JJ, Burrows PE, Graff GL, Gross ME, Martin PM, Hall M, Mast E, Bonham C, Bennett W, Zumhoff M (2002) Organic light-emitting devices with extended operating lifetimes on plastic substrates. Appl Phys Lett 81:2929
15. Groner MD, George SM, McLean SM, Carcia PF (2006) Gas diffusion barriers on polymers using Al_2O_3 atomic layer deposition. Appl Phys Lett 88:051907-1–051907-3
16. Carcia PF, McLean RS, Reilly MH, Gorner MD, George SM (2006) Ca test of Al_2O_3 gas diffusion barriers grown by atomic layer deposition on polymers. Appl Phys Lett 89:031915-1–031915-3
17. Dunkel R, Bujas R, Klein A, Horndt V (2005) Method of measuring ultralow water vapor permeation for OLED displays. Proc IEEE 93:1478–1482
18. a. Mocon Inc. www.mocon.com, Accessed Mar 2011. b. http://www.mocon.com/permeation.php Accessed Mar 2011
19. Ranade A, D'Souza NA, Wallace RM, Gnade BE (2005) High sensitivity gas permeability measurement system for thin plastic films. Rev Sci Instrum 76:013902-1–013902-5
20. Nilsson PO, Forssell G (1977) Optical properties of calcium. Phys Rev B 16:3352–3358
21. Nisato G, Bouten PCP, Slikkerveer PJ, Bennett WD, Graff GL, Rutherford N, Wiese L (2001) Evaluating high performance diffusion barriers: the calcium test. Proc Int Display Workshop/Asia Display 1435–1438
22. Nisato G, Kuilder M, Bouten P, Moro L, Philips O, Rutherford N (2003) Thin film encapsulation for Oleds: evaluation of multi-layer barriers using a Ca test. Proc Soc Info Display Symp Digest Tech Papers 34:550–553
23. Paetzold R, Winnacker A, Henseler D, Cesari V, Heuser K (2003) Permeation rate measurements by electrical analysis of calcium corrosion. Rev Sci Instrum 74:5147–5150
24. Jitanu A, Klein LC (2009) Sol-Gel Hybrids for electronic applications: hermetic coatings for microelectronics and energy storage. In: Merhari L (ed) Hybrid nanocomposites for nanotechnology: electronic optical magnetic and bio/medical applications. Springer, Berlin, pp 429–453
25. Sanchez C, Ribot F (1994) Chemical design of hybrid organic-inorganic materials synthesized via sol-gel. New J Chem 10:1007–1040
26. Jitianu A, Britchi A, Deleanu C, Badescu V, Zaharescu M (2003) Comparative study of the sol-gel processes starting with different substituted Si-alkoxide. J Non-Cryst Solids 319: 263–279
27. Unger B, Jancke H, Hahnert M, Stade H (1994) The early stages of the sol-gel processing of TEOS. J Sol-Gel Sci Tech 2:51
28. Wojcik AB, Klein LC (1995) Transparent inorganic/organic copolymers by the sol-gel process: copolymers of tetraethyl orthosilicate (TEOS), vinyl triethoxysilane (VTES) and (meth)acrylate monomers. J Sol-Gel Sci Tech 4:57–66
29. Wojcik AB, Klein LC (1995) Transparent inorganic/organic copolymers by the sol-gel process: Thermal behavior of copolymers of tetraethyl orthosilicate (TEOS), vinyl triethoxysilane (VTES) and (meth)acrylate monomers. J Sol-Gel Sci Tech 5:77–82
30. Wojcik AB, Klein LC (1997) Organic/inorganic hybrids by the sol-gel process: classification of synthesis methods. Appl Organomet Chem 11:129–135
31. Avnir D, Klein LC, Levy D, Schubert U, Wojcik AB (1998) Organo-silica Sol-Gel materials. In: Rappoport Z, Apeloig Y (eds) The chemistry of organosilicon compounds, vol 2. Wiley, London, pp 2317–2362 (Chapter 40)

32. Jackson A, Jitianu A, Klein LC (2006) Development of hermetic barrier using vinyl triethoxysilane (VTEOS) and Sol-Gel processing. Mater Matters (Sigma-Aldrich) 1:11–12
33. Amberg-Schwab S, Hoffmann M, Bader H (1996) Barrierschichten für Verpackungsmaterialien; Barriereschichten aus anorganish-organischen Hybridopolymeren. Kunststoffe 86:660–664
34. Amberg-Schwab S, Hoffmann M, Bader H, Gessler M (1998) Inorganic-Organic polymers with barrier for water vapor, oxygen and flavors. J Sol-Gel Sci Tech 13:141–146
35. Amberg-Schwab S, Katschorek H, Weber U, Hoffmann M, Burger A (2000) Barrier properties of inorganic-organic polymers: influence of starting compounds, curing, conditions and storage-scaling-up to industrial application. J Sol-Gel Sci Tech 19:125–129
36. Hoffmann M, Amberg-Schwab S (1998) Barrier properties of inorganic-organic polymers. Mat Res Soc Symp Proc 519:309–314
37. Amberg-Schwab S, Katschorek H, Weber U, Burger A (2003) Inorganic-organic polymers as migration barriers against liquid and volatile compounds. J Sol-Gel Sci Tech 26:699–703
38. Amberg-Schwab S, Weber U, Burger A, Nique S, Xalter R (2006) Development of passive and active barrier coatings on basis of inorganic-organic polymers. Monatshefte für Chemie (Chemical Monthly) 137:657–666
39. Sanchez C, Julian B, Belleville P, Popall M (2005) Applications of hybrid organic-inorganic nanocomposites. J Mat Chem 15:3559–3592
40. Dahmouche K, Atik M, Mello NC, Bonagamba TJ, Panepucci H, Aegerter MA, Judeinstein P (1997) Investigation of new ion-conducting ORMOLYTES: structure and properties. J Sol-Gel Sci Technol 8:711
41. Jerome R, Mazurek M (1997) Synthesis and characterisation of the molecular structure. In: Tant MR, Mauritz KA, Wilkes GL (eds) Ionomers: synthesis, structure, properties and applications. Blackie Academic and Professional, London, p 3
42. Watanabe H (2002) Patent No. JP 2002011116
43. Jitianu A, Amatucci G, Klein LC (2008) Organic-inorganic Sol-Gel thick films for humidity barriers. J Mater Res 23:2084–2090
44. Rondinella VV, Matthewson MJ (1993) Effect of loading mode and coating on dynamic fatigue of optical fiber in two-point bending. J Am Ceram Soc 76:139–144
45. Wei TS, Skutnik BJ (1988) Effect of coating on fatigue behavior of optical fiber. J Non-Cryst Solids 102:100–105
46. Nishiyama N, Horie K, Sakura T (1988) Adsorption behavior of a silane coupling agent onto a colloidal silica surface studies by 29Si NMR spectroscopy. J Colloid Interface Sci 129: 113–119
47. Jitianu A, Doyle J, Amatucci G, Klein LC (2008) Methyl-modified melting gels for hermetic barrier coatings, In Proceedings MS&T enabling surface coating systems: multifunctional coatings (CD-ROM), Pittsburgh, PA, pp 2171–2182
48. Jitianu A, Amatucci G, Klein LC (2008) Phenyl-substituted siloxane hybrid gels that soften below 140°C. J Am Ceram Soc 92:36–40
49. Jitianu A, Doyle J, Amatucci G, Klein LC (2010) Methyl modified siloxane melting gels for hydrophobic films. J Sol-Gel Sci Tech 53:272–279
50. Klein LC, Jitianu A (2010) Organic-inorganic hybrid melting gels. J Sol-Gel Sci Tech 55: 86–93
51. Matsuda A, Sasaki T, Hasegawa K, Tatsumisago M, Minami T (2001) Thermal softening behavior and application to transparent thick films of poly(benzylsilsesquioxane) particles prepared by sol-gel process. J Am Cer Soc 84:775–780
52. Masai H, Tokuda Y, Yoko T (2005) Gel-melting method for preparation of organically modified siloxane low-melting glasses. J Mater Res 20:1234–1241
53. Kakiuchida H, Takahashi M, Tokuda Y, Masai H, Kuniyoshi M, Yoko T (2006) Viscoelastic and structural properties of the phenyl-modified polysiloxane system with a three-dimensional structure. J Phys Chem B 110:7321–7327
54. Kakiuchida H, Takahashi M, Tokuda Y, Masai H, Kuniyoshi M, Yoko T (2007) Effects of organic groups on structure and viscoelastic properties of organic-inorganic polysiloxane. J Phys Chem B 111:982–988

55. Jitianu A, Lammers K, Arbuckle GA, Amatucci GG, Klein LC (2011) Thermal analysis of organically modified siloxane melting gels. Accepted in J Therm Anal
56. Kumar RS, Auch M, Ou E, Ewald G, Jin CS (2002) Low moisture permeation measurement through polymer substrates for organic light emitting devices. Thin Solid Films 417:120–126
57. Wojcik AB, Klein LC, Rondinella VV, Matthewson MJ, Foy PR (1993) Organically modified silicate coatings for optical fibers. SPIE Proc 2074:135–141

Index

M. Aparicio et al. (eds.), *Sol-Gel Processing for Conventional and Alternative Energy*, Advances in Sol-Gel Derived Materials and Technologies,
DOI: 10.1007/978-1-4614-1957-0, © Springer Science+Business Media New York 2012